PRINCIPLES OF APPLIED MATHEMATICS

Transformation and Approximation

James P. Keener

Department of Mathematics
University of Utah

The Advanced Book Program

CRC Press
Taylor & Francis Group
Boca Raton London New York

CRC Press is an imprint of the
Taylor & Francis Group, an **informa** business

A CHAPMAN & HALL BOOK

First published 1988 by Perseus Books Publishing, L.L.C.

Published 2018 by CRC Press
Taylor & Francis Group
6000 Broken Sound Parkway NW, Suite 300
Boca Raton, FL 33487-2742

© 1988, 1995 by Taylor & Francis Group, LLC
CRC Press is an imprint of Taylor & Francis Group, an Informa business

No claim to original U.S. Government works

ISBN 13: 978-0-201-48363-5 (pbk)
ISBN 13: 978-0-201-15674-4 (hbk)

Visit the Taylor & Francis Web site at
http://www.taylorandfrancis.com

and the CRC Press Web site at
http://www.crcpress.com

Library of Congress Cataloging-in-Publication Data
Keener, James P.
    Principles of applied mathematics : transformation and
approximation / James P. Keener.
        p.   cm.
    Includes index.

    1. Transformations (Mathematics) 2. Asymptotic expansions.
I. Title.
QA601.K4 1987
515.7'23–dc19   87-18630
                                                  CIP

This book was typeset in *Textures* using a Macintosh SE computer.

To my three best friends, Kristine, Samantha, and Justin, who have adjusted extremely well to having a mathematician as husband and father.

PREFACE

Applied mathematics should read like good mystery, with an intriguing beginning, a clever but systematic middle, and a satisfying resolution at the end. Often, however, the resolution of one mystery opens up a whole new problem, and the process starts all over. For the applied mathematical scientist, there is the goal to explain or predict the behavior of some physical situation. One begins by constructing a mathematical model which captures the essential features of the problem without masking its content with overwhelming detail. Then comes the analysis of the model where every possible tool is tried, and some new tools developed, in order to understand the behavior of the model as thoroughly as possible. Finally, one must interpret and compare these results with real world facts. Sometimes this comparison is quite satisfactory, but most often one discovers that important features of the problem are not adequately accounted for, and the process begins again.

Although every problem has its own distinctive features, through the years it has become apparent that there are a group of tools that are essential to the analysis of problems in many disciplines. This book is about those tools. But more than being just a grab bag of tools and techniques, the purpose of this book is to show that there is a systematic, even esthetic, explanation of how these classical tools work and fit together as a unit.

Much of applied mathematical analysis can be summarized by the observation that we continually attempt to reduce our problems to ones that we already know how to solve.

This philosophy is illustrated by an anecdote (apocryphal, I hope) of a mathematician and an engineer who, as chance would have it, took a gourmet cooking class together. As this course started with the basics, the first lesson was on how to boil a pot of water. The instructor presented the students with two pots of water, one on the counter, one on a cold burner, while another burner was already quite hot, but empty.

In order to test their cooking aptitudes, the instructor first asked the engineer to demonstrate to the rest of the class how to boil the pot of water that was already sitting on the stove. He naturally moved the pot carefully from the cold burner to the hot burner, to the appreciation of his classmates. To be sure the class understood the process, the pot was moved back to its original spot on the cold burner at which time the mathematician was asked to demonstrate how to heat the water in the pot sitting on the counter. He promptly and confidently exchanged the position of the two pots, placing the pot from the counter onto the cold burner and the pot from the burner onto the counter. He then stepped back from the stove, expecting an appreciative response from his mates. Baffled by his actions, the instructor asked him to explain what he had done, and he replied naturally, that he had simply reduced his problem to one that everyone already knew how to solve.

We shall also make it our goal to reduce problems to those which we already know how to solve.

We can illustrate this underlying idea with simple examples. We know that to solve the algebraic equation $3x = 2$, we multiply both sides of the equation with the inverse of the "operator" 3, namely $1/3$, to obtain $x = 2/3$. The same is true if we wish to solve the matrix equation $Ax = f$ where

$$A = \begin{pmatrix} 3 & 1 \\ 1 & 3 \end{pmatrix}, \qquad f = \begin{pmatrix} 1 \\ 2 \end{pmatrix}.$$

Namely, if we know A^{-1}, the inverse of the matrix operator A, we multiply both sides of the equation by A^{-1} to obtain $x = A^{-1}f$. For this problem,

$$A^{-1} = \frac{1}{8} \begin{pmatrix} 3 & -1 \\ -1 & 3 \end{pmatrix}, \qquad \text{and} \qquad x = \frac{1}{8} \begin{pmatrix} 1 \\ 5 \end{pmatrix}.$$

If we make it our goal to invert many kinds of linear operators, including matrix, integral, and differential operators, we will certainly be able to solve many types of problems. However, there is an approach to calculating the inverse operator that also gives us geometrical insight. We try to transform the original problem into a simpler problem which we already know how to solve. For example, if we rewrite the equation $Ax = f$ as $T^{-1}AT\left(T^{-1}x\right) = T^{-1}f$ and choose

$$T = \begin{pmatrix} 1 & -1 \\ 1 & 1 \end{pmatrix}, \qquad T^{-1} = \frac{1}{2} \begin{pmatrix} 1 & 1 \\ -1 & 1 \end{pmatrix},$$

we find that

$$T^{-1}AT = \begin{pmatrix} 4 & 0 \\ 0 & -2 \end{pmatrix}.$$

With the change of variables $y = T^{-1}x$, $g = T^{-1}f$, the new problem looks like two of the easy algebraic equations we already know how to solve, namely $4y_1 = 3/2$, $-2y_2 = 1/2$. This process of separating the coupled equations into uncoupled equations works only if T is carefully chosen, and exactly how this is done is still a mystery. Suffice it to say, the original problem has been transformed, by a carefully chosen change of coordinate system, into a problem we already know how to solve.

This process of changing coordinate systems is very useful in many other problems. For example, suppose we wish to solve the boundary value problem $u'' - 2u = f(x)$ with $u(0) = u(\pi) = 0$. As we do not yet know how to write down an inverse operator, we look for an alternative approach. The single most important step is deciding how to represent the solution $u(x)$. For example, we might try to represent u and f as polynomials or infinite power series in x, but we quickly learn that this guess does not simplify the solution process much. Instead, the "natural" choice is to represent u and f as trigonometric series,

$$u(x) = \sum_{k=1}^{\infty} u_k \sin kx, \qquad f(x) = \sum_{k=1}^{\infty} f_k \sin kx.$$

Using this representation (i.e., coordinate system) we find that the original differential equation reduces to the infinite number of separated algebraic equations $(k^2 + 2)\, u_k = -f_k$. Since these equations are separated (the kth equation depends only on the unknown u_k), we can solve them just as before, even though there are an infinite number of equations. We have managed to simplify this problem by transforming into the correctly chosen coordinate system.

For many of the problems we encounter in the sciences, there is a natural way to represent the solution that transforms the problem into a substantially easier one. All of the well-known special functions, including Legendre polynomials, Bessel functions, Fourier series, Fourier integrals, etc., have as their common motivation that they are natural for certain problems, and perfectly ridiculous in others. It is important to know when to choose one transform over another.

Not all problems can be solved exactly, and it is a mistake to always look for exact solutions. The second basic technique of applied mathematics is to reduce hard problems to easier problems by ignoring small terms. For example, to find the roots of the polynomial $x^2 + x + .0001 = 0$, we notice that the equation is very close to the equation $x^2 + x = 0$ which has roots $x = -1$, and $x = 0$, and we suspect that the roots of the original polynomial are not too much different from these. Finding how changes in parameters affect the solution is the goal of perturbation theory and asymptotic analysis, and in this example we have a regular perturbation problem, since the solution is a regular (i.e., analytic) function of the "parameter" .0001 .

Not all reductions lead to such obvious conclusions. For example, the polynomial $.0001x^2 + x + 1 = 0$ is close to the polynomial $x + 1 = 0$, but the first has two roots while the second has only one. Where did the second root go? We know, of course, that there is a very large root that "goes to infinity" as ".0001 goes to zero," and this example shows that our naïve idea of setting all small parameters to zero must be done with care. As we see, not all problems with small parameters are regular, but some have a singular behavior. We need to know how to distinguish between regular and singular approximations, and what to do in each case.

This book is written for beginning graduate students in applied mathematics, science, and engineering, and is appropriate as a one-year course in applied mathematical techniques (although I have never been able to cover all of this material in one year). We assume that the students have studied at an introductory undergraduate level material on linear algebra, ordinary and partial differential equations, and complex variables. The emphasis of the book is a working, systematic understanding of classical techniques in a modern context. Along the way, students are exposed to models from a variety of disciplines. It is hoped that this course will prepare students for further study of modern techniques and in-depth modeling in their own specific discipline.

One book cannot do justice to all of applied mathematics, and in an effort to keep the amount of material at a manageable size, many important topics were not included. In fact, each of the twelve chapters could easily be expanded into an entire book. The topics included here have been selected, not only for their scientific importance, but also because they allow a logical flow to the development of the ideas and techniques of transform theory and asymptotic analysis. The theme of transform theory is introduced for matrices in Chapter one, and revisited for integral equations in Chapter three, for ordinary differential equations in Chapters four and seven, for partial differential equations in Chapter eight, and for certain nonlinear evolution equations in Chapter nine.

Once we know how to solve a wide variety of problems via transform theory, it becomes appropriate to see what harder problems we can reduce to those we know how to solve. Thus, in Chapters ten, eleven, and twelve we give a survey of the three basic areas of asymptotic analysis, namely asymptotic analysis of integrals, regular perturbation theory and singular perturbation theory.

Here is a summary of the text, chapter by chapter. In Chapter one, we review the basics of spectral theory for matrices with the goal of understanding not just the mechanics of how to solve matrix equations, but more importantly, the geometry of the solution process, and the crucial role played by eigenvalues and eigenvectors in finding useful changes of coordinate systems. This usefulness extends to pseudo-inverse operators as well as operators in Hilbert space, and so is a particularly important piece of background information.

In Chapter two, we extend many of the notions of finite dimensional vector spaces to function spaces. The main goal is to show how to represent objects in a function space. Thus, Hilbert spaces and representation of functions in a Hilbert space are studied. In this context we meet classical sets of functions such as Fourier series and Legendre polynomials, as well as less well-known sets such as the Walsh functions, Sinc functions, and finite element bases.

In Chapter three, we explore the strong analogy between integral equations and matrix equations, and examine again the consequences of spectral theory. This chapter is more abstract than others as it is an introduction to functional analysis and compact operator theory, given under the guise of Fredholm integral equations. The added generality is important as a framework for things to come.

In Chapter four, we develop the tools necessary to use spectral decompositions to solve differential equations. In particular, distributions and Green's functions are used as the means by which the theory of compact operators can be applied to differential operators. With these tools in place, the completeness of eigenfunctions of Sturm-Liouville operators follows directly.

Chapter five is devoted to showing how many classical differential equations can be derived from a variational principle, and how the eigenvalues of a differential operator vary as the operator is changed.

Chapter six is a pivotal chapter, since all chapters following it require a substantial understanding of analytic function theory, and the chapters preceding it require no such knowledge. In particular, knowing how to integrate and differentiate analytic functions at a reasonably sophisticated level is indispensable to the remaining text. Section 6.3 (applications to Fluid Flow) is included because it is a classically important and very lovely subject, but it plays no role in the remaining development of the book, and could be skipped if time constraints demand.

Chapter seven continues the development of transform theory and we show that eigenvalues and eigenfunctions are not always sufficient to build a transform, and that operators having continuous spectrum require a generalized construction. It is in this context that Fourier, Mellin, Hankel, and Z transforms, as well as scattering theory for the Schrödinger operator are studied.

In Chapter eight we show how to solve linear partial differential and difference equations, with special emphasis on (as you guessed) transform theory. In this chapter we are able to make specific use of all the techniques introduced so far and solve some problems with interesting applications.

Although much of transform theory is rather old, it is by no means dead. In Chapter nine we show how transform theory has recently been used to solve certain nonlinear evolution equations. We illustrate the inverse scattering transform on the Korteweg-deVries equation and the Toda lattice.

In Chapter ten we show how asymptotic methods can be used to approximate the horrendous integral expressions that so often result from transform techniques.

In Chapter eleven we show how perturbation theory and especially the study of nonlinear eigenvalue problems uses knowledge of the spectrum of a linear operator in a fundamental way. The nonlinear problems in this chapter all have the feature that their solutions are close to the solutions of a nearby linear problem.

Singular perturbation problems fail to have this property, but have solutions that differ markedly from the naïve simplified problem. In Chapter twelve, we give a survey of the three basic singular perturbation problems (slowly varying oscillations, initial value problems with vastly different time scales, and boundary value problems with boundary layers).

This book has the lofty goal of reaching students of mathematics, science, and engineering. To keep the attention of mathematicians, one must be systematic, and include theorems and proofs, but not too many explicit calculations. To interest an engineer or scientist, one must give specific examples of how to solve meaningful problems but not too many proofs or too much abstraction. In other words, there is always someone who is unhappy.

In an effort to minimize the total displeasure and not scare off too many members of either camp, there are some proofs and some computations. Early in the text, there are disproportionately more proofs than computations because as the foundations are being laid, the proofs often give important insights. Later, after the bases are established, they are invoked in the problem solving process, but proofs are deemphasized. (For example, there are no proofs in Chapters eleven and twelve.) Experience has shown that this approach is, if not optimal, at least palatable to both sides.

This is intended to be an applied mathematics book and yet there is very little mention of numerical methods. How can this be? The answer is simple and, I hope, satisfactory. This book is primarily about the principles that one uses to solve problems and since these principles often have consequence in numerical algorithms, mention of numerical routines is made when appropriate. However, FORTRAN codes or other specific implementations can be found in many other good resources and so (except for a code for fast Walsh transforms) are omitted. On the other hand, many of the calculations in this text should be done routinely using symbolic manipulation languages such as REDUCE or MACSYMA. Since these languages are not generally familiar to many, included here are a number of short programs written in REDUCE in order to encourage readers to learn how to use these remarkable tools. (Some of the problems at the end of several chapters are intended to be so tedious that they force the reader to learn one of these languages.)

This text would not have been possible were it not for the hard work and encouragement of many other people. There were numerous students who struggled through this material without the benefit of written notes while the course was evolving. More recently, Prof. Calvin Wilcox, Prof. Frank Hoppensteadt, Gary De Young, Fred Phelps, and Paul Arner have been very influential in the shaping of this presentation. Finally, the patience of Annetta Cochran and Shannon Ferguson while typing from my illegible scrawling was exemplary.

In spite of all the best efforts of everyone involved, I am sure there are still typographical errors in the text. It is disturbing that I can read and reread a section of text and still not catch all the erors. My hope is that those that remain are both few and obvius and will not lead to undue confusion.

CONTENTS

1

FINITE DIMENSIONAL VECTOR SPACES

1.1 Linear Vector Spaces

In any problem we wish to solve, the goal is to find a particular object, chosen from among a large collection of contenders, which satisfies the governing constraints of the problem. The collection of contending objects is often a VECTOR SPACE, and although individual elements of the set can be called by many different names, it is typical to call them VECTORS. Not every set of objects constitutes a vector space. To be specific, if we have a collection of objects S, we must first define addition and scalar multiplication for these objects. The operations of addition and scalar multiplication are defined to satisfy the usual properties: If $x, y, z \in S$, then

1. $x + y = y + x$ (commutative law)
2. $x + (y + z) = (x + y) + z$ (associative law)
3. $0 \in S$, $0 + x = x$ (additive identity)
4. $-x \in S$, $-x + x = 0$ (additive inverse)

and if $x \in S$, $\alpha, \beta \in \mathbb{R}$ (or $\mathbb{C}$) then

1. $\alpha(\beta x) = (\alpha\beta)x$

1

2. $(\alpha + \beta)x = \alpha x + \beta x$

3. $\alpha(x + y) = \alpha x + \alpha y$

4. $1x = x$, $0x = 0$.

Once the operations of addition and scalar multiplication are defined, we define a vector space as follows:

Definition

A set of objects S is called a LINEAR VECTOR SPACE if two properties hold:

1. If $x, y \in S$, then $x + y \in S$ (closure under vector addition),

2. If $\alpha \in \mathbb{R}$ (or $\alpha \in \mathbb{C}$) and $x \in S$, then $\alpha x \in S$ (closure under scalar multiplication).

If the scalars are real, S is a real vector space, while if the scalars are complex, S is a complex vector space.

Examples

1. The set of real ordered pairs (x, y) (denoted $\mathbb{R}^2$) is a linear vector space if addition is defined by $(x_1, y_1) + (x_2, y_2) = (x_1 + x_2, y_1 + y_2)$, and scalar multiplication is defined by $\alpha(x, y) = (\alpha x, \alpha y)$.

2. The set of real n-tuples $(x_1, x_2, x_3, \ldots, x_n)$ (denoted $\mathbb{R}^n$) forms a linear vector space if addition and scalar multiplication are defined component wise as in the above example. Similarly the set of complex n-tuples (denoted $\mathbb{C}^n$) forms a linear vector space with the complex scalars.

3. The set of all polynomials of degree n forms a linear vector space, if addition and scalar multiplication are defined in the usual way.

4. The set of all continuous functions defined on some interval $[a, b]$ forms a linear vector space.

There are many sets that are not linear vector spaces. For example, the set of all ordered pairs of the form $(x, 1)$ is not closed under addition or scalar multiplication and so does form a vector space. On the other hand, ordered pairs of the form $(x, 3x)$ do comprise a linear vector space. Similarly, the set of continuous functions $f(x)$ defined on the interval $[a, b]$ with $f(a) = 1$ does not form a vector space, while if instead the constraint $f(a) = 0$ is imposed, the properties of closure hold.

In a vector space, we can express one element of S in terms of additions and scalar multiplications of other elements. For example, if $x_1, x_2, x_3, \ldots, x_n,$

are elements of S, then $x = \alpha_1 x_1 + \alpha_2 x_2 + \ldots + \alpha_n x_n$ is also in S, and x is said to be a LINEAR COMBINATION of $x_1, x_2, \ldots, x_n$. If there is some linear combination with $\alpha_1 x_1 + \alpha_2 x_2 + \ldots + \alpha_n x_n = 0$, and not all of the scalars α_j are zero, then the set $x_1, x_2, \ldots, x_n$ is said to be LINEARLY DEPENDENT. On the other hand, if the only linear combination of $x_1, x_2, \ldots, x_n$ which is zero has all α_i equal to zero, the set $\{x_1, x_2, \ldots, x_n\}$ is said to be LINEARLY INDEPENDENT.

Examples

1. The single nontrivial vector $x \in \mathbb{R}^2$, $x \neq 0$ forms a linearly independent set because the only way to have $\alpha x = 0$ is if $\alpha = 0$.

2. A linear combination of two vectors in $\mathbb{R}^2$, $\alpha_1(x_1, y_1) + \alpha_2(x_2, y_2)$, is the zero vector whenever $x_1 = -\beta x_2$ and $y_1 = -\beta y_2$ where $\beta = \alpha_2/\alpha_1$. Geometrically, this means that the vector (x_1, y_1) is collinear with the vector (x_2, y_2), i.e., they have the same direction in $\mathbb{R}^2$.

An important example of linearly independent vectors is the polynomials. For example, the powers of x, $\{1, x, x^2, \ldots, x^n\}$ form a linearly independent set. An easy proof of linearly independence is to define $f(x) = a_0 + a_1 x + a_2 x^2 + \cdots + a_n x^n$ and to determine when $f(x) \equiv 0$. Clearly, $f(0) = a_0$ and $f(x) = 0$ implies $a_0 = 0$. The kth derivative of f at $x = 0$,

$$\frac{d^k f(x)}{dx^k}\bigg|_{x=0} = k! a_k$$

is zero if and only if $a_k = 0$. Therefore, $f(x) \equiv 0$ if and only if $a_j = 0$ for $j = 0, 1, 2, \ldots, n$.

It is often convenient to represent an arbitrary element of the vector space as a linear combination of a predetermined set of elements of S. For example, we might want to represent the continuous functions as a linear combination of certain polynomials or of certain trigonometric functions. We ask the question, given a subset of S, when can any element of S be written as a linear combination of elements of the specified subset?

Definition

A set of vectors $T \in S$ is a SPANNING SET if every $x \in S$ can be written as a linear combination of the elements of T.

Definition

If a spanning set $T \in S$ is linearly independent then T is a BASIS for S.

One can easily show that every basis of S has the same number of elements. If a basis T is finite, the number of elements is called the DIMENSION of the space.

Examples

1. In $\mathbb{R}^2$, any three vectors are linearly dependent, but any two non-collinear vectors form a basis.

2. In $\mathbb{R}^3$ the vectors $\{(1,0,0),\ (0,1,0),\ (0,0,1),\ (1,1,0),\ (0,1,1)\}$ form a spanning set, but the last two vectors are not necessary to form a spanning set. The smallest spanning set is the basis and the dimension is 3. From this set, there are eight different ways to choose a subset which is a basis. In $\mathbb{R}^n$, the so-called "natural" basis is the set of vectors $\{e_1, e_2, \ldots, e_n\}$ where e_k has entries $e_{kj} = \delta_{kj}$ where $\delta_{kj} = 0$ if $j \neq k$ and $\delta_{kk} = 1$. δ_{kj} is called the Kronecker delta.

3. The set $\{1, x, x^2, x^2 - 1\}$ is linearly dependent. However, any of the subsets $\{1, x, x^2\}$, $\{1, x, x^2 - 1\}$, or $\{x, x^2, x^2 - 1\}$ forms a basis for the set of quadratic polynomials.

4. The set of all polynomials does not have finite dimension, since any linear combination of polynomials of degree n or less cannot be a polynomial of degree $n + 1$. The set of all continuous functions is also infinite since the polynomials are a subset of this set. We will show later that the polynomials form a "basis" for the continuous functions, but this requires more technical information than we have available now.

When you first learned about vectors in an undergraduate Physics or Mathematics course, you were probably told that vectors have DIRECTION and MAGNITUDE. This is certainly true in $\mathbb{R}^3$, but in more complicated vector spaces the concept of direction is a bit hard to visualize. The direction of a vector is always given relative to some other reference vector by the angle between the two vectors. To get a concept of direction and angles in general vector spaces we introduce the notion of an INNER PRODUCT.

Definition

If $x, y \in S$, then $\langle x, y \rangle$ is called the INNER PRODUCT (also called scalar product or dot product) if $\langle \cdot, \cdot \rangle$ is a bilinear operator $\langle \cdot, \cdot \rangle : S \times S \to \mathbb{R}$, (or $\mathbb{C}$ if S is a complex vector space) with the properties:

1. $\langle x, y \rangle = \overline{\langle y, x \rangle}$, (overbar denotes complex conjugate)
2. $\langle \alpha x, y \rangle = \alpha \langle x, y \rangle$,
3. $\langle x + y, z \rangle = \langle x, z \rangle + \langle y, z \rangle$,

4. $\langle x, x \rangle > 0$ if $x \neq 0$, $\langle x, x \rangle = 0$ if and only if $x = 0$.

A linear vector space with an inner product is called an INNER PRODUCT SPACE.

Examples

1. In $\mathbb{R}^n$, suppose $x = (x_1, x_2, \ldots, x_n)$ and $y = (y_1, y_2, \ldots, y_n)$ then

$$\langle x, y \rangle = \sum_{k=1}^{n} x_k y_k$$

is the usual EUCLIDEAN INNER PRODUCT. If $x, y \in \mathbb{C}^n$, then

$$\langle x, y \rangle = \sum_{k=1}^{n} x_k \bar{y}_k.$$

2. For a real continuous function $f(x)$ defined for $x \in [0, 1]$, we discretize $f(x)$ to define a vector in $\mathbb{R}^n$, $F = (f(x_1), f(x_2), \ldots, f(x_n))$ where $x_k = k/n$, and it makes sense to define

$$\langle F, G \rangle_n = \frac{1}{n} \sum_{k=1}^{n} f(x_k) g(x_k).$$

(This is not an inner product for the continuous functions. Why?) Taking the limit $n \to \infty$, we obtain

$$\langle f, g \rangle = \int_0^1 f(x) g(x) dx.$$

It is an easy matter to verify that this defines an inner product on the vector space of continuous functions.

3. There are other interesting ways to define an inner product for functions. For example, if the functions f and g are continuously differential on the interval $[0, 1]$, we might define

$$\langle f, g \rangle = \int_0^1 \left(f(x) \overline{g(x)} + f'(x) \overline{g'(x)} \right) dx.$$

More generally, if f and g are n times continuously differentiable on $[0, 1]$ we might define

$$\langle f, g \rangle = \int_0^1 \left(\sum_{j=0}^{n} \frac{d^j f(x)}{dx^j} \frac{\overline{d^j g(x)}}{dx^j} \right) dx$$

as an inner product. Indeed, one can show that these satisfy the required properties of inner products.

A vector in a linear vector space has a NORM (also called amplitude, magnitude, or length) if for $x \in S$ there is a function $\|\cdot\| : S \to \mathbb{R}^+$ (the nonnegative real numbers) with the properties:

1. $\|x\| > 0$ if $x \neq 0$ and $\|x\| = 0$ implies $x = 0$.
2. $\|\alpha x\| = |\alpha| \, \|x\|$ for $\alpha \in \mathbb{R}$ (or $\alpha \in \mathbb{C}$ if S is a complex vector space).
3. $\|x + y\| \leq \|x\| + \|y\|$ (TRIANGLE INEQUALITY).

For example, in $\mathbb{R}^n$ we can take $\|x\|^2 = \sum_{i=1}^{n} |x_i|^2$ or more generally, $\|x\|^p = \sum_{i=1}^{n} |x_i|^p$. For $p = 2$ we have the usual Euclidean norm, however, for differing values of p the meaning of size changes. For example, in $\mathbb{R}^2$ the "unit sphere" $\|x\| = 1$ with $p = 2$ is the usual sphere $x_1^2 + x_2^2 = 1$ while with $p = 1$ the "unit sphere" $|x| + |y| = 1$ is a diamond (a square with vertices on the axes). If we take $p = \infty$ and $\|x\| = \max_i |x_i|$, then the "unit sphere" $\|x\| = 1$ is a square.

A space with a norm is called a NORMED VECTOR SPACE. Although there are many ways to define norms, one direct way to find a norm is if we have an inner product space. If $x \in S$ and if $\langle x, y \rangle$ is an inner product for S then $\|x\| = \sqrt{\langle x, x \rangle}$ is called the INDUCED NORM for S. For example, in $\mathbb{R}^n$,

$$\|x\| = \left(\sum_{i=1}^{n} |x_i|^2 \right)^{1/2},$$

is the induced norm and for real continuous functions with inner product $\langle f, g \rangle = \int_a^b f(x)g(x)dx$, the induced norm is

$$\|f\| = \left(\int_a^b |f(x)|^2 \, dx \right)^{1/2}.$$

For n times continuously differentiable complex valued functions, the inner product

$$\langle f, g \rangle = \int_a^b \sum_{j=0}^{n} f^{(j)}(x)\overline{g^{(j)}(x)}dx$$

induces the norm

$$\|f\| = \left(\int_a^b \sum_{j=0}^{n} \left| f^{(j)}(x) \right|^2 dx \right)^{1/2}.$$

For an induced norm, $\|x\| > 0$ if $x \neq 0$, and $\|\alpha x\| = |\alpha| \, \|x\|$ for scalars α. To see that the triangle inequality also holds, we need to first prove an important result:

Theorem 1.1 (Schwarz Inequality)

For an inner product space,

$$|\langle x, y \rangle|^2 \leq \|x\|^2 \cdot \|y\|^2.$$

Proof

For $x, y \in S, \alpha$ a scalar,

$$0 \leq \|x - \alpha y\|^2 = \|x\|^2 - 2 \operatorname{Re} \langle x, \alpha y \rangle + |\alpha|^2 \|y\|^2.$$

If $y \neq 0$, we make $\|x - \alpha y\|^2$ as small as possible by picking $\alpha = \frac{\langle x, y \rangle}{\|y\|^2}$, from which the Schwarz inequality is immediate.

Using the Schwarz inequality, we see that the triangle inequality holds for any induced norm, since

$$\|x + y\|^2 = \|x\|^2 + 2 \operatorname{Re} \langle x, y \rangle + \|y\|^2$$

$$\leq \|x\|^2 + 2\|x\| \|y\| + \|y\|^2 = (\|x\| + \|y\|)^2$$

or

$$\|x + y\| \leq \|x\| + \|y\|.$$

One can show that in $\mathbb{R}^3$, $\cos \theta = \frac{\langle x, y \rangle}{\|x\| \|y\|}$ where $\langle x, y \rangle$ is the usual Euclidean inner product and θ is the angle between the vectors x, y. In other real inner product spaces, such as the space of continuous functions, we take this as our definition of $\cos \theta$. The law of cosines and Pythagoras' Theorem are immediate. For example,

$$\|x - y\|^2 = \|x\|^2 - 2\|x\| \|y\| \cos \theta + \|y\|^2$$

is the law of cosines, and if $\cos \theta = 0$ then

$$\|x + y\|^2 = \|x\|^2 + \|y\|^2 \qquad \text{(PYTHAGORAS)}.$$

Following this definition of $\cos \theta$, we say that the vectors x and y are ORTHOGONAL if $\langle x, y \rangle = 0$.

Examples

1. $(1, 0)$ is orthogonal to $(0, 1)$ in $\mathbb{R}^2$ using the Euclidean inner product since $\langle (0, 1), (1, 0) \rangle = 0$.

2. With inner product $\langle f, g \rangle = \int_0^{2\pi} f(x)g(x)dx$, $\sin x$ and $\sin 2x$ are orthogonal on $[0, 2\pi]$ since

$$\int_0^{2\pi} \sin x \sin 2x \, dx = 0.$$

3. With inner product $\langle f, g \rangle = \int_0^1 f(x)g(x)dx$, the angle between the function 1 and x on $[0, 1]$ is $30°$ since $\cos\theta = \frac{(1,x)}{\|1\|\cdot\|x\|} = \sqrt{3}/2$.

Orthogonal vectors are nice for a number of reasons. If the set of vectors

$$\{\phi_1, \phi_2, \phi_3, \ldots, \phi_n\} \in S, \quad \phi_i \neq 0,$$

are mutually orthogonal, that is, $\langle \phi_i, \phi_j \rangle = 0$ for $i \neq j$, then they form a linearly independent set. To see this suppose we can find $\alpha_1, \alpha_2, \ldots, \alpha_n$ so that

$$\alpha_1\phi_1 + \alpha_2\phi_2 + \cdots + \alpha_n\phi_n = 0.$$

The inner product of this expression with ϕ_j is

$$\langle \phi_j, \alpha_1\phi_1 + \alpha_2\phi_2 + \cdots + \alpha_n\phi_n \rangle = \langle \phi_j, \alpha_j\phi_j \rangle = \bar{\alpha}_j \|\phi_j\|^2 = 0$$

so that $\alpha_j = 0$ for $j = 1, 2, \ldots, n$, provided ϕ_j is nontrivial.

If the vectors $\{\phi_1, \phi_2, \ldots, \phi_n\}$ form a basis for S, we can represent any element $f \in S$ as a linear combination of ϕ_j,

$$f = \sum_{j=1}^n \alpha_j\phi_j.$$

Taking the inner product of f with ϕ_i we find a matrix equation for the coefficients α_j, $B\alpha = \eta$ where $B = (b_{ij})$, $b_{ij} = \langle \phi_i, \phi_j \rangle$, $\alpha = (\alpha_i)$, $\eta_i = \langle f, \phi_i \rangle$. This matrix problem is always uniquely solvable since the $\{\phi_i\}$ are linearly independent. However, the solution is simplified enormously when $\{\phi_i\}$ are mutually orthogonal, since then B is a diagonal matrix and easily inverted to yield

$$\alpha_i = \frac{\langle f, \phi_i \rangle}{\|\phi_i\|^2}.$$

The coefficient α_i carries with it an appealing geometrical concept in $\mathbb{R}^n$. If we want to "project" a vector f onto ϕ we might imagine shining a light onto ϕ and measuring the shadow cast by f. From simple trigonometry, the length of this shadow is $\|f\| \cos\theta$ where θ is the angle in the plane defined by f and ϕ, between the vectors f and ϕ. This length is exactly $\frac{\langle f, \phi \rangle}{\|\phi\|}$ and the vector of the shadow, $\frac{\langle f, \phi \rangle \phi}{\|\phi\|^2}$, is called the projection of f onto ϕ. Notice that in the proof of the Schwarz inequality given above, we made the quantity $\|x - \alpha y\|$ as small as possible by picking αy to be the projection of x onto y.

The GRAM-SCHMIDT ORTHOGONALIZATION PROCEDURE is an inductive technique to generate a mutually orthogonal set from any linearly independent set of vectors. Given the vectors $x_1, x_2, x_3, \ldots, x_n$, we set $\phi_1 = x_1$. If we take $\phi_2 = x_2 - \alpha\phi_1$, the requirement $\langle \phi_2, \phi_1 \rangle = 0$ determines

$$\alpha = \frac{\langle x_2, \phi_1 \rangle}{\|\phi_1\|^2}$$

so

$$\phi_2 = x_2 - \frac{\langle x_2, \phi_1 \rangle}{\|\phi_1\|^2}\phi_1.$$

In other words, we get ϕ_2 by subtracting from x_2 the projection of x_2 onto ϕ_1. Proceeding inductively, we find

$$\phi_n = x_n - \sum_{j=1}^{n-1} \frac{\langle x_n, \phi_j \rangle \, \phi_j}{\langle \phi_j, \phi_j \rangle}$$

from which it is clear that $\langle \phi_k, \phi_n \rangle = 0$ for $k < n$. We are left with mutually orthogonal vectors which have the same span as the original set.

Example

Consider the set of continuous functions $C[a, b]$, and the powers of x, $\{1, x, x^2, x^3, \ldots, x^n\}$, which on any interval $[a, b]$ $(a \neq b)$ form a linearly independent set.

Take $[a, b] = [-1, 1]$ and the inner product

$$\langle f, g \rangle = \int_{-1}^{1} f(x)g(x)dx.$$

Using the Gram-Schmidt procedure, we find

$$\phi_0 = 1$$

$$\phi_1 = x$$

$$\phi_2 = x^2 - 1/3$$

$$\phi_3 = x^3 - 3x/5$$

and so on. The functions thus generated are called the LEGENDRE POLYNOMIALS. Other sets of orthogonal polynomials are found by changing the domain or the definition of the inner product. We will see more of these in Chapter 2.

1.2   Spectral Theory for Matrices

In a vector space, there are many different choices for a basis. The main message of this section and, in fact, of this text, is that by a careful choice of basis, many problems can be "diagonalized." This is, in fact, the main idea behind TRANSFORM THEORY.

Suppose we want to solve the matrix problem

$$Ax = y$$

where A is an $n \times n$ matrix. We view the entries of the vectors x and y as the coordinates of vectors relative to some basis, usually the natural basis. That is,

$$x = (x_1, x_2, \ldots, x_n)^T = \sum_{j=1}^{n} x_j e_j^T.$$

The same vector x would have different coordinates if it were expressed relative to a different basis.

Suppose the matrix equation $Ax = y$ is expressed in coordinates relative to a basis $\{\psi_1, \psi_2, \ldots, \psi_n\}$ and we wish to re-express the problem relative to some new basis $\{\phi_1, \phi_2, \ldots, \phi_n\}$. What does this change of basis do to the original representation of the matrix problem?

Since $\{\psi_j\}$ forms a basis, there are numbers c_{ji} for which $\phi_i = \sum_{j=1}^{n} c_{ji}\psi_j$. The numbers c_{ji} are the coordinates of the vector ϕ_i relative to the original basis $\{\psi_j\}$. In the case that $\{\psi_j\}$ is the natural basis, the matrix $C = (c_{ij})$ is exactly the matrix with vectors ϕ_i in its ith column. (Notice that the convention for the indices of C, namely $\phi_i = \sum_{j=1}^{n} c_{ji}\psi_j$ is not the same as for matrix multiplication.)

If $\hat{x}_i$ and x_i' are coordinates of a vector x relative to the basis $\{\psi_i\}$ and $\{\phi_i\}$, respectively, then $x = \sum_{j=1}^{n} \hat{x}_j\psi_j = \sum_{i=1}^{n} x_i'\phi_i = \sum_{i=1}^{n}\sum_{j=1}^{n} x_i'c_{ji}\psi_j$ so that $\hat{x}_j = \sum_{i=1}^{n} c_{ji}x_i'$. Written in vector notation, $\hat{x} = Cx'$ where $C = (c_{ij})$ is the matrix of coefficients which expresses the basis $\{\phi_i\}$ in terms of the basis $\{\psi_i\}$.

Now the original problem $A\hat{x} = \hat{y}$ becomes $A'x' = C^{-1}ACx' = y'$, where $A' = C^{-1}AC$, $\hat{x}, \hat{y}$ are the coordinates of x, y relative to the original basis $\{\psi_i\}$, and x', y' are the coordinates of x, y relative to the new basis $\{\phi_i\}$. The transformation $A' = C^{-1}AC$ is called a SIMILARITY TRANSFORMATION, and we see that all similarity transformations are equivalent to a change of basis and vice versa. Two matrices are said to be EQUIVALENT if there is a similarity transformation between them.

This transformation can be made a bit clearer by examining what happens when the original basis is the natural basis. If A_E is the representation of A with respect to the natural basis and $\{\phi_i\}$ is the new basis, then C_1 is the matrix whose columns are the vectors ϕ_i. The representation of A with respect to $\{\phi_i\}$ is $A_1 = C_1^{-1} A_E C_1$. Similarly, if C_2 has as its columns the basis vectors $\{\psi_i\}$, then the representation of A with respect to $\{\psi_i\}$ is $A_2 = C_2^{-1} A_E C_2$. Thus, if we are given A_2, the representation of A with respect to $\{\psi_i\}$, and wish to find A_1, the representation of A with respect to $\{\phi_i\}$, we find that

$$A_1 = C_1^{-1} C_2 A_2 C_2^{-1} C_1$$

$$= \left(C_2^{-1} C_1\right)^{-1} A_2 \left(C_2^{-1} C_1\right).$$

In other words, the matrix C that takes A_2 to A_1 is $C = C_2^{-1} C_1$ where C_1 has ϕ_i as columns and C_2 has ψ_i as columns.

Under what conditions is there a change of basis (equivalently, a similarity transformation) that renders the matrix A diagonal? Apparently we must find a matrix C so that $AC = C\Lambda$ where Λ is diagonal, that is, the column vectors of C, say y_i, must satisfy $Ay_i = \lambda_i y_i$, where λ_i is the ith diagonal element of Λ.

Definition

An EIGENPAIR of A is a pair (λ, x), $\lambda \in \mathbb{C}$, $x \in \mathbb{C}^n$ satisfying

$$Ax = \lambda x, \qquad x \neq 0.$$

The vector x is called the EIGENVECTOR and λ is called the EIGENVALUE. A number λ is an eigenvalue of A if and only if it is a root of the polynomial $\det(A - \lambda I) = 0$, called the CHARACTERISTIC POLYNOMIAL for A.

Our motivation here for finding eigenvectors is that they provide a way to represent a matrix operator as a diagonal operator. Another important interpretation of eigenvectors is geometrical. If we view the matrix A as a transformation that transforms one vector into another, then the equation $Ax = \lambda x$ expresses the fact that some vector, when it is transformed by A, has its direction unchanged, even though the length of the vector may have changed. For example, a rigid body rotation (in three dimensions) always has an axis about which the rotation takes place, which therefore is the invariant direction. A similar statement about invariance under transformation shows why the function $f(x) = e^x$ is such an important function, since e^x is invariant under the operation of differentiation.

Theorem 1.2

Suppose A is an $n \times n$ matrix

1. If A has n linearly independent eigenvectors, there is a change of basis in $\mathbb{C}^n$ so that relative to the new basis A is diagonal.

2. If A is real and has n linearly independent real eigenvectors, there is a real change of basis in $\mathbb{R}^n$ so that relative to the new basis A is diagonal.

3. If C is the matrix whose columns are the eigenvectors of A, then $C^{-1}AC = \Lambda$ is the diagonal matrix of eigenvalues.

Proof

To prove this theorem, suppose $x_1, x_2, \ldots, x_n$ are linearly independent eigenvectors of A, $Ax_i = \lambda_i x_i$. Let C be the matrix with columns x_i. Then we have

$$
\begin{aligned}
AC &= A(x_1, \ldots, x_n) = (Ax_1, \ldots, Ax_n) = (\lambda_1 x_1, \ldots, \lambda_n x_n) \\
&= (x_1, \ldots, x_n) \begin{bmatrix} \lambda_1 & & & \\ & \lambda_2 & & 0 \\ & & \ddots & \\ 0 & & & \lambda_n \end{bmatrix} = C\Lambda,
\end{aligned}
$$

where Λ is diagonal, or

$$
C^{-1}AC = \Lambda.
$$

Examples

1. The matrix

$$
A = \begin{pmatrix} 3 & 1 \\ 1 & 3 \end{pmatrix}
$$

has eigenpairs $\lambda_1 = 4$, $x_1 = (1,1)^T$, and $\lambda_2 = 2$, $x_2 = (1,-1)^T$. The vectors x_1, x_2 are real and linearly independent. The matrix

$$
C = \begin{pmatrix} 1 & 1 \\ 1 & -1 \end{pmatrix}
$$

gives the required change of basis and

$$
C^{-1}AC = \begin{pmatrix} 4 & 0 \\ 0 & 2 \end{pmatrix}.
$$

2. The ROTATIONAL MATRIX

$$
A = \begin{pmatrix} \sin\theta & \cos\theta \\ -\cos\theta & \sin\theta \end{pmatrix}
$$

has eigenvalues $\lambda = \sin\theta \pm i\cos\theta$ and eigenvectors $(\mp i, 1)^T$. There is no real change of basis which diagonalizes A, although it can be diagonalized using complex basis vectors.

3. The matrix

$$A = \begin{pmatrix} 0 & 1 \\ 0 & 0 \end{pmatrix}$$

has characteristic polynomial $\lambda^2 = 0$, so there are two eigenvalues $\lambda = 0$, counting multiplicity. However, there is only one eigenvector $x_1 = (1, 0)^T$. There is no change of basis which diagonalizes A.

We need to determine when there are n linearly independent eigenvectors. The problem is that, even though the polynomial equation

$$\det(A - \lambda I) = 0$$

always has n roots, counting multiplicity, (called the ALGEBRAIC MULTI-PLICITY OF λ) there need not be an equal number of eigenvectors (called the GEOMETRIC MULTIPLICITY of λ). It is this possible deficiency that is of concern. In general, the geometric multiplicity is no greater than the algebraic multiplicity for an eigenvalue λ. For any eigenvalue λ, the geometric multiplicity is at least one, since $\det(A - \lambda I) = 0$ implies there is at least one nontrivial vector x for which $(A - \lambda I)x = 0$. In fact, every square matrix A has at least one eigenpair (although it may be complex).

Theorem 1.3

If A has n distinct eigenvalues, then it has n linearly independent eigen-vectors.

The proof of this statement is by induction. For each eigenvalue λ_k, we let x_k be the corresponding eigenvector. For $k = 1$, the single vector $x_1 \neq 0$ forms a linearly independent set. For the induction step, we suppose $x_1, x_2, \ldots, x_{k-1}$ are linearly independent. We try to find scalars $\alpha_1, \alpha_2, \ldots, \alpha_k$ so that

$$\alpha_1 x_1 + \alpha_2 x_2 + \cdots + \alpha_k x_k = 0.$$

If we multiply this expression by A and use that $Ax_i = \lambda_i x_i$, we learn that

$$\alpha_1 \lambda_1 x_1 + \alpha_2 \lambda_2 x_2 + \cdots + \alpha_k \lambda_k x_k = 0.$$

If we multiply instead by λ_k, we obtain

$$\lambda_k \alpha_1 x_1 + \lambda_k \alpha_2 x_2 + \cdots + \lambda_k \alpha_k x_k = 0.$$

Comparing these two expressions, we find that

$$\alpha_1 (\lambda_1 - \lambda_k) x_1 + \cdots + \alpha_{k-1} (\lambda_{k-1} - \lambda_k) x_{k-1} = 0.$$

It follows that $\alpha_1 = \alpha_2 = \cdots = \alpha_{k-1} = 0$, since $x_1, x_2, \ldots, x_{k-1}$ are assumed to be linearly independent and λ_k is distinct from $\lambda_1, \lambda_2, \ldots, \lambda_{k-1}$, and finally $\alpha_k = 0$ since x_k is assumed to be nontrivial.

There is another important class of matrices for which diagonalization is always possible, namely the SELF-ADJOINT MATRICES.

Definition

For any matrix A, the ADJOINT of A is defined as the matrix A^* where $\langle Ax, y \rangle = \langle x, A^*y \rangle$, for all x, y in $\mathbb{C}^n$. A matrix A is SELF-ADJOINT if $A^* = A$.

Definition

If $A = (a_{ij})$, the TRANSPOSE of A is $A^T = (a_{ji})$.

To find the adjoint matrix A^* explicitly, note that (using the Euclidean inner product)

$$\langle Ax, y \rangle = \sum_{i=1}^{n} \sum_{j=1}^{n} a_{ij} x_j \bar{y}_i = \sum_{j=1}^{n} x_j \left(\sum_{i=1}^{n} a_{ij} \bar{y}_i \right)$$

so that $A^* = \bar{A}^T$. Important subclasses of matrices are the REAL SYMMETRIC matrices which have $A = A^T$, and the HERMITIAN matrices which satisfy $A = \bar{A}^T$. Both of these are self-adjoint since $A = A^*$.

Theorem 1.4

If A is self-adjoint the following statements are true:

1. $\langle Ax, x \rangle$ is real for all x,
2. All eigenvalues are real,
3. Eigenvectors of distinct eigenvalues are orthogonal,
4. The eigenvectors form an n-dimensional basis,
5. The matrix A can be diagonalized.

The proof of this theorem is as follows:

1. If $A = A^*$, then $\langle Ax, x \rangle = \langle x, A^*x \rangle = \langle \overline{Ax, x} \rangle$ so $\langle Ax, x \rangle$ is real.
2. If $Ax = \lambda x$, then $\langle Ax, x \rangle = \langle \lambda x, x \rangle = \lambda \langle x, x \rangle$. Since $\langle Ax, x \rangle$ and $\langle x, x \rangle$ are real, λ must be real.
3. Consider the eigenpairs (λ, x), (μ, y) with $\lambda \neq \mu$. Then

$$\lambda \langle x, y \rangle = \langle \lambda x, y \rangle = \langle Ax, y \rangle = \langle x, Ay \rangle = \langle x, \mu y \rangle = \mu \langle x, y \rangle$$

so that

$$(\lambda - \mu) \langle x, y \rangle = 0.$$

Since $\lambda \neq \mu$, then $\langle x, y \rangle = 0$.

The proof of item 4 requires more background on linear manifolds.

Definition

1. A LINEAR MANIFOLD $M \subset S$ is a subset of S which is closed under vector addition and scalar multiplication.
2. An INVARIANT MANIFOLD M for the matrix A is a linear manifold $M \subset S$ for which $x \in M$ implies $Ax \in M$.

Examples

1. $N(A)$, the null space of A, is the set of all x for which $Ax = 0$. If x and y satisfy $Ax = 0$ and $Ay = 0$, then $A(\alpha x + \beta y) = 0$ so that $N(A)$ is a linear manifold. Since $A0 = 0$ belongs to the manifold $N(A)$, $Ax = 0$ is in $N(A)$ as well, hence $N(A)$ is invariant. As an example, the null space of the matrix

$$A = \begin{pmatrix} 0 & 1 \\ 0 & 0 \end{pmatrix}$$

is spanned by the vector $(1,0)^T$.

2. $R(A)$, the range of A, is the set of all x for which $Ay = x$ for some y. Clearly $R(A)$ is a linear manifold and it is invariant since if $x \in R(A)$ then surely $Ax \in R(A)$.

We observe that if M is an invariant manifold over the complex scalar field for some matrix A, there is at least one vector $x \in M$ with $Ax = \lambda x$ for some $\lambda \in \mathbb{C}$. To verify this, notice that since M lies in an n-dimensional space, it has a basis, say $\{x_1, x_2, \ldots, x_k\}$. For any $x \in M$, $Ax \in M$ so x and Ax have a representation relative to the basis $\{x_1, x_2, \ldots, x_k\}$. In particular, take $x = \sum_{i=1}^{k} \alpha_i x_i$ and $Ax_i = \sum_{j=1}^{k} \beta_{ji} x_j$. To solve $Ax - \lambda x = 0$, we must have

$$Ax = \sum_{i=1}^{k} \alpha_i (Ax_i) = \lambda \sum_{i=1}^{k} \alpha_i x_i,$$

so that

$$\sum_{i=1}^{k} \alpha_i \left(\sum_{j=1}^{k} \beta_{ji} x_j - \lambda x_i \right) = 0,$$

or

$$\sum_{i=1}^{k} \alpha_i \sum_{j=1}^{k} (\beta_{ji} - \lambda \delta_{ij}) x_j = 0.$$

Changing the order of summation and observing that $\{x_j\}$ are linearly independent, we have

$$\sum_{i=1}^{k} (\beta_{ji} - \lambda \delta_{ij}) \alpha_i = 0, \qquad j = 1, 2, \ldots, k,$$

which in matrix notation is $(B - \lambda I)\alpha = 0$ where $B = (\beta_{ij})$ is a $k \times k$ matrix. Of course, we noted earlier that every square matrix has at least one eigenpair.

We are now ready to prove item 4, namely that the eigenvectors of an $n \times n$ self-adjoint matrix form an n-dimensional orthogonal basis. Suppose $A = A^*$. Then there is at least one eigenpair (λ_1, x_1) with $Ax_1 = \lambda_1 x_1$, and λ_1 real. If, in addition, A is real, then the eigenvector x_1 is real and it forms a linearly independent set (of one element). For the induction step, suppose we have found $k - 1$ mutually orthogonal eigenvectors x_i, $Ax_i = \lambda x_i$ with λ_i real for $i = 1, 2, \ldots, k - 1$. We form the linear manifold

$$M_k = \left\{ x \mid \langle x, x_j \rangle = 0, \quad j = 1, 2, \ldots, k-1 \right\},$$

called the orthogonal complement of the $k - 1$ orthogonal eigenvectors x_1, $x_2, \ldots, x_{k-1}$. This manifold is invariant for A since

$$\langle Ax, x_j \rangle = \langle x, Ax_j \rangle = \lambda_j \langle x, x_j \rangle = 0 \ .$$

for $j = 1, 2, \ldots, k - 1$. Therefore, M_k contains (at least) one eigenvector x_k corresponding to a real eigenvalue λ_k and clearly $\langle x_k, x_j \rangle = 0$ for $j < k$, since $x_k \in M_k$. The eigenvalue λ_k is real since all eigenvalues of A are real.

In summary, we can state the main result of this section:

Theorem 1.5 (SPECTRAL DECOMPOSITION THEOREM FOR SELF-ADJOINT MATRICES)

If A is an $n \times n$ self-adjoint matrix, there is an orthogonal basis

$$\{x_1, x_2, \ldots, x_n\}$$

for which

1. $Ax_i = \lambda_i x_i$ with λ_i real.
2. $\langle x_i, x_j \rangle = \delta_{ij}$ (orthogonality).
3. The matrix Q with x_j as its jth column vector is unitary, that is,

$$Q^{-1} = Q^*.$$

4. $Q^* A Q = \Lambda$ where Λ is diagonal with real diagonal entries λ_i.

Suppose we wish to solve $Ax = y$ where $A = A^*$. We now know that relative to the basis of eigenvectors, this is a diagonal problem. That is, setting $x = Qx'$, $y = Qy'$ we find that $Q^*AQx' \equiv \Lambda x' = y'$. If all the eigenvalues are nonzero, the diagonal matrix Λ is easily inverted for $x' = \Lambda^{-1}y'$. Re-expressing this in terms of the original basis we find that $x = Q\Lambda^{-1}Q^*y$. These operations can be summarized in the following commuting diagram.

$$
\begin{array}{ccccccc}
 & Ax & = & y & \xrightarrow{\text{find } A^{-1} \text{ directly}} & x & = & A^{-1}y \\
\text{transform} & \left. \begin{array}{c} x = Qx' \\ y = Qy' \end{array} \right\downarrow & & & & \left\uparrow \begin{array}{cc} x' = Q^*x & \text{inverse} \\ y' = Q^*y & \text{transform} \end{array} \right. \\
 & \Lambda x' & = & y' & \xrightarrow{\Lambda^{-1} \text{ diagonal system}} & x' & = & \Lambda^{-1}y'
\end{array}
$$

The solution of $Ax = y$ can be found directly by finding the inverse of A (the top of the diagram) or indirectly in three steps by making a change of coordinate system to a diagonal problem (the leftmost vertical descent), solving the diagonal problem (the bottom of the diagram), and then changing coordinates back to the original coordinate system (the rightmost vertical ascent).

The beauty of diagonalization can be illustrated on a piece of graph paper. Suppose we wish to solve $Ax = b$ where

$$A = \begin{pmatrix} 4 & -1 \\ 2 & 1 \end{pmatrix}.$$

We first calculate the eigenvalues and eigenvectors of A and find $\lambda_1 = 2$,

$$y_1 = \begin{pmatrix} 1 \\ 2 \end{pmatrix}$$

and $\lambda_2 = 3$,

$$y_2 = \begin{pmatrix} 1 \\ 1 \end{pmatrix}.$$

We plot these vectors on a piece of rectangular graph paper. Any vector b can be represented uniquely as a linear combination of the two basis vectors y_1, y_2. To find the solution of $Ax = b$ we take half of the first coordinate of b and one third of the second coordinate of b with respect to the basis of eigenvectors. In this way, for example, if $b = (4, 2)$ we find graphically (see Fig. 1.1), that $x = (1, 0)$ solves $Ax = b$ relative to the natural basis.

As was suggested in the preface, this diagonalization procedure is the basis on which most transform methods work, and will be reiterated throughout this text. Hopefully, this commuting diagram will become firmly emblazoned in your mind before the end of this text.

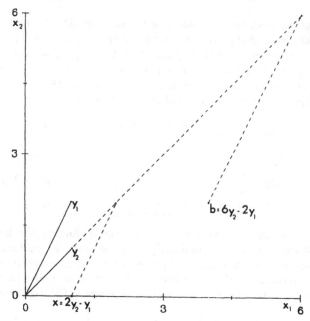

FIGURE 1.1   Graphical solution of the 2×2 matrix equation

$$\begin{pmatrix} 4 & -1 \\ 2 & 1 \end{pmatrix} \begin{pmatrix} x_1 \\ x_2 \end{pmatrix} = \begin{pmatrix} 4 \\ 2 \end{pmatrix},$$

using the eigenvectors of the matrix A as coordinates.

1.3   Geometrical Significance of Eigenvalues— The Minimax Principle

Aside from providing a natural basis in $\mathbb{R}^n$ in which to represent linear problems, eigenvectors and eigenvalues have significance relating to the geometry of the linear transformation A.

For a real symmetric matrix A, the quadratic form $Q(x) = \langle Ax, x \rangle$ produces a real number for every vector x in $\mathbb{R}^n$. Since $Q(x)$ is a continuous function in $\mathbb{R}^n$, it attains a maximum on the closed, bounded set of vectors $\|x\| = 1$. Suppose the maximum is attained at $x = x_1$. Then for every unit vector x orthogonal to x_1, $Q(x) \le Q(x_1)$. But, on the subset $\langle x, x_1 \rangle = 0$, $\|x\| = 1$, $Q(x)$ again attains a maximum at, say, $x = x_2$. Continuing inductively in this way we produce a set of mutually orthognal unit vectors, x_1, $x_2, \ldots, x_n$ at which the local extremal values of $Q(x)$ are attained on the set $\|x\| = 1$. We shall show that the x_i are in fact the eigenvectors of A and that the values $Q(x_i)$ are the corresponding eigenvalues.

Theorem 1.6 (Maximum Principle)

If A is a real symmetric matrix and $Q(x) = \langle Ax, x \rangle$, the following statements hold:

1. $\lambda_1 = \max_{\|x\|=1} Q(x) = Q(x_1)$ is the largest eigenvalue of the matrix A and x_1 is the eigenvector corresponding to eigenvalue λ_1.

2. (inductive statement). Let $\lambda_k = \max Q(x)$ subject to the constraints

   a. $\langle x, x_j \rangle = 0$, $j = 1, 2, \ldots, k-1$,

   b. $\|x\| = 1$.

   Then $\lambda_k = Q(x_k)$ is the kth eigenvalue of A, $\lambda_1 \geq \lambda_2 \geq \cdots \geq \lambda_k$ and x_k is the corresponding eigenvector of A.

Example

Consider the matrix

$$A = \begin{pmatrix} 3 & 1 \\ 1 & 3 \end{pmatrix},$$

then $Q(x) = \langle Ax, x \rangle = 3y_1^2 + 2y_1 y_2 + 3y_2^2$, where

$$x = \begin{pmatrix} y_1 \\ y_2 \end{pmatrix}.$$

We change to polar coordinates by setting $y_1 = \cos\theta$, $y_2 = \sin\theta$ so that $Q(\theta) = 3 + \sin 2\theta$. Clearly $\max Q(\theta) = 4$ occurs at $\theta = \pi/4$ and $\min Q(\theta) = 2$ at $\theta = -\pi/4$. It follows that $\lambda_1 = 4$, $x_1 = (1,1)$ and $\lambda_2 = 2$, $x_2 = (1,-1)$.

Example

In differential geometry, one finds that the curvature tensor of a surface is a symmetric 2×2 matrix. As a result, the directions of maximal and minimal curvature on any two-dimensional surface are orthogonal. You can check this out by a careful look at the shape of your forearm, for example.

To understand the geometrical significance of the maximum principle, we view the quadratic form $Q(x) = \langle Ax, x \rangle$ as a radial map of the sphere. That is, to each point on the sphere $\|x\| = 1$, associate $Q(x)$ with the point in $\mathbb{R}^n$, $Q(x)$ units from the origin in the same radial direction as the vector x. In this way, if A is positive definite, (a positive definite matrix A is one for which $\langle Ax, x \rangle > 0$ for all $x \neq 0$), $Q(x)$ looks like an ellipsoid (or football) in $\mathbb{R}^n$. (It is not exactly an ellipsoid, although the level surface $Q(x) = 1$ is an ellipsoid.) The maximum of $Q(x)$ occurs along the major axis of the football-shaped object. If we intersect it with a plane orthogonal to the major axis, then restricted to this plane, the maximum of $Q(x)$ occurs along the semi-major axis, and so on.

To verify statement 1 we apply the method of Lagrange multipliers to $Q(x)$. Specifically, we seek to maximize the function $P(x) = Q(x) - \mu(\langle x, x \rangle - 1)$. If the maximum occurs at x_1, then $P(x_1 + h) - P(x_1) \leq 0$ for any h. Expanding $P(x_1 + h)$ we find that

$$P(x_1 + h) - P(x_1) = 2(\langle Ax_1, h \rangle - \mu \langle x_1, h \rangle) + \langle Ah, h \rangle - \mu \langle h, h \rangle.$$

Notice that $\langle Ax_1, h \rangle - \mu \langle x_1, h \rangle$ is linear in h and $\langle Ah, h \rangle - \mu \langle h, h \rangle$ is quadratic in h. The quadratic expression has fixed sign, and can be chosen arbitrarily small. However, unless the linear expression is identically zero, its sign is arbitrary and for all sufficiently small h, determines the sign of $P(x_1 + h) - P(x_1)$. Thus, we must have $\langle Ax_1 - \mu x_1, h \rangle = 0$ for all h. Taking $h = Ax_1 - \mu x_1$ we conclude that

$$Ax_1 - \mu x_1 = 0,$$

so that x_1 is the eigenvector of A. Furthermore, $Q(x_1) = \langle Ax_1, x_1 \rangle = \mu \langle x_1, x_1 \rangle = \mu$, since $\|x_1\| = 1$ so that $\max Q(x) = \mu$ and the vector which maximizes $Q(x)$ is an eigenvector of A.

To verify the inductive step, we again use Lagrange multipliers and maximize

$$P(x) = \langle Ax, x \rangle - \mu(\langle x, x \rangle - 1) - \sum_{j=1}^{k-1} \alpha_j \langle x, x_j \rangle,$$

where $x_1, x_2, \ldots, x_{k-1}$ are eigenvectors of A corresponding to known eigenvalues $\lambda_1 \geq \lambda_2 \ldots \geq \lambda_{k-1}$. Note that $\frac{\partial P(x)}{\partial \mu} = 0$ implies that $\langle x, x \rangle = 1$ and $\frac{\partial P(x)}{\partial \alpha_j} = 0$ implies that $\langle x, x_j \rangle = 0$. Suppose x_k maximizes $P(x)$ so that $P(x_k + h) - P(x_k) \leq 0$. Using that $\langle x, x_j \rangle = 0$ for $j = 1, 2, \ldots, k-1$, we find that,

$$P(x_k + h) - P(x_k) = 2(\langle Ax_k, h \rangle - \mu \langle x_k, h \rangle) + \langle Ah, h \rangle - \mu \langle h, h \rangle.$$

Using the same argument as before, we require $\langle Ax_k - \mu x_k, h \rangle = 0$ for all h. In other words, $Ax_k - \mu x_k = 0$ and x_k is an eigenvector of A. Finally, $Q(x_k) = \langle Ax_k, x_k \rangle = \mu \langle x_k, x_k \rangle = \mu$ since $\|x_k\| = 1$ so that $\max Q(x)$ is the eigenvalue and the vector which maximizes $Q(x)$ is an eigenvector of A. Clearly $\lambda_k \leq \lambda_{k-1} \leq \cdots \leq \lambda_1$.

From the maximum principle we can find the kth eigenvalue and eigenvector only after the previous $k-1$ eigenvectors are known. It is useful to have a characterization of λ_k that makes no reference to other eigenpairs. This is precisely the motivation of the

Theorem 1.7 (Courant Minimax Principle)

For any real symmetric matrix A,

$$\lambda_k = \min_C \max_{\substack{\|x\|=1 \\ Cx=0}} \langle Ax, x \rangle,$$

where C is any $(k-1) \times n$ matrix.

The geometrical idea of this characterization is as follows. For any non-trivial matrix C, the constraint $Cx = 0$ corresponds to the requirement that x lie on some $n - k + 1$ dimensional hyperplane in $\mathbb{R}^n$. We find λ_k by first maximizing $Q(x)$ with $\|x\| = 1$ on the hyperplane and then we vary the hyperplane until this maximum is as small as possible.

To prove the minimax principle we note that since A is symmetric, there is a unitary matrix Q that diagonalizes A, so that $A = Q\Lambda Q^T$ where Λ is the diagonal matrix of eigenvalues $\lambda_1 \geq \lambda_2 \geq \cdots \geq \lambda_n$. It follows that

$$\langle Ax, x \rangle = \langle Q\Lambda Q^T x, x \rangle = \langle \Lambda y, y \rangle = \sum_{i=1}^n \lambda_i y_i^2,$$

where $y = Q^T x$ and $Cx = 0$ if and only if $CQy = By = 0$, where $B = CQ$. We define μ by

$$\mu = \min_C \left(\max_{\substack{\|x\|=1 \\ Cx=0}} \langle Ax, x \rangle \right) = \min_B \left\{ \max_{\substack{\|y\|=1 \\ By=0}} \sum_{i=1}^n \lambda_i y_i^2 \right\}.$$

If we pick the matrix B so that $By = 0$ implies $y_1 = y_2 = \cdots = y_{k-1} = 0$, then $\mu \leq \max_{\|y\|=1} \sum_{i=k}^n \lambda_i y_i^2 = \lambda_k$. On the other hand, since the matrix B has rank $\leq k-1$, the solution of $By = 0$ is not unique, and we can still satisfy the equation $By = 0$ with the $n - k$ additional restrictions $y_{k+1} = y_{k+2} = \cdots = y_n = 0$. With these restrictions, $By = 0$ becomes a system of $k - 1$ equations in k unknowns which always has nontrivial solutions. For this restricted set of vectors, the maximum for each B is possibly reduced. Therefore,

$$\mu \geq \min_B \left\{ \max_{\substack{\|y\|=1 \\ By=0}} \sum_{i=1}^k \lambda_i y_i^2 \right\} \geq \min_B \left\{ \max_{\substack{\|y\|=1 \\ By=0}} \lambda_k \sum_{i=1}^k y_i^2 \right\} \geq \lambda_k.$$

Since $\mu \geq \lambda_k$ and $\mu \leq \lambda_k$, it must be that $\mu = \lambda_k$, as proposed.

Using the minimax principle we can estimate the eigenvalues of the matrix A.

Theorem 1.8

Suppose the quadratic form $Q(x) = \langle Ax, x \rangle$ is constrained by k linear constraints $Bx = 0$ to the quadratic form $\hat{Q}$ in $n - k$ variables. The relative extremal values of Q, denoted $\lambda_1 \geq \lambda_2 \geq \cdots \geq \lambda_n$ and the relative extremal values of $\hat{Q}$, denoted $\hat{\lambda}_1 \geq \hat{\lambda}_2 \geq \cdots \geq \hat{\lambda}_{n-k}$ satisfy

$$\lambda_j \geq \hat{\lambda}_j \geq \lambda_{j+k}, \qquad j = 1, 2, \ldots, n - k.$$

The proof of this statement is as follows. According to the minimax principle, for any quadratic form Q, the relative maxima on the unit sphere are given by

$$\lambda_j = \min_C \left\{ \max_{\substack{\|x\|=1 \\ Cx=0}} Q(x) \right\},$$

where $Cx = 0$ contains $j - 1$ linear constraints on x. If $\hat{x}_i$, $i = 1, 2, \ldots, j - 1$ are the vectors for which the relative maxima of $\hat{Q}(x)$ are attained, then the minimum is attained by taking the columns of C to be $\hat{x}_i$, and

$$\hat{\lambda}_j = \min_C \left\{ \max_{Cx=0} \hat{Q}(x) \right\} = \max_{\substack{(x, \hat{x}_i)=0 \\ i=1,2,\ldots,j-1 \\ \|x\|=1}} \hat{Q}(x) = \max_{\substack{(x, \hat{x}_i)=0 \\ Bx=0 \\ \|x\|=1}} Q(x) \geq \lambda_{j+k},$$

since there are now $j + k - 1$ constraints on $Q(x)$. Similarly,

$$\hat{\lambda}_j = \min_{y_i} \left\{ \max_{\substack{(x, y_i)=0 \\ \|x\|=1 \\ i=1,2,\ldots,j-1}} \hat{Q}(x) \right\} \leq \max_{\substack{(x, x_i)=0 \\ \|x\|=1}} \hat{Q}(x) \leq \max_{\substack{(x, x_i)=0 \\ \|x\|=1}} Q(x) = \lambda_j,$$

where x_i, $i = 1, 2, \ldots, j - 1$ are those vectors at which the first $j - 1$ relative maxima of $Q(x)$ are attained.

The importance of these observations is that we can estimate the eigenvalues of a symmetric matrix A by comparing it with certain smaller matrices $\hat{A}$ corresponding to a constrained system $\hat{Q}$.

For a geometrical illustration of these results, consider a 3×3 positive definite matrix A and its associated quadratic form $Q(x) = \langle Ax, x \rangle$. If we view $Q(x)$ as the "radial" map of the sphere $\|x\| = 1$, the surface thus constructed is football shaped with three principal axis, the major axis, the semi-major axis, and the minor axis with lengths $\lambda_1 \geq \lambda_2 \geq \lambda_3$, respectively. If we cut this football-shaped surface with any plane through the origin, the intersection of $Q(x)$ with the plane is "elliptical" with a major and minor axis with lengths $\hat{\lambda}_1 \geq \hat{\lambda}_2$, respectively. It is an immediate consequence of our preceding discussion that

$$\lambda_1 \geq \hat{\lambda}_1 \geq \lambda_2 \geq \hat{\lambda}_2 \geq \lambda_3.$$

To illustrate the usefulness (and requisite artistry) of these results, suppose we wish to estimate the eigenvalues of the symmetric matrix

$$A = \begin{pmatrix} 1 & 2 & 3 \\ 2 & 3 & 6 \\ 3 & 6 & 8 \end{pmatrix}$$

using the minimax principle. We first notice that the matrix

$$\hat{A} = \begin{pmatrix} 1 & 2 & 3 \\ 2 & 4 & 6 \\ 3 & 6 & 9 \end{pmatrix}$$

is of rank 1 and therefore has two zero eigenvalues. The range of $\hat{A}$ is spanned by the vector $x = (1, 2, 3)^T$ and $\hat{A}$ has positive eigenvalue $\lambda_1 = 14$. It follows (see Problem 1.3.2) that the eigenvalues of A satisfy $\lambda_1 \leq 14$, $\lambda_2 \leq 0$, $\lambda_3 \leq 0$.

To use the minimax principle further we impose the single constraint $x_1 = 0$. With this constraint the extremal values of $Q(x)$ are the eigenvalues λ of the diagonal submatrix

$$\begin{pmatrix} 3 & 6 \\ 6 & 8 \end{pmatrix}.$$

For this matrix $\hat{\lambda}_1 = 12$, $\hat{\lambda}_2 = -1$ so that $\lambda_3 \leq -1 \leq \lambda_2 \leq 0$, $\lambda_1 \geq 12$.
To improve our estimate of λ_1 we observe that $Q(x) = 12$ for

$$x = \begin{pmatrix} 0 \\ 2 \\ 3 \end{pmatrix}.$$

We try

$$x = \begin{pmatrix} 1 \\ 2 \\ 3 \end{pmatrix}$$

and find that $Q(x) = \frac{183}{14} = 13.07$ so that $13.07 \leq \lambda_1 \leq 14$.

An improved estimate of λ_2 is found by imposing the single constraint $x_2 = 0$ for which the extremal values are $\lambda = \frac{9 \pm \sqrt{85}}{2}$. It follows that $-0.1098 \cong \frac{9-\sqrt{85}}{2} \leq \lambda_2 \leq 0$. These estimates are actually quite good since the exact eigenvalues are $\lambda = -1$, -0.076, and 13.076.

To understand further the significance of eigenvalues, it is useful to think about vibrations in a lattice. By a lattice we mean a collection of objects (balls or molecules, for example) which are connected by some restoring force

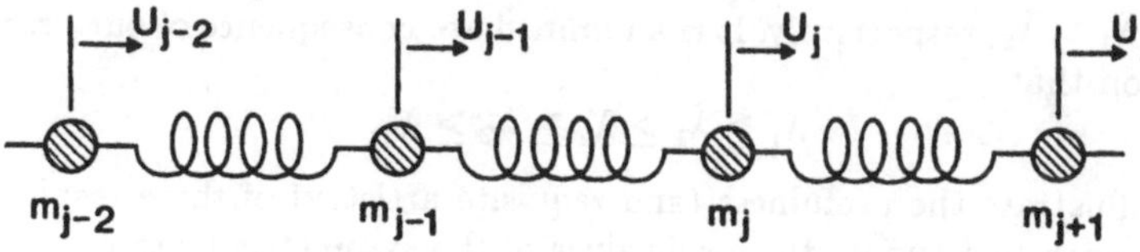

FIGURE 1.2   One dimensional lattice of balls connected by springs.

(such as springs or molecular forces). As a simple example consider a one-dimensional lattice of balls of mass m_j connected by linear springs (Fig. 1.2). Let u_j denote the displacement from equilibrium of the jth mass. The equation of motion of the system is given by

$$m_j \frac{d^2 u_j}{dt^2} = k_j \left(u_{j+1} - u_j \right) + k_{j-1} \left(u_{j-1} - u_j \right).$$

We have assumed that the restoring force of the jth spring is linearly proportional to its displacement from equilibrium with constant of proportionality k_j. Suppose that the masses at the ends of the lattice are constrained to remain fixed. This system of equations can be written as

$$\frac{d^2 u}{dt^2} = Au,$$

where u is the n-vector with components $(u_1, u_2, \ldots, u_n)$ and the $n \times n$ matrix is tridiagonal with entries

$$a_{jj} = -\frac{k_j + k_{j-1}}{m_j},$$

$$a_{j,j-1} = \frac{k_{j-1}}{m_j},$$

$$a_{j,j+1} = \frac{k_j}{m_j},$$

and all other entries are zero. Of course, one could envision a more complicated lattice for which connections are not restricted to nearest neighbor interactions in which case the matrix A need not be tridiagonal.

The first thing to try with the system $\frac{d^2 u}{dt^2} = Au$ is to assume that the solution has the form $u = u_0 e^{i\omega t}$ in which case we must have

$$Au_0 = -\omega^2 u_0.$$

In other words, the eigenvalues $\lambda = -\omega^2$ of the matrix A correspond to the natural frequencies of vibration of this lattice, and the eigenvectors determine the shape of the vibrating modes, and are called the natural modes or *normal modes* of vibration for the system.

The matrix A for the simple lattice here is symmetric if and only if the masses m_j are identical. If the masses are identical we can use the minimax principle to draw conclusions about the vibrations of the lattice.

Suppose $m_j = m$ for $j = 1, 2, \ldots n$. We calculate that

$$\langle Au, u \rangle = \sum_{i=1}^{n} \sum_{j=1}^{n} a_{ij} u_i u_j$$

$$= -\frac{k_0}{m} u_1^2 - \frac{k_n}{m} u_n^2 - \frac{1}{m} \sum_{j=1}^{n-1} k_j \left(u_j - u_{j+1} \right)^2 .$$

Since the constants k_j are all positive, we see that A is a negative definite matrix. We see further that increasing any k_j decreases the quantity $\langle Au, u \rangle$ and hence, the eigenvalues of A are decreased. From the minimax principle we have some immediate consequences.

1. If the mass m is increased, all of the eigenvalues of A are increased, that is, made less negative, so that the frequencies of vibration ω are decreased.

2. If any of the spring constants k_j are increased then the eigenvalues of A are decreased (made more negative) so that the frequencies of vibration are increased.

The natural frequencies of vibration of a lattice are also called the *resonant frequencies* because, if the system is forced with a forcing function vibrating at that frequency, the system will resonate, that is, experience large amplitude oscillations which grow (theoretically without bound) as time proceeds. This observation is the main idea behind a variety of techniques used in chemistry (such as infrared spectroscopy, electron spin resonance spectroscopy and nuclear magnetic resonance spectroscopy) to identify the structure of molecules in liquids, solids, and gasses.

It is also the idea behind microwave ovens. If an object containing a water molecule is forced with microwaves at one of its resonant frequencies the water molecule will vibrate at larger and larger amplitudes, increasing the temperature of the object. As you may know, microwave ovens do not have much effect on objects whose resonant frequencies are much different than that of water, but on objects containing water, a microwave oven is extremely efficient. Microwave ovens are set at frequencies in the 10^{11}hz range to resonate with the rotational frequencies of a water molecule.

1.4 Fredholm Alternative Theorem

The FREDHOLM ALTERNATIVE THEOREM is arguably the most important theorem used in applied mathematics as it gives specific criteria for when solutions of linear equations exist.

Suppose we wish to solve the matrix equation $Ax = b$ where A is an $n \times m$ matrix (not necessarily square). We want to know if there is a solution, and if so, how many solutions are possible.

Theorem 1.9 (Uniqueness)

The solution of $Ax = b$ (if it exists) is unique if and only if the only solution of $Ax = 0$ is $x = 0$.

Theorem 1.10 (Existence)

The equation $Ax = b$ has a solution if and only if $\langle b, v \rangle = 0$ for every vector v satisfying $A^*v = 0$.

The latter of these two statements is known as the Fredholm Alternative theorem. We have juxtaposed the two to show the importance of the homogeneous matrix problem in determining characteristics of the inhomogeneous problem.

To prove the first statement, suppose $Ax = 0$ for some $x \neq 0$. If $Ay_0 = b$ then $y_1 = y_0 + \alpha x$ also solves $Ay_1 = b$ for any choice of α so that the solution is not unique. Conversely, if solutions of $Ax = b$ are not unique then there are vectors y_1 and y_2 with $x = y_1 - y_2 \neq 0$ satisfying $Ay_1 = Ay_2 = b$. Clearly $Ax = A(y_1 - y_2) = Ay_1 - Ay_2 = 0$.

The first half of the second statement is easily proven. If $A^*v = 0$ then observe that

$$\langle v, b \rangle = \langle v, Ax \rangle = \langle A^*v, x \rangle = \langle 0, x \rangle = 0.$$

To prove the last half we seek a contradiction. Suppose $\langle v, b \rangle = 0$ for all v with $A^*v = 0$, but that $Ax = b$ has no solution. Since b is not in the range of A we write $b = b_r + b_0$ where b_r is the component of b in the range of A and b_0 is the component of b orthogonal to the range of A. Then $\langle b_0, Ax \rangle = 0$ for all x, so that $A^*b_0 = 0$. By our assumption on b, we conclude that $\langle b_0, b \rangle = 0$. Expanding $b = b_r + b_0$ we find $\langle b_0, b_r + b_0 \rangle = \langle b_0, b_r \rangle + \langle b_0, b_0 \rangle = 0$. Since $\langle b_0, b_r \rangle = 0$ it must be that $b_0 = 0$ so that $b = b_r$ is in the range of A after all.

To illustrate these theorems in a trivial way, consider the matrix

$$A = \begin{pmatrix} 1 & 2 \\ 3 & 6 \end{pmatrix}.$$

The null space of A is spanned by the vector

$$x = \begin{pmatrix} 2 \\ -1 \end{pmatrix}$$

so that solutions, if they exist, are not unique. Since the null space of A^* is spanned by the vector

$$v = \begin{pmatrix} 3 \\ -1 \end{pmatrix},$$

solutions $Ax = b$ exist only if b is of the form

$$b = \alpha \begin{pmatrix} 1 \\ 3 \end{pmatrix},$$

which is no surprise since the second row of A is three times the first row.

One relevant restatement of the Fredholm alternative is that the null space of A^* is the orthogonal complement of the range of A and that together they span $\mathbb{R}^m$. Stated another way, any vector b in $\mathbb{R}^m$ can be written uniquely as $b = b_r + b_0$ where b_r is in the range of A and b_0 is in the null space of A^*, and b_r is orthogonal to b_0.

1.5 Least Squares Solutions—Pseudo Inverses

Armed with only the Fredholm alternative, a mathematician is actually rather dangerous. In many situations the Fredholm alternative may tell us that a system of equations has no solution. But to the engineer or scientist who spent lots of time and money collecting data, this is a most unsatisfactory answer. Surely there must be a way to "almost" solve a system.

Typical examples are overdetermined systems from curve fitting. Suppose the data points (x_i, y_i), $i = 1, 2, \ldots, n$, are believed to be linearly related so that $y_i = \alpha(x_i - \bar{x}) + \beta$ with α and β to be determined, where $\bar{x} = \frac{1}{n} \sum_{i=1}^{n} x_i$. This leads to the system of n linear equations in two unknowns

$$\begin{pmatrix} x_1 - \bar{x} & 1 \\ x_2 - \bar{x} & 1 \\ \vdots & \vdots \\ x_n - \bar{x} & 1 \end{pmatrix} \begin{pmatrix} \alpha \\ \beta \end{pmatrix} = \begin{pmatrix} y_1 \\ y_2 \\ \vdots \\ y_n \end{pmatrix}.$$

If there are more than two data points this system of equations will not in general have an exact solution, if only because collected data always contain error and the fit is not exact. Since we cannot solve $Ax = b$ exactly, our goal is to find an x that minimizes $\|Ax - b\|$.

This minimal solution is norm dependent. One natural, and common, choice of norm is the Euclidean norm

$$\|x\| = \left(\sum_{i=1}^{n} x_i^2 \right)^{1/2},$$

but other choices such as $\|x\| = \max_i |x_i|$ or $\|x\| = \sum_{i=1}^{n} |x_i|$ are useful as well, but lead to different approximation schemes. The LEAST SQUARES SOLUTION is the vector x that minimizes the Euclidean norm of $Ax - b$.

To find the least squares solution, recall from the Fredholm alternative that b can always be written as $b_r + b_0 = b$, where b_r is in the range of A and b_0 is orthogonal to the range of A. Since $Ax - b_r$ is in the range of A then

$$\|Ax - b\|^2 = \|Ax - b_r\|^2 + \|b_0\|^2 .$$

We have control only over the selection of x, so the minimum is attained when we minimize $\|Ax - b_r\|^2$. Since b_r is in the range of A, $Ax = b_r$ always has a solution and $\min \|Ax - b\|^2 = \|b_0\|^2$. We know that $b = b_r + b_0$, where b_r is the projection of b onto the range of A and b_0 is in the orthogonal complement of the range of A. Therefore, $0 = \langle b_0, Ax \rangle = \langle A^*b_0, x \rangle$ for all x so that $A^*b_0 = 0$. Now $Ax = b_r$ is equivalent to $Ax = b - b_0$ so that $Ax - b$ must be in the null space of A^*, which is true if and only if $A^*Ax = A^*b$. Thus the least squares solution of $Ax = b$ is any vector x satisfying $A^*Ax = A^*b$. One such x always exists.

Another derivation of this same equation uses ideas of calculus. We wish to find the minimum of $\|Ax - b\|^2 = \langle Ax - b, Ax - b \rangle$. We let $x = x_0 + \alpha y$, where x_0 is the minimizing vector and y is an arbitrary perturbation. Since x_0 renders $\|Ax - b\|^2$ a minimum, $\langle A(x_0 + \alpha y) - b, A(x_0 + \alpha y) - b \rangle \geq \langle Ax_0 - b, Ax_0 - b \rangle$ for αy nonzero. Expanding this expression, we find that we must require

$$\alpha \langle y, A^*(Ax_0 - b) \rangle + \alpha \langle A^*(Ax_0 - b), y \rangle + \alpha^2 \langle Ay, Ay \rangle \geq 0.$$

This last expression is quadratic in αy, and unless the linear terms vanish, for sufficiently small α, a change of sign of α will produce a change of sign of the entire expression, which is not permitted. Thus we require $\langle y, A^*(Ax_0 - b) \rangle = 0$ for all y. That is, we require

$$A^*Ax_0 = A^*b.$$

We know from our first derivation that $A^*Ax = A^*b$ always has a solution. Another check of this is to apply the Fredholm alternative theorem. We require $\langle v, A^*b \rangle = 0$ for all v with $A^*Av = 0$. (Notice that A^*A is a square, self-adjoint matrix). If $A^*Av = 0$ then Av is in the null space of A^* and also in the range of A. Since these two subspaces are orthogonal, Av must be 0, the only element common to both subspaces. Thus $\langle v, A^*b \rangle = \langle Av, b \rangle = 0$, as required.

As an example of the least squares solution, recall the curve fitting problem described earlier. We seek a least squares solution of

$$A \begin{pmatrix} \alpha \\ \beta \end{pmatrix} = \begin{pmatrix} x_1 - \bar{x} & 1 \\ x_2 - \bar{x} & 1 \\ \vdots & \vdots \\ x_n - \bar{x} & 1 \end{pmatrix} \qquad \begin{pmatrix} \alpha \\ \beta \end{pmatrix} = \begin{pmatrix} y_1 \\ \vdots \\ y_n \end{pmatrix} = b.$$

Premultiplying by A^*, we find (use that $\sum_{i=1}^{n}(x_i - \bar{x}) = 0$)

$$\alpha \sum_{i=1}^{n}(x_i - \bar{x})^2 = \sum_{i=1}^{n} y_i(x_i - \bar{x})$$

$$n\beta = \sum_{i=1}^{n} y_i$$

which is always solvable for α and β.

Other types of curve fitting are possible by changing the assumptions. For example, one could try a quadratic fit $y_i = \alpha x_i^2 + \beta x_i + \gamma$. Other choices of basis functions (such as will be discussed in Chapter 2) can also be used, but in all cases we obtain a linear least squares problem. For the exponential fit $y_i = \alpha e^{\beta x_i}$, the parameters α, β occur nonlinearly. However, the equation $\ln y_i = \ln \alpha + \beta x_i$, is again linear in $\ln \alpha$ and β and leads to a linear least squares problem.

The least squares solution always exists, however, it is unique if and only if $A^*Ax = 0$ has only the trivial solution $x = 0$. If $A^*(Ax) = 0$ then $Ax \in N(A^*)$, and $Ax \in R(A)$ so that $Ax = 0$. Thus, A^*A is invertible if and only if $Ax = 0$ has no nontrivial solutions, that is, if A has linearly independent columns. If A has linearly independent columns, then the least squares solution of $Ax = b$ is given by $x = (A^*A)^{-1}A^*b$, called the MOORE-PENROSE LEAST SQUARE SOLUTION. The matrix $A' = (A^*A)^{-1}A^*$ is called the MOORE-PENROSE PSEUDO-INVERSE of A. Notice that when A is invertible, A^* is also invertible and $(A^*A)^{-1}A^* = A^{-1}A^{*-1}A^* = A^{-1}$ so that the pseudo-inverse of A is precisely the inverse of A. Notice also that the linear fit discussed earlier has a unique solution if there are at least two distinct data points x_i.

If A does not have linearly independent columns, the least squares problem does not have a unique solution. Suppose x_p is one solution of the least squares problem $A^*Ax = A^*b$. Since A has linearly dependent columns, there are vectors w that satisfy $Aw = 0$, and $x = x_p + w$ is also a least squares solution. One reasonable way to specify the solution uniquely is to seek the smallest possible solution of $A^*Ax = A^*b$. As a result, we seek x so that $\langle x, w \rangle = 0$ for all w with $Aw = 0$. That is, we want x to be orthogonal to the null space of A, and therefore, in the range of A^*.

Definition

The LEAST SQUARE PSEUDO-INVERSE of A is the matrix A' for which $x = A'b$ satisfies:

1. $A^*Ax = A^*b$.

2. $\langle x, w \rangle = 0$ for every w satisfying $Aw = 0$.

The geometrical meaning of the pseudo-inverse can be illustrated with the simple 2×2 example

$$A = \begin{pmatrix} 1 & 1 \\ .5 & .5 \end{pmatrix}.$$

For this matrix A,

$$R(A) = \left\{ \begin{pmatrix} 1 \\ .5 \end{pmatrix} \right\},$$

$$R(A^*) = \left\{ \begin{pmatrix} 1 \\ 1 \end{pmatrix} \right\}.$$

The pseudo-inverse of A must first project the vector

$$b = \begin{pmatrix} b_1 \\ b_2 \end{pmatrix}$$

into a vector in $R(A)$ and then find the inverse image of this vector in $R(A^*)$. (See Figure 1.3.) That is, x must have the form

$$x = \alpha \begin{pmatrix} 1 \\ 1 \end{pmatrix}$$

and

$$Ax = \begin{pmatrix} b_1 \\ b_2 \end{pmatrix} - \frac{1}{5} \begin{pmatrix} 1 \\ -2 \end{pmatrix} \left\langle \begin{pmatrix} 1 \\ -2 \end{pmatrix}, \begin{pmatrix} b_1 \\ b_2 \end{pmatrix} \right\rangle = \frac{1}{5} \begin{pmatrix} 4 & 2 \\ 2 & 1 \end{pmatrix} \begin{pmatrix} b_1 \\ b_2 \end{pmatrix}.$$

Since

$$A \begin{pmatrix} 1 \\ 1 \end{pmatrix} = \begin{pmatrix} 2 \\ 1 \end{pmatrix},$$

it follows that

$$x = \left(\frac{2}{5}b_1 + \frac{1}{5}b_2 \right) \begin{pmatrix} 1 \\ 1 \end{pmatrix} = \begin{pmatrix} 2/5 & 1/5 \\ 2/5 & 1/5 \end{pmatrix} \begin{pmatrix} b_1 \\ b_2 \end{pmatrix}$$

and that

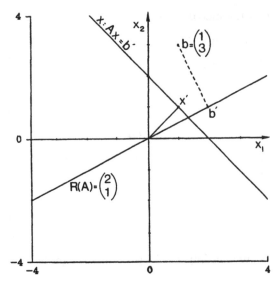

FIGURE 1.3 Graphical solution of the 2 × 2 least square matrix problem.

$$\begin{pmatrix} 1 & 1 \\ .5 & .5 \end{pmatrix} \begin{pmatrix} x_1 \\ x_2 \end{pmatrix} = \begin{pmatrix} 1 \\ 3 \end{pmatrix}.$$

$$A' = \begin{pmatrix} 2/5 & 1/5 \\ 2/5 & 1/5 \end{pmatrix} \quad \text{is the pseudo-inverse of } A.$$

There are a number of different algorithms one can use to calculate the pseudo-inverse A'. In most situations it is not a good idea to directly calculate $(A^*A)^{-1}$ since, even if A has linearly independent columns, the calculation of $(A^*A)^{-1}$ is numerically poorly posed. The algorithms that we describe here are included because they illustrate geometrical concepts, and also, in the case of the last two, because of their usefulness for computer calculations.

Method 1 (Gaussian Elimination)

The first idea is to mimic GAUSSIAN ELIMINATION and find A' via EL-EMENTARY ROW OPERATIONS. Recall that Gaussian elimination is the process of reducing the augmented matrix $[A, I]$ to $[I, A^{-1}]$ by means of elementary row operations. The goal of reduction is to subtract appropriate multiples of one row from lower rows of the matrix so that only zeros occur in the column directly below the diagonal element. For example, one step of the row reduction of the matrix

$$
A = \begin{pmatrix}
a_{11} & a_{12} & & \cdots & & a_{1n} \\
 & a_{22} & \cdots & & \cdots & a_{2n} \\
 & & \cdot & & & \\
 & & & \cdot & & \\
 & & a_{kk} & \cdots & & a_{kn} \\
 & & a_{k+1,k} & & & a_{k+1,n} \\
 0 & & & & & \\
 & & \vdots & & & \vdots \\
 & & a_{nk} & & & a_{nn}
\end{pmatrix}
$$

is accomplished by premultiplying A by the matrix (only nonzero elements are displayed)

$$
M_k = \begin{pmatrix}
1 & & & & & & \\
 & 1 & & & & & \\
 & & \cdot & & & & \\
 & & & \cdot & & & \\
 & & & 1 & & & \cdot \\
 & & & -\sigma_{k+1} & 1 & & \\
 & & & -\sigma_{k+2} & & & \\
 & & & \vdots & & \cdot & \\
 & & & -\sigma_n & & & 1
\end{pmatrix}
$$

where $\sigma_j = a_{jk}/a_{kk}$. Specifically, the nonzero entries of M_k are ones on the diagonal and $-\sigma_j$ in the kth column below the diagonal. The numbers σ_j are called multipliers for the row reduction. The result of a sequence of such row operations is to triangularize A, so that $M_n M_{n-1} \cdots M_1 A = U$ is an upper triangular matrix. The product of elementary row operations $M_n M_{n-1} \cdots M_1 = L^{-1}$ is a lower triangular matrix which when inverted gives $A = LU$. This representation of A is called an LU decomposition.

By analogy, our goal is to find some way to augment the matrix A with another matrix, say P, so that the row reduction of $[A, P]$ yields the identity and the pseudo-inverse $[I, A']$. The least squares solution we seek must satisfy $Ax = b_r$, subject to the conditions $\langle w_i, x \rangle = 0$ where $\{w_i\}$ span the null space of A, and the vector b_r is the projection of the vector b onto the range of A. We can represent b_r as $b_r = Pb = b - \sum_{i=1}^k \langle u_i, b \rangle u_i = \left(I - \sum_{i=1}^k u_i u_i^T \right) b$, where the k vectors $\{u_j\}$ form an orthonormal basis for the null space of A^*. Thus, the projection matrix P is given by $P = I - \sum_{i=1}^k u_i u_i^T$. Notice that if the null space of A^* is empty, then $k = 0$ and $P \equiv I$. If the null space of A is empty we row reduce the augmented matrix $[A, P]$, but if not, we append the additional constraints $\langle w_i, x \rangle = 0$ and row reduce the augmented matrix

$$\begin{bmatrix} A & P \\ w_i^T & 0 \end{bmatrix},$$

discarding all zero rows, which leads to $[I, A']$.

To see that this can always be done, notice that the projection P is guaranteed by the Fredholm alternative to be row consistent with the matrix A. Furthermore, the rows of A span the range of A^* and the vectors $\{w_i\}$ span the null space of A, so that, from the Fredholm alternative $\mathbb{R}^n = R(A^*) + N(A)$, the rows of A with $\{w_i^T\}$ span $\mathbb{R}^n$ and hence can be row reduced to an identity matrix (with possibly some leftover zero rows).

As an example, consider the matrix

$$A = \begin{pmatrix} 2 & 4 & 6 \\ 1 & 2 & 3 \end{pmatrix}.$$

Since A has linearly dependent columns, A has no Moore-Penrose pseudo-inverse. The null space of A is spanned by the two vectors

$$w_1 = \begin{pmatrix} 1 \\ 1 \\ -1 \end{pmatrix} , \quad w_2 = \begin{pmatrix} 2 \\ -1 \\ 0 \end{pmatrix},$$

and the null space of A^* is spanned by the single vector

$$u = \frac{1}{\sqrt{5}} \begin{pmatrix} 1 \\ -2 \end{pmatrix}.$$

As a result,

$$P = I - \frac{1}{5} \begin{pmatrix} 1 & -2 \\ -2 & 4 \end{pmatrix} = \frac{1}{5} \begin{pmatrix} 4 & 2 \\ 2 & 1 \end{pmatrix}.$$

We row reduce the augmented matrix

$$\begin{pmatrix} A & P \\ w_1^T & 0 \\ w_2^T & 0 \end{pmatrix} = \begin{pmatrix} 2 & 4 & 6 & 4/5 & 2/5 \\ 1 & 2 & 3 & 2/5 & 1/5 \\ 1 & 1 & -1 & 0 & 0 \\ 2 & -1 & 0 & 0 & 0 \end{pmatrix}.$$

Elementary row operations reduce this matrix to

$$\begin{pmatrix} 1 & 0 & 0 & 2/70 & 1/70 \\ 0 & 1 & 0 & 4/70 & 2/70 \\ 0 & 0 & 1 & 6/70 & 3/70 \end{pmatrix}$$

(we have discarded one identically zero row) so that

$$A' = \frac{1}{70} \begin{pmatrix} 2 & 1 \\ 4 & 2 \\ 6 & 3 \end{pmatrix}.$$

It is easy to check that $A^* A A' = A^*$.

Method 2 (LU Decomposition)

Any matrix A can be written as $A = LU$ where L is lower triangular and U is upper triangular (with the proviso that some row interchanges may be necessary). It is generally not recommended to row reduce A to a diagonal matrix, but simply to triangular form. The decomposition is unique if we require L to have ones on the diagonal and U to have linearly independent rows. To see why the LU decomposition is desirable, suppose we want to solve $Ax = b$ and we know that $A = LU$. We let $Ux = y$ so that $Ly = b$. Solving $Ly = b$ involves only forward substitution, and solving $Ux = y$ involves only back substitution. Thus, if both L and U are full rank, the solution vector x is easily determined.

The LU decomposition is found by direct elimination techniques. The idea is to reduce the augmented matrix $[A, I]$ to the form $[U, L^{-1}]$ by lower triangular elementary row operations. Alternately, one can simply store the multipliers from the row reduction process in the subdiagonal element that is being zeroed by the row operation (see Problem 1.5.11). The resulting lower diagonal matrix is the lower diagonal part of L and the diagonal elements of L are all ones. If A has linearly independent rows, then zero rows of U and corresponding columns of L can be harmlessly eliminated.

Example

Consider

$$A = \begin{pmatrix} 1 & 2 \\ 3 & 4 \\ 5 & 6 \end{pmatrix}.$$

By row reduction, storing multipliers below the diagonal, we find that A reduces to

$$\begin{pmatrix} 1 & 2 \\ 3 & -2 \\ 5 & 2 \end{pmatrix}$$

so that

$$U = \begin{pmatrix} 1 & 2 \\ 0 & -2 \\ 0 & 0 \end{pmatrix}$$

$$L = \begin{pmatrix} 1 & 0 & 0 \\ 3 & 1 & 0 \\ 5 & 2 & 1 \end{pmatrix}.$$

Eliminating the last row of U and last column of L (which are superfluous) we determine that

$$A = LU = \begin{pmatrix} 1 & 0 \\ 3 & 1 \\ 5 & 2 \end{pmatrix} \begin{pmatrix} 1 & 2 \\ 0 & -2 \end{pmatrix}.$$

Notice that since L has linearly independent columns, the null space of A is the same as the null space of U and similarly, the null space of A^* is the null space of L^*. The range of A is spanned by the columns of L and the range of A^* is spanned by the columns of U^*. If $A = A^*$, then $U = L^*$.

Once the LU decomposition is known, the pseudo-inverse of A is easily found.

Theorem 1.11

$$A' = U^*(UU^*)^{-1}(L^*L)^{-1}L^*.$$

This statement can be verified by substituting $x = A'b$ into $A^*Ax = A^*b$. Since U has linearly independent rows and L has linearly independent columns, UU^* and L^*L are both invertible. Notice also that if $Aw = 0$ then $Uw = 0$ as well and

$$\langle w, A'b \rangle = \langle w, U^*(UU^*)^{-1}(L^*L)^{-1}L^*b \rangle$$

$$= \langle Uw, (UU^*)^{-1}(L^*L)^{-1}L^*b \rangle = 0,$$

so that x is the smallest possible least squares solution. If A is a 2×2 matrix there is an interesting geometrical interpretation for A'. Suppose the dimensions of $N(A)$ and $R(A)$ are both one. Then L and U^* are vectors in $\mathbb{R}^2$, and UU^* and L^*L are the squared lengths of the vectors U^* and L respectively. The operation $A'b$ represents finding the length of the projection of b onto L^*, rescaling this projection and finally taking a vector of this length in the direction of U^*.

Method 3 (Orthogonal Transformations)

The beauty of orthonormal (unitary) matrices Q is that for any vector x, $\|Qx\| = \|x\|$, since $\langle Qx, Qx \rangle = \langle x, Q^*Qx \rangle = \langle x, x \rangle$. The advantage of triangular systems is that if the diagonal elements are nonzero, they can be solved by forward or backward substitution, which is nearly as nice as a diagonal system.

One important way to solve least squares problems is to reduce the matrix A to upper triangular form using carefully chosen orthogonal matrices. Another way of saying this is that we want to represent A as $A = QR$ where Q is unitary and R is upper triangular. This can always be done because this is exactly what the Gram-Schmidt procedure does, namely, it represents any vector as a linear combination of previously determined orthogonal vectors plus a new mutually orthogonal vector. In this case, the column vectors of A are the set of vectors with which we start, and the columns of Q are the orthogonal basis that results from applying the Gram-Schmidt procedure to A. The matrix R indicates how to express the columns of A as linear combinations of the orthogonal basis Q.

To actually carry this out we use a method that looks similar to Gaussian elimination. However, instead of using elementary row operations as in Gaussian elimination, we use an object called an elementary reflector, or more commonly, HOUSEHOLDER TRANSFORMATION.

Definition

An elementary reflector (Householder transformation) is a matrix of the form
$$U = I - \frac{2uu^*}{\|u\|^2} \quad , \quad \text{provided} \quad u \neq 0.$$
It is easy to verify that $U = U^*$ and $U^*U = I$.

The important property of elementary reflectors is that, just as with elementary row operations, they can be chosen to introduce zeros into a column vector.

Theorem 1.12

For $x \in \mathbb{R}^n$, let $\sigma = \pm\|x\|$, and suppose $x \neq -\sigma e_1$. If $u = x + \sigma e_1$, then $U = I - \frac{2uu^*}{\|u\|^2}$ is an elementary reflector whose action on x is $Ux = -\sigma e_1$ (e_1 is the first natural basis vector).

Proof

Since $x \neq -\sigma e_1$, $u \neq 0$ and U is an elementary reflector. Since $\sigma^2 = \|x\|^2$,
$$\|u\|^2 = 2\sigma^2 + 2\sigma x_1,$$

where x_1 is the first component of the vector x, and

$$Ux = x - \frac{uu^*x}{\sigma^2 + \sigma x_1} = x - u = -\sigma e_1.$$

It is wise to choose σ to have the same sign as x_1, so that $x \neq -\sigma e_1$ unless $x = 0$ already, in which case, the transformation is not needed.

We now see that we can put zeros in the first column of A, below the diagonal, by premultiplying by the appropriate Householder transformation. To carry out the transformation on the entire matrix, we do not actually calculate the matrix U but rather note that for any column vector y

$$Uy = y - \frac{2uu^*y}{u^*u} = y - \frac{2\langle u, y \rangle}{\langle u, u \rangle} u,$$

so that Uy can be viewed as a linear combination of two column vectors with a multiplier related to the inner product of the two vectors. Premultiplying by a sequence of Householder transformations we sequentially triangularize the matrix A. We first apply the Householder transformation that places zeros in the first column below the diagonal. Then we apply the Householder transformation that leaves the first row and first column undisturbed and places zero in the second column below the diagonal, and so on sequentially. Note that if

$$U_k = I_k - \frac{2u_k u_k^*}{\|u_k\|^2}$$

is a $k \times k$ Householder transformation that places zeros in the first column below the diagonal of a $k \times k$ matrix, then the $n \times n$ matrix

$$U_n = I_n - \frac{2u_n u_n^*}{\|u_k\|^2},$$

where $u_n^T = (0, 0, 0, \ldots, 0, u_k^T)$ ($k - 1$ zero entries) is the Householder transformation for an $n \times n$ matrix that transforms only the lower diagonal $k \times k$ submatrix and leaves the remaining matrix unscathed.

If $U_1, U_2 \cdots U_n$ are the chosen sequence of transformation matrices, then with $Q^* = U_n U_{n-1} \cdots U_1$ we have

$$\|Ax - b\| = \|Q^*Ax - Q^*b\| = \|Rx - Q^*b\|$$

where R is upper triangular. If the columns of A are linearly independent (the same condition as required for the Moore-Penrose pseudo-inverse) the matrix R has nonzero diagonal elements and is therefore invertible.

If we denote the vector

$$Q^*b = \begin{pmatrix} b_1 \\ b_2 \end{pmatrix}$$

where the vector b_1 has the same number of elements as R has rows, then the least squares solution is found by solving $Rx = b_1$ and for this choice of x, $\|Ax - b\| = \|b_2\|$. It follows that the pseudo-inverse of A is given by $A' = R'Q^*$, where R' is the pseudo-inverse of R. The vector b_2 is the residual for this least squares solution, and cannot be made smaller. It is often the case that the matrix A has linearly independent columns so that R is uniquely invertible and the solution of $Rx = b_1$ is found by back substitution. If, however, R is not invertible, one must find the smallest solution of $Rx = b_1$, which usually cannot be found by back substitution.

There is another important use for Householder transformations that should be mentioned here. As is well known, there is no way to find the eigenvalues of a matrix larger than 4×4 with a finite step algorithm. The only hope for larger matrices is an iterative procedure, and one of the best is based on the QR decomposition.

Two matrices have the same eigenvalues if they are related through a similarity transformation. The goal of this algorithm is to find an infinite sequence of similarity transformations that converts the given matrix M_0 into its diagonal matrix of eigenvalues.

The method is to use Householder transformations to decompose the matrix M_n into the product of an orthogonal matrix and an upper triangular matrix $M_n = Q_n R_n$. To make this into a similarity transformation we reverse the order of matrix multiplication and form $M_{n+1} = R_n Q_n$ so that $M_{n+1} = Q_n^{-1} M_n Q_n$. Now the amazing fact is that if M_0 has only real eigenvalues, then M_n converges to the diagonal matrix of eigenvalues. Of course, if the eigenvalues of M_0 are not all real, this cannot happen, since the orthogonal matrices Q_n are always real. However, if there are complex eigenvalues, M_n converges to a matrix with 2×2 real matrices on the diagonal, corresponding to the 2×2 matrices with complex eigenvalues.

Householder transformations provide a very efficient way to find the QR decomposition of a matrix. Good software packages to do these calculations on a computer are readily available.

Method 4 (Singular Value Decomposition (SVD))

The final method by which A' can be calculated is the SINGULAR VALUE DECOMPOSITION. This method extends the idea of the spectral decomposition of a matrix to nonsquare matrices and is often used in applications.

The singular value decomposition of a matrix A is similar to the spectral representation of a square matrix $A = T\Lambda T^{-1}$. If A is self-adjoint, then $A = Q\Lambda Q^{-1}$ where Q is orthogonal. Once the spectral decomposition is known, the inverse is easily found to be

$$A^{-1} = T\Lambda^{-1}T^{-1}$$

provided the diagonal matrix Λ has only nonzero diagonal entries. If A is symmetric, the pseudo-inverse of A is $A' = Q\Lambda'Q^{-1}$ where Λ' is the pseudo-inverse of the diagonal matrix Λ. What is needed is an analogous result for nonsymmetric matrices. First we will state the theorem and then show why it is true.

Theorem 1.13

Suppose A is an $m \times n$ matrix.

1. A can be factored as

$$A = Q_1 \Sigma Q_2^*$$

where $Q_1(m \times m)$ and $Q_2(n \times n)$ are unitary matrices that diagonalize AA^* and A^*A respectively, and Σ is an $m \times n$ diagonal matrix (called the matrix of SINGULAR VALUES) with diagonal entries $\sigma_{ii} = \sqrt{\lambda_i}$, where $\lambda_i, i = 1, 2, \ldots, \min(m, n)$, are the eigenvalues of AA^* (see Problem 1.2.3).

2. The pseudo-inverse of A is

$$A' = Q_2 \Sigma' Q_1^*,$$

where Σ', the pseudo-inverse of Σ, is an $n \times m$ diagonal matrix with diagonal entries $\sigma'_{ii} = \frac{1}{\sigma_{ii}} = \frac{1}{\sqrt{\lambda_i}}$ provided $\sigma_{ii} \neq 0$, and 0 otherwise.

The factorization $A = Q_1 \Sigma Q_2^*$ is called the singular value decomposition of A. To show that this decomposition always exists, notice that the matrix A^*A is a symmetric $n \times n$ matrix with non-negative eigenvalues $\lambda_1, \lambda_2, \ldots, \lambda_n$. Let Q_2 be the $n \times n$ matrix which diagonalizes A^*A. The columns of Q_2 are the orthogonal eigenvectors $x_1, x_2, \ldots, x_n$ of A^*A, normalized to have length one. Then

$$A^*A = Q_2 \Lambda Q_2^*.$$

Suppose A^*A has k positive eigenvalues. Then $\lambda_1, \lambda_2, \ldots, \lambda_k > 0$, $\lambda_{k+1} = \lambda_{k+2} = \cdots = \lambda_n = 0$. Let $y_i = \frac{Ax_i}{\sqrt{\lambda_i}}$, $i = 1, 2, \ldots, k$, and notice that

$$
\begin{aligned}
\langle y_i, y_j \rangle &= \frac{1}{\sqrt{\lambda_i}\sqrt{\lambda_j}} \langle Ax_i, Ax_j \rangle = \frac{1}{\sqrt{\lambda_i}\sqrt{\lambda_j}} \langle x_i, A^*Ax_j \rangle \\
&= \frac{\sqrt{\lambda_j}}{\sqrt{\lambda_i}} \langle x_i, x_j \rangle = \delta_{ij}.
\end{aligned}
$$

We use the Gram-Schmidt procedure to extend the y_i's to form an orthonormal basis for $\mathbb{R}^m$, $\{y_i\}_{i=1}^m$. Let $Q_1 = [y_1, y_2, \cdots, y_m]$. The vectors y_i are the orthonormal eigenvectors of AA^*. We now examine the entries of $Q_1^*AQ_2$. For $i \leq k$, the ij^{th} entry of $Q_1^*AQ_2$ is

$$\langle y_i, Ax_j \rangle = \frac{1}{\sqrt{\lambda_i}} \langle Ax_i, Ax_j \rangle = \frac{1}{\sqrt{\lambda_i}} \langle x_i, A^*Ax_j \rangle$$

$$= \frac{\lambda_j}{\sqrt{\lambda_i}} \langle x_i, x_j \rangle = \sqrt{\lambda_j} \delta_{ij}.$$

For $i > k$, note that $AA^*y_i = 0$ so that A^*y_i is simultaneously in the null space of A and the range of A^*. Therefore $A^*y_i = 0$ and $\langle y_i, Ax_j \rangle = \langle A^*y_i, x_j \rangle = 0$. It follows that $Q_1^*AQ_2 = \Sigma$ where $\Sigma = (\sigma_{ij})$, $\sigma_{ij} = \sqrt{\lambda_j}\delta_{ij}$, $i = 1, 2, \ldots, m$, $j = 1, 2, \ldots, n$.

It is noteworthy that the orthogonal vectors $\{x_1, x_2, \ldots, x_k\}$ span the range of A^* and the vectors $\{x_{k+1}, \ldots, x_n\}$ span the null space of A. Similarly, the vectors $\{y_1, y_2, \ldots y_k\}$ span the range of A and $\{y_{k+1}, \ldots, y_m\}$ span the null space of A^*. Thus the orthogonal matrices Q_1 and Q_2 provide an orthogonal decomposition of $\mathbb{R}^m$ and $\mathbb{R}^n$ into $\mathbb{R}^m = R(A) \oplus N(A^*)$ and $\mathbb{R}^n = R(A^*) \oplus N(A)$, respectively. Further, the singular value decomposition $A = Q_1\Sigma Q_2^*$ shows the geometry of the transformation of x by A. In words, Q_2^*x decomposes x into its orthogonal projections onto $R(A^*)$ and $N(A)$, then ΣQ_2^*x rescales the projection of x onto $R(A^*)$, discarding the projection of x onto $N(A)$. Finally, multiplication by Q_1 places ΣQ_2^*x onto the range of A.

To find A', the pseudo-inverse of A, we want to minimize

$$\|Ax - b\|^2 = \|Q_1\Sigma Q_2^*x - b\|^2.$$

For any unitary matrix Q,

$$\|Qx\|^2 = \langle Qx, Qx \rangle = \langle x, Q^*Qx \rangle = \langle x, x \rangle = \|x\|^2$$

since $Q^*Q = I$. Therefore, $\|Ax - b\|^2 = \|Q_1\Sigma Q_2^*x - b\| = \|\Sigma v - Q_1^*b\|$ where $v = Q_2^*x$. It is an easy matter to verify that the pseudo-inverse of an $m \times n$ diagonal matrix D with entries $d_{ij} = \sigma_i\delta_{ij}$ is an $n \times m$ diagonal matrix D' with entries $d_{ij}' = \frac{1}{\sigma_i}\delta_{ij}$ whenever $\sigma_i \neq 0$ and $d_{ij}' = 0$ otherwise. We therefore take $y = \Sigma'Q_1^*b$, and y is the smallest least squares solution. The vector

$$x = Q_2\Sigma'Q_1^*b$$

has the same norm as y and therefore gives the smallest least squares solution of $Ax = b$. Hence $A' = Q_2\Sigma'Q_1^*$, as stated earlier.

We can illustrate the geometrical meaning of this pseudo-inverse first for symmetric matrices. If $A = Q\Lambda Q^{-1}$, with Q unitary, we know that Q represents the basis relative to which A has a diagonal representation. If one of the diagonal elements of Λ is zero, then of course, the inverse of A cannot be found. To find $A'b$, we form $Q^{-1}b$ which re-expresses b relative to the new coordinate system. The expression $\Lambda'Q^{-1}b$ projects $Q^{-1}b$ onto the range of Λ

and determines the smallest element whose image under Λ is $Q^{-1}b$. Finally $Q(\Lambda'Q^{-1}b)$ re-expresses this vector in terms of the original coordinate system (see Problem 1.5.6).

The change of coordinates Q identifies for us the range and null space of A. What we have actually done is to project the vector b onto the range of A and then find another vector in the range of A whose image is this projection. Since $A = A^*$, this is the correct thing to do.

This construction of A' can work only for symmetric matrices where $R(A) = R(A^*)$. Unfortunately, this construction fails for nonsymmetric matrices for two important reasons (see Problem 1.5.8). First since $A \neq A^*$, $R(A) \neq R(A^*)$ so projecting onto $R(A)$ does not also put us on $R(A^*)$. Furthermore, if $A \neq A^*$, the basis provided by the diagonalization of A is not an orthogonal basis and hence projections are not orthogonal.

The correct sequence of operations is provided by the singular value decomposition $A' = Q_2\Sigma'Q_1^*$. To find the least squares solution of $Ax = b$, we first use Q_1 to decompose b into its projection onto $R(A)$ and $N(A^*)$. The part on $R(A)$ is rescaled by Σ' and then transformed onto $R(A^*)$ by Q_2.

This again illustrates nicely the importance of transform methods. As we know, the least squares solution of $Ax = b$ can be found by direct techniques. However, the change of coordinates provided by the orthogonal matrices Q_1, Q_2 transforms the problem $Ax = b$ into $\Sigma y = \hat{b}$ where $y = Q_2^*x$, $\hat{b} = Q_1^*b$, which is a diagonal (separated) problem. This least squares problem is easily solved and transforming back we find, of course, $x = Q_2\Sigma'Q_1^*b$. This transform process is illustrated in the commuting diagram

$$
\begin{array}{ccc}
Ax = b & \xrightarrow{\quad A' \quad} & x = A'b = Q_2\Sigma'Q_1^*b \\
\begin{array}{c} y = Q_2^*x \\ \hat{b} = Q_1^*b \end{array} \Big\downarrow & & \Big\uparrow \begin{array}{c} x = Q_2y \\ b = Q_1\hat{b} \end{array} \\
\Sigma y = \hat{b} & \xrightarrow{\quad \Sigma' \quad} & y = \Sigma'\hat{b}
\end{array}
$$

where the top line represents the direct solution process, while the two vertical lines represent transformation into and out of the appropriate coordinate representation, and the bottom line represents solution of a diagonal system.

The singular value decomposition also has the important feature that it allows one to stabilize the inversion of unstable (or ill-conditioned) matrices. To illustrate this fact, consider the matrix

$$
A = \begin{pmatrix} 1 + \dfrac{2\epsilon}{\sqrt{10}} & 1 - \dfrac{2\epsilon}{\sqrt{10}} \\ 2 - \dfrac{\epsilon}{\sqrt{10}} & 2 + \dfrac{\epsilon}{\sqrt{10}} \end{pmatrix}
$$

where we take ϵ to be a small number. We can visualize what is happening by viewing the equation $Ax = b$ as describing the intersection of two straight lines in the

$$x = \begin{pmatrix} x_1 \\ x_2 \end{pmatrix}$$

plane and by noting that for ϵ small the lines are almost parallel. Certainly small changes in b give large changes to the location of the solution. Thus, it is apparent that the inversion of A is very sensitive to numerical error.

We calculate the singular value decomposition of A to be

$$A = \frac{1}{\sqrt{10}} \begin{pmatrix} 1 & 2 \\ 2 & -1 \end{pmatrix} \begin{pmatrix} \sqrt{10} & 0 \\ 0 & \epsilon \end{pmatrix} \begin{pmatrix} 1 & 1 \\ 1 & -1 \end{pmatrix}$$

and the inverse of A to be

$$A^{-1} = \frac{1}{\sqrt{10}} \begin{pmatrix} 1 & 1 \\ 1 & -1 \end{pmatrix} \begin{pmatrix} 1/\sqrt{10} & 0 \\ 0 & 1/\epsilon \end{pmatrix} \begin{pmatrix} 1 & 2 \\ 2 & -1 \end{pmatrix}$$

$$= \begin{pmatrix} .1 + 2/\epsilon\sqrt{10} & .2 - 1/\epsilon\sqrt{10} \\ .1 - 2/\epsilon\sqrt{10} & .2 + 1/\epsilon\sqrt{10} \end{pmatrix}.$$

It is clear that the singular value ϵ makes the matrix A^{-1} very large and the solution $x = A^{-1}b$ is unstable.

If instead of using the full inverse, we replace ϵ by zero and use the singular value decomposition to find the pseudo-inverse,

$$A' = 1/\sqrt{10} \begin{pmatrix} 1 & 1 \\ 1 & -1 \end{pmatrix} \begin{pmatrix} 1/\sqrt{10} & 0 \\ 0 & 0 \end{pmatrix} \begin{pmatrix} 1 & 2 \\ 2 & -1 \end{pmatrix} = \frac{1}{10} \begin{pmatrix} 1 & 2 \\ 1 & 2 \end{pmatrix},$$

the solution is now stable. That is, small changes in the vector b translate into only small changes in the solution vector $x = A'b$ and, in fact, x varies only in the direction orthogonal to the nearly parallel straight lines of the original problem. In other words, the singular value decomposition identified the stable and unstable directions for this problem, and by purposely replacing ϵ by zero, we were able to keep the unstable direction from contaminating the solution.

In a more general setting, one can use the singular value decomposition to filter noise (roundoff error) and stabilize inherently unstable calculations. This is done by examining the singular values of A and determining which of the singular values are undesirable, and then setting these to zero. The resulting pseudo-inverse does not carry with it the destabilizing effects of the noise and roundoff error. It is this feature of the singular value decomposition that makes it the method of choice whenever there is a chance that a system of equations has some potential instability.

The singular value decomposition is always recommended for curve fitting. The SVD compensates for noise and roundoff error propagation as well as the fact that solutions may be overdetermined or underdetermined. In a least squares fitting problem it is often the case that there is simply not enough experimental evidence to distinguish between certain basis functions and the resulting least squares matrix is (nearly) underdetermined. The SVD signals this deficiency by having correspondingly small singular values. The user then has the option of setting to zero small singular values. The decision of how small is small is always up to the user. However, with a little experimentation one can usually learn what works well.

1.6   Numerical Considerations

One cannot help but have tremendous respect for the applied mathematicians and scientists that preceded us. People like Legendre, Bessel, Cauchy, Euler, Fourier, the Bernoullis, and many more, were not only very smart but also diligent. They could do long tedious calculations, and catch their sign errors, all without so much as a caclulator! Now that computers are a part of everyday life, we use computers to perform calculations that in an earlier day would have been done by hand, or not done at all.

It seems that every applied mathematics text that is written these days has a list of explicit algorithms or specific computer programs or else it contains disks with software programs to be used on your personal computer. In this chapter we discussed Gaussian elimination, LU decompositions, QR decompositions, Householder transformations, and singular value decompositions, all of which can and should be implemented on a computer. Our primary reason for discussing these techniques was to help you understand the theoretical aspects and implications of bases, transformations, eigenvalues, eigenvectors and the like, since all of these objects have major uses in more complicated settings. It is not the primary goal of this text to teach you numerical analysis. However, if you are not comfortable with these procedures as numerical algorithms, you should know that there are excellent packages readily available for use. There is little need to write your own routines.

For matrices that are not too large, the interactive program MATLAB has all of the capabilities you should need. It will find the LU, QR, or SV decompositions of a matrix quickly and easily. MATLAB makes use of the more extensive packages EISPACK and LINPACK. EISPACK is a package of over 70 state of the art Fortran subroutines useful for solving and analyzing linear systems of equations and related matrix problems. The advantage of these last two packages is that they take advantage of special structures of larger sparse systems and hence are more efficient for large problems than MATLAB.

These packages are generally available on mainframe computers. For personal computers, reliable software to do linear algebra is rapidly becoming available. There is a growing number of packages that will fit on a PC. If you cannot find a package to purchase, then the book *Numerical Recipes* certainly contains the Fortran or Pascal program of your dreams, and these will work quite well on your PC.

The literature on numerical linear algebra is in relatively good shape and this book certainly does not need to reproduce well-known algorithms. There is, however, one significant form of computer calculation that is omitted from most books, namely the use of algebraic symbol manipulation languages, such as REDUCE or MACSYMA, which operate on mainframe computers, and smaller versions such as muMATH or MAPLE that can be used on personal computers and work stations. These languages are very useful because they do algebra and calculus calculations exactly, symbolically, and quickly, and they do them without making sign errors.

It is strongly recommended that you become familiar with one of these packages. To help orient you to doing explicit calculations in this way, we include here a computer program, but this one is written in the language REDUCE. Other REDUCE programs and routines will be included in later sections of the text to remind you that many explicit, tedious, calculations can be done much more quickly this way. Later on, we also include some exercises that are intended to be too tedious to do by hand, but which can be done readily with one of these languages.

This first routine generates orthogonal polynomials. If you did Problem 1.1.6, you already know that the Gram-Schmidt procedure, while easy to describe, is tedious to actually carry out. Why not let the computer do the dirty work? Here, then, is a REDUCE procedure which generates the first $N+1$ orthogonal polynomials on the interval A to B with respect to the inner product $\langle f,g \rangle = \int_A^B f(x)g(x)W(x)dx$ (with no sign errors!):

```
PROCEDURE GRAM(A,B,W,N);
BEGIN
OPERATOR PHI, NORMSQ;
PHI(0):=1;
TF:=INT(W,X);
NORMSQ(0):=SUB(X=B,TF)-SUB(X=A,TF);
FOR J:=1:N DO
BEGIN
PHI(J):=X**J;
JS:=J-1;
FOR K:=0:JS DO
BEGIN
TF:=INT(X**J*PHI(K)*W,X);
TA:=SUB(X=B,TF)-SUB(X=A,TF);
PHI(J):=PHI(J)-TA*PHI(K)/NORMSQ(K);
END;
TF:=INT(W*PHI(J)**2,X);
NORMSQ(J):=SUB(X=B,TF)-SUB(X=A,TF);
```

```
WRITE PHI (J);
NORMSQ(J);
END;
END;
```

The statement GRAM(-1,1,1,10); will generate the first eleven Legendre polynomials.

For this routine one must have both A and B finite. Unfortunately, this routine will not generate the Chebyshev polynomials since REDUCE cannot integrate $w(x) = (1 - x^2)^{-1/2}$. However, this deficiency can be remedied by adding instructions on how to integrate functions like $w(x)$ or by first making the obvious trigonometric substitution $x = \sin\theta$.

Further Reading

There are numerous linear algebra books that discuss the introductory material of this chapter. Three that are recommended are:

1. P. R. Halmos, *Finite Dimensional Vector Spaces*, Van Nostrand Reinhold, New York, 1958.

2. J. N. Franklin, *Matrix Theory*, Prentice-Hall, Englewood Cliffs, New Jersey, 1968.

3. G. Strang, *Linear Algebra and its Applications*, Academic Press, New York, 1980.

The geometrical and physical meaning of eigenvalues is discussed at length in the classic books

4. R. Courant and D. Hilbert, *Methods of Mathematical Physics*, Volume I, Wiley-Interscience, New York, 1953.

5. J. H. Wilkinson, *The Algebraic Eigenvalue Problem*, Oxford University Press, London, 1965.

Numerical aspects of linear algebra are discussed, for example, in

6. G. W. Stewart, *Introduction to Matrix Computations*, Academic Press, New York, 1980,

7. G. E. Forsythe, M. A. Malcolm and C. B. Moler, *Computer Methods for Mathematical Computations*, Prentice-Hall, 1977,

and for least squares problems in

8. C. L. Lawson and R. J. Hanson, *Solving Least Squares Problems*, Prentice-Hall, Englewood Cliffs, New Jersey, 1974,

while many useable FORTRAN and PASCAL programs are described in

9. W. H. Press, B. P. Flannery, S. A. Teukolsky, and W. T. Vetterling, *Numerical Recipes: The Art of Computing,* Cambridge University Press, Cambridge, 1966.

Packaged software routines are described in various user's guides. For example,

10. C. Moler, MATLAB User's Guide, May 1981, Department of Computer Science, University of New Mexico.

11. J. J. Dongarra, J. R. Bunch, C. B. Moler, and G. W. Stewart, *LINPACK User's Guide, SIAM,* Philadelphia, 1979.

12. B. T. Smith, J. M. Boyle, J. J. Dongarra, B. S. Garbow, Y. Ikebe, V. C. Klema, C. B. Moler, *Matrix Eigensystem Routines—EISPACK Guide,* Lecture Notes in Computer Science, Volume 6, Springer-Verlag, 1976.

Finally, symbol manipulation languages are described in their user's guides.

13. A. C. Hearn, *REDUCE User's Manual,* Version 3.1, Rand Publication CP78, The Rand Corporation, Santa Monica, California 90406.

14. K. O. Geddes, G. H. Gonnet, B. W. Char, *MAPLE User's Manual,* Department of Computer Science, University of Waterloo, Waterloo, Ontario, Canada N2L 3G1.

15. *muMATH-83 Reference Manual,* The Soft Warehouse, P. O. Box 11174 Honolulu, Hawaii 96828.

16. R. H. Rand, *Computer Algebra in Applied Mathematics: An Introduction to MACSYMA,* Pitman, Boston, 1984.

Problems for Chapter 1

Section 1.1

1. Show that in any inner product space

$$\|x + y\|^2 + \|x - y\|^2 = 2\|x\|^2 + 2\|y\|^2.$$

Interpret this geometrically in $\mathbb{R}^2$.

2. **a.** Verify that, in an inner product space,

$$\text{Re}\,\langle x, y \rangle = \frac{1}{4}\left(\|x + y\|^2 - \|x - y\|^2\right).$$

b. Show that in any real inner product space there is at most one inner product which generates the same induced norm.

c. In $\mathbb{R}^n$, show that $\|x\| = (\sum_{k=1}^n |x_k|^p)^{1/p}$ can be induced by an inner product if and only if $p = 2$.

3. Show that any set of mutually orthogonal vectors is linearly independent.

4. a. Show that $\mathbb{R}^n$ with norm $\|x\|_\infty = \max_i\{|x_i|\}$ is a normed linear vector space.

b. Show that $\mathbb{R}^n$ with norm $\|x\|_1 = \sum_{k=1}^n |x_i|$ is a normed linear vector.

5. Verify that the choice $\alpha = \frac{(x,y)}{\|y\|^2}$ makes $\|x - \alpha y\|^2$ as small as possible. Show that $|\langle x, y \rangle|^2 = \|x\|^2\|y\|^2$ if and only if x and y are linearly dependent.

6. Starting with the set $\{1, x, x^2, \ldots, x^k, \ldots\}$, use the Gram-Schmidt procedure and the inner product

$$\langle f, g \rangle = \int_a^b f(x)g(x)w(x)dx \quad , \quad w(x) > 0$$

to find the first four orthogonal polynomials when

a. $a = -1, b = 1, w = 1$ (Legendre polynomials)

b. $a = -1, b = 1, w = (1 - x^2)^{-1/2}$ (Chebyshev polynomials)

c. $a = 0, b = \infty, w(x) = e^{-x}$ (Laguerre polynomials)

d. $a = -\infty, b = \infty, w(x) = e^{-x^2}$ (Hermite polynomials)

7. Starting with the set $\{1, x, x^2, \ldots, x^n, \ldots\}$ use the Gram-Schmidt procedure and the inner product

$$\langle f, g \rangle = \int_{-1}^1 (f(x)g(x) + f'(x)g'(x))\, dx$$

to find the first five orthogonal polynomials.

Section 1.2

1. a. Represent the matrix

$$A = \begin{pmatrix} 1 & 1 & 2 \\ 2 & 1 & 3 \\ 1 & 0 & 1 \end{pmatrix} \quad \text{(relative to the natural basis)}$$

relative to the basis $\{(1, 1, 0), (0, 1, 1), (1, 0, 1)\}$

b. If the representation of A with respect to the basis

$$\{(1, 1, 2), (1, 2, 3), (3, 4, 1)\} \quad \text{is} \quad A = \begin{pmatrix} 1 & 1 & 1 \\ 1 & 2 & 3 \\ 3 & 2 & 1 \end{pmatrix}$$

find the representation of A with respect to the basis $\{(1, 0, 0), (0, 1, -1), (0, 1, 1)\}$.

2. a. Prove that two symmetric matrices are equivalent if and only if they have the same eigenvalues.

  b. Show that if A and B are equivalent, then

$$\det A = \det B.$$

  c. Is the converse true?

3. a. Show that if A is an $n \times m$ matrix and B an $m \times n$ matrix, then AB and BA have the same nonzero eigenvalues.

  b. Show that the eigenvalues of AA^* are real and non-negative.

4. Show that the eigenvalues of a real skew-symmetric $(A = -A^T)$ matrix are imaginary.

5. Find a basis for the range and null space of the following matrices and their adjoints

  a.

$$A = \begin{pmatrix} 1 & 2 \\ 1 & 3 \\ 2 & 5 \end{pmatrix}$$

  b.

$$A = \begin{pmatrix} 1 & 2 & 3 \\ 3 & 1 & 2 \\ 1 & 1 & 1 \end{pmatrix}.$$

6. Find the matrices T, Λ so that $A = T\Lambda T^{-1}$ for the following matrices A

  a.

$$\begin{pmatrix} 1 & 0 & 0 \\ 1/4 & 1/4 & 1/2 \\ 0 & 0 & 1 \end{pmatrix}$$

  b.

$$\begin{pmatrix} 0 & 1 \\ -1 & 0 \end{pmatrix}$$

  c.

$$\begin{pmatrix} 1 & 0 & 0 \\ 1 & 2 & 0 \\ 2 & 1 & 3 \end{pmatrix}$$

  d.

$$\begin{pmatrix} 1/2 & 1/2 & \sqrt{3}/6 \\ 1/2 & 1/2 & \sqrt{3}/6 \\ \sqrt{3}/6 & \sqrt{3}/6 & 5/6 \end{pmatrix}.$$

7. Find the spectral representation of the matrix

$$A = \begin{pmatrix} 7 & 2 \\ -2 & 2 \end{pmatrix}.$$

Illustrate how $Ax = b$ can be solved geometrically using the appropriately chosen coordinate system on a piece of graph paper.

8. Suppose P is the matrix that projects (orthogonally) any vector onto a manifold M. Find all eigenvalues and eigenvectors of P.

9. The sets of vectors $\{\phi_i\}_{i=1}^n$, $\{\psi_i\}_{i=1}^n$ are said to be *biorthogonal* if $\langle \phi_i, \psi_j \rangle = \delta_{ij}$. Suppose $\{\phi_i\}_{i=1}^n$ and $\{\psi_i\}_{i=1}^n$ are biorthogonal.

   a. Show that $\{\phi_i\}_{i=1}^n$ and $\{\psi_i\}_{i=1}^n$ each form a linearly independent set.

   b. Show that any vector in $\mathbb{R}^n$ can be written as a linear combination of $\{\phi_i\}$ as

   $$x = \sum_{i=1}^n \alpha_i \phi_i$$

   where $\alpha_i = \langle x, \psi_i \rangle$.

   c. Re-express b) in matrix form; that is, show that

   $$x = \sum_{i=1}^n P_i x$$

   where P_i are projection matrices with the properties that $P_i^2 = P_i$ and $P_i P_j = 0$ for $i \neq j$. Express the matrix P_i in terms of the vectors ϕ_i and ψ_i.

10. a. Suppose the eigenvalues of A are distinct. Show that the eigenvectors of A and the eigenvectors of A^* form a biorthogonal set.

   b. Suppose $A\phi_i = \lambda_i \phi_i$ and $A^* \psi_i = \bar{\lambda}_i \psi_i$, $i = 1, 2, \ldots, n$ and that $\lambda_i \neq \lambda_j$ for $i \neq j$. Prove that $A = \sum_{i=1}^n \lambda_i P_i$ where $P_i = \phi_i \psi_i^*$ is a projection matrix.
   Remark: This is an alternate way to express the Spectral Decomposition Theorem for a matrix A.

   c. Express the matrices C and C^{-1} where $A = C\Lambda C^{-1}$, in terms of ϕ_i and ψ_i.

Section 1.3

1. Use the minimax principle to show that the matrix

$$\begin{pmatrix} 2 & 4 & 5 & 1 \\ 4 & 2 & 1 & 3 \\ 5 & 1 & 60 & 12 \\ 1 & 3 & 12 & 48 \end{pmatrix}$$

has an eigenvalue $\lambda_4 < -2.1$ and an eigenvalue $\lambda_1 > 67.4$.

2. a. Prove an inequality relating the eigenvalues of a symmetric matrix before and after one of its diagonal elements is increased.

   b. Show that the smallest eigenvalue of

$$A = \begin{pmatrix} 8 & 4 & 4 \\ 4 & 8 & -4 \\ 4 & -4 & 7 \end{pmatrix}$$

   is smaller than $-1/3$.

3. Show that the intermediate eigenvalue λ_2 of

$$\begin{pmatrix} 1 & 2 & 3 \\ 2 & 2 & 4 \\ 3 & 4 & 3 \end{pmatrix}$$

   is not positive.

4. The moment of inertia of any solid object about an axis along the unit vector x is defined by

$$I(x) = \int_R d_x^2(y)\rho\, dV,$$

   where $d_x(y)$ is the perpendicular distance from the point y to the axis along x, ρ is the density of the material, and R is the region occupied by the object. Show that $I(x)$ is a quadratic function of x, $I(x) = x^T A x$ where A is a symmetric 3×3 matrix.

Section 1.4

1. Under what conditions do the matrices of Problem 1.2.5 have solutions $Ax = b$? Are they unique?

2. Suppose P projects vectors in $\mathbb{R}^n$ (orthogonally) onto a linear manifold M. What is the solvability condition for the equation $Px = b$?

3. Show that the matrix $A = (a_{ij})$ where $a_{ij} = \langle \phi_i, \phi_j \rangle$ is uniquely invertible if and only if the vectors ϕ_i are linearly independent.

4. A square matrix A (with real entries) is positive-definite if $\langle Ax, x \rangle > 0$ for all $x \neq 0$. Use the Fredholm alternative to prove that a positive definite matrix is uniquely invertible.

Section 1.5

1. Find the least squares pseudo-inverse for the following matrices:

   a.
   $$\begin{pmatrix} 1 & 0 & 0 \\ 0 & 1 & 0 \\ 1 & 1 & 0 \end{pmatrix}$$

   b.
   $$\begin{pmatrix} 3 & -1 \\ 1 & 1 \\ -1 & 3 \end{pmatrix}$$

   c.
   $$\begin{pmatrix} 3 & 1 & 2 \\ -1 & 1 & -2 \end{pmatrix}$$

   d.
   $$\begin{pmatrix} -1 & 0 & 1 \\ -1 & 2/3 & 2/3 \\ -1 & -2/3 & 7/3 \end{pmatrix}.$$

2. Verify that the least squares pseudo-inverse of an $m \times n$ diagonal matrix D with $d_{ij} = \sigma_i \delta_{ij}$ is the $n \times m$ diagonal matrix D' with $d'_{ij} = \frac{1}{\sigma_i}\delta_{ij}$ whenever $\sigma_i \neq 0$ and $d'_{ij} = 0$ otherwise.

3. a. For any two vectors x, y in $\mathbb{R}^n$ with $\|x\| = \|y\|$, find the Householder (orthogonal) transformation U that satisfies $Ux = y$.

   b. Verify that a Householder transformation U satisfies $U^* = U^{-1}$.

4. Use the Gram-Schmidt procedure to find the QR representation of the matrix
   $$A = \begin{pmatrix} 2 & 1 & 3 \\ 4 & 2 & 1 \\ 9 & 1 & 2 \end{pmatrix}.$$

5. For the matrix
   $$A = \begin{pmatrix} 5 & -3 \\ 0 & 4 \end{pmatrix},$$
   illustrate on a piece of graph paper how the singular value decomposition $A = Q_1 \Sigma Q_2^*$ transforms a vector x onto Ax. Compare this with how $A = T\Lambda T^{-1}$ transforms a vector x onto Ax.

6. For the matrix
   $$A = \begin{pmatrix} 1 & 2 \\ 2 & 4 \end{pmatrix},$$
   illustrate on a piece of graph paper how the least squares pseudo-inverse $A' = Q^{-1}\Lambda Q$ transforms a vector b into the least squares solution of $Ax = b$.

7. For the matrices

$$A = \begin{pmatrix} 1 & 1 \\ 0 & 0 \end{pmatrix}$$

and

$$A = \begin{pmatrix} 1.002 & 0.998 \\ 1.999 & 2.001 \end{pmatrix},$$

illustrate on a piece of graph paper how the least squares pseudo-inverse $A' = Q_2\Sigma'Q_1^*$ transforms a vector b onto the least squares solution of $Ax = b$. For the second of these matrices, show how setting the smallest singular value to zero stabilizes the inversion process.

8. For a nonsymmetric matrix $A = T^{-1}\Lambda T$, it is not true in general that $A' = T^{-1}\Lambda'T$ is the pseudo-inverse. Find a 2×2 example which illustrates geometrically what goes wrong.

9. Find the singular value decomposition and pseudo-inverse of

$$A = \begin{pmatrix} 2\sqrt{5} & -2\sqrt{5} \\ 3 & 3 \\ 6 & 6 \end{pmatrix}.$$

10. Find the least squares solution of the n linear equations

$$a_i x + b_i y = c_i \quad i = 1, 2, \ldots, n.$$

If $r_j, j = 1, 2, \ldots, n(n-1)/2$ are the solutions of all possible pairs of such equations, show that the least squares solution

$$r = \begin{pmatrix} x \\ y \end{pmatrix}$$

is a convex linear combination of r_j, specifically

$$r = \sum_{j=1}^{k} p_i r_j \quad, \quad k = n(n-1)/2$$

where

$$p_j = \frac{D_j^2}{\sum_{i=1}^{k} D_i^2}$$

and D_j is the determinant of the ith 2-equation subsystem. Interpret this result goemetrically.

11. The matrix M_k represents elementary row operations if it has elements

$$m_{ij} = 1 \qquad \text{if} \quad i = j$$

$$m_{ik} = -a_{ik} \qquad \text{if} \quad i > k$$

$$m_{ij} = 0 \qquad \text{otherwise.}$$

Let L have entries $l_{ij} = 1$ if $i = j$, $l_{ij} = a_{ij}$ if $i > j$, $l_{ij} = 0$ otherwise. Show that $(M_n M_{n-1} \cdots M_1)^{-1} = L$.

12. Find the unitary matrix A that minimizes $\sum_{i=1}^{n} \|x_i - A y_i\|^2$ where the vectors $\{x_i\}$, $\{y_i\}$, $i = 1, 2, \ldots n$ are specified. Give a geometrical interpretation of the meaning of this problem. Hint: Let X, Y be matrices with columns $\{x_i\}$, $\{y_i\}$, respectively. Use that $\sum_{i=1}^{n} \langle x_i, x_i \rangle = tr\left(X^T X\right)$ and that $Y X^T$ has a singular value decomposition $Y X^T = U \Sigma V^T$.

Section 1.6

1. Generalize the REDUCE procedure GRAM to find orthogonal polynomials with respect to the inner product

$$\langle f, g \rangle = \int_a^b \left(f(x) g(x) + f'(x) g'(x) \right) dx.$$

1.7 Appendix: Jordan Canonical Form

Not all matrices can be diagonalized. If a matrix is not symmetric and has some multiple eigenvalues, the theorems of Chapter 1 on diagonalization may fail. For example, these theorems fail for the matrix

$$A = \begin{pmatrix} 0 & 1 \\ 0 & 0 \end{pmatrix}$$

which has $\lambda = 0$ as an eigenvalue of multiplicity 2, but only one eigenvector. It is natural to inquire about what can be done in the case that a matrix cannot be diagonalized.

The goal is to find a set of basis vectors for which the representation of the matrix A is, if not diagonal, then as "close to diagonal" as possible. To the extent that it is possible we will use the eigenvectors of the matrix A and then complete this set with other appropriately chosen vectors.

The key observation is provided by the example

$$A = \begin{pmatrix} 0 & 1 \\ 0 & 0 \end{pmatrix}.$$

For this matrix we already know the eigenvector

$$x_1 = \begin{pmatrix} 1 \\ 0 \end{pmatrix}$$

satisfies $Ax_1 = 0$. The second basis vector

$$x_2 = \begin{pmatrix} 0 \\ 1 \end{pmatrix}$$

is, of course, linearly independent of x_1, and $Ax_2 = x_1$, so that $A^2 x_2 = 0$.

Theorem 1.14

If $x_0 \neq 0$ satisfies $A^k x_0 = 0$ and $A^{k-1} x_0 \neq 0$ for some $k > 1$, then the set of vectors $x_0, Ax_0, \ldots, Ax_0^{k-1}$ is linearly independent.

To prove this theorem, we seek constants $a_0, a_1, \ldots, a_{k-1}$ so that $a_0 x_0 + a_1 A x_0 + \cdots + a_{k-1} A^{k-1} x_0 = 0$. Since $A^k x_0 = 0$ but $A^{k-1} x_0 \neq 0$, we multiply by A^{k-1} and learn that $a_0 = 0$. If we multiply instead by A^{k-2} we can conclude that $a_1 = 0$ and so on inductively. That is, if we suppose that $a_0 = a_1 = \cdots = a_{j-1} = 0$ we multiply by A^{k-j-1} to deduce that $a_j = 0$ as well, which completes the proof.

If $A^k x_0 = 0$ for some $k > 0$, we examine the null space of A^k denoted $N_k = N(A^k)$. If x is in N_k, then $A^k x = 0$ so that $A^{k+1} x = 0$ and as a result x is in N_{k+1} as well. Thus, N_k is a subset of N_{k+1} which is a subset of N_{k+j} for $j > 1$. The dimension of N_k cannot grow without bound, so we let q be the smallest integer for which $N_q = N_{q+1}$. Then $N_{q+j} = N_q$ for all $j \geq 0$. Indeed, if x is in N_{q+j}, $A^{q+j} x = A^{q+1} A^{j-1} x = 0$. Since $A^{j-1} x$ is in $N_{q+1} = N_q$, it follows that $0 = A^q A^{j-1} x = A^{q+j-1} x$ so that x is in N_{q+j-1}, and so on.

Let $R_k = R(A^k)$ be the range of A^k.

Theorem 1.15

Any vector x in $\mathbb{R}^n$ can be uniquely represented as $x = x_r + x_n$ where x_r is in R_q and x_n is in N_q. Furthermore, A is invertible on the subspace R_q.

For any $n \times n$ square matrix A, the rank of A is the dimension of $R(A)$, the nullity is the dimension of $N(A)$ and the $\text{rank}(A) + \text{nullity}(A) = n$. Since $R(A^q)$ and $N(A^q)$ span $\mathbb{R}^n$ we need only show the uniqueness of the decomposition. The decomposition will be unique if the only element common to N_q and R_q is the zero vector.

If x is in N_q then $A^q x = 0$ and if x is in R_q then $A^q y = x$ for some y and $A^{2q} y = 0$. Since $N_{2q} = N_q$, y is in N_q, and $x = A^q y = 0$.

To see that A is invertible on R_q we need to show that the solution of $Ax = y$ is unique whenever y is in R_q. Recall that uniqueness is guaranteed if and only if the only solution of $Ax = 0$ is $x = 0$. Suppose there is an x in R_q for which $Ax = 0$. Since x is in R_q there is a y with $A^q y = x$ and $A^{q+1} y = 0$. However, $N_{q+1} = N_q$ so $x = A^q y = 0$ which completes the proof.

If N_q is not empty, we can split the space on which A acts into two complementary invariant subspaces N_q and R_q. Our goal is to find a basis for N_q.

Since $N_q \neq N_{q-1}$, there are a finite number of linearly independent vectors $x_1, x_2, \ldots, x_{p_1}$ that span the difference between N_{q-1} and N_q so that x_j is in N_q but not in N_{q-1}. Thus, $A^q x_j = 0$ and $A^{q-1} x_j \neq 0$ for $j = 1, 2, \ldots, p_1$. Notice also that $A^k x_j$ belongs to N_{q-k} but not to N_{q-k-1}. The set of vectors $x_i, Ax_i, \ldots, A^q x_i$ is called a MAXIMAL CHAIN. We also note that if $x_1, x_2, \ldots, x_p$ are linearly independent and lie in N_s but not N_{s-1}, then $Ax_1, Ax_2, \ldots, Ax_p$ form a linearly independent set in N_{s-1}. If not, there are constants $\alpha_1, \ldots, \alpha_p$ so that $0 = \alpha_1 Ax_1 + \cdots + \alpha_p Ax_p = A(\alpha_1 x_1 + \cdots + \alpha_p x_p) = 0$ and $\alpha_1 x_1 + \cdots + \alpha_p x_p$ lies in N_1. Since $x_1, x_2, \ldots, x_p$ lie in N_s and not N_{s-1} (or N_1), $\alpha_1 x_1 + \alpha_2 x_2 + \cdots + \alpha_p x_p = 0$, implying that $\alpha_1 = \alpha_2 = \cdots = \alpha_p = 0$.

Proceeding inductively, suppose some set of vectors $y_1, y_2, \ldots, y_r$ span the deficiency between N_{q-j} and N_{q-j+1}. If $Ay_1, Ay_2, \ldots, Ay_r$ span the deficiency between N_{q-j-1} and N_{q-j} we can continue on, but if not we begin new chains with linearly independent vectors $x_{p_j+1}, \ldots, x_{p_{j+1}}$ so that the deficiency between N_{q-j-1} and N_{q-j} is now spanned by $Ay_1, Ay_2, \ldots, Ay_r$ and $x_{p_j+1}, \ldots, x_{p_{j+1}}$. In this way, we eventually have a collection of maximal chains

$$x_1, Ax_1 \cdots \cdots A^{p_1} x_1$$

$$x_2, Ax_2 \cdots \cdots A^{p_2} x_2$$

$$x_r, Ax_r \cdots \cdots A^{p_r} x_r$$

which forms a basis for N_q.

We are now in a position to apply this machinery to the matrix $A - \lambda_1 I$. We suppose λ_1 is an eigenvalue of A. There is a smallest integer q_1 so that $N_{q_1}(\lambda_1) = N((A - \lambda_1 I)^{q_1})$ satisfies $N_{q_1}(\lambda_1) = N_{q_1+1}(\lambda_1)$. We split the space $\mathbb{R}^n$ into $N_{q_1}(\lambda_1)$ and R_{q_1} and form a basis for $N_{q_1}(\lambda_1)$ with the set of maximal chains for $A - \lambda_1 I$.

Since R_{q_1} is invariant for $A - \lambda_1 I$, A restricted to R_{q_1} has another eigenvalue $\lambda_2 \neq \lambda_1$. But restricted to R_{q_1}, there is again a smallest integer q_2 so that $N_{q_2}(\lambda_2) = N(A - \lambda_2 I)^{q_2}$ satisfies $N_{q_2}(\lambda_2) = N_{q_2+1}(\lambda_2)$, and this subspace can be spanned by the set of maximal chains for $A - \lambda_2 I$. Proceeding inductively we eventually run out of eigenvalues, and conclude that we have spanned all of $\mathbb{R}^n$ with the collection of maximal chains of $A - \lambda_i I$.

To see what A looks like relative to this basis, notice first that if $x_j, (A - \lambda_j)x_j, (A - \lambda_j)^2 x_j \cdots (A - \lambda_j)^{p_j-1} x_j = y_{p_j}, y_{p_j-1}, \ldots, y_1$, is a maximal chain, then $Ay_i = \lambda_j y_i + y_{i-1}$ for $i = 2, 3, \ldots, p_j$. The vectors y_i are called GENERALIZED EIGENVECTORS of A and the first element of a chain is always a genuine eigenvector.

We form a matrix T whose columns consist of all maximal chains of vectors.

Theorem 1.16

The matrix $\hat{A} = T^{-1} AT$ is the matrix with elements $\hat{a}_{ij}$, where $\hat{a}_{ii} = \lambda_i$ are the eigenvalues of A and $\hat{a}_{i,i+1} = \sigma_i$ and all other elements are zero. The element σ_i is either zero or one, and σ_i is one if the ith basis vector is a generalized eigenvector, and not the first vector in a maximal chain.

The proof of this theorem is by direct calculation. The matrix $\hat{A}$ is called the *Jordan Canonical Form* for the matrix A. In general, two matrices are equivalent if and only if they have the same Jordan canonical form.

Examples

1. The matrix

$$A = \begin{pmatrix} 1 & 1 \\ 0 & 1 \end{pmatrix}$$

has a double eigenvalue $\lambda = 1$. The matrix

$$A - I = \begin{pmatrix} 0 & 1 \\ 0 & 0 \end{pmatrix}$$

and has one eigenvector

$$x_1 = \begin{pmatrix} 1 \\ 0 \end{pmatrix}$$

which is the last vector in a maximal chain. The solution of $(A - I)x_2 = x_1$ is

$$x_2 = \begin{pmatrix} 0 \\ 1 \end{pmatrix} + \begin{pmatrix} \alpha \\ 0 \end{pmatrix}$$

where α is arbitrary. For any α, the vectors x_1, x_2 form a basis for $\mathbb{R}^2$, and relative to this basis, A has the representation

$$\hat{A} = \begin{pmatrix} 1 & 1 \\ 0 & 1 \end{pmatrix}.$$

2. The matrix

$$A = \begin{pmatrix} -3 & 1 & 2 \\ -8 & 3 & 4 \\ -4 & 1 & 3 \end{pmatrix}$$

has the characteristic polynomial $\det(A - \lambda I) = (\lambda - 1)^3$, so that A has an eigenvalue $\lambda = 1$ of multiplicity 3. The matrix $A - I$ is

$$A - I = \begin{pmatrix} -4 & 1 & 2 \\ -8 & 2 & 4 \\ -4 & 1 & 2 \end{pmatrix}$$

and has a null space spanned by two vectors $x_1 = (4, 8, 4)^T$, $x_3 = (1, 4, 0)^T$. We form a maximal chain from x_1 by setting $(A-I)x_2 = x_1$, $x_2 = (-1, 0, 0)^T$. Relative to the basis x_1, x_2, x_3, the matrix A has the representation

$$\hat{A} = \begin{pmatrix} 1 & 1 & 0 \\ 0 & 1 & 0 \\ 0 & 0 & 1 \end{pmatrix}.$$

Problems

Find Jordan Canonical forms and the corresponding bases for the following matrices:

1.

$$\begin{pmatrix} -3 & -1 & 0 \\ 1 & 0 & 1 \\ 5 & -2 & 0 \end{pmatrix}$$

2.

$$\begin{pmatrix} 3 & 3 & 4 \\ -10 & -16 & -24 \\ 7 & 12 & 18 \end{pmatrix}$$

3.

$$\begin{pmatrix} 3 & 3 & 4 & 5 \\ 2 & 6 & 8 & 10 \\ -13 & -20 & -30 & -40 \\ 9 & 13 & 20 & 27 \end{pmatrix}$$

4.

$$\begin{pmatrix} 13 & -3 & -2 & -4 \\ 16 & -3 & -2 & -6 \\ 16 & -4 & -2 & -5 \\ 16 & -4 & -3 & -4 \end{pmatrix}$$

5.

$$\begin{pmatrix} 5 & -1 & 1 & 1 & 0 & 0 \\ 1 & 3 & -1 & -1 & 0 & 0 \\ 0 & 0 & -1 & -1 & 0 & 0 \\ 0 & 0 & 0 & 4 & -1 & -1 \\ 0 & 0 & 0 & 0 & 3 & 1 \\ 0 & 0 & 0 & 0 & 1 & 3 \end{pmatrix}$$

2

FUNCTION SPACES

2.1 Complete Metric Spaces

So far we have dealt only with finite dimensional vector spaces. In this chapter we will extend the ideas of Chapter 1 to vector spaces without finite dimension. We begin by defining a linear vector space, as before, to be a collection of objects, S, which are closed under addition and scalar multiplication. That is, if x and y are in S, and α is a real (or complex) number, then $x + y$ is in S and αx is in S.

Definition

A METRIC d in the vector space S is a measure of distance between two elements of S, $d(\cdot, \cdot) : S \times S \to \mathbb{R}^+$ for which

1. $d(x, y) = d(y, x)$ (symmetry)
2. $d(x, y) \geq 0$ and $d(x, y) = 0$ if and only if $x = y$.
3. $d(x, y) \leq d(x, z) + d(z, y)$ (TRIANGLE INEQUALITY).

In applications it is simply naive to assert that we can or should look for exact solutions. Even if we can solve a model problem exactly, there are always approximations made. In the modeling process we always ignore terms, make simplifying assumptions, and never know parameter values exactly. Of course, numerical solutions are inexact. A more relevant desire is to find a solution that is close to the "true" solution, and then to sequentially improve the model and solution techniques until the obtained solution is in some sense "close enough."

For this reason, it is quite relevant to concern ourselves with sequences and their convergence. Suppose we have a sequence of vectors $\{x_n\}$ in S.

Definition

A sequence $\{x_n\}$ in S is said to have a limit x in S or to converge to x in S (denoted $\lim_{n \to \infty} x_n = x \in S$) if for any $\epsilon > 0$ there is an integer N, (depending on the size of ϵ) so that for any $n \geq N$, $d(x_n, x) < \epsilon$.

A concept of closeness that does not require knowledge of the limiting vector is as follows:

Definition

A sequence $\{x_n\}$ in S is called a CAUCHY SEQUENCE if for any $\epsilon > 0$ there is an integer N (depending on ϵ) so that for every $m, n \geq N$, $d(x_n, x_m) < \epsilon$.

It is an easy matter (using the triangle inequality) to show that a convergent sequence is always a Cauchy sequence. As we shall see, the converse is not necessarily true.

Definition

The vector space S is called a NORMED SPACE if there is a function (called a norm) denoted $\| \cdot \| : S \to \mathbb{R}^+$ which for all x in S satisfies

1. $\|x\| \geq 0$ and $\|x\| = 0$ implies $x = 0$
2. $\|\alpha x\| = |\alpha| \cdot \|x\|$
3. $\|x + y\| \leq \|x\| + \|y\|$.

(This is the same definition as given in Chapter 1.)

If there is a norm, we automatically have a metric by taking $d(x, y) = \|x - y\|$, but the opposite is not necessarily true. In Chapter 1 we saw simple examples of normed linear spaces.

Examples of INFINITE DIMENSIONAL SPACES are SEQUENCE SPACES, that is the set of all sequences of numbers $\{x_n\}$ with appropriate additional restrictions. For example, the set of sequences $\{x_n\}$ with

$$\sum_{n=1}^{\infty} |x_n|^2 < \infty$$

is a normed vector space (denoted ℓ^2), if we take

$$\|x\| = \left(\sum_{n=1}^{\infty} |x_n|^2 \right)^{1/2}.$$

Other sequence spaces, with $\sum_{n=1}^{\infty} |x_n|^p < \infty$, are normed vector spaces with

$$\|x\| = \left(\sum_{n=1}^{\infty} |x_n|^p \right)^{1/p}, \qquad p \geq 1$$

(called ℓ^p spaces). To see that these spaces are not finite dimensional, let e_k be the sequence $e_k = \{\delta_{ik}\}$, $i = 1, 2, \ldots$, and note that no nontrivial, finite, linear combination of e_k can be the zero sequence, so that any finite collection of the e_k is linearly independent. Therefore the sequence space has an infinite number of linearly independent elements.

Other examples of infinite dimensional vector spaces are FUNCTION SPACES. For example, the set of continuous functions defined on the closed interval $[a, b]$, denoted $C[a, b]$, is a vector space. Norms can be defined in a variety of ways. For example,

$$\|f\| = \max_{x \in [a,b]} |f(x)|$$

is called the SUPREMUM or UNIFORM NORM, and

$$\|f\| = \left(\int_a^b |f|^2 dx \right)^{1/2}$$

is called the L^2 (pronounced "L-two") norm. More generally, the L^p norm is

$$\|f\| = \left(\int_a^b |f|^p dx \right)^{1/p}.$$

The only hard part of verifying that these are norms is to show that the triangle inequality holds. In L^2 one can use the Schwarz inequality. For L^p the proof is harder; can you supply it?

Function spaces are actually quite a bit bigger than we have so far suggested, since for a particular norm, $\|\cdot\|$, the space is defined as the set of all f with $\|f\| < \infty$. So, for example, the function $f(x) = x^{-1/3}$ belongs to $L^1[0, 1]$ and $L^2[0, 1]$ but not to $C[0, 1]$ nor to $L^p[0, 1]$, if $p \geq 3$.

Definition

A normed linear space S is complete if every Cauchy sequence in S is convergent in S.

The key idea here is that the convergence must be in S so that the limiting object, if it exists, must be in the same space as the elements of the sequence.

As a simple example, a sequence of rational numbers need not converge to a rational number, but may converge to an irrational number. Thus the set of rational numbers is not complete, but the real numbers, because they contain all possible limits of sequences, are complete.

Not all normed linear vector spaces are complete. As an example, we will demonstrate that the set of continuous functions is complete with respect to the uniform norm but not complete with respect to the L^2 norm. First suppose that $\{f_n(t)\}$ is a Cauchy sequence of continuous functions with respect to the uniform norm. For each fixed point t, the sequence is a Cauchy sequence in $\mathbb{R}$ and therefore convergent in $\mathbb{R}$. Let $f(t)$ be the pointwise limit of the sequence $\{f_n(t)\}$. For any $\epsilon > 0$ pick N so that

$$\max_{t \in [a,b]} |f_n(t) - f_m(t)| < \epsilon/3$$

for all $n, m \geq N$. Since $\{f_n\}$ is a Cauchy sequence, this can always be done. Furthermore, since $f_n(t)$ is continuous, for $\epsilon > 0$ there is a $D > 0$ so that for $\delta \leq D$, $|f_k(t + \delta) - f_k(t)| < \epsilon/3$ for some fixed $k \geq N$. It follows that

$$|f(t) - f(t+\delta)| \leq |f(t) - f_k(t)| + |f_k(t) - f_k(t+\delta)| + |f_k(t+\delta) - f(t+\delta)| < \epsilon$$

whenever $\delta \leq D$, so that $f(t)$, the pointwise limit, is a continuous function. Thus, the limit is defined and belongs to the set of continuous functions. Notice that convergence is indeed uniform (i.e., convergence is with respect to the uniform norm), since for any $n, m \geq N$ and any $t \in [a, b]$, $|f_n(t) - f_m(t)| < \epsilon$. Taking the limit $m \to \infty$, we obtain that

$$\lim_{m \to \infty} |f_n(t) - f_m(t)| = |f_n(t) - \lim_{m \to \infty} f_m(t)| = |f_n(t) - f(t)| < \epsilon$$

for all $n \geq N$ and $t \in [a, b]$. (Can you justify taking the limit inside the absolute value?)

This is not the case for the continuous functions using the L^p norm. To see this consider the sequence of continuous functions

$$f_n(t) = \begin{cases} 0 & 0 \leq t < 1/2 - 1/n \\ 1/2 + n/2(t - 1/2) & 1/2 - 1/n \leq t \leq 1/2 + 1/n \\ 1 & 1/2 + 1/n \leq t \leq 1 \end{cases}$$

(A sketch of $f_n(t)$ is shown in Figure 2.1). The sequence $\{f_n(t)\}$ is not a Cauchy sequence using the uniform norm, but with any of the L^p norms ($1 \leq p < \infty$), it is a Cauchy sequence. (Can you show this?) However, it is

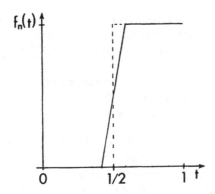

FIGURE 2.1   Sketch of the function $f_n(t)$.

easy to see that the L^p limit of the sequence $\{f_n(t)\}$ is the function

$$f(t) = \begin{cases} 0 & 0 \le t < 1/2 \\ 1/2 & t = 1/2 \\ 1 & 1/2 < t \le 1 \end{cases}$$

and this is certainly not a continuous function. Thus, the Cauchy sequence $\{f_n(t)\}$ is not convergent in the space of continuous functions, measured by the L^2 norm. That is, this space is not complete.

This observation presents a most serious difficulty. We certainly want to work only with complete function spaces and so we seize upon the idea of trying to complete the space of continuous functions with norm L^2, say. To do this we might guess that we should start with the set of continuous functions and then add as many functions as we can that arise as limits of continuous functions. For example, the function $f(t) = 0$ for $x \ne \frac{1}{2}$, $f\left(\frac{1}{2}\right) = 1$ should be in the set, and the function $g(t) = 0$ for $x \ne 1/3$, $2/3$, $g\left(\frac{1}{3}\right) = g\left(\frac{2}{3}\right) = 1$ should be in the set. Continuing in this way, we quickly see that the function $\chi(t) = 0$ for t irrational, $\chi(t) = 1$ for t rational should also be in the set. But here is the problem:

$$\|\chi\| = \left(\int_0^1 \chi^2(t)dt \right)^{1/2}$$

does not make sense, at least when we think of Riemann integration the way we were taught in Calculus. Remember that, in the Riemann theory, we approximate an integral by the sum $\sum f(x_i)\Delta x_i$, where the x_i are chosen points inside small intervals of length Δx_i. Notice that for the function $\chi(t)$, if we happen to pick only irrational points x_i, our approximate integral is zero, whereas if we happen to pick only rational points x_i, our approximate integral is one. Other choices for the points x_i can give an approximate answer anywhere between zero and one, no matter how large the number of points becomes. Thus, we cannot use Riemann integration to evaluate this integral. We conclude that the space of all functions with $\|f\|^2 = \int_0^1 |f|^2 dt < \infty$ cannot be a complete space.

This is where the theory of Lebesgue integration comes to the rescue. Rather than abandoning all hope of making L^2 complete, we redefine what we mean by integration in such a way as to make sense of $\int_0^1 |\chi(t)|^2 dt$, for example.

This explains the importance of the Lebesgue integral. For purposes of this text, it is only necessary to acquire a simple understanding of the Lebesgue integral and some of the concepts related to it. For a more detailed study of the ideas, please consult an analysis text such as listed in the references.

Here, then, are the properties you need to absorb in order to understand what Lebesgue integration is all about:

0. A set of real numbers is a SET OF MEASURE ZERO if for every $\epsilon > 0$, the set can be covered by a collection of open intervals whose total length is less than ϵ. For example, any finite set of numbers has measure zero. In fact, any countable set of numbers has measure zero. If we can count (or sequentially order) a set, then we surround the nth point x_n with an open interval of length $\epsilon/2^n$, and the total length of the intervals is no greater than ϵ.

   For notation, let $\int_R$ denote the Riemann integral and let $\int_L$ denote the Lebesgue integral.

1. If $\int_R f(t)dt$ exists, then $\int_L f(t)$ exists and $\int_R f(t)dt = \int_L f(t)$

2. If $\int_L f(t)$ and $\int_L g(t)$ exist and α is a scalar, then

$$\int_L \alpha f = \alpha \int_L f \quad \text{and} \quad \int_L (f+g) = \int_L f + \int_L g.$$

3. If $\int_L f^2(t)$ and $\int_L g^2(t)$ exist, then $\int_L f(t)g(t)$ and $\int_L (f+g)^2$ exist (i.e., they are finite).

4. If f and g are equal, except on a set of measure zero, then

$$\int_L (f-g) = 0 \quad , \quad \int_L (f-g)^2 = 0.$$

These four properties tell us how to evaluate the Lebesgue integral of many functions. Of course, if we can evaluate the Riemann integral, the Lebesgue integral is the same (property 1) or if we can change the definition of the function on at most a set of measure zero to become a Riemann integrable function, we also know its Lebesgue integral (properties 1 and 4). For all the functions we will see in this text, this suffices as a definition of Lebesgue integration. However, there are functions (rather complicated and pathological to be sure) for which this definition is not sufficient. For example, suppose a set C is uncountable, has positive measure, but contains no intervals, and suppose $f(x) = 1$ for x in C, $f(x) = 0$ for x not in C. The Lebesgue integral of this f cannot be evaluated by simply relating it to a Riemann integrable function g which differs from f on at most a set of measure zero, since no such g exists. For this f, another definition of Lebesgue integration is needed.

Property 4 also means that in L^2 the distance between two functions which differ only on a set of measure zero is zero and we must identify the two functions as the same function in order that the L^2 norm have the property that $\|x\| = 0$ if and only if $x = 0$. From now on in this text, unless noted otherwise, the statement $f = g$ means simply that $\|f - g\| = 0$. For the uniform norm, this also implies that $f(t) = g(t)$ at each point t, but for other norms, such as the L^2 norm, it means that f may differ from g on a set of measure zero.

5. (Lebesgue Dominated Convergence Theorem) Suppose $\{f_n(t)\}$ is a sequence of integrable functions on $[a, b]$ (i.e., $f_n(t) \in L^1[a, b]$) and $f_n(t)$ converges to $f(t)$ pointwise almost everywhere (that is, except on a set of measure zero). If there is an integrable function $g(t)$ so that for every $n \geq N$, $|f_n(t)| \leq g(t)$, the pointwise limit $f(t)$ is integrable and

$$\lim_{n \to \infty} \int_a^b f_n(t) = \int_a^b \lim_{n \to \infty} f_n(t) = \int_a^b f(t).$$

In other words, the limit process and integration may be interchanged without harm.

6. $L^2[a, b]$ is a complete space.

7. Continuous functions are dense in $L^2[a, b]$. In other words, inside any ϵ neighborhood of any function in $L^2[a, b]$ there is a continuous function.

8. (Fubini) Suppose M and M' are two closed intervals and that f is defined on the product of these intervals $M \times M'$.

   a. The integral $\int_{M \times M'} |f|$ exists and is finite (i.e., f is integrable) if and only if one of the iterated integrals $\int_M \left(\int_{M'} |f| \right)$ or $\int_{M'} \left(\int_M |f| \right)$ exists (and is finite).

   b. If f is integrable, then

$$\int_{M \times M'} f = \int_M \left(\int_{M'} f \right) = \int_{M'} \left(\int_M f \right).$$

   In other words, exchange of the order of integration is permissible.

In Chapter 1 we defined an inner product space to be a linear vector space for which there is an inner product. For an infinite dimensional space S the same definition of inner product holds.

Definition

An inner product $\langle \ , \ \rangle$ is a bilinear operation on $S \times S$ into the real (or complex) numbers with the properties

1. $\langle x, y \rangle = \overline{\langle x, y \rangle}$
2. $\langle \alpha x, y \rangle = \alpha \langle x, y \rangle$
3. $\langle x + y, z \rangle = \langle x, z \rangle + \langle y, z \rangle$
4. $\langle x, x \rangle \geq 0, \langle x, x \rangle = 0$ if and only if $x = 0$.

In the same way as before, the Schwarz inequality

$$|\langle x, y \rangle|^2 \leq \langle x, x \rangle \langle y, y \rangle$$

holds. Likewise, orthogonality and the Gram-Schmidt procedure work as in Chapter 1.

Examples of inner products are, for the sequence space ℓ^2,

$$\langle x, y \rangle = \sum_{i=1}^{\infty} x_i \bar{y}_i$$

or for the function space $L^2[a, b]$, $\langle f, g \rangle = \int_a^b f(t)\bar{g}(t)dt$. To be sure that this inner product exists for all f, g in $L^2[a, b]$ one must invoke property (3) of the Lebesgue integral. Can you show that the inner product $\langle x, y \rangle = \sum_{i=1}^{\infty} x_i \bar{y}_i$ is defined (i.e. bounded) for every sequence x, y in ℓ^2?

If a space S has defined on it an inner product, then $(\langle x, x \rangle)^{1/2}$ is a norm for the space S. The triangle inequality follows since

$$\begin{aligned} \|x + y\|^2 &= \langle x + y, x + y \rangle = \langle x, x \rangle + 2 \operatorname{Re} \langle x, y \rangle + \langle y, y \rangle \\ &\leq \|x\|^2 + 2\|x\| \|y\| + \|y\|^2 = \left(\|x\| + \|y\|^2\right). \end{aligned}$$

A complete inner product space is called a HILBERT SPACE (where the norm is the natural norm) and a complete normed space (without an inner product) is called a BANACH SPACE. Throughout the rest of this text we will make heavy use of Hilbert Spaces, especially the two most important examples, ℓ^2 and $L^2[a, b]$ with some reference to Sobolev spaces, which have not yet been defined. Examples of Banach Spaces are $L^p[a, b]$ for $p \neq 2$, but in this text we will not use these spaces.

2.1.1   Sobolev Spaces

The Hilbert space $L^2[a, b]$ with inner product $\langle f, g \rangle = \int_a^b f(x)\overline{g(x)}dx$ is a complete inner product space. To define a SOBOLEV SPACE, we start with all functions in $L^2[a, b]$ and then restrict attention further to those functions whose derivatives can be defined and are also in $L^2[a, b]$. (We do not yet know how to take the derivative of an arbitrary L^2 function, but we shall rectify this deficiency in Chapter 4 when we study distributions.) The collection of all such functions is called the Sobolev space $H^1[a, b]$.

The inner product for $H^1[a, b]$ is given by

$$\langle f, g \rangle = \int_a^b \left(f(x)\overline{g(x)} + f'(x)\overline{g'(x)} \right) dx$$

(complex conjugation is needed only if f and g are complex valued), and the norm is

$$\|f\| = \left(\int_a^b \left(|f(x)|^2 + |f'(x)|^2 \right) dx \right)^{1/2}.$$

One need not stop with first derivatives. For any function in L^2 we can define the nth order derivative using the theory of distributions. Those functions whose derivatives through order n are in L^2 comprise the Sobolev space $H^n[a, b]$. The inner product for this space is

$$\langle f, g \rangle = \int_a^b \left(\sum_{j=0}^{n} f^{(j)}(x)\overline{g^{(j)}}(x) \right) dx,$$

and the norm is

$$\|f\| = \left(\int_a^b \left(\sum_{j=0}^{n} \left| f^{(j)}(x) \right|^2 \right) dx \right)^{1/2}.$$

Two facts about Sobolev spaces (which we will not prove) make them quite important. First, Sobolev spaces are complete. Thus, a Sobolev space is a Hilbert space and all of the subsequent theory about operators on a Hilbert space applies. Second, we have the SOBOLEV INEQUALITIES:

Theorem 2.1

1. If $f \in H^1[a, b]$, then f is equivalent to a continuous function, and if M is any closed and bounded set containing x, then there is a constant $c \geq 0$ for which

$$|f(x)| \leq c \left(\int_M \left(|f(x)|^2 + |f'(x)|^2 \right) dx \right)^{1/2}.$$

2. If $f \in H^{k+1}[a, b]$, then f is equivalent to a C^k function.

In higher spatial dimensions, one must use higher order derivatives to obtain the same amount of smoothness. That is, if f is defined on some subset of $\mathbb{R}^n$, and if $f \in H^{n+k}(U)$ for every bounded open set U, then f is equivalent to a C^k function.

As you might imagine, the principal usefulness of Sobolev spaces is when additional smoothness is desired. Thus, they are especially nice for approximation procedures, such as the Galerkin methods or finite element methods, where convergence to smooth functions is desired.

The disadvantage of Sobolev spaces is that many of the operators that are important in applications are not self adjoint with respect to the Sobolev inner product. As we have seen with matrix operators, and as we shall see with operators on a Hilbert space, self adjointness is crucial to the development of spectral theory and transform methods. For this reason, we shall not make extensive use of Sobolev spaces in this text, but at the same time, you should be aware of their tremendous importance in certain contexts.

2.2   Approximation in Hilbert Spaces

2.2.1   Fourier Series and Completeness

Since we have abandoned the goal of finding "exact" solutions to problems, what we need is some way to find good approximate solutions, and this leads us to the question of how best to approximate a function in a real Hilbert Space. Suppose we have a favorite set of functions $\{\phi_i\}_{i=1}^n$ with which we wish to approximate the function f in H, the Hilbert Space. To approximate f in the best possible way, we want a linear combination of $\{\phi_i\}_{i=1}^n$ which is as close as possible, in terms of the norm of H, to f. In other words, we want to minimize

$$\left\| f - \sum_{i=1}^n \alpha_i \phi_i \right\|^2 .$$

Now

$$\left\| f - \sum_{i=1}^n \alpha_i \phi_i \right\|^2 = \|f\|^2 - 2 \sum_{i=1}^n \alpha_i \langle f, \phi_i \rangle + \sum_{i=1}^n \sum_{j=1}^n \alpha_i \alpha_j \langle \phi_i, \phi_j \rangle .$$

Differentiating this expression with respect to α_k and setting the result to zero, we find that the minimum will occur only if we choose

$$\Phi \alpha = \beta$$

where $\alpha = (\alpha_i)$, $\beta = (\langle f, \phi_i \rangle)$ and the matrix $\Phi = (\phi_{ij})$ has entries $\phi_{ij} = \langle \phi_i, \phi_j \rangle$. If the vectors ϕ_i are linearly independent, the matrix equation $\Phi \alpha = \beta$

can be solved by Gaussian elimination for the coefficients α. However, notice that our problem is much easier if the matrix Φ is the identity, i.e., if $\{\phi_i\}$ is an orthonormal set. Observe further that if we choose to increase the size of the set by adding more linearly independent vectors to the approximating set, all of components of α will in general be modified by the higher order approximation. This will not, however, be a problem if the ϕ_i are chosen to be orthonormal. Thus, since we know how to orthonormalize any set, (via the Gram-Schmidt procedure) we assume from the start that the set $\{\phi_i\}_{i=1}^n$ is orthonormal.

If the set $\{\phi_i\}_{i=1}^n$ is orthonormal, then

$$\left\| f - \sum_{i=1}^n \alpha_i \phi_i \right\|^2 = \sum_{i=1}^n (\alpha_i - \langle f, \phi_i \rangle)^2 + \|f\|^2 - \sum_{i=1}^n \langle f, \phi_i \rangle^2 .$$

This expression is minimized for arbitrary f by taking $\alpha_i = \langle f, \phi_i \rangle$. This choice of α_i minimizes the norm, and our approximation for f is $f \sim \sum_{i=1}^n \langle \phi_i, f \rangle \phi_i$ which we call a (generalized) FOURIER SERIES. The coefficients α_i are called FOURIER COEFFICIENTS. With this choice of α_i, the error of our approximation is

$$\left\| f - \sum_{i=1}^n \alpha_i \phi_i \right\|^2 = \|f\|^2 - \sum_{i=1}^n \langle f, \phi_i \rangle^2 .$$

Since the error can never be negative, it follows that

$$\sum_{i=1}^n \alpha_i^2 = \sum_{i=1}^n \langle f, \phi_i \rangle^2 \leq \|f\|^2 < \infty,$$

which is known as BESSEL'S INEQUALITY. Since this is true for all n, if the set $\{\phi_i\}$ is infinite, we can take the limit $n \to \infty$, and conclude that $g = \sum_{i=1}^\infty \alpha_i \phi_i$ converges to some function in H. (We know this because $\{g_n\}$, $g_n = \sum_{i=1}^n \alpha_i \phi_i$ is a Cauchy sequence in H and H is complete.)

Notice that for any orthonormal set $\{\phi_i\}$, the identification $\alpha_i = \langle f, \phi_i \rangle$ induces a transformation between L^2 and the sequence space ℓ^2. That is, if $f \in L^2$, then the sequence $\{\alpha_i\}$ with $\alpha_i = \langle f, \phi_i \rangle$ is in ℓ^2 and vice versa, if $\{\alpha_i\}$ is a sequence in ℓ^2, then $g = \sum_{i=1}^\infty \alpha_i \phi_i$ is in L^2.

For any orthonormal set $\{\phi_i\}_{i=1}^\infty$, the best approximation of f is $g = \sum_{i=1}^\infty \langle f, \phi_i \rangle \phi_i$, which is the projection of f onto the space spanned by the set $\{\phi_i\}$. It is natural to ask if it is ever possible that $g = f$. It is easy to see that even if one has an infinite set $\{\phi_i\}_{i=1}^\infty$, the set may be deficient. For example, the best approximation of $\cos x$ in terms of $\{\sin nx\}_{n=1}^\infty$ for $x \in [0, 2\pi]$ with inner product $\langle f, g \rangle = \int_0^{2\pi} f(x)g(x)dx$ is $\cos x \sim 0$ which certainly is not equal.

Definition

An orthonormal set $\{\phi_i\}_{i=1}^{\infty}$ is COMPLETE if $\sum_{i=1}^{\infty} \langle f, \phi_i \rangle \phi_i = f$ for every f in the Hilbert Space H.

Do not be confused by this second use of the word complete. In the last section we said that a space was complete if all Cauchy sequences converge. Here we say that the set $\{\phi_i\}_{i=1}^{\infty}$ is complete if the projection of f onto that set is always exactly f.

Theorem 2.2

A set $\{\phi_i\}_{i=1}^{\infty}$ is complete in H if any of the following (equivalent) statements holds:

1. $f = \sum_{i=1}^{\infty} \langle \phi_i, f \rangle \phi_i$ for all f in H,

2. For $\epsilon > 0$, there is an $N < \infty$ so that for all $n \geq N$, $\|f - \sum_{i=1}^{n} \langle f, \phi_i \rangle \phi_i\| < \epsilon$,

3. $\|f\|^2 = \sum_{i=1}^{\infty} \langle f, \phi_i \rangle^2$ for all f in H (Parseval's equality).

4. If $\langle f, \phi_i \rangle = 0$ for all i, then $f = 0$.

5. There is no function $\psi \neq 0$ in H for which the set $\{\phi_i\} \cup \psi$ forms an orthogonal set.

2.2.2 Orthogonal Polynomials

What kinds of orthogonal sets are complete? Two common classes of complete orthogonal functions are the ORTHOGONAL POLYNOMIALS and the TRIGONOMETRIC FUNCTIONS. There are many others such as BESSEL FUNCTIONS and WALSH FUNCTIONS which we will discuss in due course.

To verify that orthogonal polynomial sets are complete, we first prove a fundamental theorem,

Theorem 2.3 (WEIERSTRASS APPROXIMATION THEOREM)

For any continuous function $f \in C[a, b]$ and any $\epsilon > 0$, there is a polynomial $p(x) = \sum_{j=1}^{n} \alpha_j x^j$ so that

$$\max_{a \leq x \leq b} |f(x) - p(x)| < \epsilon.$$

We will prove this statement only on the interval $x \in [0,1]$, but the generalization to the interval $[a,b]$ is straightforward. The proof makes use of the BERNSTEIN POLYNOMIALS

$$\beta_{n,j}(x) = \binom{n}{j} x^j (1-x)^{n-j} \quad , \quad x \in [0,1]$$

and we approximate the function $f(x)$ by the polynomial

$$p_n(x) = \sum_{j=1}^{n} f(j/n) \beta_{n,j}(x).$$

It is important to take note of some properties of the polynomials $\beta_{n,j}(x)$. From the binomial theorem we know that

$$\sum_{j=0}^{n} \binom{n}{j} x^j y^{n-j} = (x+y)^n,$$

$$\sum_{j=0}^{n} (j/n) \binom{n}{j} x^j y^{n-j} = x(x+y)^{n-1},$$

$$\sum_{j=0}^{n} \left(\frac{j}{n}\right)^2 \binom{n}{j} x^j y^{n-j} = \left(1-\frac{1}{n}\right) x^2 (x+y)^{n-2} + \frac{x}{n}(x+y)^{n-1}.$$

Setting $y = 1 - x$, we find that

1. $\sum_{j=0}^{n} \beta_{n,j}(x) = 1$,
2. $\sum_{j=0}^{n} \left(\frac{j}{n}\right) \beta_{n,j}(x) = x$ and
3. $\sum_{j=0}^{n} (j/n)^2 \beta_{n,j}(x) = \left(1-\frac{1}{n}\right) x^2 + \frac{x}{n}$.

Definition

For the function $f(x)$, the MODULUS of CONTINUITY of $f(x)$ is

$$\omega(f,\delta) = \begin{array}{c} \text{l.u.b.} \\ x, y \in [0,1] \\ |x-y| \le \delta \end{array} |f(x) - f(y)|.$$

(l.u.b. means least upper bound).

In words, $\omega(f, \delta)$ is the most the function $f(x)$ can vary over an interval of length δ. We will show that

$$|f(x) - p_n(x)| \le \frac{9}{4} \omega \left(f, \frac{1}{\sqrt{n}}\right).$$

We begin by subdividing the interval $[0,1]$ into n subintervals with mesh points $x_j = j/n$, $j = 0, 1, 2, \ldots, n$. Using property 1) we write

$$f(x) - p_n(x) = \sum_{j=0}^{n} (f(x) - f(x_j)) \beta_{n,j}(x), \quad x_j = j/n,$$

and for each fixed x we break the sum apart into two sums

$$= \sum_{\substack{j \\ |x-x_j| \le \delta}} (f(x) - f(x_j)) \beta_{n,j}(x) + \sum_{\substack{j \\ |x-x_j| > \delta}} (f(x) - f(x_j)) \beta_{n,j}(x)$$

$$= S_1(x) + S_2(x).$$

For the first of these two sums,

$$|S_1(x)| \le \omega(f; \delta) \left(\sum_{\substack{j \\ |x-x_j| \le \delta}} \beta_{n,j}(x) \right) \le \omega(f; \delta),$$

since $|x - x_j|$ is restricted to be smaller than δ and $\beta_{n,j}(x)$ is positive in the interval $0 \le x \le 1$. For the second sum, $|x - x_j| > \delta$, so we let k be the smallest integer for which $k + 1 \ge \left| \frac{x-x_j}{\delta} \right|$. Then

$$|f(x) - f(x_j)| \le (k+1)\omega(f; \delta) \le \left(1 + \left| \frac{x - x_j}{\delta} \right| \right) \omega(f; \delta).$$

It follows that

$$|S_2(x)| \le \omega(f; \delta) \sum_{\substack{j \\ |x-x_j| > \delta}} \left(1 + \left| \frac{x - x_j}{\delta} \right| \right) \beta_{n,j}(x)$$

$$\le \omega(f; \delta) \left(1 + \sum_{\substack{j \\ |x-x_j| > \delta}} \frac{(x - x_j)^2}{\delta^2} \beta_{n,j}(x) \right).$$

Now, from properties 1), 2) and 3)

$$\sum_{j=0}^{n} (x - x_j)^2 \beta_{n,j}(x) = \frac{x(1 - x)}{n} \le \frac{1}{4n}.$$

Combining this information we see that $|S_2(x)| \leq \left(1 + \frac{1}{4n\delta^2}\right)\omega(f;\delta)$ so that

$$|f(x) - p_n(x)| \leq \omega(f;\delta)\left(2 + \frac{1}{4n\delta^2}\right).$$

Choosing $\delta = \frac{1}{\sqrt{n}}$ we obtain the predicted estimate. Finally, for any continuous function, the modulus of continuity $\omega\left(f; \frac{1}{\sqrt{n}}\right)$ can be made as small as we wish simply by taking n sufficiently large. In this way, we obtain the stated approximation theorem.

To show that the polynomials are complete in $L^2[a, b]$, we recall that the continuous functions are dense in L^2. This means that for any f in $L^2[a, b]$ there is a continuous function g in $C[a, b]$ for which $\int_b^a (f - g)^2 dx < (\epsilon/2)^2$ for any $\epsilon > 0$. Using the Weierstrass theorem, we approximate the continuous function $g(x)$ with a polynomial $p_n(x)$ so that

$$|g(x) - p_n(x)| \leq \frac{\epsilon}{2\sqrt{a-b}} \quad \text{for all} \quad x \in [a, b],$$

and then $\|f - p_n\| \leq \|f - g\| + \|g - p_n\| \leq \epsilon/2 + \left(\int_a^b (g(x) - p_n(x))^2 \, dx\right)^{1/2} \leq \epsilon$ which shows that the polynomials are dense in L^2. If we use instead a least squares polynomial approximation of order n, the approximation error can only get smaller, verifying that completeness property 2) of Theorem 2 holds.

There are numerous examples of orthogonal polynomial sets. Some of the most well-known are in the following list: We define the inner product to be $\langle u, v \rangle = \int_a^b u(x)v(x)\omega(x)dt$ with $\omega(x) \geq 0$.

1. **LEGENDRE POLYNOMIALS:** Take $a = -1$, $b = 1$, $\omega(x) = 1$, then $P_0(x) = 1$, $P_n(x) = \frac{1}{2^n n!} \frac{d^n}{dx^n}(x^2 - 1)^n$, $n > 0$.

2. **CHEBYSHEV POLYNOMIALS:** Take $a = -1$, $b = 1$, $\omega(x) = (1 - x^2)^{-1/2}$, then
$$T_n(x) = \cos(n\cos^{-1} x).$$

3. **HERMITE POLYNOMIALS:** Take $a = -\infty$, $b = \infty$, $\omega(x) = e^{-x^2}$, then $H_n(x) = (-1)^n e^{x^2/2} \frac{d^n}{dx^n}\left(e^{-x^2/2}\right)$.

4. **LAGUERRE POLYNOMIALS:** Take $a = 0$, $b = \infty$, $\omega(x) = e^{-x}$, then
$$L_n(x) = \frac{e^x}{n!}\frac{d^n}{dx^n}\left(x^n e^{-x}\right).$$

5. **JACOBI POLYNOMIALS:** Take $a = -1$, $b = 1$, $\omega(x) = (1+x)^\alpha(1-x)^\beta$ provided $\alpha, \beta > -1$, then
$$P_n^{\alpha\beta}(x) = \frac{(-1)^n}{2^n n!}(1-x)^{-\alpha}(1+x)^{-\beta}\frac{d^n}{dx^n}\left[(1-x)^\alpha(1+x)^\beta\left(1-x^2\right)^n\right].$$

The Chebyshev and Gegenbauer polynomials are special cases of the Jacobi polynomials.

6. GEGENBAUER POLYNOMIALS: Take $a = -1, b = 1, \omega(x) = (1 - x^2)^{\gamma - 1/2}$, and $\alpha = \beta = \gamma - 1/2, \gamma > 1/2$ for Jacobi polynomials, then

$$C_n^\gamma(x) = \frac{-(-1)^n}{2^n n!}(1 - x^2)^{1/2 - \gamma}\frac{d^n}{dx^n}\left[(1 - x^2)^{\gamma - 1/2 + n}\right].$$

These polynomials are not always normalized to have norm one. For example, the Legendre polynomials are usually normalized to have leading term $P_n(x) = \frac{(2n)!}{2^n(n!)^2}x^n + \cdots$ and as a result

$$\int_{-1}^{1} P_n^2(x)dx = \frac{2}{2n + 1}.$$

In later chapters we will see why these functions are "natural" for certain problems. Notice that we have not proven completeness for the Hermite or Laguerre polynomials, since our proof assumed that we are on a finite interval. We will correct this omission later (Chapter 7).

2.2.3 Trigonometric Series

The most familiar class of non-polynomial orthogonal functions are the TRIGONO-METRIC FUNCTIONS

$$\{\sin(nx)\ ,\ \cos(nx)\}_{n=0}^\infty\ ,\ x \in [0, 2\pi]$$

which are orthogonal relative to the inner product

$$\langle u, v \rangle = \int_0^{2\pi} u(x)v(x)dx.$$

To show that trigonometric functions are complete on $L^2[0, 2\pi]$ we suppose $f(\theta)$ is any continuous, 2π-periodic function. Define $\psi(x, y) = \rho f(\theta)$ where $x = \rho\cos\theta$, $y = \rho\sin\theta$, and apply the Weierstrass approximation theorem to the unit square $-1 \leq x, y \leq 1$. (We have not actually proven the validity of the Weierstrass theorem in more than one spatial dimension. Can you supply the proof?). There is a polynomial $Q_n(x, y)$ for which $|\psi(x, y) - Q_n(x, y)| < \epsilon/2$ on the unit square. Set $\rho = 1$ to obtain $|f(\theta) - Q_n(\cos\theta, \sin\theta)| < \epsilon/2$. From trigonometric identities, it follows that $Q_n(\cos\theta, \sin\theta)$ is a trigonometric series of the form

$$P_n(\theta) = Q_n(\cos\theta, \sin\theta) = \sum_{j=0}^{n}(\alpha_j \sin n\theta + b_j \cos n\theta).$$

For any continuous function $g(\theta)$, $0 \leq \theta \leq 2\pi$ (but not necessarily periodic) there is a 2π periodic continuous $f(\theta)$ which approximates $g(\theta)$ in L^2,

$\int_0^{2\pi} (f(\theta) - g(\theta))^2 \, d\theta \leq \epsilon^2/4$. The easiest way to construct $f(\theta)$ is to take $f(\theta) = g(\theta)$ on the interval $[0, 2\pi - \delta]$ and then add a thin boundary layer to $f(\theta)$ to make it periodic and continuous. It follows from the triangle inequality that

$$\|g - P_n(\theta)\| \leq \|g - f\| + \|f - P_n(\theta)\| \leq \epsilon.$$

The function $P_n(\theta)$ is not necessarily the "best" n^{th} order trigonometric series. Therefore, we might get a better approximation (but no worse) if we replace $P_n(\theta)$ by the least squares trigonometric approximation, the nth order FOURIER SERIES

$$g_n(\theta) = \sum_{j=0}^{n} (a_j \sin(n\theta) + b_j \cos(n\theta))$$

where

$$a_j = \frac{1}{\pi} \int_0^{2\pi} g(\theta) \sin(j\theta) d\theta, \quad j > 0,$$

$$b_j = \frac{1}{\pi} \int_0^{2\pi} g(\theta) \cos(j\theta) d\theta, \quad j > 0,$$

$$a_0 = 0 \quad , \quad b_0 = \frac{1}{2\pi} \int_0^{2\pi} g(\theta) d\theta.$$

Even though the trigonometric Fourier series converge in the sense of L^2, it is tempting to use Fourier series in a pointwise manner. For reference we state here an important theorem about the pointwise convergence of Fourier series.

Theorem 2.4 (Fourier Convergence Theorem)

Suppose $f(x)$ is piecewise $C^1[0, 2\pi]$ (i.e., is continuous and has continuous derivative except possibly at a finite number points at which there is a jump discontinuity, at which left and right derivatives exist). The Fourier series of f converges to $\frac{1}{2}[f(x^+) + f(x^-)]$ for every x in $(0, 2\pi)$. At $x = 0$ and $x = 2\pi$ the Fourier series of f converges to $\frac{1}{2}[f(0^+) + f(2\pi^-)]$.

The proof of this theorem fits more naturally into our discussion of distributions in Chapter 4, so we will defer it until then. Although it is tempting to use Fourier series in a pointwise way, if $f(x)$ is merely continuous, and not piecewise C^1, $f(x)$ and its Fourier series may differ at an infinite number of points, even though the Fourier series converges to $f(x)$ in L^2. This illustrates in an important way the difference between convergence in L^2 and convergence in the uniform norm.

There are other useful representations of the trigonometric Fourier series. The representation in terms of sines and cosines listed above always converges to $f(x)$ on the interval $[0, 2\pi]$ and to the 2π-periodic extension of $f(x)$ on the infinite interval $-\infty < x < \infty$. By a simple change of scale it follows that

$$f(x) \quad = \quad \sum_{j=0}^{\infty} \left(a_j \sin \frac{2j\pi x}{L} + b_j \cos \frac{2j\pi x}{L} \right)$$

$$a_j \quad = \quad \frac{2}{L} \int_0^L f(x) \sin \frac{2j\pi x}{L} dx,$$

$$b_j \quad = \quad \frac{2}{L} \int_0^L f(x) \cos \frac{2j\pi x}{L} dx,$$

$$j = 1, 2, \ldots$$

$$a_0 \quad = \quad 0 \quad , \quad b_0 = \frac{1}{L} \int_0^L f(x) dx,$$

converges in L^2 to $f(x)$ on the interval $[0, L]$, and to the periodic extension of f on the infinite interval. One can use only sines or only cosines with the result that the series reduce to

$$f(x) \quad = \quad \sum_{j=1}^{\infty} a_j \sin \frac{j\pi x}{L}$$

$$a_j \quad = \quad \frac{2}{L} \int_0^L f(x) \sin \frac{j\pi x}{L} dx$$

or

$$f(x) \quad = \quad \sum_{j=0}^{\infty} b_j \cos \frac{j\pi x}{L}$$

$$b_j \quad = \quad \frac{2}{L} \int_0^L f(x) \cos \frac{j\pi x}{L} dx \quad , \quad j \geq 1$$

$$b_0 \quad = \quad \frac{1}{L} \int_0^L f(x) dx$$

and the series converge to the $2L$ periodic odd or even (respectively) extensions of $f(x)$ on $-\infty < x < \infty$. These formulae follow from the full Fourier series applied to odd or even functions.

Finally, it is often useful to represent Fourier series in terms of complex exponentials

$$f(x) \quad = \quad \sum_{j=-\infty}^{\infty} a_j e^{2\pi ijx/L}$$

$$a_j \quad = \quad \frac{1}{L} \int_0^L f(x) e^{-2\pi ijx/L} dx,$$

which, again, follows directly from our earlier expressions.

2.2.4   Discrete Fourier Transforms

In applications, such as signal processing, it is often the case that a function is not known for every value of its argument but only at certain discrete evenly spaced points of the independent variable t. If the n values $f = \{f_j\}_{j=0}^{n-1}$ are given, we define the DISCRETE FOURIER TRANSFORM of f to be the values $\{g_j\}_{j=0}^{n-1}$ given by

$$g_j = \sum_{k=0}^{n-1} f_k e^{2\pi ikj/n} \quad , \quad j = 0, 1, 2, \ldots, n-1.$$

It is easy to show that the inverse of this transform is

$$f_k = \frac{1}{n} \sum_{j=0}^{n-1} g_j e^{-2\pi ijk/n} \quad , \quad k = 0, 1, 2, \ldots, n-1.$$

If one uses the data $\{f_j\}_{j=0}^{n-1}$ to generate the values $\{g_j\}_{j=0}^{n-1}$ in a direct way, namely, using the formula exactly as stated, one must make n exponential evaluations, n multiplications and n additions to evaluate one of the values g_k. Even if the complex exponentials are tabulated in advance, one must perform n^2 multiplications. However, in the process of finding the values g_k, many operations are superfluously repeated.

By a careful reexamination of the formulae many calculations that are superfluous can be eliminated. For example, if n is an even number, we can write

$$g_j = \sum_{k=0}^{n/2-1} f_{2k} e^{2\pi i \frac{2kj}{n}} + e^{\frac{2\pi ij}{n}} \left(\sum_{k=0}^{n/2-1} f_{2k+1} e^{2\pi i \frac{2kj}{n}} \right).$$

The important observation here is that the expression

$$\sum_{k=0}^{n/2-1} f_{2k+\ell} e^{\frac{2\pi ijk}{n/2}} \quad , \quad \ell = 0, 1$$

represents the Discrete Fourier Transform of the two data sets $\{f_{2k+\ell}\}, \ell = 0, 1$ for which there are only $n/2$ points. Another way to write this is to denote $\left\{ g_j^{(n)} \right\}$ as a Discrete Fourier Transform of n points. Then

$$g_j^{(n)} = g_{j,1}^{(n/2)} + e^{\frac{2\pi ij}{n}} g_{j,2}^{(n/2)}$$

$$j = 1, 2, \ldots, n/2$$

$$g_{j+n/2}^{(n)} = g_{j,1}^{(n/2)} - e^{\frac{2\pi ij}{n}} g_{j,2}^{(n/2)}$$

where $\{g_{j,1}^{(n/2)}\}$, $\{g_{j,2}^{(n/2)}\}$ are the Discrete Fourier Transforms of $n/2$ interlaced points. Each of the $n/2$ transforms requires only $n^2/4$ multiplications (as they stand) so to construct the set $\{g_j^{(n)}\}$ requires only $n^2/2 + n$ multiplications, a significant saving over n^2, if n is large.

However, if n is a multiple of 4, we need not stop here since the Discrete Fourier Transform of the two data sets $\{f_{2k+\ell}\}, \ell = 0, 1$ can be split up again for another significant computational saving. This further splitting requires that we calculate transforms of the four sets of points $\{f_{4k+\ell}\}, \ell = 0, 1, 2, 3$.

If $n = 2^p$, the best thing to do is to use this simplification recursively, and find the two point transform of all pairs of points which are spaced 2^{p-1} points apart. From these we construct the transform of all sets of four points spaced 2^{p-2} points apart, and continue to build up from there until the desired transform is found.

Suppose the number of multiplications required for this procedure to produce an n point transform is $K(n)$. From our recursive formula we realize that

$$K(2n) = 2K(n) + 2n$$

multiplications are required for $2n$ points. The solution of this equation is

$$K(n) = n \log_2 n = p \cdot 2^p,$$

where $K(2) = 2$. Certainly, this is a significant saving over the naive calculation which required n^2 multiplications.

It is not a difficult matter to write a computer program that calculates discrete Fourier transforms in this fast way. However, most computers currently have available Fast Fourier Transform software that perform these computations in this efficient way, or if such a program is not readily available, the code given in the book by Press, et al., will work quite well.

Other versions of the discrete Fourier transform may be appropriate in certain situations. For example, with real data, one may prefer a real valued transform (the discrete Fourier transform is complex valued). The discrete Sine transform is defined for the data $\{f_k\}_{k=1}^{n-1}$ by

$$g_j = \sum_{k=1}^{n-1} f_k \sin \frac{\pi k j}{n}$$

with inverse transform

$$f_k = \frac{2}{n} \sum_{k=1}^{n-1} g_j \sin \frac{\pi k j}{n}.$$

Of course, the transform and inverse transform are the same operation except for the scale factor $\frac{2}{n}$.

Once this transform is defined, it is not hard to devise a fast Sine transform algorithm which is more efficient than using the direct calculation. Computer programs to implement this fast Sine transform are also readily available.

One further observation needs to be made here. This transform is actually a matrix multiplication by a matrix A whose entries are $a_{kj} = \sin \pi kj/n$. The columns of the matrix are mutually orthogonal and the norm of each column is $\sqrt{n/2}$. The usefulness of this transform will be enhanced if we can find matrices which are diagonalized by this transform. Indeed, just as Fourier Transforms are useful in the study of ordinary and partial differential equations, the discrete Fourier transforms, and especially the fast algorithms, are quite useful in the study of the discretized versions of differential equations, which we will see in chapters 4 and 8.

2.2.5   Walsh Functions and Walsh Transforms

WALSH FUNCTIONS are a complete set of orthonormal piecewise constant functions defined on the interval $x \in [0, 1)$, which are quite useful in the processing of digital information.

There are two equivalent definitions of the Walsh functions. The first is inductive and gives a good understanding of how they are generated and why they are orthogonal and complete. The goal is to build a set of orthogonal, piecewise constant functions. We first show that, if we do this correctly, the set of functions will be complete.

Suppose $f(x)$ is a continuous function. For some fixed n, we divide the interval [0,1] into 2^n equal subintervals, and define

$$q_n(x) = f\left(\frac{j - 1/2}{2^n}\right) \quad j = 1, 2, \ldots, 2^n$$

for $j - 1 \leq 2^n x \leq j$, that is, on each subinterval, $q_n(x)$ is the same as f at the midpoint of the interval. It follows that $|f(x) - q_n(x)| < \omega\left(f, \frac{1}{2^n}\right)$ which is smaller than ϵ for n large enough and that

$$\int_0^1 (f(x) - q_n(x))^2 \, dx < \epsilon^2.$$

Since continuous functions are dense in L^2, the piecewise constant functions are also dense in L^2.

To generate an orthonormal set of piecewise constant functions we take $\text{Wal}(0, x) = 1$, for x in [0,1] and

$$\text{Wal}(1, x) = \begin{cases} 1, & 0 \leq x < 1/2 \\ -1, & 1/2 \leq x < 1 . \end{cases}$$

Using these two functions as starters, we proceed inductively, defining

$$\text{Wal}(2n, x) = \begin{cases} \text{Wal}(n, 2x) & 0 \le x < 1/2 \\ (-1)^n \text{Wal}(n, 2x - 1), & 1/2 \le x < 1 \end{cases}$$

and

$$\text{Wal}(2n + 1, x) = \begin{cases} \text{Wal}(n, 2x) & 0 \le x < 1/2 \\ (-1)^{n+1}\text{Wal}(n, 2x - 1), & 1/2 \le x < 1. \end{cases}$$

In words, we take a known Walsh function $\text{Wal}(n, x)$ defined on $[0,1)$, contract it down by a change of scale to the interval $[0, 1/2)$ and then reuse the scaled down function on the interval $[1/2, 1)$ with multiplier ± 1 to generate two new Walsh functions. In this way $\text{Wal}(1, x)$ is used to generate $\text{Wal}(2, x)$, and $\text{Wal}(3, x)$, and these two in turn generate $\text{Wal}(k, x)$, $k = 4, 5, 6, 7$ in the next step. Notice that the 2^n Walsh functions $\text{Wal}(k, x)$, $k = 0, 1, 2, \ldots, 2^n - 1$ are constant on intervals of length no less than $1/2^n$.

A second definition of Walsh functions is a bit more direct and more useful for computer programming, although less insightful. For any $x \in [0, 1)$ let $x = \sum_{j=1}^{\infty} a_j 2^{-j}, a_j = 0$ or 1 be the BINARY EXPANSION of x. Define $a_k(x)$ to be the kth coefficient of this expansion. For example, to determine a_1 we divide the interval $[0, 1)$ into two subintervals $[0, 1/2)$ and $[1/2, 1)$. If $x < 1/2$, we take $a_1 = 0$ whereas if $x \ge 1/2$, we take $a_1 = 1$. Proceeding inductively, we suppose $a_1, a_2, \ldots, a_k$ are known. Let

$$x_k = \sum_{j=1}^{k} a_j 2^{-j}.$$

If $x \ge x_k + \frac{1}{2^{k+1}}$ we take $a_{k+1} = 1$ whereas if $x < x_k + \frac{1}{2^{k+1}}$ we take $a_{k+1} = 0$. In this way the a_k are uniquely specified. For example, in this representation, $x = 3/8$ has $a_1 = 0$, $a_2 = 1$, $a_3 = 1$, and $a_k = 0$ for $k \ge 4$.

With x given in its binary expansion we define

$$\text{Wal}(2^k - 1, x) = (-1)^{a_k(x)}.$$

The Walsh functions $\text{Wal}(2^k - 1, x)$ flip alternately between ± 1 at the points $x_j = j/2^k$, $j = 1, 2, \ldots, 2^k - 1$. To determine the remaining Walsh functions, we define the "exclusive or" binary operation $\oplus$ (also called add without carry) using the table

$\oplus$	0	1
0	0	1
1	1	0.

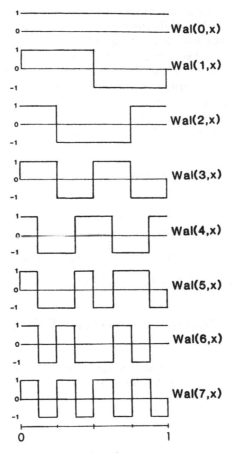

FIGURE 2.2 The first eight Walsh functions.

For integers i, j, we define $i \oplus j$ as the operation $\oplus$ applied to the binary representations of i and j. Since $\oplus$ is its own additive inverse, $i \oplus i = 0$ and $(i \oplus j) \oplus j = i$. The definition of the remaining Walsh functions follows from

$$\text{Wal}(i \oplus j, x) = \text{Wal}(i, x) \cdot \text{Wal}(j, x).$$

To make some sense of this definition, we note that we can uniquely decompose any integer n into a $\oplus$ sum of the integers $2^k - 1$, as

$$n = \sum_{j=0}^{N} b_j 2^j = \sum_{j=1}^{N} \oplus (b_{j-1} \oplus b_j)(2^j - 1) \oplus b_N (2^{N+1} - 1).$$

For example, $4 = 3 \oplus 7$, and $10 = 1 \oplus 3 \oplus 7 \oplus 15$. The way to view this decomposition operationally is via simple shifts. For example, to decompose the integer 13, write its binary representation 1101 (since $8+4+1=13$) and $\oplus$ add this to its shift 0110 (shift the entries 1101 one position to the right to get 0110) as $1101 \oplus 0110 = 1011$. The resulting zeros and ones are coefficients of integers $2^j - 1$. In this case $13 = 15 \oplus 3 \oplus 1$. Another way to view this decomposition uses the fact that $\oplus$ is its own additive inverse. As a result, $13 \oplus 15 = 1101 \oplus 1111 = 0010$ so $13 \oplus 15 \oplus 3 = 0010 \oplus 11 = 1$ and $13 \oplus 15 \oplus 3 \oplus 1 = 0$ so that $13 = 15 \oplus 3 \oplus 1$ as found before.

The proof of this identity follows easily by induction on N. Suppose the binary representation of a number is $b_{N+1} b_N b_{N-1} \cdots b_1 b_0$ and suppose that,

$$b_N b_{N-1} \cdots b_1 b_0 = \sum_{j=1}^{N} \oplus (b_{j-1} \oplus b_j)(2^j - 1) \oplus b_N (2^{N+1} - 1).$$

It is certainly true that $b_0 = b_0(2^1 - 1)$. We use that the binary representation of $2^j - 1$ is the string of j ones, $11 \ldots 11$ and then

$$b_{N+1} b_N b_{N-1} \cdots b_1 b_0 = b_N b_{N-1} \cdots b_1 b_0 \oplus b_{N+1} 00 \cdots 00$$

$$= b_N b_{N-1} \cdots b_1 b_0 \oplus b_{N+1} \left((2^{N+2} - 1) \oplus (2^{N+1} - 1) \right)$$

$$= \sum_{j=1}^{N+1} \oplus (b_{j-1} \oplus b_j)(2^j - 1) \oplus b_{N+1} (2^{N+2} - 1).$$

Now that n is decomposed as a linear combination of integers $2^j - 1$ and $\text{Wal}(2^n - 1, x) = (-1)^{a_j(x)}$, it follows that

$$\text{Wal}(n, x) = (-1)^{\left(\sum_{j=1}^{N} a_j(b_j + b_{j-1}) + a_{N+1} b_N \right)}$$

or

$$\text{Wal}(n, x) = (-1)^{\left(\sum_{j=1}^{N} (a_j + a_{j+1}) b_j + a_1 b_0 \right)}$$

where $x = \sum_{j=1}^{\infty} a_j 2^{-j}$. Notice that replacing $\oplus$ by $+$ in these formulas is not a misprint. Why?

To verify that this definition is equivalent to our first inductive definition, we note that if $n = \sum_{j=0}^{N} b_j 2^j$, then $2n = \sum_{j=0}^{N+1} b_j' 2^{j-1}$ where $b_j' = b_{j-1}$, $b_0' = 0$ and $2n + 1 = \sum_{j=0}^{N+1} b_j' 2^{j-1}$ where $b_j' = b_{j-1}$ with $b_0' = 1$. Thus for $m = 2n$ or $2n + 1$

$$\begin{aligned}
\text{Wal}(m, x) &= (-1)^{\left(\sum_{j=1}^{N+1} (a_j + a_{j+1}) b_j' + a_1 b_0' \right)} \\
&= (-1)^{\left(\sum_{j=1}^{N} (a_j' + a_{j+1}') b_j + a_1' b_0 + a_1 (b_0 + b_0') \right)} \\
&= \text{Wal}(n, x')(-1)^{a_1(b_0 + b_0')}
\end{aligned}$$

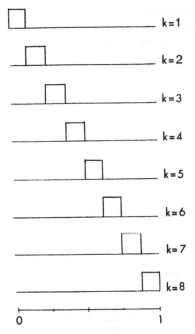

FIGURE 2.3   The "natural basis" of piecewise constant functions.

where $a'_j = a_{j+1}$, $x' = (2x) \bmod 1$. When $0 \le x < 1/2$, $a_1 = 0$ and when $1/2 \le x < 1$, $(-1)^{b_0} = (-1)^n$, so the two definitions agree.

It is clear (by induction) that for $n \ge 1$, $\int_0^1 \mathrm{Wal}(n, x)dx = 0$, whereas $\int_0^1 \mathrm{Wal}(0, x)dx = 1$. To show orthogonality of the set, note that

$$\int_0^1 \mathrm{Wal}(i, x)\mathrm{Wal}(j, x)dx = \int_0^1 \mathrm{Wal}(i \oplus j, x)dx = \delta_{ij},$$

since $i \oplus i = 0$. For $j \le 2^n - 1$, the Walsh functions $\mathrm{Wal}(j, x)$ are constant on intervals of length no less than $1/2^n$. Since there are 2^n such functions and they are orthogonal, hence linearly independent, they span the same set as the "natural" basis functions

$$g_k(x) = \begin{cases} 1 & k - 1 \le 2^n x \le k \\ & \qquad\qquad\qquad k = 1, 2, \ldots, 2^n \\ 0 & \text{elsewhere} . \end{cases}$$

It follows that the Walsh functions are complete. The natural basis functions are shown for $n = 3$ in Figure 2.3.

Even numbered Walsh functions are symmetric and odd numbered ones are antisymmetric about the point $x = 1/2$. In light of the analogy with the trigonometric functions, the notation

$$\text{cal}(k, x) \;\; = \;\; \text{Wal}(2k, x)$$

$$\text{sal}(k, x) \;\; = \;\; \text{Wal}(2k - 1, x)$$

has been adopted. Corresponding to this notation, the identities

$$\text{cal}(i, x)\text{cal}(j, x) \;\; = \;\; \text{cal}(i \oplus j, x)$$

$$\text{cal}(i, x)\text{sal}(j, x) \;\; = \;\; \text{sal}\left((i \oplus (j - 1)) + 1, x\right)$$

$$\text{sal}(i, x)\text{sal}(j, x) \;\; = \;\; \text{cal}\left((i - 1) \oplus (j - 1), x\right)$$

can be shown to hold.

Since the Walsh functions are an orthonormal, complete, set of functions in $L^2[0, 1]$, we can represent any function in $L^2[0, 1]$ as

$$f(x) = \sum_{n=1}^{\infty} a_n \text{Wal}(n, x)$$

where

$$a_n = \int_0^1 f(t)\text{Wal}(n, t)dt$$

which is the WALSH TRANSFORM PAIR for $f(x)$. Note that this integral is easy to evaluate since $\text{Wal}(n, x)$ is ± 1. If the function $f(x)$ is not continuous but consists rather of evenly spaced (digitized) sample points f_i, the DISCRETE WALSH TRANSFORM pair is defined as

$$f_i \;\; = \;\; \sum_{n=0}^{N-1} \alpha_n \text{Wal}(n, x_i)$$

$$i = 0, 1, 2, \ldots, N - 1$$

$$\alpha_i \;\; = \;\; \frac{1}{N} \sum_{n=0}^{N-1} f_n \text{Wal}(i, x_n) \;, \;\; x_i = \frac{i + 1/2}{N}$$

for $N = 2^p$, some power of 2. One can show that, except for the normalizing factor $\frac{1}{N}$, the forward and inverse Walsh transform are the same operation (see problem 2.2.19).

Numerical routines that perform the discrete Walsh transforms are extremely fast compared even to the Fast Fourier transform. If $N = 2^p$ for some

p, notice that

$$\alpha_i = \frac{1}{N} \sum_{k=0}^{N-1} f_k \, \mathrm{Wal}(i, x_k)$$

$$= \frac{1}{2M} \sum_{k=0}^{M-1} \left(f_{2k} \, \mathrm{Wal}\left(i, x_{2k}\right) + f_{2k+1} \, \mathrm{Wal}\left(i, x_{2k+1}\right) \right)$$

where $M = N/2$. We now use that $\mathrm{Wal}(i, x_k) = \mathrm{Wal}(k, x_i)$ (see problem 2.2.19) to write that

$$\alpha_i = \frac{1}{2M} \sum_{k=0}^{M-1} \left(f_{2k} \, \mathrm{Wal}(2k, x_i) + f_{2k+1} \, \mathrm{Wal}(2k + 1, x_i) \right)$$

and finally, we use the inductive definition of the Walsh functions and find

$$\alpha_i = \frac{1}{M} \sum_{k=0}^{M-1} \left(\frac{f_{2k} + f_{2k+1}}{2} \right) \mathrm{Wal}(k, \hat{x}_i)$$

$$0 \leq i \leq M - 1$$

$$\alpha_{i+M} = \frac{1}{M} \sum_{k=0}^{M-1} (-1)^k \left(\frac{f_{2k} - f_{2k+1}}{2} \right) \mathrm{Wal}(k, \hat{x}_i)$$

where $\hat{x}_i = \frac{i+1/2}{M}$. The obvious observation to make is that the first M coefficients α_i, $i = 1, \ldots, M - 1$ are the Walsh transform of the average of consecutive data points and the coefficients α_i, $i = M, M+1, \ldots, 2M - 1$ are the Walsh transform of half the difference of consecutive data points.

This formula applied recursively enables us to evaluate the Walsh transform without ever evaluating a Walsh function, but merely by transforming the data by adding or subtracting successive pairs of data points. The following FORTRAN code calculates the Walsh transform of 2^p data points by exploiting this formula to the fullest extent. In the code, the division by two is not done until the very end, when it becomes division by 2^p. This step saves time and gives an operation count of $p2^p$ adds or subtracts, and 2^p divisions. This is indeed faster than the fastest Fast Fourier Transform code.

```
        SUBROUTINE FWT(F,G,N)
        DIMENSION F(1),G(1)
C
C       F...INPUT VECTOR CONTAINING N DATA POINTS
C
```

```
C          G...WORK VECTOR WITH LENGTH AT LEAST N
C
C          N...LENGTH OF INPUT VECTOR, ASSUMED TO BE  POWER OF 2
C             IF N IS NOT A POWER OF 2, THE SUBROUTINE USES TH
C             LARGEST POWER OF TWO LESS THAN N
C
C          EXCEPT FOR THE LOOP DO 60 I=..., THIS CODE CAN BE
C          INTEGERIZED. THAT IS, ALL VARIABLES CAN BE USED AS
C          INTEGERS
C
C          CALCULATE P WHERE N.GE.2**P
            IP=0
            ITP=1
   10       ITP=2*ITP
            IF(ITP.LE.N)THEN
            IP=IP+1
            GO TO 10
            ENDIF
            ITP=ITP/2
            NN=ITP
            WRITE(6,*)NN,IP
C
C          FIND THE WALSH TRANSFORM OF THE VECTOR F
            DO 50 I=1,IP
            JS=NN/ITP
            ITP=ITP/2
            DO 50 J=1,JS
            JST=1+2*(J-1)*ITP
            CALL COMBIN(F(JST),G(JST),ITP)
   50       CONTINUE
            DN=FLOAT(NN)
            DO 60 I=1,NN
            F(I)=F(I)/NN
   60       CONTINUE
            RETURN
            END
C
C          THIS ROUTINE ADDS AND SUBTRACTS CONSECUTIVE PAIRS
C          OF ELEMENTS OF F AND PUTS THEM BACK INTO THE
C          APPROPRIATE POSITION IN F.
            SUBROUTINE COMBIN(F,G,NS)
            DIMENSION F(1),G(1)
            A=-1.
            DO 100 I=1,NS
```

```
      G(I)=F(2*I-1)-F(2*I))
      A=-A
      G(I+NS)=A*(F(2*I-1)-F(2*I))
100   CONTINUE
      IST=2*NS
      DO 200 I=1,IST
      F(I)=G(I)
200   CONTINUE
      RETURN
      END
```

2.2.6 Finite Elements

As we know, mutually orthogonal basis functions are useful because they give least squares approximations of a function f using the Fourier coefficients $f = \sum_{n=1}^{\infty} \alpha_k \phi_k(x)$, $\alpha_k = \langle f, \phi_k \rangle$. FINITE ELEMENTS are functions which are not mutually orthogonal but nonetheless give useful approximations to functions in a Hilbert space.

Suppose the interval [0,1] is subdivided into N subintervals of length $h = 1/N$ and $x_j = j/N$, $j = 0, 1, 2, \ldots, N$ are the endpoints of the intervals.

Definition

A FINITE ELEMENT SPACE $S^h(k, r)$ is the set of all functions $\phi(x)$ defined on [0,1] satisfying

1. $\phi(x)$ is a polynomial of degree less than or equal to k, on the interval $[x_j, x_{j+1}]$,

2. $\phi(x)$ is r times continuously differentiable on [0,1].

If $r = 0$, the functions ϕ are continuous but not differentiable and if $\phi(x)$ is allowed to be discontinuous we set $r = -1$.

The finite element space $S^h(k, r)$ is a finite dimensional vector space of dimension $N(k - r) + r + 1$ and as a result has a finite set of basis functions. The simplest example is the set of piecewise constant functions

$$\phi_j(x) = \begin{cases} 1 & x_{j-1} \leq x \leq x_j \\ 0 & \text{otherwise} \end{cases}$$

denoted $S^h(0, -1)$. This is the set of functions from which we generated the Walsh functions in the previous section.

The next reasonable choice of basis functions are the piecewise linear

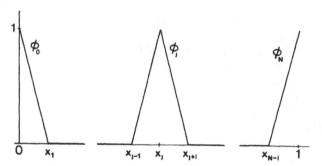

FIGURE 2.4 Piecewise linear finite element basis functions.

"tent" functions, denoted

$$\phi_j(x) = \begin{cases} (x - x_{j-1})/h & x_{j-1} \leq x \leq x_j \\ (x_{j+1} - x)/h & x_j \leq x \leq x_{j+1} \\ 0 & \text{otherwise} \end{cases}$$

for $j = 0, 1, 2, \ldots, N$, depicted in Figure 2.4. Since these functions are continuous but not continuously differentiable, they span the finite element space $S^h(1, 0)$.

The third example of a finite element space $S^h(3, 1)$, are the piecewise cubic Hermite functions, or more commonly, cubic splines. The basis for $S^h(3, 1)$ is spanned by two sets of cubic polynomials

$$\phi_j(x) = \begin{cases} (|x - x_j| - h)^2(2|x - x_j| + h)/h^3 & x_{j-1} \leq x \leq x_{j+1} \\ 0 & \text{otherwise} \end{cases}$$

$$\psi_j(x) = \begin{cases} (x - x_j)(|x - x_j| - h)^2/h^2 & x_{j-1} \leq x \leq x_{j+1} \\ 0 & \text{otherwise} \end{cases}$$

for $j = 0, 1, 2, \ldots, N$, depicted in Figure 2.5. The important thing to note is that the functions $\phi_j(x)$, $\psi_j(x)$ satisfy the conditions

$$\phi_j(x_k) = \delta_{kj}, \quad \phi_j'(x_k) = 0$$
$$\psi_j(x_k) = 0, \quad \psi_j'(x_k) = \delta_{kj}$$

from which they are uniquely determined.

There are a number of ways that finite elements are used. In the first, we can easily interpolate data points. For example, a continuous piecewise linear function with values $f(x_j) = f_j$ is given by

$$f(x) = \sum_{i=0}^{N} f_i \phi_i(x)$$

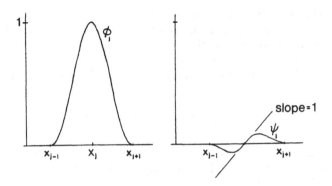

FIGURE 2.5 Piecewise cubic finite element basis functions.

where $\phi_i(x)$, $i = 0, 1, 2, \ldots, N$ are the piecewise linear tent functions that span the space $S^h(1,0)$. A continuously differentiable function with values $f(x_j) = f_j$ and derivative values $f'(x_j) = f'_j$ is given by

$$f(x) = \sum_{i=0}^{N} f_i \phi_i(x) + \sum_{i=0}^{N} f'_i \psi_i(x)$$

where $\phi_i(x)$ and $\psi_i(x)$ are the piecewise cubic polynomial functions that span $S^h(3,1)$.

For these interpolation schemes, the coefficients of the basis functions are uniquely and easily determined. However, we may want to find an interpolate of the data points $f(x_j) = f_j$ that satisfies other properties. For example, we may want the interpolating function to be a twice continuously differentiable piecewise cubic polynomial, in which case the coefficients of the functions ψ_i, $i = 0, 1, 2, \ldots, N$ are determined by a system of linear equations (see problem 2.2.24).

Other interpolation schemes frequently involve choosing the coefficients to minimize some positive object. For example, we may try to approximate a known function $f(x)$ in a least squares sense by minimizing

$$\left\| f(x) - \sum_{j=1}^{n} \alpha_j \chi_j(x) \right\|,$$

where $\{\chi_j(x)\}$ are the chosen set of finite element functions. We already know that this expression is minimized if we pick α_j to satisfy the matrix equation

$$\Phi \alpha = \beta$$

where

$$\phi_{ij} = \langle \chi_i, \chi_j \rangle \quad , \quad \beta_i = \langle f, \chi_i \rangle .$$

Except for the case $S^h(0, -1)$, the basis functions are not mutually orthogonal, so the matrix Φ is not diagonal. However, for $S^h(1, 0)$ and $S^h(3, 1)$, most of the inner products are zero, so that the matrix has a banded structure, and the matrix equation is still reasonably easy to solve.

We can define a "smooth" interpolation of the data points $f(x_i) = f_i$ by setting

$$f(x) = \sum_{i=0}^{N} f_i \phi_i(x) + \sum_{i=0}^{N} \alpha_i \psi_i(x)$$

with f_i specified and the α_i chosen to minimize $\int_0^1 (f''(x))^2 \, dx$. This reduces to the least squares problem of minimizing $\|g + h\|^2$ where

$$g = \sum_{i=0}^{N} f_i \phi_i''(x)$$

and

$$h = \sum_{i=0}^{N} \alpha_i \psi_i''(x).$$

The solution is given by

$$\Phi \alpha = \beta$$

where

$$\phi_{ij} = \langle \psi_i'', \psi_j'' \rangle \quad , \quad \beta_j = -\sum_{i=0}^{N} f_i \langle \phi_i'', \psi_j'' \rangle.$$

The matrix Φ is tridiagonal and the terms $\langle \phi_i'', \psi_j'' \rangle$ are nonzero only if i and j are consecutive, which simplifies the computation somewhat. One can show (problem 2.2.25) that the resulting interpolate is twice continuously differentiable.

If a function is unknown, but described by a differential equation, we can also use finite elements. For example, suppose we wish to solve the differential equation $u'' = 0$ subject to boundary conditions $u(0) = a$, $u(1) = b$, using a cubic spline approximation. The first step is to reformulate the differential equation as a least squares problem. In fact, using the Calculus of Variations, it can be shown that the solution of this differential equation is equivalent to minimizing

$$\int_0^1 u'^2 dx$$

subject to the boundary conditions $u(0) = a$, $u(1) = b$. We represent the function $u(x)$ approximately as

$$u = \sum_{j=0}^{N} \alpha_j \phi_j(x) + \sum_{j=0}^{N} \beta_j \psi_j(x)$$

where $\alpha_0 = a$ and $\alpha_N = b$ are specified, but the remaining coefficients are yet to be determined. The problem is now to minimize

$$\|f + g\|^2$$

where we can identify f and g as

$$f = a\phi_0'(x) + b\phi_N'(x)$$

$$g = \sum_{j=1}^{N} \alpha_j \phi_j'(x) + \sum_{j=0}^{N} \beta_j \psi_j'(x).$$

It is a straightforward, but tedious, calculation to show that the least squares solution is given by

$$u(x) = \sum_{j=0}^{N} ((b-a)x_j + a)\,\phi_j(x) + \sum_{j=0}^{N} (b-a)\psi_j(x)$$

$$= bx + a(1-x),$$

which is not surprising since this is the exact solution of the problem.

This method of using some set of functions to minimize a functional is called the Rayleigh-Ritz technique. In Chapter 5 it will become more evident how to find an equivalent object to minimize. However, if we do not know or cannot find an equivalent object to minimize, a technique, known as the Galerkin method, should work. The idea is to pick a function for which the differential operator is orthogonal to as many linearly independent functions as possible. Said another way, we project the equation onto a finite dimensional subspace of the Hilbert Space and solve the resulting finite dimensional problem. For example, to solve $u'' = 0$ with $u(0) = a$, $u(1) = b$ using the Galerkin technique and cubic splines, we set

$$u = \sum_{j=0}^{N} \alpha_j \phi_j(x) + \sum_{j=0}^{N} \beta_j \psi_j(x)$$

and then choose α_j and β_j so that

$$\langle u'', \phi_j \rangle = 0$$

and

$$\langle u'', \psi_j \rangle = 0$$

for $j = 0, 1, 2, \ldots, N$. For this equation, as well as many others, the Rayleigh-Ritz and Galerkin techniques are equivalent. Keep in mind that the answer one gets from an approximation scheme such as this depends in a nontrivial way on how the inner product is defined. Thus, the Sobolev inner product gives a different answer than the L^2 inner product.

For most practical purposes, finite elements cannot be used without the aid of a computer, since the matrices that must be inverted are not diagonal. However, with a computer, the potential use can be extended to more than one independent variable, where they can be used to design and build surfaces or to solve partial differential equations in an approximate sense. But this vast subject is best deferred to another text.

2.2.7 Sinc Functions

The finite element functions work well for approximations on finite domains, but are not at all useful on an infinite domain such as the real line. For approximations on the real line, the Whittaker-Cardinal functions, or Sinc functions, as they are commonly known, are quite useful.

Suppose $h > 0$ is a fixed mesh width (or step size). We define

$$S_0(x) = \frac{\sin(\pi x/h)}{(\pi x/h)}$$

and

$$S_k(x) = S_0(x - kh) \quad , \quad -\infty < k < \infty.$$

The functions $S_k(x)$ have the useful property that $S_k(jh) = \delta_{jk}$ and so provide for easy interpolation of functions on the real line. To interpolate the function $f(x)$ on the real line, we define the Whittaker Cardinal function

$$C(f)(x) = \sum_{k=-\infty}^{\infty} f(kh)S_k(x)$$

whenever this series converges.

Now, the amazing fact about the function $C(f)(x)$ is that for a large class of functions, $C(f)(x)$ is exactly $f(x)$ and for many other functions, $C(f)(x)$ gives a very accurate approximation to $f(x)$.

The principal use of the Sinc functions is as a basis for a Galerkin approximation scheme. However, to make proper use of the Sinc functions, we need to understand more of their properties and that requires more knowledge of complex variables and analytic function theory than we have so far assumed. Thus, we defer further discussion of Sinc functions to Chapter 6.

Further Reading

In this chapter we introduced some basic concepts of function spaces, Lebesgue integration, and so on. A more careful development of these concepts can be found in any good analysis book, for example,

1. W. Rudin, *Principles of Mathematical Analysis*, McGraw-Hill, New York, 1964.

An introductory book on Fourier Series is

2. R.V. Churchill, *Fourier Series and Boundary Value Problems*, McGraw-Hill, New York, 1963,

and more advanced discussions of orthogonal functions are given in

3. H. Hochstadt, *The Functions of Mathematical Physics*, Wiley-Interscience, New York, 1971,

4. R. Askey, *Orthogonal Polynomials and Special Functions*, SIAM, Philadelphia, 1975.

Walsh functions are not commonly included in standard applied mathematics literature. They were first discussed in 1923 in

5. J.L. Walsh, "A closed set of normal orthogonal functions," Am. J. Math. **45**, 5-43 (1923).

A good introduction to Walsh functions is found in

6. The Hatfield Polytechnic, *Symposium on Theory and Applications of Walsh Functions*, 1971,

as well as

7. K.G. Beauchamp, *Applications of Walsh and Related Functions*, Academic Press, London, 1984,

8. H.F. Harmuth, *Transmission of Information by Orthogonal Functions*, Springer-Verlag, New York, 1972.

Finally, a good source of information on splines, including FORTRAN codes is

9. C. deBoor, "A Practical Guide to Splines," Applied Mathematical Sciences, Vol.27, Springer-Verlag, New York, 1978.

More extensive resources on how to use finite elements to solve differential equations are

10. R. Wait and A.R. Mitchell, *Finite Element Analysis and Applications*, Wiley and Sons, New York, 1968,

11. G. Strang and G. Fix, *An Analysis of the Finite Element Method*, Prentice-Hall, 1973.

Problems for Chapter 2

Section 2.1

1. Verify that ℓ^2 is a normed vector space. Show that for all sequences $x, y \in \ell^2$, the inner product $\langle x, y \rangle = \sum_{i=1}^{\infty} x_i \bar{y}_i$ is defined and satisfies the requisite properties.

2. Show that $\mathbb{R}^n$ is a complete normed linear vector space with the norm $\|x\| = \max\{|x_1|, |x_2|, \ldots, |x_n|\}$.

3. Show that the sequence $\{x_n\}$, $x_n = \sum_{k=1}^{n} \frac{1}{k!}$ is a Cauchy sequence using the measure of distance $d(x, y) = |x - y|$.

4. Show that the continuous functions on $[0,1]$ form an infinite dimensional vector space. Find a set of linearly independent vectors which is not finite.

5. Show that the sequence of functions $\{f_n(t)\}$

$$f_n(t) = \begin{cases} 0 & 0 \le t < 1/2 - 1/n \\ 1/2 + n(t - 1/2)/2 & 1/2 - 1/n \le t \le 1/2 + 1/n \\ 1 & 1/2 + 1/n \le t \le 1 \end{cases}$$

is a Cauchy sequence in $L^2[0, 1]$ but not in $C[0, 1]$ (i.e., using the uniform norm).

6. a. Suppose that for any function $f(t)$, $\| |f(t)| \| = \|f(t)\|$ and that if $0 < f \le g$, then $\|f\| \le \|g\|$. If $\{g_n(t)\}$ is a sequence of functions for which $\lim_{n \to \infty} g_n(t) = g(t)$ (i.e., $\lim_{n \to \infty} \|g_n(t) - g(t)\| = 0$, show that $\lim_{n \to \infty} |g_n(t)| = |g(t)|$.
   Hint: Show that $| |u| - |v| | \le |u - v|$.

   b. Give some examples of norms for which the two assumptions hold.

7. a. Prove that the set of numbers $\{\frac{1}{n}\}$ $n = 1, 2, \ldots$, forms a set of measure zero.

   b. Prove that the set of rational numbers p/q and $0 \le p/q \le 1$ forms a set of measure zero.
   Hint: To do this, one must find a way to sequentially list the rational numbers and be sure that all rational numbers are included in the list.

8. a. Use property 4) of Lebesgue integration to evaluate

$$\int_0^1 \chi(t) \quad \text{where} \quad \chi(t) = \begin{cases} 1 & t \quad \text{rational} \\ 0 & t \quad \text{irrational} \end{cases}$$

   b. Use property 5) of Lebesgue integration to evaluate $\int_0^1 \chi(t)$.

9. Prove Theorem 2.2.

10. Show that the Lebesgue dominated convergence theorem fails for the sequence of functions $f_n(x) = n^2 x e^{-nx}$ for $x \in [0, 1]$.

11. For $f(x, y) = \frac{x^2 - y^2}{(x^2 + y^2)^2}$, show that

$$\int_0^1 \left(\int_0^1 f(x, y) dx \right) dy \neq \int_0^1 \left(\int_0^1 f(x, y) dy \right) dx.$$

What went wrong?

Section 2.2

1. Find the best quadratic polynomial fit to $f(x) = |x|$ on the interval $[-1, 1]$ relative to the inner product $\langle u, v \rangle = \int_{-1}^1 u(t) v(t) w(t) dt$ in the cases $w(t) = 1$, $w(t) = (1 - t^2)^{1/2}$ and $w(t) = (1 - t^2)^{-1/2}$. What are the differences between these approximations?

2. Find the best L^1 linear approximation of e^x on $[0,1]$, i.e., minimize

$$\int_0^1 |e^x - \alpha - \beta x| dx.$$

Describe in general how to minimize $\int_0^1 |f(x) - \alpha - \beta x| dx$ for any monotone function $f(x)$. What additional assumptions on $f(x)$ do you need?

3. a. Find the best least squares polynomial approximation of degree $n \leq 5$ for the function $f(x) = \sin \pi x$ in the interval $x \in [-1, 1]$ using the inner product $\langle f, g \rangle = \int_{-1}^1 f(x) g(x) dx$.
   Remark: The first 5 Legendre polynomials are

$$
\begin{aligned}
P_0(x) &= 1 \\
P_1(x) &= x \\
P_2(x) &= (3x^2 - 1)/2 \\
P_3(x) &= (5x^3 - 3x)/2 \\
P_4(x) &= (35x^4 - 30x^2 + 3)/8 \\
P_5(x) &= (63x^5 - 70x^3 + 15x)/8
\end{aligned}
$$

(Why not use REDUCE for this problem?)

   b. Solve the same problem using the Sobolev inner product $\langle f, g \rangle = \int_{-1}^1 (f(x) g(x) + f'(x) g'(x)) dx$.

   c. Discuss the differences between these two approximations. In general, what differences do you expect to see between use of the L^2 norm and the Sobolev norm?

4. Show that the trigonometric functions form a complete orthogonal basis for any Sobolev space $H^n[0, 2\pi]$ whose functions $f(x)$ satisfy $f(o) = f(2\pi)$. What is the difference between n terms of an L^2 and an H^1 Fourier approximation of some piecewise C^1 function? Illustrate the difference with a specific example of your choosing.

5. Define $T_n(x) = \cos(n \cos^{-1} x)$, $n \geq 1$, $T_0 = 1$. Show that

   a. $T_n(x)$ is an nth order polynomial.

   b. $\int_{-1}^{1} T_n(x) T_k(x)(1 - x^2)^{-1/2} dx = 0$ if $k \neq n$.

   c. The sequence $\{T_n(x)\}$ is generated (up to a scale factor) by the Gram-Schmidt procedure starting with the set $\{1, x, x^2, \ldots, \}$ using the weight function $w = (1 - x^2)^{-1/2}$ and $\langle f, g \rangle = \int_{-1}^{1} f(x)g(x)w(x)dx$.

6. Suppose $\{\phi_n(x)\}_{n=0}^{\infty}$ is a set of orthonormal polynomials relative to the L^2 inner product with positive weight function $w(x)$ on the domain $[a, b]$, and suppose $\phi_n(x)$ is a polynomial of degree n with leading coefficient $k_n x^n$.

   a. Show that $\phi_n(x)$ is orthogonal to any polynomial of degree less than n.

   b. Show that the polynomials satisfy a recurrence relation

   $$\phi_{n+1}(x) = (A_n x + B_n)\phi_n(x) + C_n \phi_{n-1}(x)$$

   where $A_n = \frac{k_{n+1}}{k_n}$. Express B_n and C_n in terms of A_n, A_{n-1} and ϕ_n.

   c. Evaluate A_n, B_n, and C_n for the Legendre and Chebyshev polynomials.

7. Show that the functions

   $$P_n(x) = \frac{1}{w(x)} \frac{d^n}{dx^n} \left(w(x)(1 - x^2)^n \right), \quad n = 0, 1, 2, \ldots$$

   are orthogonal to every polynomial of degree less than n with respect to the inner product $\langle f, g \rangle = \int_{-1}^{1} f(x)g(x)w(x)dx$.

8. Prove that $P_n(x) = \frac{(-1)^n}{2^n n!} \frac{d^n}{dx^n} \left((1 - x^2)^n\right)$ is generated (up to scalar multiple) from the set $1, x, x^2, \ldots$ by the Gram-Schmidt procedure using the L^2 inner product with weight function $w(x) = 1$ on the interval $[-1, 1]$.

9. Show that the Chebyshev polynomials can be expressed as

   $$T_n(x) = \frac{(-1)^n 2^n}{(2n)!}(1 - x^2)\frac{d^n}{dx^n}(1 - x^2)^{n-1/2}.$$

10. Show that $H_{n+1}(x) - 2x H_n(x) + 2n H_{n-1}(x) = 0$, $n \geq 1$ where $H_n(x)$ is the nth order Hermite polynomial.

11. Show that $L_{n+1}(x) - (2n + 1 - x)L_n(x) + n^2 L_{n-1}(x) = 0$, $n \geq 1$, where $L_n(x)$ is the nth order Laguerre polynomial.

12. Suppose $f(t)$ and $g(t)$ are 2π periodic functions with Fourier series representations

$$f(t) = \sum_{k=-\infty}^{\infty} f_k e^{ikt} \quad , \quad g(t) = \sum_{k=-\infty}^{\infty} g_k e^{ikt}.$$

Find the Fourier series of

$$h(t) = \int_0^{2\pi} f(t - x)g(x)dx.$$

13. Suppose $\{f_n\}$ and $\{g_n\}$ are both periodic sequences, i.e., $f_{k+N} = f_k$ and $g_{k+N} = g_k$ for all k, with N fixed. Find the discrete Fourier transform of the sequence $\{h_n\}$ where $h_n = \sum_{j=1}^{N} f_j g_{N-j}$.

14. Show that the discrete Fourier Transform $\{f_n\} \rightarrow \{g_n\}$ can be viewed as a change of coordinate system in $\mathbb{R}^n$. What is the matrix T corresponding to this change of basis?

15. a. Verify that $\frac{1}{n}\sum_{j=0}^{n-1} e^{\frac{-2\pi ijk}{n}} \sum_{l=0}^{n-1} f_l e^{\frac{-2\pi ijl}{n}} = f_k$.

 b. Verify that $\frac{2}{n}\sum_{j=1}^{n-1} \sin\frac{\pi jl}{n} \sin\frac{\pi jk}{n} = \delta_{kl}$.

16. Represent the integer 23 as a $\oplus$ sum of integers $2^n - 1$.

17. Show that $\text{Wal}(n, x)$ is even about $x = 1/2$ if n is even and is odd about $n = 1/2$ if n is odd.

18. Verify the identities

$$\text{cal}(i, x)\text{cal}(j, x) = \text{cal}(i \oplus j, x)$$

$$\text{cal}(i, x)\text{sal}(j, x) = \text{sal}(i \oplus (j - 1) + 1, x)$$

$$\text{sal}(i, x)\text{sal}(j, x) = \text{cal}((i - 1) \oplus (j - 1), x)$$

19. a. Let $x_j = \frac{i+1/2}{N}$, $N = 2^p$ for some fixed p. Show that

$$\text{Wal}(k, x_j) = \text{Wal}(j, x_k) \quad \text{for} \quad 0 \leq j, k \leq N - 1.$$

 b. Show that, except for the factor $\frac{1}{N}$, the forward and inverse discrete Walsh transform correspond to the same matrix multiplication.

20. Suppose $\{x_j\}$ and $\{y_j\}$ are sequences with discrete Walsh transforms $\{A_j\}$, $\{B_j\}$ respectively. Define the dyadic convolution of x and y as $z = x * y$ where $z_s = \frac{1}{N}\sum_{r=0}^{N-1} x_r y_{r \oplus s}$. Show that if $N = 2^p$ for some integer p, then the discrete Walsh transform of $\{z_s\}$ is $\{A_s B_s\}$.
Remark: This is the "dyadic" convolution theorem for Walsh transforms.

21. Prove "Parseval's Theorem" for Walsh functions: If the discrete Walsh Transform of $\{x_j\}$ is $\{A_k\}$, then $\frac{1}{N}\sum_{j=0}^{N-1} x_j^2 = \sum_{k=0}^{N-1} A_k^2$.

22. Find the cubic spline $f(x)$ satisfying $f(x_j) = f_j$, $j = 0, 1, 2$, $x_j = j$ which minimizes $\int_0^2 f''^2(x)dx$.

23. Suppose the functions $\phi_i(x)$, $\psi_i(x)$, $i = 0, 1, 2, \ldots, N$ are the piecewise cubic polynomial basis functions for $S^h(3, 1)$.

    a. Verify that, for $i \neq 0, N$

    $$\int_0^1 \phi_i^2(x)dx = 26h/35 , \quad \int_0^1 \psi_i^2(x)dx = 2h^3/105$$

    $$\int_0^1 \phi'^2_i(x)dx = 12h/5h , \quad \int_0^1 \psi'^2_i(x)dx = 4h/15$$

    $$\int_0^1 \phi''^2_i(x)dx = 24/h^3 , \quad \int_0^1 \psi''^2_i(x)dx = 8/h$$

    b. Verify that

    $$\int_0^1 \phi_i(x)\phi_{i+1}(x)dx = 9h/70 \qquad \int_0^1 \psi_i(x)\psi_{i+1}(x)dx = -h^3/140$$

    $$\int_0^1 \phi_i'(x)\phi_{i+1}'(x)dx = -6/5h , \qquad \int_0^1 \psi_i'(x)\psi_{i+1}'(x)dx = -h/30$$

    $$\int_0^1 \phi_i''(x)\phi_{i+1}''(x)dx = -12/h^3 , \qquad \int_0^1 \psi_i''(x)\psi_{i+1}''(x)dx = 2/h$$

    c. Verify that

    $$\int_0^1 \psi_i(x)\phi_{i+1}(x)dx = 13h^2/420$$

    $$\int_0^1 \psi_i'(x)\phi_{i+1}'(x)dx = -1/10$$

    $$\int_0^1 \psi_i''(x)\phi_{i+1}''(x)dx = -6/h^2$$

Remark: Here is a good example of a problem that makes mathematics texts tedious and boring. Instead of doing all of these integrals by hand, why not use this exercise to polish your REDUCE skills? For example, the following sequences of commands will evaluate

$$I_1 = \int_0^1 \phi_i^2(x)dx \quad \text{and} \quad I_2 = \int_0^1 \phi'^2_i(x)dx :$$

```
F:  =  (x-h)**2*(2*x+h)/h**3;
FP: =  DF(F,X);
S1: =  INT(F**2,x);
S2: =  INT(FP**2;x);
I1: =  2*SUB(x=h,S1);
I2: =  2*SUB(x=h,S2);
```

24. Suppose the functions $\phi_i(x)$, $\psi_i(x)$ $i = 0, 1, 2, \ldots, N$ are the piecewise cubic polynomial basis functions for $S^h(3,1)$. The function $\sum_{i=0}^{N}(f_i\phi_i(x) + \alpha_i\psi_i(x))$ interpolates the data points $f(x_i) = f_i$. Show that each of the following specifications requires solution of a tridiagonal matrix equation for the coefficients α_i. Determine explicitly the matrix equation to be solved.

    a. Minimize $\int_0^1 (f''(x))^2 \, dx$.

    b. Minimize $\int_0^1 (f'(x))^2 \, dx$.

    c. Require that $f(x)$ be twice continuously differentiable with $f'(0) = \alpha_0$, $f'(1) = \alpha_N$ given.

    d. With $g(x)$ known, $f_i = g(x_i)$ minimize

$$\int_0^1 (g(x) - f(x))^2 \, dx.$$

25. Suppose $f(x)$ is the piecewise cubic polynomial interpolate of the data points $f(x_i) = f_i$ and is chosen to minimize $\int_0^1 (f''(x))^2 \, dx$. Show that $f(x)$ is twice continuously differentiable.
Hint: Show that the equations for $\alpha_1, \ldots, \alpha_{N-1}$ from problem 24.a and c are identical.

3

INTEGRAL EQUATIONS

3.1   Introduction

The first application of the theory developed in the previous two chapters will be to INTEGRAL EQUATIONS. As we shall see, integral equations have a number of features that are strikingly similar to those of matrices, even though integral equations typically act on infinite dimensional Hilbert spaces. Integral equations occur as models in many different physical contexts.

Example   Population Dynamics

Suppose $u(t)$ represents the number of individuals in some population. The simplest way to describe the time evolution of the population is to assume that the net of per capita births and deaths is a fixed constant A and that $f(t)$ is the net rate of influx of individuals due to migration. Then

$$\frac{du}{dt} - Au = f(t)$$

describes the rate of change of the population $u(t)$. If we multiply both sides of this equation by the integrating factor e^{-At} and integrate, we find

$$u(t) = u_0 e^{At} + \int_0^t e^{A(t-\tau)} f(\tau) \, d\tau.$$

This is an example of an integral equation of the form

$$u(t) = \int_a^b k(t, \tau) f(\tau) \, d\tau$$

which is called a FREDHOLM INTEGRAL EQUATION of the FIRST KIND. The special case where $k(t, \tau) = 0$ for $\tau > t$ is called a VOLTERRA INTEGRAL EQUATION. Even though this example is somewhat trivial, it is interesting because it represents the solution of a differential equation as an integral. That is, the inverse of the differential operator is shown to be an integral operator.

Example Renewal Processes

Suppose we want to make a more careful analysis of birth and death processes taking into account the age structure of the population. We want to find $B(t)$, the rate of births at time t. The current births are only from individuals born previously who are now bearing offspring, so that

$$B(t) = \int_0^t \beta(a) s(a) B(t - a) \, da + f(t)$$

where $B(t - a)$ is the rate of birth a units of time previous, $s(a)$ is the probability density of survival to age a, β is the fertility of individuals who are t units of age and $f(t)$ is the rate of births from individuals more than t units of age. With the change of variable $\tau = t - a$ this equation becomes

$$B(t) = \int_0^t B(\tau) \beta(t - \tau) s(t - \tau) \, d\tau + f(t)$$

which is an equation of the form

$$u(t) = \int_a^b k(t, \tau) u(\tau) \, d\tau + f(t)$$

known as a FREDHOLM INTEGRAL EQUATION of the SECOND KIND.

Differential equations can often be converted into integral equations. For example, integrating by parts, one can show that the solution of $\frac{d^2 u}{dx^2} = f(x)$, $u(0) = 0$, $u(1) = 0$ is given by

$$u(x) = \int_0^1 k(x, y) f(y) \, dy$$

where

$$k\left(x,y\right) = \begin{cases} y\left(x-1\right) & 0 \leq y < x \leq 1 \\ x\left(y-1\right) & 0 \leq x < y \leq 1 \end{cases}$$

Therefore, the boundary value problem

$$\frac{d^2u}{dx^2} + \omega\left(x\right)u = g\left(x\right), u\left(0\right) = u\left(1\right) = 0$$

is equivalent to the integral equation

$$u\left(x\right) = G\left(x\right) - \int_0^1 k\left(x,y\right)\omega\left(y\right)u\left(y\right)dy$$

where $G\left(x\right) = \int_0^1 k\left(x,y\right)g\left(y\right)dy$, a Fredholm integral equation of the second kind.

One feature of these integral equations is that the kernel $k\left(x,y\right)$ in each case satisfies

$$\int_a^b \int_a^b k^2\left(x,y\right)dx\,dy < \infty.$$

As we shall see, this is an important class of kernels (called Hilbert-Schmidt kernels), about which much can be said.

Another important class of integral equations is the SINGULAR INTEGRAL EQUATIONS.

The problem of Abel is a classical example of a singular integral equation. The goal is to determine the shape of a hill given only the descent time of a sliding frictionless bead as a function of its initial height. Suppose the bead begins its descent from height x with zero initial speed. When the bead is at height y the sum of its kinetic energy $mv^2/2$ and potential energy $-mg\left(x-y\right)$ is constant so that

$$\frac{ds}{dt} = \sqrt{2g\left(x-y\right)}$$

where $\frac{ds}{dt} = v$ is the speed of the bead. It follows that the descent time is

$$T\left(x\right) = \int_0^x \frac{f\left(y\right)}{\sqrt{x-y}}dy$$

where

$$f\left(y\right) = \frac{1}{\sqrt{2g\,dy/ds}}$$

and $y = y\left(s\right)$ is the height of the hill as a function of the arc length coordinate s. This equation is known as ABEL'S INTEGRAL EQUATION.

Another example of a singular integral equation relates to X-RAY TOMOGRAPHY. Using x-rays, one would like to construct a picture of the two-dimensional cross section of some object without cutting (surgery).

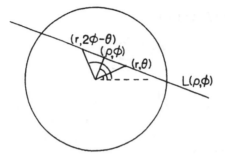

FIGURE 3.1   Sketch of the line $L(\rho, \phi)$ along which the absorption of an x-ray is integrated.

We suppose that when a high frequency electromagnetic wave penetrates an object, it travels along a straight line without bending or spreading, but the intensity decreases as it travels because of absorption. We assume that the x-ray is absorbed at a linear rate g which is a fixed, unknown function of position. Thus, if the initial intensity of the beam is I_0, after passing through the object the intensity is

$$I = I_0 \exp\left[-\int_L g \, ds\right]$$

where the integral is a line integral along the straight path of travel. We set $f(\rho, \phi) = \int_{L(\rho, \phi)} g(r, \theta) \, ds = \ln(I_0/I)$ where $\ln(I_0/I)$, which is a measured quantity, varies as ρ, ϕ are changed. The notation $L(\rho, \phi)$ denotes that the line L is closest to the origin at the point (ρ, ϕ), expressed in polar coordinates (see Figure 3.1).

We represent $g(r, \theta)$ in its Fourier series

$$g(r, \theta) = \sum_{n=-\infty}^{\infty} g_n(r) e^{in\theta}, \quad g_n(r) = \frac{1}{2\pi} \int_0^{2\pi} g(r, \theta) e^{-in\theta} \, d\theta.$$

To go from the point (ρ, ϕ) to the boundary of the circle along the arc L, one can travel in either of two directions, one which increases and one which decreases the angle θ (see Figure 3.1). As a result, along an increment of arc the contribution to the integral is

$$df = (g(r, \theta) + g(r, 2\phi - \theta)) \, ds$$

$$= 2 \sum_{n=-\infty}^{\infty} g_n(r) e^{in\phi} \cos n(\phi - \theta) \frac{r \, dr}{(r^2 - \rho^2)^{1/2}}.$$

Now $\cos(\phi - \theta) = \rho/r$ so that

$$f = \sum_{n=-\infty}^{\infty} f_n(\rho) e^{in\phi}$$

where

$$f_n(\rho) = 2 \int_\rho^1 g_n(r) \frac{\cos\left[n \cos^{-1}(\rho/r)\right] r \, dr}{(r^2 \doteq \rho^2)^{1/2}}.$$

Recall from Chapter 2 that the Chebyshev polynomials are given by $T_n(x) = \cos(n \cos^{-1} x)$ so we have

$$f_n(\rho) = 2 \int_\rho^1 \frac{g_n(r) T_n(\rho/r) r \, dr}{(r^2 - \rho^2)^{1/2}}.$$

Since $f_n(\rho)$ can in principle be measured, the goal is to find $g_n(r)$ knowing $f_n(\rho)$.

The solution of these singular integral equations in 1963 allowed A.M. Cormack to build the first CAT (Computer Aided Tomography) scan device, a device which is now common in hospitals and medical centers. In 1979 Professor Cormack was awarded the Nobel prize in physiology (shared with G.N. Hounsfield) for his accomplishments.

3.2   Bounded Linear Operators in Hilbert Space—The Fredholm Alternative

Definition

A BOUNDED LINEAR OPERATOR $L : H \to H$ is a mapping which takes one function in H into another function in H and satisfies the two properties:

1. L is **linear**: $L(\alpha f + \beta g) = \alpha L f + \beta L g$ for all real numbers α, β and all f, g in H.

2. L is **bounded** (or **continuous**): There is a constant $K > 0$ so that
$$\|Lf\| \leq K\|f\| \text{ for all } f \text{ in } H.$$

(Here $\| \cdot \|$ is the natural norm of H.)

The norm of the operator L is

$$\|L\| = \underset{u \neq 0}{\text{l.u.b.}} \frac{\|Lu\|}{\|u\|},$$

That is, $\|L\|$ is the smallest constant K for which $\|Lu\| \leq K\|u\|$.

Examples

1. $Lf = f$, the identity operator, is clearly linear and bounded (with $\|L\| = 1$).

2. $Lf = \int_0^1 f(x)\,dx$ is certainly linear. To show boundedness in $L^2[0, 1]$ note that

$$\|Lf\|^2 = \int_0^1 (Lf)^2\,dt = \int_0^1 \left(\int_0^1 f(x)\,dx \right)^2 dt$$

$$= \left(\int_0^1 (f(x) \cdot 1)dx \right)^2 \leq \int_0^1 f^2(x)\,dx = \|f\|^2.$$

by the Schwarz inequality. More generally, the integral operator

$$Ku = \int_0^1 k(x, y)\, u(y)\, dy$$

is linear and bounded in $L^2[0, 1]$ if $\int_0^1 \int_0^1 k^2(x, y)\, dy\, dx < \infty$. The proof is basically the same, since

$$\|Ku\|^2 = \int_0^1 \left(\int_0^1 k(x, y)\, u(y)\, dy \right)^2 dx$$

$$\leq \int_0^1 \left(\int_0^1 k^2(x, y)dy \right) \left(\int_0^1 u^2(y)\, dy \right) dx$$

$$= \left(\int_0^1 \int_0^1 k^2(x, y)dx\, dy \right) \|u\|^2.$$

3. $Lf = \frac{df}{dx}$ is a linear operator, but it is not bounded in L^2. If we take $f = \sin nx$ for x in $[0, 2\pi]$, then $\|Lf\|^2 = n^2 \int_0^{2\pi} \cos^2 nx\, dx = n^2\pi$. However, $\|f\|^2 = \int_0^{2\pi} \sin^2 nx\, dx = \pi$, so that $\|Lf\| = n\|f\|$. Clearly, there is no constant K which is bounded yet bigger than every integer n.

Linear operators in Hilbert space are the natural generalization of matrices in $\mathbb{R}^n$. As with matrices, we define the DOMAIN, RANGE and NULL SPACE of L as

$D(L) = $ Domain of $L = \{f \in H \mid Lf$ is defined$\}$
$R(L) = $ Range of $L = \{f \in H \mid f = Lg$ for some $g \in H\}$
$N(L) = $ Nullspace of $L = \{f \in H \mid Lf = 0\}$

If our analogy with matrices is correct, we expect the operator L to have an adjoint L^* and that adjoint should be involved in a statement about the existence of solutions of $Lu = f$.

Definition

The adjoint of the operator L is that operator L^* (if it exists) for which $\langle Lf, g \rangle = \langle f, L^*g \rangle$ for all f, g in H.

For real matrices $A = (a_{ij})$, the adjoint matrix is $A^* = (a_{ij}^*)$ where $a_{ij}^* = a_{ji}$. For a real integral operator $(Ku)(x) = \int_a^b k(x, y)u(y)dy$ on $L^2[a, b]$,

$$
\begin{aligned}
\langle Ku, v \rangle &= \int_a^b \left(\int_a^b k(x, y)u(y)dy \right) v(x)dx \\
&= \int_a^b u(y) \left(\int_a^b k(x, y)v(x)dx \right) dy \\
&= \int_a^b u(x) \left(\int_a^b k(y, x)v(y)dy \right) dx = \langle u, K^*v \rangle
\end{aligned}
$$

so that

$$
(K^*v)(x) = \int_a^b k(y, x)v(y)dy
$$

is the adjoint integral operator.

Adjoint operators as defined here need not always exist. For example, using this definition, the differential operator $Lf = \frac{df}{dx}$ does not have an adjoint. If we try

$$
\begin{aligned}
\langle Lf, g \rangle &= \int_0^1 \frac{df}{dx} g\, dx = fg \Big|_0^1 \\
&- \int_0^1 f \frac{dg}{dx} dx = f(1)g(1) - f(0)g(0) + \left\langle f, \frac{-dg}{dx} \right\rangle,
\end{aligned}
$$

we are tempted to conclude that $L^*g = \frac{-dg}{dx}$, but this is not correct since this definition of adjoint does not hold for *all* choices of f and g, but only for those for which $f(1)g(1) - f(0)g(0) = 0$. When we come to differential operators in Chapter 4 we will be forced to remedy this deficiency by adding a statement about the domain of the operator to the definition of the adjoint.

Under what conditions are we assured that an adjoint operator exists? To answer this question we need some new notation.

Definition

A subspace $M \subset H$ (called a LINEAR MANIFOLD) is *closed* if every sequence in M which is convergent in H is also convergent in M, (that is, if M is itself a Hilbert space).

One important example of a closed linear manifold is the null space of a bounded linear operator. If $\{u_n\}$ is a sequence of elements in the null space of L which converges to $u^* \in H$, then

$$\lim_{n \to \infty} \|Lu_n - Lu^*\| = \lim_{n \to \infty} \|L(u_n - u^*)\| \le \lim_{n \to \infty} K\|u_n - u^*\| = 0$$

and

$$Lu^* = \lim_{n \to \infty} Lu_n = \lim_{n \to \infty} 0 = 0,$$

so that u^* is also in $N(L)$.

As another example, suppose M is any linear manifold in H. Define the ORTHOGONAL COMPLEMENT of M to be

$$M' = \{f \in H \mid \langle f, g \rangle = 0 \quad \text{for all} \quad g \in M\}.$$

To see that M' is a closed linear manifold, suppose $\{u_n\}$ is a sequence in M', and $\lim_{n \to \infty} u_n = u^*$ is in H. For every g in M

$$\langle u^*, g \rangle = \left\langle \lim_{n \to \infty} u_n, g \right\rangle = \lim_{n \to \infty} \langle u_n, g \rangle = 0.$$

The interchange of limits here is again valid since the linear operator $\langle u, g \rangle$ is a bounded linear operator for any fixed g in M.

Any element f in H has a unique decomposition as $f = f_1 + f_2$ where f_1 is in the closure of M (if it is not closed), and f_2 is in M' (which is always closed).

Definition

1. A LINEAR FUNCTIONAL $T : H \to \mathbb{R}$ is a linear map of H into the real numbers satisfying

$$T(\alpha f + \beta g) = \alpha T f + \beta T g$$

   for α, β real numbers and f, g in H.

2. A linear functional $T : H \to \mathbb{R}$ is *bounded* if there is a positive real number K for which $|T(u)| \le K\|u\|$ for all $u \in H$.

An obvious example of a linear functional is the inner product $Tf = \langle f, g \rangle$ for any fixed g in H. In fact, according to the following theorem all bounded linear functionals can be represented as inner products.

Theorem 3.1 (Riesz Representation Theorem)

For any bounded linear functional $T : H \rightarrow \mathbb{R}$, there is a unique g in H so that $Tf = \langle f, g \rangle$.

Proof

Let $N = \{ f \epsilon H \mid Tf = 0 \}$ be the null space of the linear functional T. If $N = H$ set $g = 0$ and we are done.

If N is a proper subset of H, then N', the orthogonal complement of N, is nontrivial and closed. Take g_0 to be any element of N' with $\|g_0\| = 1$ and set $g = T(g_0) g_0$.

To verify that $T(f) = \langle f, g \rangle$ notice that $y = T(f) g_0 - f T(g_0)$ is in the null space of T, so that $\langle y, g_0 \rangle = 0$. As a result $T(f) \langle g_0, g_0 \rangle = \langle f, T(g_0) g_0 \rangle$ so that $T(f) = \langle f, g \rangle$.

The function g is unique, since if there is another function $\hat{g}$, say, for which $Tf = \langle f, g \rangle = \langle f, \hat{g} \rangle$ for all f in H, then, $\langle f, g - \hat{g} \rangle = 0$ for all f, and in particular for $f = g - \hat{g}$. But then $\|g - \hat{g}\| = 0$ so $g = \hat{g}$.

The Riesz Representation theorem fails for unbounded linear functionals. To see why this is true consider the example $Tf = f(0)$, which is not a bounded linear functional using the L^2 norm. As we shall discuss in more detail in Chapter 4, there is no function g which when integrated against $f(x)$ gives $f(0)$. So what went wrong?

Quite simply, since the null space of T includes any function in L^2 with $f(0) = 0$, the closure of the null space is all of L^2 and the orthogonal complement of the null space contains only the function $g = 0$ (in L^2, this means $g = 0$ almost everywhere) and so the proof fails. In contrast, the null space of a bounded linear operator is always closed, so this problem does not occur. Another way to express the failure is to note that in L^2, the functional $Tf = f(0)$ is not even well defined since any two functions which differ only at $x = 0$ are identified in L^2 as the same function. As a result there are functions f and g for which $Tf \neq Tg$ even though $f = g$ in L^2. In fact, this functional makes sense only for functions which are continuous near 0.

We are now able to come to the conclusion that we have been seeking:

Theorem 3.2

If L is a bounded linear operator in H, then

1. L^* exists.

2. L^* is a bounded, linear operator.

Proof

Since L is a bounded linear operator, $Tu = \langle Lu, v \rangle$ is a bounded linear functional for each fixed v in H. Linearity of T is obvious. T is bounded since $|Tu| = |\langle Lu, v \rangle| \leq \|Lu\| \|v\| \leq (K\|v\|) \|u\|$. By the Riesz Representation Theorem we can pick g so that $Tu = \langle u, g \rangle$ and then $L^*v = g$ defines the adjoint operator of L. To see that L^* is linear let $L^*v_1 = g_1, L^*v_2 = g_2$ and $L^* (\alpha v_1 + \beta v_2) = g_3$. Then

$$\begin{aligned}
\langle u, g_3 \rangle &= \langle u, L^* (\alpha v_1 + \beta v_2) \rangle = \langle Lu, \alpha v_1 + \beta v_2 \rangle \\
&= \overline{\alpha} \langle u, g_1 \rangle + \overline{\beta} \langle u, g_2 \rangle = \\
&= \langle u, \alpha g_1 + \beta g_2 \rangle \text{ for all } u \text{ in } H.
\end{aligned}$$

To see that L^* is bounded, note that

$$\|L^*v\|^2 = \langle L^*v, L^*v \rangle = \langle L (L^*v), v \rangle \leq K\|L^*v\| \|v\|$$

so that $\|L^*v\| \leq K\|v\|$ where K is the bound for L.

With this machinery in place we can finally state

Theorem 3.3

FREDHOLM ALTERNATIVE THEOREM: If L is a bounded linear operator in H with closed range, the equation $Lf = g$ has a solution if and only if $\langle g, v \rangle = 0$ for every v in the null space of L^*.

The proof of this theorem is a word for word repeat of the proof given in Chapter 1 for matrices. We use the fact that the range of L is closed to ensure that any vector g in H can be written as $g = g_r + g_n$ where g_r belongs to the range of L and g_n belongs to the null space of L^*. In general, if the range of L is not closed we can only conclude that g_r belongs to the closure of the range of L. The end result can be stated in shorthand form as

$$H = \bar{R}(L) + N(L^*)$$

where $\bar{R}(L)$ is the closure of the range of L and $N(L^*)$ is the null space of the operator L^*. The decomposition of H into $\bar{R}(L)$ and $N(L^*)$ is an orthogonal decomposition called a direct sum, implying that any vector g in H can be expressed uniquely as $g = g_r + g_n$ where $\langle g_r, g_n \rangle = 0, g_r \in \bar{R}(L), g_n \in N(L^*)$.

Example

To illustrate the Fredholm alternative theorem, consider the linear integral equation

$$Lu = u(t) + \lambda \int_0^1 su(s)\,ds = f(t).$$

The null space of L is empty unless $\lambda = -2$ in which case it is spanned by $u = $ constant. The adjoint operator is

$$L^*v = v(t) + \lambda t \int_0^1 v(s)\,ds$$

whose null space is empty unless $\lambda = -2$, in which case it is spanned by $v(t) = t$. The Fredholm alternative implies that for $\lambda \neq -2$ a unique solution of $Lu = f$ exists for all f but when $\lambda = -2$ a (nonunique) solution exists if and only if $\int_0^1 tf(t)\,dt = 0$. (Notice that we cannot be sure that the Fredholm alternative applies here since we have not yet shown that the range of the integral operator is closed. This we will do in Section 3.5.)

The integral equation is solved exactly by

$$u(t) = f(t) - \frac{\lambda}{1 + \lambda/2} \int_0^1 sf(s)\,ds = Rf$$

which verifies the above existence statement. The operator R is called the RESOLVENT OPERATOR (i.e. INVERSE INTEGRAL OPERATOR) for the operator L. We shall see more of resolvent operators in later sections.

3.3   Compact Operators—Hilbert Schmidt Kernels

Boundedness of an operator is not quite enough to conclude that it is similar to a matrix. The important property of integral operators is that they are *compact*, which is a stronger statement than boundedness.

Definition

A set S in H is COMPACT (or sequentially compact) if any sequence $\{x_n\}$ chosen from S contains a convergent subsequence.

From this definition, any bounded set in a finite dimensional space is compact, although bounded sets in L^2 are not necessarily compact. For example, the sequence $\{\sin nx\}_{n=1}^\infty$ which consists of mutually orthogonal elements in $L^2[0, \pi]$ cannot possibly have a convergent subsequence since the distance between any two elements of the sequence is $\sqrt{2}$. (This type of compactness is often called *sequential compactness* or *precompactness* to distinguish it from the common definition found in most analysis texts, that a compact set is closed and bounded.)

Compact sets are, however, bounded. To see this, suppose we have an unbounded compact set S. Choose a sequence of points $\{x_n\}$ in S with $\|x_n\| \geq n$. Now it is possible to choose a subsequence of $\{x_n\}$ say $\{y_n\}$, with the property that $\|y_j - y_k\| > 1$ for all $j < k$. This follows since if $\{y_j\}_{j=1}^{k-1}$ are chosen, they form a bounded set and since $\|y_k - y_j\| \geq |\,\|y_k\| - \|y_j\|\,|$, $\|y_k\|$ can be chosen as large as we need. Now the sequence $\{y_k\}_{k=1}^{\infty}$ has no convergent subsequences although any subset of a compact set is a compact set, leading to an obvious contradiction.

Definition

An operator K is *compact* if it transforms bounded sets into compact sets.

A compact operator is automatically a bounded operator since it maps the bounded set $\|u\| = 1$ into a compact, hence bounded, set. The least upper bound of this image set is the bound of the operator. If K is a bounded linear operator and if $\{\phi_n\}$ is an infinite set of orthonormal functions, then

$$\lim_{n \to \infty} K\phi_n = 0.$$

If not, there must be an infinite subsequence of $\{\phi_n\}$, say $\{\psi_n\}$, for which $\|K\psi_n\| \geq \varepsilon$ for every n. Since $\{\psi_n\}$ is bounded, the sequence $\{K\psi_n\}$ has a convergent subsequence, say $\{K\chi_n\}$, for which

$$\lim_{n \to \infty} K\chi_n = f \neq 0,$$

so that

$$\lim_{n \to \infty} \langle K\chi_n, f \rangle = \|f\|^2 \neq \epsilon.$$

However, since the functions χ_n are mutually orthonormal,

$$\lim_{n \to \infty} \langle K\chi_n, f \rangle = \lim_{n \to \infty} \langle \chi_n, K^* f \rangle = 0$$

by Bessel's inequality, which is a contradiction. Notice that since $\lim_{n \to \infty} K\phi_n = 0$, the inverse of the operator K, if it exists, is unbounded.

A simple example of a compact operator on $L^2[a, b]$ is the integral operator $Ku = \int_a^b k(x, y)\, u(y)dy$ where $k(x, y) = \sum_{i=1}^{n} p_i(x)q_i(y)$ with $\int_a^b p_i^2(x)dx < \infty$ and $\int_a^b q_i^2(x)dx < \infty$. The kernel $k(x, y)$ is called a SEPARABLE or DEGENERATE kernel and K is clearly a compact operator since it maps bounded sets into bounded, finite dimensional sets, which are compact.

To find other compact operators we use the following

Theorem 3.4

If a linear operator K can be approximated (in the operator norm) by a sequence of compact linear operators K_n, with $\lim_{n \to \infty} \|K_n - K\| = 0$, then K is a compact operator.

To prove this fact we need to show that for any bounded sequence $\{u_n\}$, the sequence $\{Ku_n\}$ has a convergent subsequence. Since K_1 is compact, there is a subsequence $\{u_n^{(1)}\}$ of $\{u_n\}$ for which $\{K_1 u_n^{(1)}\}$ is convergent. Since K_2 is compact, there is a subsequence $\{u_n^{(2)}\}$ of $\{u_n^{(1)}\}$ for which $\{K_2 u_n^{(2)}\}$ is convergent. Proceeding inductively there is a subsequence $\{u_n^{(j)}\}$ of $\{u_n^{(j-1)}\}$ for which $\{K_j u_n^{(j)}\}$ is convergent. Define the sequence $\{y_n\}$ to be the "diagonal" subsequence of $\{u_n\}$, namely $y_j = u_j^{(j)}$. Then

$$
\begin{aligned}
\|K y_i - K y_j\| & \leq \|K y_i - K_n y_i\| \\
& + \|K_n y_i - K_n y_j\| \\
& + \|K_n y_j - K y_j\| \\
& \leq \|K - K_n\|(\|y_i\| + \|y_j\|) + \|K_n y_i - K_n y_j\|
\end{aligned}
$$

Now $\{K_n y_i\}$ is convergent for all fixed n, the sequence $\{y_i\}$ is bounded, and $\|K - K_n\|$ is as small as we wish for all large n. Therefore, $\{K y_i\}$ is a Cauchy sequence, which, being in H, converges.

The main impact of this result is that, as we shall see, a large class of integral operators are compact operators.

Definition

If $k(x, y)$ satisfies $\int_a^b \int_a^b k^2(x, y) dx\, dy < \infty$, the kernel $k(x, y)$ is called a HILBERT-SCHMIDT KERNEL and the corresponding integral operator is called a HILBERT-SCHMIDT OPERATOR.

Throughout this chapter we shall assume that the kernel $k(x, y)$ is real, but all of what follows can easily be extended to allow complex kernels.

Hilbert-Schmidt operators are compact. We will show this in two steps. First we will show that a Hilbert-Schmidt operator is bounded, and then, that it can be approximated by a sequence of degenerate operators.

First, to see that the Hilbert-Schmidt operator K is bounded suppose that $g = Kf = \int_a^b k(x, y)f(y) dy$. Then, using the L^2 norm,

$$
\|g\|^2 = \int_a^b g^2(x) dx = \int_a^b \left(\int_a^b k(x, y) f(y) dy \right)^2 dx
$$

$$\leq \int_a^b \left(\int_a^b k^2(x,y)dy \right) \left(\int_a^b f^2(y)dy \right) dx$$

$$= \|f\|^2 \int_a^b \int_a^b k^2(x,y)dx\, dy.$$

Before we continue to the second step, we must demonstrate an important fact about orthonormal bases.

Theorem 3.5

If the sets $\{\Phi_n(x)\}_{n=1}^\infty, \{\Psi_n(x)\}_{n=1}^\infty$ are complete and orthonormal on $L^2[a,b]$ and $L^2[c,d]$ respectively, then the set of all products

$$\{\Phi_n(x)\Psi_m(y)\}_{m,n=1}^\infty$$

is a complete orthonormal set on L^2 of the product of the two intervals, $x \in [a,b], y \in [c,d]$.

To prove this fact, recall that one criterion for completeness is that a set $\{\Phi_n\}$ is complete if and only if $\langle \Phi_n, f \rangle = 0$ for all n implies $f = 0$. Suppose

$$\int_a^b \int_c^d f(x,y)\Phi_i(x)\Psi_j(y)dy\, dx = 0$$

for all i, j. Then

$$\int_a^b \left(\int_c^d f(x,y)\Psi_j(y)dy \right) \Phi_i(x)dx = 0$$

for all i, so that $\int_c^d f(x,y)\psi_j(y)dy = 0$ for all j and almost everywhere in x, since $\{\Phi_i\}$ forms a complete set. Since $\{\Psi_i\}$ also forms a complete set, this in turn implies that $f(x,y) = 0$ almost everywhere on the rectangle $[a,b] \times [c,d]$ which verifies completeness. It is an easy matter to verify that $\{\Phi_n(x)\Psi_m(y)\}_{m,n=1}^\infty$ is an orthonormal set.

Suppose $\{\phi_i(x)\}$ is a complete orthonormal set on $L^2[a,b]$. We can represent the kernel $k(x,y)$ as

$$k(x,y) = \sum_{i=1}^\infty \sum_{j=1}^\infty a_{ij}\phi_i(x)\phi_j(y)$$

where $a_{ij} = \int_a^b \int_a^b k(x,y)\phi_i(x)\phi_j(y)dx\, dy$ are the Fourier coefficients for the kernel $k(x,y)$. From Parseval's equality,

$$\sum_{i=1}^\infty \sum_{j=1}^\infty |a_{ij}|^2 = \int_a^b \int_a^b k^2(x,y)dx\, dy < \infty.$$

We define the degenerate kernel $k_n(x, y)$ by

$$k_n(x, y) = \sum_{i=1}^{n} \sum_{j=1}^{n} a_{ij} \phi_i(x) \phi_j(y)$$

which approximates $k(x, y)$ in $L^2([a, b] \times [a, b])$. As a result the operator K_n with kernel $k_n(x, y)$ approximates K since

$$\|Ku - K_n u\|^2 = \left\| \int_a^b \left(\sum_{i=n+1}^{\infty} \sum_{j=1}^{\infty} + \sum_{i=1}^{n} \sum_{j=n+1}^{\infty} \right) a_{ij} \phi_i(x) \phi_j(y) u(y) dy \right\|^2$$

$$\leq \left(\sum_{i=n+1}^{\infty} \sum_{j=1}^{\infty} + \sum_{i=1}^{n} \sum_{j=n+1}^{\infty} \right) |a_{ij}|^2 \|u\|^2$$

by the Schwarz inequality. Certainly this becomes arbitrarily small for large enough n.

3.4   Spectral Theory for Compact Operators

Degenerate operators are equivalent to matrix operators. To see this, suppose we wish to solve $u(x) = f(x) + \lambda K_n u$ where K_n is the degenerate operator

$$K_n u(x) = \int_a^b \sum_{i=1}^{n} \phi_i(x) \psi_i(y) u(y) dy.$$

From the form of this equation it is apparent that $u(x) = f(x) + \lambda \sum_{i=1}^{n} \alpha_i \phi_i(x)$ where $\alpha_i = \int_a^b \psi_i(y) u(y) dy$ are unknown constants. Taking the inner product of u with $\psi_j(x)$ we obtain

$$\alpha_j = \langle u, \psi_j \rangle = \langle f, \psi_j \rangle + \lambda \sum_{i=1}^{n} \alpha_i \langle \phi_i, \psi_j \rangle .$$

If we let α be the vector $\alpha = (\alpha_i)$, $A = (a_{ij})$ be the matrix with entries $a_{ij} = \langle \phi_j, \psi_i \rangle$ and $f = (f_i)$, $f_j = \langle f(x), \psi_j \rangle$, then this equation has the matrix representation

$$(I - \lambda A) \alpha = f,$$

which, if we can solve for α, gives us the solution $u(x)$.

From Chapter 1, we know all about solutions of this matrix problem and in particular, we know the important role that eigenvalues of A play. The main goal of this section is to determine how to generalize the spectral theory for matrices to infinite dimensional linear compact operators. The reason this can be done for compact operators is that, as we have already seen, the compacting action of the operator tends to minimize the effect of "most" of the possible independent directions in the vector space. As a result the difference between the action of a finite dimensional operator and the action of a compact operator is "small."

Definition

An eigenvalue, eigenvector pair is the pair (μ, u), μ a scalar, $u \in H, u \neq 0$, satisfying the equation $Ku = \mu u$.

For a compact linear operator K, eigenvalues and eigenvectors have a number of important properties:

1. *The multiplicity of any eigenvalue $\mu \neq 0$ is finite.* That is, the number of linearly independent eigenfunctions for any eigenvalue $\mu \neq 0$ is finite.
   If not, there is an infinite set of functions $\{\phi_i\}_{i=1}^{\infty}$ which can be made orthonormal (via the Gram-Schmidt procedure) for which $K\phi_i = \mu\phi_i$. Since K is compact, $\lim_{i\to\infty} K\phi_i = 0$, but if $\mu \neq 0$ the set $\{\mu\phi_i\}$ is not convergent, so we have a contradiction.

2. Since the compact linear operator K is a bounded operator, *the adjoint operator K^* exists and is bounded.* We say that an operator K is *self adjoint* if $Ku = K^*u$ for all u. For example, if K is a Hilbert-Schmidt operator, the adjoint (using the real L^2 inner product with weight function $\omega(x) = 1$) is given by $K^*u(x) = \int_a^b k(y, x)u(y)dy$, and is self-adjoint if and only if $k(x, y) = k(y, x)$.

3. *If K is self-adjoint, the eigenvalues are real.* We use the complex inner product and note that

$$\begin{aligned} \mu\langle u, u\rangle &= \langle Ku, u\rangle = \overline{\langle u, Ku\rangle} \\ &= \overline{\langle K^*u, u\rangle} = \overline{\langle Ku, u\rangle} = \bar{\mu}\langle u, u\rangle. \end{aligned}$$

4. *If K is self-adjoint, eigenfunctions corresponding to distinct eigenvalues are orthogonal.* Suppose $Ku = \mu_1 u$ and $Kv = \mu_2 v$ with $\mu_1 \neq \mu_2$. Then $\mu_1\langle u, v\rangle = \langle Ku, v\rangle = \langle u, Kv\rangle = \mu_2\langle u, v\rangle$ which implies that $\langle u, v\rangle = 0$ if $\mu_1 \neq \mu_2$.

5. *If K is self-adjoint, the number of distinct nonzero eigenvalues is either finite or $\lim_{n\to\infty} \mu_n = 0$.* For each distinct μ_n there are at most a finite number of eigenfunctions. Let $u_n, \|u_n\| = 1$ be one of the possibly many eigenfunctions for μ_n. The sequence $\{u_n\}$ is an orthonormal sequence since $\mu_n \neq \mu_m$ for $n \neq m$. Since K is compact, if there are an infinite number of μ_n, $0 = \lim_{n\to\infty} \|Ku_n\|^2 = \lim_{n\to\infty} \mu_n^2\|u_n\|^2 = \lim_{n\to\infty} \mu_n^2$.

6. (*Maximum Principle*) *Every nontrivial self-adjoint compact operator has a nontrivial eigenpair* (μ_1, ϕ_1) *where*

$$|\mu_1| = \max_{\|u\|=1} |\langle Ku, u \rangle| = \|K\|,$$

and the maximum is attained by ϕ_1.

We define $M = \underset{\|u\| \neq 0}{l.u.b.} \frac{|\langle Ku, u \rangle|}{\|u\|^2}$. We will show first that $M \leq \|K\|$ and then that $M \geq \|K\|$. To establish an upper bound on M, note that, by the Schwarz inequality, $|\langle Ku, u \rangle| \leq \|Ku\| \cdot \|u\|$ so that $\frac{|\langle Ku, u \rangle|}{\|u\|^2} \leq \frac{\|Ku\|}{\|u\|} \leq \|K\|$. It follows that $M \leq \|K\|$. For Hilbert-Schmidt operators, recall that $\|K\|^2 \leq \int_a^b \int_a^b k^2(x, y) dx \, dy < \infty$.

To prove that $M \geq \|K\|$ we observe that

$$\langle K(u+v), u+v \rangle - \langle K(u-v), u-v \rangle = 2 \langle Ku, v \rangle + 2 \langle Kv, u \rangle.$$

Since $\langle K(u+v), u+v \rangle \leq M\|u+v\|^2$ and $\langle K(u-v), (u-v) \rangle \geq -M\|u-v\|^2$, it follows that

$$2 \langle Ku, v \rangle + 2 \langle Kv, u \rangle \leq M \left(\|u+v\|^2 + \|u-v\|^2 \right) = 2M \left(\|u\|^2 + \|v\|^2 \right).$$

If we set $v = \frac{Ku}{\|Ku\|} \|u\|$ and substitute, we find that $4\|Ku\| \, \|u\| \leq 4M\|u\|^2$ so that $\frac{\|Ku\|}{\|u\|} \leq M$ for all $u \neq 0$. In other words $\|K\| \leq M$. Comparing this with the previous inequality, we conclude that $\|K\| = M$.

Since $M = \|K\|$, there is a sequence of normalized functions $\{z_n\}, \|z_n\| = 1$ for which $\lim_{n \to \infty} |\langle Kz_n, z_n \rangle| = \|K\|$, and there is therefore a subsequence $\{x_n\}$ for which $\langle Kx_n, x_n \rangle$ converges to either $\|K\|$ or $-\|K\|$. Define $\mu_1 = \lim_{n \to \infty} \langle Kx_n, x_n \rangle$. Then

$$\begin{aligned} \|Kx_n - \mu_1 x_n\|^2 &= \|Kx_n\|^2 - 2\mu_1 \langle Kx_n, x_n \rangle + \mu_1^2 \|x_n\|^2 \\ &\leq \|K\|^2 - 2\mu_1 \langle Kx_n, x_n \rangle + \mu_1^2 \to 0 \end{aligned}$$

as $n \to \infty$. This means that $\lim_{n \to \infty} (Kx_n - \mu_1 x_n) = 0$.

Since K is compact, Kx_n contains a convergent subsequence, say $\{Ky_n\}$, which converges to some function. Since $Ky_n - \mu_1 y_n$ converges to zero, it follows that y_n converges to some function ϕ_1, and since K is continuous, $\mu_1 \phi_1 = \lim_{n \to \infty} Ky_n = K \lim_{n \to \infty} y_n = K\phi_1$. (The interchange of limits is valid for a continuous operator. See Exercise 3.4.5.)

7. *A compact self adjoint operator* K *is either degenerate or else it has an infinite number of mutually orthogonal eigenfunctions.* To see this we proceed inductively, and suppose we have found $n-1$ eigenpairs of $K, (\mu_i, \phi_i), i = 1, 2, \ldots, n-1$ with the ϕ_i mutually orthonormal. We define a new compact operator

$$K_n u = Ku - \sum_{k=1}^{n-1} \mu_k \phi_k(x) \langle u, \phi_k \rangle.$$

If K is an integral operator, the kernel corresponding to K_n is $k_n(x, y) = k(x, y) - \sum_{k=1}^{n-1} \mu_k \phi_k(x) \phi_k(y)$. The following properties of K_n are immediate:

a. If $\langle u, \phi_i \rangle = 0$ for $i = 1, 2, \ldots, n-1$, then $K_n u = K u$.

b. If $u = \sum_{i=1}^{n-1} \alpha_i \phi_i$, then $K_n u = 0$, since $K \phi_i = \mu_i \phi_i$.

c. Since $\langle K_n u, \phi_i \rangle = 0, i = 1, 2, \ldots, n-1$, the range of K_n is orthogonal to the span of $\{\phi_i\}, i = 1, 2, \ldots, n-1$.

d. If $K_n \psi = \mu \psi$ then ψ is in the range of K_n, orthogonal to ϕ_i for $i = 1, 2, \ldots, n-1$ and $K \psi = K_n \psi = \mu \psi$ so that (μ, ψ) is an eigenpair for K as well.

We can now apply property 6 (the maximum principle) to the reduced operator K_n to find another eigenpair. We learn that if K_n is nontrivial, it has a nontrivial eigenpair (μ_n, ϕ_n) where

$$|\mu_n| = \max_{u \neq 0} \frac{|\langle K_n u, u \rangle|}{\|u\|^2} = \|K_n\|$$

and the maximum is attained by $u = \phi_n$. Furthermore, the new eigenfunction ϕ_n is orthogonal to the previous eigenfunctions $\phi_i, i = 1, 2, \ldots n-1$, and, of course, it can be normalized.

e. (Maximum Principle for self-adjoint compact operators.) The nth eigenpair (μ_n, ϕ_n) of K can be characterized by

$$|\mu_n| = \max_{\substack{u \neq 0 \\ \langle u, \phi_j \rangle = 0 \\ j=1,2,\ldots,n-1}} \frac{|\langle K u, u \rangle|}{\|u\|^2} \leq |\mu_{n-1}|.$$

The maximum is attained by $u = \phi_n$ and $\langle \phi_n, \phi_i \rangle = 0$ for $i = 1, 2, \ldots, n-1$. Thus one of two possibilities occurs. Either K is degenerate, in which case

$$K u = \sum_{i=1}^{n} \mu_i \phi_i(x) \langle u, \phi_i(x) \rangle$$

for some n, or else K has an infinite orthonormal set of eigenfunctions $\{\phi_i\}$ with corresponding eigenvalues $\mu_i, |\mu_1| \geq |\mu_2| \geq \ldots$, and $\lim_{n \to \infty} \mu_n = 0$.

Now comes the main result of this section.

Theorem 3.6

If K is a compact self adjoint linear operator, the set $\{\phi_i\}$ of its orthonormal eigenfunctions is complete over the range of K. That is, every function in the range of K can be expanded in a Fourier series in terms of the set $\{\phi_i\}$.

To see this, suppose $u_n = f - \sum_{i=1}^{n-1} \alpha_i \phi_i$, $\alpha_i = \langle f, \phi_i \rangle$ for any f in H. Since $\langle u_n, \phi_i \rangle = 0$ for $i = 1, 2, \ldots n - 1$,

$$\|K u_n\|^2 = \|K_n u_n\|^2 \le |\mu_n|^2 \|u_n\|^2 \le |\mu_n|^2 \|f\|^2,$$

so either $K u_n = 0$ for n large enough or else $\lim_{n \to \infty} K u_n = 0$.

If g is in the range of K, there is an f so that $Kf = g$, and

$$0 = \lim_{n \to \infty} \left\| K \left(f - \sum_{i=1}^{n-1} \alpha_i \phi_i \right) \right\| = \lim_{n \to \infty} \left\| g - \sum_{i=1}^{n-1} \alpha_i \mu_i \phi_i \right\|.$$

It follows that

$$g = \sum_{i=1}^{\infty} \beta_i \phi_i \quad \text{where} \quad \beta_i = \mu_i \langle f, \phi_i \rangle = \langle f, K\phi_i \rangle = \langle Kf, \phi_i \rangle = \langle g, \phi_i \rangle,$$

i.e. g is represented as a Fourier series of ϕ_i.

Example

For any 2π-periodic function $k(t)$ in $L^2[0, 2\pi]$, let $Ku = \int_0^{2\pi} k(x-t)u(t)dt$. The eigenfunctions and eigenvalues of K are easily found. We try $\phi_n(x) = e^{inx}$ and observe that

$$K\phi_n = \int_0^{2\pi} k(x - t)e^{int}dt = e^{inx} \int_{x-2\pi}^{x} k(s)e^{-ins}ds = \mu_n \phi_n$$

where

$$\mu_n = \int_0^{2\pi} k(s)e^{-ins}ds$$

because of the periodicity of $k(s)$. We already know that the functions $\{e^{inx}\}_{n=-\infty}^{\infty}$ are complete for functions in $L^2[0, 2\pi]$. It is interesting to note, however, that K is self adjoint only if $k(s) = k(-s)$ for all s, although its eigenfunctions are complete even when K is not self adjoint.

Example

From Section 3.1 we know that the boundary value problem $u'' + \lambda u = 0$ with $u(0) = u(1) = 0$ is equivalent to the integral equation

$$u(x) = -\lambda \int_0^1 k(x,y)u(y)dy = \lambda K u$$

where

$$k(x,y) = \begin{cases} y(x-1) & 0 \le y < x \le 1 \\ x(y-1) & 0 \le x < y \le 1 \end{cases}$$

is a symmetric, continuous kernel. The eigenfunctions of this operator are the functions $\{\sin n\pi x\}_{n=1}^{\infty}$ and are therefore complete over the range of K. The range of K in this case is the set of continuous functions $u(x)$ with $u(0) = u(1) = 0$.

This example illustrates the technique by which we will show, in Chapter 4, that eigenfunctions of many differential equations form complete sets.

3.5   Resolvent and Pseudo-Resolvent Kernels

If it exists, a RESOLVENT OPERATOR is the inverse operator for an integral equation

$$u(x) - \lambda \int_a^b k(x,t)u(t)dt = f(x)$$

or more generally, for $(I - \lambda K)u = f$, where K is a compact operator. The Fredholm Alternative theorem tells us when the resolvent operator exists.

Theorem 3.7

If K is a compact linear operator, then $(I - \lambda K)u = f$ has a solution if and only if f is orthogonal to the null space of $(I - \lambda K)^*$.

Recall that if a bounded linear operator L has closed range, then $H = R(L) + N(L^*)$, or equivalently, the equation $Lf = g$ has a solution if and only if $\langle g, v \rangle = 0$ for every v in H with $L^*v = 0$.

To show that the Fredholm alternative theorem holds for operators of the form $L = I - \lambda K$ where K is compact, we must show that L has closed range (the operator L is clearly bounded). This we do in two steps: First we show that $\|Lu\| \ge c\|u\|$ for some fixed $c > 0$ for every u in the orthogonal complement of $N(L)$, then we show that this implies the range of L is closed.

First, suppose there are functions $\{u_i\}$, with u_i in the orthogonal complement of $N(L)$, $\|u_i\| = 1$, for which $\lim_{i \to \infty} \|Lu_i\| = 0$. Then $Lu_i = u_i - \lambda K u_i$ converges to zero. Since K is compact, $K u_i$ has a convergent subsequence, say $\{K v_i\}$, and of course $\{v_i\}$ converges since $K v_i - v_i/\lambda \to 0$. If $\lim_{i \to \infty} v_i = w$, then $\lim_{i \to \infty} K v_i = K w$ and $K w = w/\lambda$ so that w is in $N(L)$. The orthogonal complement of $N(L)$ is a closed set so $\lim_{i \to \infty} v_i = w$ is in both $N(L)$ and its orthogonal complement, i.e. $w = 0$. But $\|v_i\| = 1$ so that $\|w\| = 1$, a contradiction, and $\frac{Lu}{\|u\|}$ must be bounded away from zero.

To see that the range of L is closed, we let $\{f_n\}$ be a sequence in $R(L)$ which converges to some element $f \in H$. Since f_n is in the range of L, there are functions u_n so that $Lu_n = f_n$. Let v_n be the projection of u_n onto the orthogonal complement of $N(L)$. The sequence $\{v_n\}$ is a Cauchy sequence since $Lv_n = f_n$ and

$$\|v_n - v_m\| \leq \frac{1}{c}\|L(v_n - v_m)\| = \frac{1}{c}\|f_n - f_m\|.$$

and therefore $\{v_n\}$ converges in H, $\lim_{n \to \infty} v_n = v$. We now know that $Lv_n \to f$ and $v_n \to v$, and since L is a bounded operator $Lv = f$ so that f is in the range of L, and the range of L is therefore closed.

For a self-adjoint operator, we can solve $(I - \lambda K)u = f$ explicitly in terms of the eigenfunctions of K. If K is self adjoint and the normalized eigenfunctions of K are $\{\phi_n\}$, since $g(x) = u(x) - f(x)$ must be in the range of K, we can expand $g(x)$ in terms of the eigenfunctions of K as

$$g(x) = \sum_{i=1}^{\infty} \alpha_i \phi_i, \quad \alpha_i = \langle g, \phi_i \rangle.$$

Using that $u = g + f$, we find from the integral equation that

$$\sum_{i=1}^{\infty} \alpha_i (1 - \lambda \mu_i)\phi_i(x) = \lambda K f.$$

Since $K f$ is in the range of K, it has an expansion

$$K f = \sum_{i=1}^{\infty} \beta_i \phi_i \quad \text{where} \quad \beta_i = \mu_i \langle f, \phi_i \rangle.$$

It follows that $\alpha_i (1 - \lambda \mu_i) = \lambda \beta_i = \lambda \mu_i \langle f, \phi_i \rangle$, whence

$$
\begin{aligned}
u(x) &= f(x) + \lambda \sum_{i=1}^{\infty} \frac{\mu_i \phi_i(x) \langle f, \phi_i \rangle}{1 - \lambda \mu_i} \\
&= f(x) + \lambda \int_a^b r(x, t; \lambda) f(t)\, dt = f + \lambda R f.
\end{aligned}
$$

The kernel

$$r(x,t;\lambda) = \sum_{i=1}^{\infty} \frac{\mu_i \phi_i(x)\phi_i(t)}{1 - \lambda\mu_i}$$

is called the RESOLVENT KERNEL and the operator R is called the RESOLVENT OPERATOR. One can show (Problem 3.4.6) that $u(x)$ is in $L^2[a,b]$ if $f(x)$ is in $L^2[a,b]$.

There is an important point to make here that may have slipped by unnoticed. When we make the guess $u = f + \sum_{i=1}^{\infty} \alpha_i \phi_i(x)$ we are introducing a change of representation for u, or equivalently a change of coordinate system. The net result of this change of coordinates is that in the new system, the equation for the component α_i of ϕ_i makes no reference to other components α_j for $j \neq i$. In other words, we have diagonalized an infinite dimensional problem by using as coordinates the eigenfunctions of the operator, which we know form a complete set, or basis, for the range space. This is another illustration of the commuting diagram of transform theory shown first in Chapter 1. The only difference between this transform and those of Chapter 1 is that now the bases are infinite.

We see (as is also true for matrices) that the "best" coordinate system for a linear operator is the set of eigenfunctions for that operator, and, if they form a complete set, they diagonalize the operator equation into separated algebraic equations.

Notice that the existence of the resolvent operator R is dependent upon the value of λ. If $\lambda = \frac{1}{\mu_k}$ for some k, the resolvent operator fails to exist. This should come as no surprise, since $\lambda = \frac{1}{\mu_k}$ implies that the homogeneous problem $f = 0$ has the nontrivial solution $u = \phi_k$. We also see an immediate verification of the Fredholm Alternative, namely if $\lambda = \frac{1}{\mu_k}$ for some k, then a solution exists if and only if $\langle \phi_k, f \rangle = 0$, in which case the kth term (or terms) drop out of the sum for the solution. Of course, if $\lambda = \frac{1}{\mu_k}$, the inverse is not unique since arbitrary amounts of the null space may be added to the solution.

In general, the least squares solution is that function $u(x)$ which is orthogonal to the null space of the operator and which has as its image the orthogonal projection of f onto the range of the operator.

From this definition of the least squares solution it is clear how to find a pseudo-inverse operator for a self-adjoint integral equation. In case $\lambda = \frac{1}{\mu_k}$ for some k, the kth coefficient α_k cannot be determined and the best one can do is to solve exactly the projected equation

$$u(x) = f(x) - \sum_{\mu_j \stackrel{=}{} \mu_k} \langle f, \phi_j \rangle \, \phi_j(x) + \lambda K u.$$

in which case the smallest solution is

$$u(x) = f(x) + \sum_{\mu_j \neq \mu_k} \frac{\mu_j \phi_j(x) \langle f, \phi_j \rangle}{\mu_k - \mu_j} - \sum_{\mu_j = \mu_k} \phi_j(x) \langle f, \phi_j \rangle$$

$$= f(x) + \lambda \hat{R} f$$

The operator $\hat{R}$ is the "pseudo"-resolvent operator for this problem, and it exists since we have taken care to project out those eigenfunctions corresponding to eigenvalue μ_k.

Example

Find the pseudo-resolvent operator for the equation

$$Lu = u - 2 \int_0^1 su(s)ds = f(t).$$

This integral operator is not self adjoint. The null space of L is spanned by the function $\phi = $ constant and the null space of L^* is spanned by $\psi = t$. The least square solution must satisfy

$$u - 2 \int_0^1 su(s)ds = f(t) - 3t \int_0^1 sf(s)ds$$

and

$$\int_0^1 u\,ds = 0.$$

It is a straightforward calculation to show that

$$u(t) = f(t) + \int_0^1 \left(\frac{3s}{2} - 3ts - 1 \right) f(s)ds$$

is the pseudo-resolvent solution.

3.6   Approximate Solutions

In practice it is often the case that the eigenfunctions of nondegenerate operators K are not known or are difficult to find, or, if the operator is not self adjoint, the completeness of the eigenfunctions (if they even exist, see Problem 3.4.2) may not be guaranteed. The only reasonably general solution technique for nondegenerate kernels is to try to find a differential equation that is equivalent to the integral equation and then to find eigenfunctions of the differential equation. For example, by direct differentiation (using Leibniz' rule) one can easily show that

$$\lambda u(x) = \int_0^\pi \min(x,t)u(t)dt$$

is equivalent to the differential equation $u'' = -\frac{1}{\lambda}u$ with boundary conditions $u(0) = u'(\pi) = 0$, from which one can find the eigenpairs.

When explicit calculations fail an ITERATIVE SOLUTION TECHNIQUE can be quite useful. The idea is to guess an approximate solution and then try to improve it. For example, a reasonable, but crude, guess is that the solution of $u(x) = f(x) + \lambda(Ku)(x)$ looks like $u_0(x) = f(x)$. Of course, this is too crude to be acceptable but we might try to correct it by taking

$$u_1(x) = f(x) + \lambda K f = f(x) + \lambda K u_0.$$

Continuing in this way we take

$$u_n = f + \lambda K u_{n-1}.$$

If this scheme is convergent we can stop at some point and have a reasonable approximate solution. These iterates are generally referred to as NEUMANN ITERATES.

To see that these iterates actually converge to a solution of our problem we must prove

Theorem 3.8 (Contraction Mapping Principle)

If M is a continuous (not necessarily linear) operator on H with the properties that

1. M maps a closed and bounded set S into itself,

2. $\|Mx - My\| \leq \delta \|x - y\|$ for all $x, y \in S$, and for some δ, $0 \leq \delta < 1$, and if the sequence $\{x_n\}$ is generated by the iteration $x_{n+1} = Mx_n$, $x_0 \in S$, then $\lim_{n\to\infty} x_n$ exists and the limiting object $x = \lim_{n\to\infty} x_n$ satisfies the equation $x = Mx$.

To prove the Contraction Mapping Principle, we show that the sequence $\{x_n\}$ is a Cauchy sequence. We examine $\|x_m - x_n\|$ with $m > n$

$$\|x_m - x_n\| = \left\|\sum_{k=1}^{m-n}(x_{n+k} - x_{n+k-1})\right\| \leq \sum_{k=1}^{m-n}\|x_{n+k} - x_{n+k-1}\|$$

$$\leq \sum_{k=0}^{m-n-1}\delta^k\|x_{n+1} - x_n\| \leq \frac{\delta^n}{1-\delta}\|x_1 - x_0\|.$$

Since x_0 and x_1 both lie in S, they are bounded by α, say, and

$$\|x_m - x_n\| \leq \frac{2\alpha\delta^n}{1 - \delta}$$

which is arbitrarily small with n sufficiently large and $\delta < 1$. Since $\{x_n\}$ is a Cauchy sequence it has a limit in the space H (assumed to be complete), call it x.

Since M is a continuous operator

$$\|Mx_n - Mx\| \leq \|M\|\,\|x_n - x\| \to 0$$

as $n \to \infty$ so that

$$x = \lim_{n\to\infty} x_{n+1} = \lim_{n\to\infty} Mx_n = Mx$$

as required.

To invoke the contraction mapping principle for Neumann iterates, we must show that iterates are uniformly bounded and contracting. First, notice that since $u_n = f + \lambda K u_{n-1}$,

$$\|u_n\| \leq \|f\| + |\lambda|\,\|K\|\,\|u_{n-1}\|$$

and if $\|u_{n-1}\| \leq \alpha$ then $\|u_n\| \leq \alpha$ provided $\alpha = \frac{\|f\|}{1-|\lambda|\|K\|} > 0$. Since $u_0 = f, \|u_0\| = \|f\| < \alpha$, it follows that $\|u_n\| \leq \alpha$ for all n, provided $|\lambda|\,\|K\| < 1$.

Second, since $u_n - u_{n-1} = \lambda K(u_{n-1} - u_{n-2})$,

$$\|u_n - u_{n-1}\| \leq |\lambda|\,\|K\|\,\|u_{n-1} - u_{n-2}\| \leq \delta\|u_{n-1} - u_{n-2}\|$$

where $\delta = |\lambda|\,\|K\|$. Thus, if $|\lambda|\,\|K\| < 1$, we have a contraction map, and the iterates converge to the solution with geometric rate no worse than $|\lambda|\,\|K\|$. Furthermore, we have actually proven that solutions of the integral equation exist for all λ with $|\lambda| < 1/\|K\|$.

The restriction $|\lambda| < 1/\|K\|$ is a guarantee that $1/\lambda$ is larger than all the eigenvalues of K. If some of the eigenfunctions of K are known, this requirement can be weakened by projecting out those eigenfunctions. In general, however, $1/\lambda$ should be kept far away from the eigenvalues of K for an iterative technique to work with reasonable speed.

As a simple example of the usefulness of Neumann iterates consider the integral equation

$$u(x) = f(x) + \int_0^x u(t)dt.$$

The most direct way to solve this equation is to differentiate it and solve the resulting differential equation. However, to see that Neumann iterates also work, we take $u_0(x) = f(x)$ and

$$u_n(x) = f(x) + \int_0^x u_{n-1}(t)dt.$$

It follows that

$$u_1(x) = f(x) + \int_0^x f(t)dt$$

$$u_2(x) = f(x) + \int_0^x f(t)dt + \int_0^x (x-t)f(t)dt$$

(using integration by parts). We are led to guess that

$$u_n(x) = f(x) + \int_0^x \sum_{k=0}^{n-1} \frac{(x-t)^k}{k!} f(t)dt$$

which we then verify to be correct. That is,

$$
\begin{aligned}
u_{n+1}(x) &= f(x) + \int_0^x f(t)dt + \int_0^x \left(\int_0^t \sum_{k=0}^{n-1} \frac{(t-s)^k}{k!} f(s)ds \right) dt \\
&= f(x) + \int_0^x f(t)dt + \int_0^x \left(\int_s^x \sum_{k=0}^{n-1} \frac{(t-s)^k}{k!} dt \right) f(s)ds \\
&= f(x) + \int_0^x \sum_{k=0}^{n} \frac{(x-t)^k}{k!} f(t)dt.
\end{aligned}
$$

Of course, letting n approach infinity, we have

$$u(x) = \lim_{n \to \infty} u_n(x) = f(x) + \int_0^x e^{(x-t)} f(t)dt$$

which is therefore the resolvent operator for this equation.

The second class of approximate solution techniques are the Galerkin schemes. We suppose P_n is a bounded linear projection (P_n is a projection if $P_n^2 = P_n$) that projects the Hilbert space H onto an n-dimensional subspace M_n of H. The Galerkin approximation for the equation $u - \lambda K u = f$ is given by

$$(I - \lambda P_n K)u_n = P_n f.$$

That is, we project f and the compact operator K onto the subspace M_n and solve the resulting finite dimensional problem for u_n.

The Galerkin approximation is useful if we can be assured it converges. To that end we have

Theorem 3.9

If $\{P_n\}$ is a sequence of projections with the property that $\|P_n u - u\| \to 0$ as $n \to \infty$ for every $u \in H$, and if $(I - \lambda K)^{-1}$ exists, then u_n, the solution of $(I - \lambda P_n K) u_n = P_n f$, converges to the solution of $(I - \lambda K)u = f$ as $n \to \infty$.

We omit the proof of this theorem although it is not difficult to give. The key step is showing that the operator $(I - \lambda P_n K)^{-1}$ exists for all n sufficiently large and that $(I - \lambda P_n K)^{-1}$ is close to $(I - \lambda K)^{-1}$ in the operator norm.

As an example of how to make use of this theorem we consider finding an approximate solution of the integral equation

$$u(x) + \int_0^1 k(x, \xi) u(\xi) d\xi = f(x)$$

using the piecewise linear finite elements. We define $\phi_k(x)$ to be the piecewise linear continuous function with $\phi_k(x_j) = \delta_{kj}$ and linear on all the intervals $[x_j, x_{j+1}]$ where $x_j = j/N$. The projection of any continuous function $f(x)$ onto this set of basis functions is given by

$$(P_N f)(x) = \sum_{j=0}^N f(x_j) \phi_j(x).$$

Since the continuous functions are dense in $L^2[0, 1]$, we are assured that $\|P_N f - f\| \to 0$ for every $f \in L^2[0, 1]$ as $N \to \infty$. Now we write the Galerkin approximation as

$$u_n(x) + \sum_{j=0}^N \left(\int_0^1 k(x_j, \xi) u_n(\xi) d\xi \right) \phi_j(x) = \sum_{j=0}^N f(x_j) \phi_j(x),$$

and the resulting equation has a degenerate kernel which can, in principle, be solved using simple linear algebra. The approximate solution $u_n(x)$ will, of course, be piecewise linear, but as N gets large, we are assured the solution will get "close" to the exact solution. Once again it is clear that the quality of the approximation depends strongly on the choice of projection operator. The choices of projection operator are vast, ranging from the finite element projections to orthogonal projections based on the L^2 or Sobolev inner products, using orthogonal sets of basis functions, such as described in Chapter 2.

3.7 Singular Integral Equations

Although they are not central to this text, the singular integral equations derived in Section 3.1 represent an important class of integral equations. We will sketch briefly how to solve some of these equations.

We consider first Abel's equation

$$T(x) = \int_0^x \frac{f(y) dy}{\sqrt{x - y}}.$$

The trick to solving this equation is to multiply both sides of the equation by $\frac{dx}{\sqrt{\eta-x}}$ and integrate from $x = 0$ to $x = \eta$. If we do this, the right-hand side becomes

$$\int_0^\eta \frac{dx}{\sqrt{\eta-x}} \int_0^x \frac{f(y)dy}{\sqrt{x-y}} = \int_0^\eta f(y) \left(\int_y^\eta \frac{dx}{\sqrt{\eta-x}\sqrt{x-y}} \right) dy,$$

after interchanging the order of integration. We can evaluate

$$I = \int_y^\eta \frac{dx}{\sqrt{\eta-x}\sqrt{x-y}}$$

by making the clever trigonometric substitution $x = y \cos^2\theta + \eta \sin^2\theta$, from which we learn that $I = \pi$. As a result our equation has become

$$\int_0^\eta \frac{T(x)dx}{\sqrt{\eta-x}} = \pi \int_0^\eta f(x)dx.$$

It follows that

$$f(x) = \frac{1}{\pi} \frac{d}{dx} \int_0^x \frac{T(y)}{\sqrt{x-y}} dy.$$

This is a valid solution provided $T(x)$ is continuous, $T(0) = 0$ (why?) and $T'(x)$ has at most a finite number of jump discontinuities.

A similar technique can be used to solve the tomography problem

$$f_n(\rho) = 2 \int_\rho^1 \frac{g_n(r)T_n(\rho/r)r\,dr}{(r^2-\rho^2)^{1/2}}.$$

We assume that $f_n(\rho)$ is known and seek to determine $g_n(r)$. We multiply both sides of this equation by

$$T_n(\rho/z)(z/\rho)/(\rho^2-z^2)^{1/2}$$

and integrate from $\rho = z$ to $\rho = 1$ to obtain

$$\int_z^1 z\frac{T_n(\rho/z)f_n(\rho)}{\rho\sqrt{\rho^2-z^2}} d\rho = 2 \int_z^1 g_r(r) \left(\int_z^r \frac{zrT_n(\rho/r)T_n(\rho/z)d\rho}{\rho\sqrt{r^2-\rho^2}\sqrt{\rho^2-z^2}} \right) dr$$

after having interchanged the order of integration.

The problem now is to evaluate

$$I_n(r,z) = 2 \int_z^r \frac{zrT_n(\rho/r)T_n(\rho/z)d\rho}{\rho\sqrt{r^2-\rho^2}\sqrt{\rho^2-z^2}}.$$

To evaluate I_1, note that for $n = 1, T_1(x) = x$ from which it follows that

$$I_1(r,z) = 2 \int_z^r \frac{\rho\,d\rho}{\sqrt{r^2-\rho^2}\sqrt{\rho^2-z^2}}$$

which we recognize from above as having $I_1 = \pi$. To evaluate I_0, we use that $T_0(x) = 1$ so that

$$I_0(r,z) = 2 \int_z^r \frac{rz\,d\rho}{\rho\sqrt{r^2 - \rho^2}\sqrt{\rho^2 - z^2}}.$$

With the change of variables $\rho^2 = z^2 \cos^2\theta + r^2 \sin^2\theta$, we learn that

$$I_0(r,z) = 4rz \int_0^{\pi/2} \frac{d\theta}{(z^2 + r^2) + (z^2 - r^2)\cos 2\theta}$$

which can be evaluated using contour integral techniques (see Problem 6.4.19) as

$$I_0 = \pi.$$

In fact, one can use the same technique to verify that $I_n = \pi$, for all n, so that

$$\pi \int_z^1 g_n(r)\,dr = \int_z^1 \frac{zT_n(\rho/z)f_n(\rho)\,d\rho}{\rho\sqrt{\rho^2 - z^2}}$$

or

$$g_n(r) = -\frac{1}{\pi}\frac{d}{dr} \int_r^1 r\frac{T_n(\rho/r)f_n(\rho)\,d\rho}{\rho\sqrt{\rho^2 - z^2}}.$$

Further Reading

The primary reason we have included integral equations in this text is because the Hilbert-Schmidt theory and theory of compact operators has a direct analogy with matrix theory. Some books that discuss these theoretical aspects further include the classical works

1. W. V. Lovitt, *Linear Integral Equations*, Dover, New York, 1950,

2. F. G. Tricomi, *Integral Equations*, John Wiley and Sons, New York, 1957,

and the more recent works

3. A. W. Naylor and G. R. Sell, *Linear Operator Theory in Engineering and Science*, Holt, Rinehart and Winston, New York, 1971,

4. H. Hochstadt, *Integral Equations*, Wiley-Interscience, New York, 1973.

Numerical solutions of integral equations are described in

5. K. E. Atkinson, *A Survey of Numerical Methods for the Solution of Fredholm Integral Equations of the Second Kind*, SIAM, Philadelphia, 1976.

Many of the problems we discuss in later chapters are initially formulated as differential equations, but can be converted to integral equations to facilitate their analysis. Some examples of problems that begin life as integral equations are the population model and the tomography problem discussed in the first section. For further discussion of population dynamics models as integral equations see

6. F. Hoppensteadt, *Mathematical Theories of Populations: Demographics, Genetics and Epidemics*, SIAM, Philadelphia, 1975.

The papers in which Cormack published his initial work on tomography are

7. A. M. Cormack, Representation of a Function by its Line Integrals, with Some Radiological Applications, J. Appl. Phys. **34**, 2722–2727 (1963), and J. Appl. Phys. **35**, 2908–2912 (1964).

For a more extensive account of the tomography problem, see

8. F. Natterer, *The Mathematics of Computerized Tomography*, Wiley, New York, 1986.

A similar integral equation problem arises from the study of seismic wave propagation in a stratified medium, described in

9. K. Aki and P. G. Richards, *Quantitative Seismology, Theory and Methods*: Vol. 2, Freeman, 1980.

Problems for Chapter 3

Section 3.1

1. Verify that the solution of $\frac{d^2u}{dx^2} = f(x), u(0) = 0, u(1) = 0$ is given by $u(x) = \int_0^1 k(x,y)f(y)dy$ where

$$k(x,y) = \begin{cases} y(x-1) \ 0 \le y < x \le 1 \\ x(y-1) \ 0 \le x < y \le 1. \end{cases}$$

Section 3.2

1. Show that $Tf = f(0)$ is not a bounded linear functional on the space of continuous functions measured with the L^2 norm, but it is a bounded linear functional if measured using the uniform norm.

2. Show that the null space of a bounded linear functional is closed.

When do the following integral equations have solutions?

3. $u(x) = f(x) + \lambda \int_0^{1/2} u(t)dt.$

4. $u(x) = f(x) + \lambda \int_0^1 xtu(t)dt.$

5. $u(x) = f(x) + \lambda \int_0^{2\pi} \sum_{j=1}^n \frac{\cos jt \cos jx}{j} u(t)dt.$

6. $u(x) = f(x) + \lambda \int_{-1}^1 \sum_{j=0}^n P_j(x)P_j(t)u(t)dt$ where $P_j(x)$ is the jth Legendre polynomial.

Section 3.3

1. Find solutions of

$$u(x) - \lambda \int_0^{2\pi} \sum_{j=1}^n \frac{\cos jt \cos jx}{j} u(t)dt = \sin^2 x \quad n \geq 2.$$

for all values of λ. Find the resolvent kernel for this equation. (Find the least squares solution if necessary.)

2. Find the solutions of

$$u(x) - \lambda \int_{-1}^1 \sum_{j=0}^n P_j(x)P_j(t)u(t)dt = x^2 + x$$

for all values of λ where $P_j(x)$ is the jth Legendre polynomial. (Find the least squares solution if necessary.)

Remark: Recall that the Legendre polynomials are normalized to have $\int_{-1}^1 P_j^2(t)dt = \frac{2}{2j+1}.$

3. Show that any subset of a (sequentially) compact set is (sequentially) compact.

Section 3.4

1. a. Find the eigenfunctions for the integral operator

$$Ku = \int_0^1 k(x,\xi)u(\xi)d\xi$$

where

$$k(x,\xi) = \begin{cases} x(1-\xi) & 0 \leq x < \xi \leq 1 \\ \xi(1-x) & 0 \leq \xi < x \leq 1. \end{cases}$$

b. Find the expansion of

$$f(x) = \begin{cases} x/2 & 0 \le x < 1/2 \\ \frac{1-x}{2} & 1/2 < x \le 1, \end{cases}$$

in terms of the eigenfunctions of K. Is there a solution of $Ku = f$?

2. Find the eigenvalues and eigenfunctions for the integral operator $Ku = \int_0^\pi k(x,\xi)u(\xi)d\xi$ where

a. $k(x,\xi) = x\xi$.

b. $k(x,\xi) = \sin x \sin \xi + \alpha \cos x \cos \xi$.

c. $k(x,\xi) = \sum_{n=1}^\infty \frac{\sin(n+1)x \sin n\xi}{n^2}$.
   Remark: The existence of eigenfunctions is only guaranteed for self-adjoint operators.

d. $k(x,\xi) = \min(x,\xi)$.

3. Find the eigenvalues and eigenfunctions for the integral operator

$$Ku = \int_{-1}^1 k(x,\xi)u(\xi)d\xi$$

$$k(x,\xi) = 1 - |x - \xi|.$$

4. Find the adjoint kernel for an integral operator when the inner product is

$$\langle u, v \rangle = \int_a^b u(x)v(x)w(x)dx, w(x) > 0.$$

5. Suppose $\{u_n\}$ is a convergent sequence in H and K is a bounded (continuous) linear operator on H. Use the definition of convergence to show that $\lim_{n\to\infty}(Ku_n) = K(\lim_{n\to\infty} u_n)$.

6. a. Suppose K is a self-adjoint compact operator with normalized eigenfunctions $\{\phi_i\}_{i=1}^\infty$, $K\phi_i = \mu_i\phi_i$. Suppose $f = \sum_{i=1}^\infty \alpha_i\phi_i, \sum_{i=1}^\infty \alpha_i^2 < \infty$. Show that $u - \lambda Ku = f$ has a solution $u = \sum_{i=1}^\infty \beta_i\phi_i$ and that $\sum_{i=1}^\infty \beta_i^2 < \infty$ provided $\lambda\mu_i \ne 1$ for $i = 1, 2, \ldots$.

   b. Suppose $f \in L^2[a, b]$ and K is a self-adjoint compact operator on $L^2[a, b]$. Show that if $u - \lambda Ku = f$, then $u \in L^2[a, b]$.

Section 3.5

Show that the following are equivalent to differential equations, find the resolvent (or pseudo-resolvent) operator, and solve the integral equation.

1. $u(x) = 1 + \int_0^x u(t)dt$.

2. $u(x) = 1 + \int_0^x (t - x) u(t) dt.$

3. $u(x) = 1 + x + \int_0^x (x - t) u(t) dt.$

Find resolvent (or pseudo-resolvent) kernels and solve the following integral equations.

4. $u(x) = x + \lambda \int_0^{1/2} u(t) dt.$

5. $u(x) = \frac{5x}{6} + \frac{1}{2} \int_0^1 xtu(t) dt.$

6. $u(t) = p(t) + \int_{t_0}^t a(t) b(s) u(s) ds.$

Section 3.6

1.   a. How many terms in the Neumann series

$$u_n = \sum_{j=0}^n \lambda^j K^j f$$

must be used to guarantee that $\|u - u_n\| \le \epsilon$, where $u = \lim_{n \to \infty} u_n$. Hint: Show that $\|u_n - u\| \le \frac{\|\lambda K\|^{n+1} \|f\|}{1 - \|\lambda K\|}$

b. Use your estimate to calculate the number of terms needed in the Neumann series to obtain $\|u - u_n\| < .1$ for the integral equation

$$u(x) + \int_0^1 x\xi u(\xi) d\xi = x.$$

Calculate the first few iterates and compare with the exact solution.

2. Under what conditions will the iteration procedure

$$x_{n+1} = x_n - f(x_n)$$

converge to a root of $f(x) = 0$?

3. When will the iteration procedure

$$x_{n+1} = x_n - \frac{f(x_n)}{f'(x_n)} \quad \text{(Newton's method)}$$

converge to a root of $f(x) = 0$?

4. Invent an iteration scheme to solve matrix equations $Ax = b$. (Write $A = D + R$ where D is diagonal.) Use Neumann iterates and estimate the error of the iterations after n iterates. Carry out the details for $A = \begin{pmatrix} 4 & 1 \\ 0 & 3 \end{pmatrix}$ for three iterates and compare these results with the exact solution.

5. Use the Neumann iterates to solve Problem 3.5.2 (sum the series).

6. a. Show that the solution of

$$u(t) = 1 + \int_0^t s \ln(s/t) u(s) ds$$

satisfies the differential equation

$$u'' + \frac{u'}{t} + u = 0 \quad u(0) = 1, u'(0) = 0.$$

b. Use Neumann iterates to show that

$$u(t) = \sum_{k=0}^{\infty} \frac{(-1)^k (t/2)^{2k}}{(k!)^2}.$$

Remark: This function is $J_0(t)$, the zeroth order Bessel function.

7. Suppose $\{\phi_j(x)\}_{j=1}^{\infty}$ are a mutually orthogonal set of basis functions for $L^2[a, b]$. What is the matrix equation that results from the Galerkin approximation for the equation $u - \lambda K u = f$ in terms of the functions ϕ_j? Does the Galerkin procedure converge? Is there any noticeable advantage or disadvantage of orthogonal polynomials over interpolating functions (finite elements)?

Section 3.7

1. The time of descent of a sliding bead on a frictionless wire is a constant independent of initial height. Find the shape of the wire.
Answer: A cycloid.

2. Find the shape of the wire if the descent time of a sliding bead is

a.

$$T(y) = \frac{1}{\alpha} \sqrt{\frac{2y}{g}}, \quad \alpha < 1$$

b.

$$T(y) = \frac{4}{3\sqrt{2g}} y^{3/2}$$

3. Snell's law states that in a medium where the speed of propagation of a wave depends only on depth, the ratio $p = \cos \theta(z)/c(z)$ along the trajectory of a wave (a ray) is constant, where $\theta(z)$ is the angle the ray makes with a horizontal line, and $c(z)$ is the speed of propagation as a function of depth z.

b. Find expressions for the horizontal distance traveled $x(p)$ and the time of travel for a wave traveling between the surface $z = 0$ and its maximum depth $z(p)$. Show that

$$\frac{x(p)}{p} = 2 \int_0^{z(p)} \frac{c(z)dz}{\sqrt{1 - c^2(z)p^2}}$$

is the point on the surface to which a wave returns after penetrating to its maximum depth.

c. Find the inverse speed function $z = z(c)$ by solving the above integral equation

Hint: Let $y = 1/c^2(z)$.

Answer:

$$z(c) = -\frac{1}{\pi} \int_{1/c_0}^{1/c} \frac{cx(p)dp}{\sqrt{c^2 p^2 - 1}}.$$

d. Suppose the arrival time on the surface as a function of distance x from the source of an acoustic wave is

$$T(x) = \frac{2}{b} \ln \left(\sqrt{\frac{b^2 x^2}{4a^2} + 1} + \frac{bx}{2a} \right).$$

Find the speed of travel as a function of depth $c = c(z)$.

Answer: $c = a + bz$.

4

DIFFERENTIAL OPERATORS AND GREEN'S FUNCTIONS

4.1 Distributions and the Delta Function

When solving matrix problems, the idea of an inverse matrix operator is very important. For a matrix A, the inverse A^{-1} is a matrix for which $AA^{-1} = I$ so that the solution of $Ax = b$ is given by $x = A^{-1}b$.

For a differential operator L, it is reasonable to expect that the inverse operator L^{-1} is an integral operator, say

$$L^{-1}\phi = \int_a^b g(x,t)\phi(t)dt.$$

For example, we already know that the solution of the differential equation $Lu \equiv u'' = \phi(x)$, with $u(0) = u(1) = 0$ is given by

$$u(x) \;=\; \int_0^1 g(x,y)\phi(y)dy$$

where

$$g(x,y) = \begin{cases} y(x-1) & 0 \le y < x \le 1 \\ x(y-1) & 0 \le x < y \le 1. \end{cases}$$

If the inverse operator is indeed an integral operator, we expect that

$$\phi(x) = L\left(L^{-1}\phi\right) = \int_a^b Lg(x,t)\phi(t)dt.$$

In other words, the kernel g should satisfy the equation $Lg = \delta(x,t)$ where the function δ satisfies

$$\int_a^b \delta(x,t)\phi(t)dt = \phi(x),$$

for all reasonable choices of $\phi(x)$.

This requirement for the function $\delta(x,t)$ is quite disheartening because we quickly realize that no such function exists. Suppose, for example, that $\delta(x,t)$ is positive and continuous on some open interval. If we pick $\phi(t)$ to be positive on this same interval except at $t = x$, and zero elsewhere, then clearly $\int_{-\infty}^{\infty} \phi(t)\delta(x,t)dt \ne \phi(x) = 0$. Thus, we expect $\delta(x,t) = 0$ except possibly at $x = t$. On the other hand, if the function $\delta(x,t)$ differs from zero only at the one point $x = t$ (a set of measure zero) then

$$\int_a^b \delta(x,t)\phi(t)dt = 0$$

for any continuous function $\phi(t)$.

It is often the case in science and engineering courses that we are told that the δ-function $\delta(x,t)$ is defined by

$$\delta(x,t) = 0 \qquad x \ne t$$

$$\int_{-\infty}^{\infty} \delta(x,t)dt = 1.$$

However, for any reasonable definition of integration there can be no such function. How is it, then, that many practicing scientists have used this definition with impunity for years?

The goal of this section is to make mathematical sense of delta-"functions." There are, traditionally, two ways to do so, the first through DELTA SEQUENCES and the second through the THEORY OF DISTRIBUTIONS.

The idea of delta sequences is to realize that, although a delta-function cannot exist, we might be able to find a sequence of functions $\{S_k(x)\}$ which in the limit $k \to \infty$ satisfies the defining equation

$$\lim_{k \to \infty} \int_{-\infty}^{\infty} S_k(t)\phi(t)dt \;=\; \phi(0)$$

for all continuous functions $\phi(x)$. If this is true, then, by a shift of origin

$$\lim_{k \to \infty} \int_{-\infty}^{\infty} S_k(t-x)\phi(t)dt \;=\; \phi(x).$$

Notice that this definition of S_k in no way implies that $\lim_{k \to \infty} S_k(x)$ exists. In fact, we are certainly *not* allowed to interchange the limit process with integration. (Recall that one criterion that allows the interchange of limit and integration is given by the Lebesgue dominated convergence theorem in Chapter 2, which will not hold for delta sequences.)

There are a number of suggestive examples of delta sequences. The choice

$$S_k(x) \;=\; \begin{cases} k & -1/2k \le x \le 1/2k \\ 0 & |x| > 1/2k \end{cases}$$

is a delta sequence, since

$$\int_{-\infty}^{\infty} S_k(t)\phi(t)dt \;=\; k \int_{-1/2k}^{1/2k} \phi(t)dt \to \phi(0) \quad \text{as} \quad k \to \infty$$

because of the mean value theorem, provided $\phi(x)$ is continuous near $x = 0$.

Another example is $S_k(x) = \frac{1}{\pi}\frac{k}{1+k^2x^2}$. In fact, if $f(x)$ is non-negative with $\int_{-\infty}^{\infty} f(t)dt = 1$, then the sequence of functions $\{S_k(x)\}$ with $S_k(x) = kf(kx)$ is a delta sequence.

To prove this fact, suppose $\phi(x)$ is a continuous, bounded function on the infinite interval $-\infty < x < \infty$. Then

$$\int_{-\infty}^{\infty} S_k(x)\phi(x)dx \;=\; k \int_{-\infty}^{\infty} f(kx)\phi(x)dx$$

$$=\; \phi(0) + \int_{-\infty}^{\infty} f(x)\eta\left(\frac{x}{k}\right) dx,$$

where $\eta(x) = \phi(x) - \phi(0)$. Now,

$$\left| \int_{-\infty}^{\infty} f(x)\eta\left(\frac{x}{k}\right) dx \right| \le \sup |\eta(x)| \quad \left[\int_{-\infty}^{-A} f(x)dx + \int_{A}^{\infty} f(x)dx \right]$$

$$+\; \left| \int_{-A}^{A} f(x)\eta\left(\frac{x}{k}\right) dx \right|$$

We can pick A sufficiently large so that $\int_A^\infty f(x)dx$ and $\int_{-\infty}^{-A} f(x)dx$ are arbitrarily small. For A fixed and large, we can pick k large enough so that $|\eta(x/k)|$ is also arbitrarily small on the interval $-A \le x \le A$. It follows, by picking both A and k large, that

$$\left| \int_{-\infty}^\infty f(x)\eta\left(\frac{x}{k}\right) dx \right|$$

is arbitrarily small, so our result is established.

Another delta sequence which is quite important in the theory of Fourier series and Fourier transforms is

$$S_k(x) = \frac{\sin kx}{\pi x}.$$

($S_k(x)$ is also related to Sinc functions.) Notice that $S_k(x)$ is of the form $kf(kx)$ although since $S_k(x)$ is not of one sign, the previous proof does not apply. To verify that $\{S_k(x)\}$ forms a delta-sequence suppose that $\phi(x)$ is Lipschitz continuous on the interval $0 \le x \le b$ and that $\lim_{x\to 0+} \phi(x) = \phi(0^+)$ exists. We will demonstrate that

$$\lim_{k\to\infty} \int_0^b \phi(x)\frac{\sin kx}{x}dx = \frac{\pi}{2}\phi(0^+).$$

We examine the two integrals,

$$I_1 = \phi(0)\int_0^b \frac{\sin kx}{x}dx,$$

$$I_2 = \int_0^b (\phi(x) - \phi(0))\frac{\sin kx}{x}dx.$$

The first integral I_1 cannot be evaluated explicitly, however, using complex variable techniques (see Chapter 6), one can show that

$$\lim_{k\to\infty} \int_0^b \frac{\sin kx}{x}dx = \lim_{k\to\infty} \int_0^{kb} \frac{\sin x}{x}dx = \int_0^\infty \frac{\sin x}{x}dx = \frac{\pi}{2}.$$

Since $\phi(x)$ is Lipschitz continuous on $0 \le x \le b$ (i.e., $|\phi(x) - \phi(y)| \le K|x-y|$) it follows that

$$\frac{\phi(x) - \phi(0^+)}{x}$$

is a bounded, hence L^2, function on $0 \le x \le b$. Because of Bessel's inequality, Fourier coefficients of an L^2 function approach zero as $k \to \infty$, and I_2 can be expressed in terms of the kth Fourier coefficient of some L^2 function. Hence, $\lim_{k\to\infty} I_2 = 0$.

With a bit more work, we conclude that if $\phi(x)$ is continuous, and Lipschitz continuous on both sides of the origin, and if the limits from the left $\phi(0^-)$ and from the right $\phi(0^+)$ exist, then

$$\lim_{k \to \infty} \frac{1}{\pi} \int_{-\infty}^{\infty} \phi(x) \frac{\sin kx \, dx}{x} = \frac{\phi(0^+) + \phi(0^-)}{2}.$$

We cannot remove the requirement of Lipschitz continuity with impunity. In fact, there are continuous functions for which

$$\lim_{k \to \infty} \int_{-\infty}^{\infty} f(x) \frac{\sin kx}{x} dx$$

is any number we want, not $f(0)$.

Delta sequences turn out to be cumbersome to actually use. The idea, if it could be implemented, would be to solve the sequence of problems

$$Lg_k(x,t) = S_k(x - t)$$

to find a sequence of approximate inverse operators. Then, taking the limit $\lim_{k \to \infty} g_k(x,t) = g(x,t)$ we could produce the correct inverse for the operator L. There are a number of significant difficulties with this, however, since we not only need to find $g_k(x,t)$, but also show that the limit exists independent of the particular choice of delta sequence. A much more useful way to discuss delta-functions is through the theory of distributions. As we shall see, distributions provide a generalization of functions and inner products.

Definition

A TEST FUNCTION is a function $\phi(x) \in C^{\infty}(-\infty, \infty)$ with compact support. That is, $\phi(x)$ is continuously differentiable infinitely often, and $\phi(x) = 0$ for all $|x|$ sufficiently large. The closure of the set of x for which $\phi(x) \neq 0$ is called the *support* of ϕ.

The set of test functions forms a linear vector space, which we will call D. One example of a test function is

$$\phi(x) = \begin{cases} \exp\left(\frac{1}{x^2-1}\right) & |x| < 1 \\ 0 & |x| \geq 1 \end{cases}$$

so the set of test functions is not empty.

Definition

A LINEAR FUNCTIONAL t on D is a real number $t(\phi)$ which is defined for all ϕ in D and is linear in ϕ.

The notation that we use for linear functionals in D is $t(\phi) \equiv \langle t, \phi \rangle$. We will refer to $\langle t, \phi \rangle$ as the action of the linear functional t on ϕ. In this notation, linearity means that

$$\langle t, \alpha\phi_1 + \beta\phi_2 \rangle = \alpha \langle t, \phi_1 \rangle + \beta \langle t, \phi_2 \rangle$$

for ϕ_1, ϕ_2 in D and α, β real scalars. The notational representation of t looks exactly like an inner product between functions t and ϕ, and this similarity is intentional although misleading, since t need not be representable as a true inner product.

The simplest examples of linear functionals are indeed inner products. Suppose $f(x)$ is a locally integrable, that is, the Lebesgue integral $\int_I |f(x)|\, dx$ is defined and bounded for every finite interval I. Then the inner product

$$\langle f, \phi \rangle = \int_{-\infty}^{\infty} f(x)\phi(x)dx$$

is a linear functional if ϕ is in D.

Definition

The set $\{\phi_n\}$ of test functions is called a ZERO SEQUENCE if

1. $\cup_n \{\text{support}\,(\phi_n)\}$ is bounded.
2. $\lim_{n\to\infty} \left(\max_x \left|\frac{\partial^k \phi_n}{\partial x^k}\right|\right) = 0$, for each $k = 0, 1, 2, \ldots$.

Definition

A linear functional is *continuous* in D if and only if $\langle t, \phi_n \rangle \to 0$ for all zero sequences $\{\phi_n\}$. A continuous linear functional is called a DISTRIBUTION.

Every locally integrable function f induces a distribution through the usual inner product. If $\langle f, \phi \rangle = \int_{-\infty}^{\infty} f(x)\phi(x)dx$, then

$$|\langle f, \phi_n \rangle| \leq \int_{-\infty}^{\infty} |f|\,|\phi_n|\, dx \leq M_n \int_I |f(x)|\, dx \to 0,$$

where $M_n = \max_x |\phi_n(x)|$ and $I = \cup_n \{\text{support}\,(\phi_n)\}$. It follows that two locally integrable functions which are the same almost everywhere, induce the same distribution.

One important distribution is the HEAVISIDE DISTRIBUTION

$$\langle H, \phi \rangle = \int_0^{\infty} \phi(x)dx$$

which is equivalent to the inner product of ϕ with the well-known Heaviside function

$$H(x) = \begin{cases} 1 & x > 0 \\ 0 & x \le 0. \end{cases}$$

For any function f that is locally integrable we can interchangeably refer to its function values $f(x)$ or to its distributional values (or action)

$$\langle f, \phi \rangle = \int_{-\infty}^{\infty} f(x)\phi(x)dx.$$

There are numerous distributions which are not representable as an inner product. The example we have already seen is the distribution

$$\langle \delta, \phi \rangle = \phi(0)$$

which we call the delta-distribution. It is clear that this is a distribution. To see that this cannot be represented as an inner product let

$$\phi_a(x) = \exp\left(\frac{a^2}{x^2 - a^2}\right), \qquad |x| < a, \qquad \text{and} \qquad \phi_a(x) = 0$$

elsewhere. Clearly $\phi_a(0) = \max_x |\phi_a(x)| = 1/e$ which is independent of a. For any locally integrable function $f(x)$

$$\left| \int_{-\infty}^{\infty} f(x)\phi_a(x)dx \right| \le \frac{1}{e}\int_{-a}^{a} |f(x)|\, dx \to 0 \qquad \text{as} \qquad a \to 0,$$

which is not $\phi_a(0)$. Therefore, there is no locally integrable function $f(x)$ for which $\int_{-\infty}^{\infty} f(x)\phi(x)dx = \phi(0)$ for every test function $\phi(x)$.

The fact that some continuous linear functionals on D exist which do not correspond to inner products should be compared with the Riesz Representation Theorem (Chapter 3) which states that any bounded linear functional on L^2 can be represented as an inner product with some L^2 function. Notice that the functional $t(u) = u(0)$ defined for all $u \in L^2$ is not a bounded functional. In fact, it is not even well defined in L^2 since $u(0)$ can be changed arbitrarily without changing the function $u(x)$ in L^2, and so the Riesz Representation Theorem does not apply here.

We say that a distribution is REGULAR if it can be represented as an inner product with some locally integrable function f, and it is SINGULAR if no such function exists. In addition to the examples already mentioned, some singular distributions are

1. $\langle \delta_\zeta, \phi \rangle = \phi(\zeta)$... DELTA DISTRIBUTION

2. $\langle \Delta, \phi \rangle = \phi'(0)$... DIPOLE DISTRIBUTION.

The dipole distribution is frequently represented as $\Delta = -\delta'$ since $\phi'(0) = \lim_{\varepsilon \to 0} \frac{\phi(\varepsilon) - \phi(-\varepsilon)}{2\varepsilon} = \lim_{\varepsilon \to 0} \frac{1}{2\varepsilon}(\langle \delta_{-\varepsilon}, \phi \rangle - \langle \delta_\varepsilon, \phi \rangle)$ but since δ_ζ is not a function it is certainly not differentiable in the usual sense. We shall define the derivative of δ momentarily.

One often sees the notation $\int_{-\infty}^{\infty} \delta(x - t)\phi(t)dt = \phi(x)$. It should always be kept in mind that this is simply a notational device to represent the distribution $\langle \delta_x, \phi \rangle$ and is in no way meant to represent an actual integral or that $\delta(x - t)$ is an actual function. The notation $\delta(x - t)$ is a "SYMBOLIC FUNCTION" for the delta distribution.

The correct way to view δ_x is as an operator on the set of test functions. We should never refer to pointwise values of δ_x since it is not a function, but an operator on functions. The operation $\langle \delta_x, \phi \rangle = \phi(0)$ makes perfectly good sense and we have violated no rules of integration or function theory to make this definition.

The fact that some operators can be viewed as being generated by functions through normal integration should not confuse the issue. δ_x is not such an operator. Another operator that is operator valued but not pointwise valued is the operator (not a linear functional) $L = d/dx$. We know that d/dx cannot be evaluated at the point $x = 3$ for example, but d/dx can be evaluated pointwise only after it has first acted on a differentiable function $u(x)$. Thus, $du/dx = u'(x)$ can be evaluated at $x = 3$, only after the operand $u(x)$ is known. Similarly, $\langle \delta_x, \phi \rangle$ can be evaluated only after ϕ is known.

Although distributions are not always representable as integrals, their properties are nonetheless always defined to be consistent with the corresponding property of inner products. The following are some properties of distributions that result from this association.

1. If t is a distribution and $f \in C^\infty$ then ft is a distribution whose action is defined by $\langle ft, \phi \rangle = \langle t, f\phi \rangle$. For example, if f is continuous $f(x)\delta = f(0)\delta$. If f is continuously differentiable at 0, $f\delta' = -f'(0)\delta + f(0)\delta'$. This follows since

$$\langle f\delta', \phi \rangle = \langle \delta', f\phi \rangle = -(f\phi)' \big|_{x=0} = -f'(0)\phi(0) - f(0)\phi'(0).$$

2. Two distributions t_1 and t_2 are said to be equal on the interval $a < x < b$ if for all test functions ϕ with support in $[a, b]$, $\langle t_1, \phi \rangle = \langle t_2, \phi \rangle$. Therefore it is often said (and this is unfortunately misleading) that $\delta(x) = 0$ for $x \neq 0$.

3. The usual rules of integration are always assumed to hold. For example, by change of scale $t(\alpha x)$ we mean

$$\langle t(\alpha x), \phi \rangle = \frac{1}{|\alpha|} \left\langle t, \phi\left(\frac{x}{\alpha}\right) \right\rangle$$

and the shift of axes $t(x - \xi)$ is taken to mean

$$\langle t(x - \xi), \phi \rangle \ = \ \langle t, \phi(x + \xi) \rangle,$$

even though pointwise values of t may not have meaning. It follows for example that $\delta(x - \xi) = \delta_\xi$ and $\delta(ax) = \delta(x)/|a|$.

4. The derivative t' of a distribution t is defined by $\langle t', \phi \rangle = -\langle t, \phi' \rangle$ for all test functions $\phi \in D$. This definition is natural since, for differentiable functions

$$\langle f', \phi \rangle = \int_{-\infty}^{\infty} f'(x)\phi(x)dx = -\int_{-\infty}^{\infty} f(x)\phi'(x)dx = -\langle f, \phi' \rangle.$$

Since $\phi(x)$ has compact support, the integration by parts has no boundary contributions at $x = \pm\infty$.

If t is a distribution, then t' is also a distribution. If $\{\phi_n\}$ is a zero sequence in D, then $\{\phi'_n\}$ is also a zero sequence, so that

$$\langle t', \phi_n \rangle \ = \ -\langle t, \phi'_n \rangle \to 0 \qquad \text{as} \qquad n \to \infty.$$

It follows that for any distribution t, the nth distributional derivative $t^{(n)}$ exists and its action is

$$\left\langle t^{(n)}, \phi \right\rangle \ = \ (-1)^n \left\langle t, \phi^{(n)} \right\rangle.$$

Thus any L^2 function has distributional derivatives of all orders.

Examples

1. The Heaviside distribution $\langle H, \phi \rangle = \int_0^\infty \phi(x)dx$ has derivative

$$\langle H', \phi \rangle = -\int_0^\infty \phi'(x)dx = \phi(0)$$

since ϕ has compact support, so that $H' = \delta_0$.

2. The derivative of the δ-distribution is

$$\langle \delta', \phi \rangle = -\langle \delta, \phi' \rangle = -\phi'(0)$$

which is the negative of the dipole distribution.

3. Suppose f is continuously differentiable except at the points $x_1, x_2, \ldots, x_n$ at which f has jump discontinuities $\Delta f_1, \Delta f_2, \ldots, \Delta f_n$, respectively. Its distribution has derivative given by

$$\langle f', \phi \rangle \ = \ -\langle f, \phi' \rangle = -\int_{-\infty}^{\infty} f(x)\phi'(x)dx$$

$$= -\left\{\int_{-\infty}^{x_1} f(x)\phi'(x)dx \right.$$

$$+ \left. \int_{x_1}^{x_2} f(x)\phi'(x)dx + \cdots + \int_{x_n}^{\infty} f(x)\phi'(x)dx \right\}$$

$$= \int_{-\infty}^{\infty} \frac{df}{dx}\phi(x)dx + \sum_{k=1}^{n} \Delta f_k \phi(x_k).$$

It follows that the distributional derivative of f is

$$f' = \frac{df}{dx} + \sum_{k=1}^{n} \Delta f_k \delta_{x_k},$$

where df/dx is the usual calculus derivative of f, wherever it exists.

4. For $f(x) = |x|$, the distributional derivative of f has action

$$\langle f', \phi \rangle = -\langle f, \phi' \rangle = -\int_{-\infty}^{\infty} |x|\phi'(x)dx$$

$$= \int_{-\infty}^{0} x\phi'(x) - \int_{0}^{\infty} x\phi'(x)dx$$

$$= -\int_{-\infty}^{0} \phi(x)dx + \int_{0}^{\infty} \phi(x)dx$$

$$= -\int_{-\infty}^{\infty} \phi(x)dx + 2\int_{0}^{\infty} \phi(x)dx$$

so that $f' = -1 + 2H(x)$, and $f'' = 2\delta_0$.

Definition

A sequence of distributions $\{t_n\}$ is said to converge to the distribution t if their actions converge in $\mathbb{R}$, that is, if

$$\langle t_n, \phi \rangle \rightarrow \langle t, \phi \rangle \quad \text{for all} \quad \phi \quad \text{in} \quad D.$$

This convergence is called convergence in the sense of distribution or WEAK CONVERGENCE.

If the sequence of distributions t_n converges to t then the sequence of derivatives t'_n converges to t'. This follows since

$$\langle t'_n, \phi \rangle = -\langle t_n, \phi' \rangle \rightarrow -\langle t, \phi' \rangle = \langle t', \phi \rangle$$

for all ϕ in D.

Example

The sequence $\{t_n\} = \left\{\frac{\cos nx}{n}\right\}$ is both a sequence of functions and a sequence of distributions. As $n \to \infty$, t_n converges to 0 both as a function (pointwise) and as a distribution. It follows that $t'_n = -\sin nx$ converges to the zero distribution even though the pointwise limit is not defined.

Using distributions, we are able to generalize the concept of function and derivative to many objects which previously made no sense in the usual definitions. It is also possible to generalize the concept of a differential equation.

Definition

The differential equation $Lu = f$ is a differential equation in the sense of distribution (i.e., in the weak sense), if f and u are distributions and all derivatives are interpreted in the sense of distribution. Such a differential equation is called the WEAK FORMULATION of the differential equation.

Examples

1. To solve the equation $du/dx = 0$ in the sense of distribution, we seek a distribution u for which $\langle u, \phi' \rangle = 0$ for all test functions $\phi \in D$. This rather mundane equation has only the constant distribution as its solutions. Although this result is quite predictable the proof is unusual. We want to determine the action $\langle u, \phi \rangle$ on any $\phi \in D$ knowing that the action of u is $\langle u, \psi \rangle = 0$ whenever ψ is the derivative of a test function. Said another way, knowing the action of u on a restricted set of test functions, we want to determine the action of u on all test functions. We first show that $\psi = \phi'$ for some $\phi \in D$ if and only if $\int_{-\infty}^{\infty} \psi \, dx = 0$. Certainly if $\psi = \phi'$ then $\int_{-\infty}^{\infty} \psi \, dx = 0$ since ϕ has compact support. In the opposite direction, if $\int_{-\infty}^{\infty} \psi \, dx = 0$, then $\phi = \int_{-\infty}^{x} \psi \, dx$ is a test function, being smooth and having compact support.
Now comes the trick. Fix $\phi_0(x)$ to be any test function with $\int_{-\infty}^{\infty} \phi_0(x) dx = 1$. Any test function ϕ can be expressed as a linear combination of ϕ_0 and the derivative of another test function by

$$\phi(x) = \phi_0(x) \int_{-\infty}^{\infty} \phi(x) dx + \left(\phi(x) - \phi_0(x) \int_{-\infty}^{\infty} \phi(x) dx \right)$$

$$= \phi_0(x) \int_{-\infty}^{\infty} \phi(x) dx + \psi(x),$$

where by design, $\int_{-\infty}^{\infty} \psi(x)dx = 0$. Since $\langle u, \psi \rangle = 0$, we can calculate the action of u on any test function, to find that $\langle u, \phi \rangle = \langle u, \phi_0 \rangle \int_{-\infty}^{\infty} \phi(x)dx$. In other words, the action of u on any test function is the same as the action of the constant $\langle u, \phi_0 \rangle$ on the test function. Hence u is the constant distribution.

2. A slightly less trivial example is the differential equation $x\frac{du}{dx} = 0$. Interpreted in the sense of distribution, we seek a distribution u whose action satisfies $\langle u, (x\phi)' \rangle = 0$ or $\langle u, \psi \rangle = 0$ for any test function of the form $\psi = (x\phi)'$. It is straightforward (see Problem 4.1.4) to show that $\psi = (x\phi)'$ if and only if $\int_{-\infty}^{\infty} \psi\, dx = 0$ and $\int_0^{\infty} \psi\, dx = 0$. Suppose we pick test functions ϕ_0 and ϕ_1 to satisfy

$$\int_0^{\infty} \phi_0 dx = 1, \qquad \int_{-\infty}^{\infty} \phi_0 dx = 0$$

and

$$\int_0^{\infty} \phi_1 dx = 0, \qquad \int_{-\infty}^{\infty} \phi_1 dx = 1.$$

We can write any test function ϕ as a linear combination of the functions ϕ_0, ϕ_1 and a remainder which has the form of ψ, as

$$\phi = \phi_0(x) \int_0^{\infty} \phi\, dx + \phi_1(x) \int_{-\infty}^{\infty} \phi\, dx + \psi(x)$$

where

$$\psi(x) = \phi(x) - \phi_0(x) \int_0^{\infty} \phi\, dx - \phi_1(x) \int_{-\infty}^{\infty} \phi\, dx.$$

Using that $\langle u, \psi \rangle = 0$ we determine the action of u on any test function ϕ to be

$$\langle u, \phi \rangle = \langle u, \phi_0 \rangle \int_0^{\infty} \phi(x)dx + \langle u, \phi_1 \rangle \int_{-\infty}^{\infty} \phi\, dx.$$

We identify the action of u as that of $u = c_1 + c_2 H(x)$ where c_1 and c_2 are arbitrary constants. Notice, now, that the weak solution of this differential equation is not the same as the strong solution, since the Heaviside function is not a solution of the strong formulation.

The distributions we have defined in this section have as their "domain" the set of C^{∞} test functions. It is not necessary to always use C^{∞} test functions but merely C^n test functions with n some non-negative integer. The domain of the resulting distributions is larger, but the new class of distributions is smaller. For example, a distribution defined on C^{∞} test functions has derivatives of all orders, however, a distribution defined on C^n test functions has its

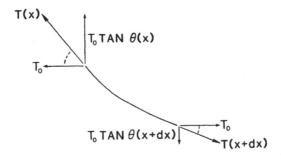

FIGURE 4.1 Forces on a string under tension.

derivative defined on the C^{n-1} test functions. Thus, we can define the delta distribution on the continuous test functions to be, as usual, $\langle \delta, \phi \rangle = \phi(0)$, which makes sense. However, if the class of test functions ϕ are not differentiable, the derivative of δ cannot be defined.

4.2   Green's Functions

We now address the question of how to find an inverse operator for a differential operator. Suppose we define the nth order linear differential operator L by $Lu = a_n(x)\frac{d^n u}{dx^n} + \cdots + a_1(x)\frac{du}{dx} + a_0(x)u$. It was suggested in the previous section that the inverse operator L^{-1} should be an integral operator whose kernel g satisfies the equation $Lg = \delta(x-t)$. We now realize that this equation must be interpreted as a differential equation in the sense of distribution. That is, we seek a function $g(x,t)$ for which the generalized operation Lg satisfies $\langle Lg, \phi \rangle = \phi(t)$. To see that this can be made to work we take as an example the problem of finding the shape of a hanging weighted string.

Suppose a string is stretched between points $x = 0$ and $x = 1$ and pinned at those points at the level $u = 0$. For each point $x \in [0,1]$ we measure the vertical displacement from zero by $u(x)$. Since the string is in static equilibrium, the vertical and horizontal forces must be in balance.

At any point x, suppose a tangent to the string makes an angle $\theta(x)$ with the horizontal. For a string, forces are transmitted only along tangential directions. Let $T(x)$ be the force in the tangential direction at the point x (see Figure 4.1). Since horizontal forces must balance, $T(x)\cos\theta(x) = T_0$, a constant independent of x. For a small segment of string between points x and $x + dx$, the vertical component of tension must match the gravitational force, i.e., $T(x+dx)\sin\theta(x+dx) - T(x)\sin\theta(x) = \rho(x)g\,ds$, where $\rho(x)$ is the density per unit length of the string, ds is the arc length of the segment between x and $x + dx$, and g is the gravitation constant. It follows that

$$T_0 \frac{\tan \theta(x + dx) - \tan \theta(x)}{dx} = \rho(x)g \frac{ds}{dx}$$

or, taking the limit $dx \to 0$,

$$T_0 \frac{d}{dx} \tan \theta(x) = \rho(x)g \frac{ds}{dx}.$$

Recognizing that $\tan \theta(x) = \frac{du}{dx}$ where $u(x)$ is the vertical displacement of the string, we have

$$T_0 \frac{d^2u}{dx^2} = \rho(x)g \left(1 + \left(\frac{du}{dx} \right)^2 \right)^{1/2}.$$

If we neglect du/dx as being small relative to 1, the shape of the string is governed by the equation

$$\frac{d^2u}{dx^2} = f(x), \qquad u(0) = u(1) = 0,$$

where $f(x) = \rho(x)g/T_0$.

It is easy enough to solve this equation by direct integration

$$\frac{du}{dx} = \int_0^x f(t)dt + \alpha,$$

and

$$u = \int_0^x (x - t)f(t)dt + \alpha x + \beta,$$

where the last expression comes after integration by parts. To satisfy boundary conditions we must choose $\beta = 0$ and $\alpha = -\int_0^1 (1 - t)f(t)dt$, so that $u(x) = \int_0^1 g(x,t)f(t)dt$ where $g(x,t) = (x - t)H(x - t) - x(1 - t)$ for $0 \leq x, t \leq 1$. Observe that we have indeed found a kernel $g(x,t)$ for our proposed inverse operator. The function g is continuous, but only piecewise differentiable. Its derivative with respect to x is, in the generalized sense,

$$g_x(x,t) = H(x - t) - (1 - t)$$

and its second derivative is $g_{xx}(x,t) = \delta(x - t)$, again in the generalized (distributional) sense. In other words, we verify that $g(x,t)$ satisfies $Lg = \delta(x - t)$ and $g(0,t) = g(1,t) = 0$ provided $0 \leq t \leq 1$, in the sense of distribution.

The function $g(x,t)$ is called a GREEN'S FUNCTION. In other texts, $g(x,t)$ is sometimes called an INFLUENCE FUNCTION because of its important physical interpretation. The formula $u(x) = \int_0^1 g(x,t)f(t)dt$ shows that $u(x)$ is the superposition of displacements resulting from applying forces

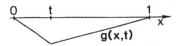

FIGURE 4.2 Plot of the Green's function $g(x,t)$ of $Lu = -u''$ with boundary conditions $u(0) = u(1) = 0$.

at each point t and summing these over all t. The function $g(x,t)$ represents the influence of a point force at position t on the displacement at position x. A sketch of $g(x,t)$ is shown in Figure 4.2.

We need a general way to find a Green's function for any linear differential operator. A strictly formal way to find a Green's function is as follows. We examine the nth order equation $Lg = \delta(x - t)$ and realize, heuristically, that if the nth derivative of g is to be like a δ-function, the $(n-1)^{\text{st}}$ derivative should be like a Heaviside function and all lower order derivatives should be continuous. If we integrate formally the equation $Lg = \delta(x - t)$ across the point $x = t$ from $x = t^-$ to $x = t^+$ we find that

$$a_n(t) \left(\frac{d^{n-1}g}{dx^{n-1}} \right) \Bigg|_{x=t^-}^{x=t^+} = 1.$$

We take as an operational definition of Green's functions

1. $Lg(x,t) = 0$ for $x \neq t$.

2. $\frac{d^k g(x,t)}{dx^k}$ continuous at $x = t$ for $k = 0, 1, 2, \ldots, n-2$.

3. $\frac{d^{n-1}g(x,t)}{dx^{n-1}} \Bigg|_{x=t^-}^{x=t^+} = \frac{1}{a_n(t)}$ (jump condition).

In addition we require $g(x,t)$ to satisfy all appropriate homogeneous boundary conditions.

To see how this formal definition actually works, we consider the second order differential operator

$$Lu = a_2(x)\frac{d^2u}{dx^2} + a_1(x)\frac{du}{dx} + a_0(x)u, \qquad x \in [a,b]$$

along with separated boundary conditions. By separated boundary conditions we mean that the boundary conditions are of the form

$$\alpha_1 u(a) + \beta_1 u'(a) = 0, \qquad \alpha_2 u(b) + \beta_2 u'(b) = 0$$

and we represent these in operator form as the boundary operators $B_1 u(a) = 0$ and $B_2 u(b) = 0$.

Suppose that $Lu = 0$ has solutions $u = u_1(x)$, $u_2(x)$ where $B_1 u_1(a) = 0$ and $B_2 u_2(b) = 0$. We expect $g(x,t)$ to be of the form

$$g(x,t) = \begin{cases} A u_1(x) u_2(t) & a \le x < t \le b \\ A u_2(x) u_1(t) & a \le t < x \le b \end{cases}$$

since g must be continuous. From the jump condition

$$\left. \frac{dg}{dx} \right|_{x=t^-}^{x=t^+} = \frac{1}{a_2(t)}$$

we require $A = \frac{1}{a_2(t)W(t)}$ where $W(t) = u_1(t)u_2'(t) - u_2(t)u_1'(t)$ is the WRON-SKIAN of the functions u_1, u_2. It follows that

$$g(x,t) = \left[u_1(x)u_2(t) + H(x - t)\left(u_2(x)u_1(t) - u_1(x)u_2(t) \right) \right] / a_2(t)W(t)$$

provided $W(t) \ne 0$. If $W(t) = 0$, we are in trouble, but we will rectify this later. For the specific example of the weighted string $Lu = u''$, $u_1(x) = x$ and $u_2(x) = x - 1$, and $W(t) \equiv 1$.

Using this expression for $g(x,t)$, we write that the solution of $Lu = f$ with homogeneous boundary conditions $B_1 u(a) = 0$ and $B_2 u(a) = 0$ is

$$u = \int_a^x \frac{u_2(x)u_1(t)f(t)}{a_2(t)W(t)} dt + \int_x^b \frac{u_1(x)u_2(t)f(t)}{a_2(t)W(t)} dt,$$

which can be verified by direct differentiation.

Since this expression for $g(x,t)$ was derived using only our formal rules, we must verify that it satisfies $Lg = \delta(x-t)$ using the theory of distributions. This equation must be satisfied in the weak sense, that is, it must be that

$$\langle Lg, \phi \rangle \equiv \left\langle g, \frac{d^2}{dx^2}(a_2\phi) - \frac{d}{dx}(a_1\phi) + a_0\phi \right\rangle = \phi(t)$$

for all test functions ϕ in D with support contained in $[a, b]$. Upon substitution of the proposed $g(x,t)$ we find that we must have

$$\int_{-\infty}^t u_1(x)u_2(t)L^*\phi \, dx + \int_t^\infty u_2(x)u_1(t)L^*\phi \, dx = a_2(t)W(t)\phi(t)$$

where $L^*\phi = (a_2\phi)'' - (a_1\phi)' + a_0\phi$. Integrating by parts, we find that this expression is identically satisfied if and only if our three operational rules are satisfied. We state this formally as a theorem:

Theorem 4.1

The function $g(x,t)$ satisfies the nth order equation $Lg = \delta(x - t)$ in the sense of distribution if and only if

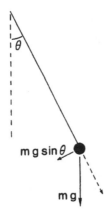

FIGURE 4.3 Forces on a pendulum.

1. $Lg = 0$ for $x \neq t$.
2. g is $n - 2$ times continuously differentiable at $x = t$.
3. $\dfrac{d^{n-1}g(x,t)}{dx^{n-1}}\bigg|_{x=t^{-}}^{x=t^{+}} = \dfrac{1}{a_n(t)}.$

Another important class of problems are the initial value problems. As a specific example consider the motion of a forced pendulum. Suppose the pendulum has mass m suspended from a weightless rod of length l and makes an angle θ with a vertical line. Balancing forces (Figure 4.3), we find that the equation of motion is

$$ml\frac{d^2\theta}{dt^2} = -mg\sin\theta + f(t),$$

where $f(t)$ is external torsional forcing and g is the gravitational constant. If we suppose that the pendulum sweeps out only a small angle then $\sin\theta \sim \theta$ and the equation of motion is

$$\ddot{\theta} + g\theta/l = \frac{f(t)}{lm}$$

which we wish to solve starting from initial data $\theta = 0$, $\dot{\theta} = 0$ at $t = 0$.

More generally suppose we wish to solve $Lu = f$ on the interval $0 \leq x < \infty$ starting from initial data $u(0) = u'(0) = 0$ (x is viewed as a time-like variable). To find a Green's function we apply the rules stated above. We require that $Lg = 0$ for $x \neq t$, that g be continuous at $x = t$ and that the jump condition $g'\big|_{x=t^{-}}^{x=t^{+}} = \frac{1}{a_2(t)}$ hold. Suppose $u_0(x)$ is a particular function that satisfies $Lu_0 = 0$ and $u_0(t) = 0$, $u_0'(t) = 1$. Since u_0 depends on both t

and x we denote $u_0 = u_0(x; t)$. From the jump condition on $g(x, t)$ we require $\frac{\partial g}{\partial x}(x, t)\, \big|_{x=t-}^{x=t^+} = \frac{1}{a_2(t)}$, and it follows that

$$g(x, t) = H(x - t)\frac{u_0(x; t)}{a_2(t)},$$

where $Lu_0 = 0$, $u_0(t; t) = 0$, $u_{0_x}(t; t) = 1$. As specific examples, the operator $Lu = u''$ has the Green's function $g(x, t) = H(x - t)(x - t)$ and the operator $Lu = u'' + \lambda^2 u$ has the Green's function $g(x, t) = \frac{1}{\lambda}\sin\lambda(x - t)H(x - t)$.

Using the proposed Green's function, we obtain

$$u(x) = \int_0^x \frac{u_0(x; t)}{a_2(t)}f(t)dt$$

as the solution of the initial value problem $Lu = f$. This formula expresses the important physical reality of CAUSALITY, namely, that the solution u at time x can be affected only by forcing before time x and can never be influenced by forcing which at time x is yet in the future.

There are two ways to verify that this solution satisfies the original equation. First, we can verify by direct differentiation that the equation $Lu = f$ is satisfied or second, we can show that $Lg = \delta(x - t)$ in the sense of distribution. Of course, both of these verifications succeed since they involve the same calculation as for the boundary value problem done previously, except now one of the two linearly independent solutions of $Lu = 0$ is chosen to be zero.

As a final example, consider the problem of finding the Green's function for the differential operator $Lu = u'' + \lambda u$ for $u \in L^2(-\infty, \infty)$ and λ a complex number. The fact that we have an infinite domain changes the details, but not the idea. The Green's function G must consist of solutions of the homogeneous problem $e^{\pm i\sqrt{\lambda}x}$. Notice that if λ is not a positive real number, then we take $\text{Im}(\sqrt{\lambda}) > 0$, so that $e^{i\sqrt{\lambda}x}$ is exponentially decaying as $x \to \infty$ and $e^{-i\sqrt{\lambda}x}$ is exponentially decaying as $x \to -\infty$. We want G to be in $L^2(-\infty, \infty)$, so we take

$$G = \begin{cases} Ae^{i\sqrt{\lambda}(x-\xi)} & x > \xi \\ Be^{-i\sqrt{\lambda}(x-\xi)} & x < \xi \end{cases}$$

and impose the continuity and jump conditions at $x = \xi$. We find $G(x, \xi; \lambda) = \frac{-i}{2\sqrt{\lambda}}e^{i\sqrt{\lambda}|x-\xi|}$, provided λ is not a positive real number.

4.3   Differential Operators

4.3.1   Domain of an Operator

Up until now we have been rather cavalier in our treatment of differential operators, since we have not stated carefully with what we are working. In this section we will define more precisely what we mean by a differential operator. The reason for this extra care is that if we hope to extend ideas of the previous chapters to differential operators then we want to define our operators L in such a way that

1. The inverse exists and is unique,

2. The adjoint operator exists,

3. The Fredholm alternative theorem holds.

The difficulty is made clear with a specific example. The differential operator $Lu = \frac{d^2u}{dx^2}$ is never uniquely invertible, since the general solution of $Lu = f$ has two arbitrary constants. As it stands, the operator L is a "many-to-one" operator, since many functions have the same second derivative. The most direct way to make the operator L invertible is to add extra conditions to the solution we seek. As a result we do not seek just any function whose second derivative is f but one which lies in a restricted subset of functions.

From now on, when we refer to a differential operator L, we shall specify not only the formal definition of how to take derivatives, say

$$Lu = a_n(x)\frac{d^nu}{dx^n} + \cdots + a_1(x)\frac{du}{dx} + a_0(x)u,$$

for an nth order operator, but also the *domain* on which it operates.

Definition

The DOMAIN of a differential operator L is given by

1. all $u \in L^2[a,b]$ for which $Lu \in L^2[a,b]$. (In this chapter the Hilbert space $L^2[a,b]$ is defined by the real inner product

$$\langle u,v \rangle = \int_a^b u(x)v(x)\omega(x)dx$$

with $\omega(x) > 0$ almost everywhere.)

2. all u which satisfy the homogeneous boundary conditions $Bu = 0$.

To specify the boundary conditions for an nth order differential operator we define the 2-point boundary operator Bu to be

$$Bu = \sum_{i=1}^{2} B_i V(x_i), \qquad V^T(x_i) = \left(u(x_i), u'(x_i), \ldots, u^{(n-1)}(x_i)\right),$$

where B_i are $n \times n$ matrices and where $x_1 = a$, $x_2 = b$. The number of independent boundary conditions contained in $Bu = 0$ is given by

$$k = \text{rank}\,[B_1, B_2].$$

For an nth order problem, the problem is well posed if $k = n$ whereas if $k > n$ or $k < n$ the problem is over conditioned or under conditioned, respectively.

In this formulation, an initial data operator has $B_1 = I$ and $B_2 = 0$. Two-point SEPARATED BOUNDARY CONDITIONS have

$$B_1 = \begin{pmatrix} \alpha_1 & \beta_1 \\ 0 & 0 \end{pmatrix} \quad \text{and} \quad B_2 = \begin{pmatrix} 0 & 0 \\ \alpha_2 & \beta_2 \end{pmatrix},$$

and the boundary conditions for which $B_1 = -B_2 = I$ are called PERIODIC BOUNDARY CONDITIONS.

The astute reader will notice a problem with this formulation. How can we specify that u satisfy boundary conditions when u is only known to be in $L^2[a, b]$ and values of u at specific points are not uniquely determined? The answer comes from the following important theorem:

Theorem 4.2

If $u \in L^2[a, b]$ and if the derivative of u (which always exists as a distribution) is also a function in $L^2[a, b]$, then u is equivalent to a continuous function on $[a, b]$.

This theorem follows from Sobolev's theorem (Chapter 2) and assures us that boundary conditions of up to order $n - 1$ can be applied for an nth order operator.

Because of the restriction $Bu = 0$, the domain of L is a submanifold of the functions for which $Lu \in L^2[a, b]$. Two operators whose only difference is the domain determined by B are nonetheless quite different operators. For example, the operator

$$Lu = \frac{d^2 u}{dx^2}$$

with the boundary conditions $u(0) = u(1) = 0$ is quite different than

$$Lu = \frac{d^2 u}{dx^2}$$

with boundary conditions $u(0) = u'(1) = 0$. To see this, notice that the solutions of $Lu = 1$ are $u = \frac{x}{2}(x - 1)$ and $u = x(x/2 - 1)$, respectively.

4.3.2 Adjoint of an Operator

Definition

The ADJOINT of a differential operator L is defined to be the operator L^* for which $\langle Lu, v \rangle = \langle u, L^*v \rangle$ for all u in the domain of L and v in domain of L^*.

This definition determines not only the operational definition of L^*, but its domain as well. Consider the example

$$Lu = \frac{du}{dx}$$

with the boundary condition $u(0) = 2u(1)$ with inner product

$$\langle u, v \rangle = \int_0^1 u(x)v(x)dx.$$

Using this inner product we have that

$$\langle Lu, v \rangle = \int_0^1 u'(x)v(x)dx = u(1)\left(v(1) - 2v(0)\right) - \int_0^1 u(x)v'(x)dx.$$

In order to make $\langle Lu, v \rangle = \langle u, L^*v \rangle$, we must take $L^*v = -\frac{dv}{dx}$ with the boundary condition $v(1) = 2v(0)$. The boundary condition $v(1) = 2v(0)$ is called the DUAL MANIFOLD for L.

In general, integration by parts yields an expression of the form $\langle Lu, v \rangle = J(u, v) + \langle u, L^*v \rangle$ where $J(u, v)$ represents boundary contributions. The domain of L^* is determined by requiring $J(u, v) = 0$ for all u in $D(L)$.

Definition

1. If the operational definitions of L and L^* agree, L is said to be FORMALLY SELF ADJOINT.

2. If L is formally self adjoint and $D(L) = D(L^*)$, L is a self-adjoint operator.

For the second order differential operator

$$Lu = a_2(x)\frac{d^2u}{dx^2} + a_1(x)\frac{du}{dx} + a_0(x)u$$

the formal adjoint is $L^*v = \frac{d^2}{dx^2}(a_2(x)v) - \frac{d}{dx}(a_1(x)v) + a_0(x)v$ where the boundary manifold $B^*v = 0$ is specified by requiring that

$$J(u, v) \equiv \left(a_2\frac{du}{dx}v - u\frac{d}{dx}(a_2v) + a_1uv \right)\Big|_a^b = 0$$

for all u satisfying $Bu = 0$.

For example, with separated boundary data $\alpha u + \beta u' = 0$ at $x = a$ or $x = b$, the adjoint condition is

$$a_2(\alpha v + \beta v') + \beta (a_2' - a_1) v = 0$$

applied at the same value of x.

The operator L is formally self adjoint if $a_2' = a_1$ and if this is the case, Lu can be written as

$$Lu = \frac{d}{dx} \left(p(x)\frac{du}{dx} \right) + q(x)u, \qquad \text{where} \qquad p = a_2, \qquad q = a_0,$$

and is called a STURM-LIOUVILLE OPERATOR. More generally, the operator

$$Lu = \frac{1}{\omega(x)} \frac{d}{dx} \left(p(x)\frac{du}{dx} \right) + q(x)u, \qquad \omega(x) > 0$$

is formally self adjoint with respect to the inner product

$$\langle u, v \rangle = \int_a^b u(x)v(x)\omega(x)dx.$$

If an operator is in the Sturm-Liouville form then the boundary contribution becomes

$$J(u, v) = p(x) \left(\frac{du}{dx}v - \frac{dv}{dx}u \right) \Big|_a^b.$$

For a formally self-adjoint second order operator there are two important classes of boundary conditions that yield self-adjoint operators. These are

1. separated boundary conditions

$$\alpha_1 u(a) + \beta_1 u'(a) = 0$$
$$\alpha_2 u(b) + \beta_2 u'(b) = 0$$

and

2. periodic boundary conditions $u(a) = u(b)$ and $u'(a) = u'(b)$ provided $p(a) = p(b)$.

We leave verification of this statement as an exercise to the reader.

4.3.3 The Extended Definition of an Operator

The differential operator L is only defined to make sense on a domain restricted by the boundary operator $Bu = 0$. It is often the case that we may want to apply L to functions other than those for which $Bu = 0$, that is, we want to extend, or generalize, the definition of L beyond the classical definition just given. To do this we follow the same steps used before when we generalized the meaning of functions to distributions. For a function u in the domain of L we know that

$$\langle Lu, v \rangle = \langle u, L^*v \rangle$$

for all v in the domain of L^*. Notice that once L^* and its domain are known, the right hand inner product makes sense for any $u \in L^2$ even if the left-hand side does not make sense.

Definition

The EXTENDED DEFINITION of L is that operator $\widehat{L}$, interpreted as a distribution, satisfying

$$\langle \widehat{L}u, v \rangle = \langle u, L^*v \rangle$$

for all $u \in L^2$ and all test functions v also in the domain of L^*.

Example

The operator $Lu = \frac{d^2u}{dx^2}$ with boundary conditions $u(0) = u(1) = 0$ is a self-adjoint operator. To find $\widehat{L}$ we write

$$
\begin{aligned}
\langle \widehat{L}u, v \rangle \;&=\; \langle u, Lv \rangle \quad \text{for all} \quad v \quad \text{with} \quad v(0) = v(1) = 0 \\[6pt]
&=\; \int_0^1 v'' u \, dx \\[6pt]
&=\; \int_0^1 u'' v \, dx + v'(1)u(1) - v'(0)u(0).
\end{aligned}
$$

Since $\widehat{L}u$ is to be interpreted as a distribution acting on the test function v, we have

$$\widehat{L}u = \frac{d^2u}{dx^2} - \delta'(x - 1)u(1) + \delta'(x)u(0).$$

This extension of u'' makes intuitive sense. Suppose $u = 0$ for $x > 1$ and $x < 0$ but experiences jump discontinuities at $x = 0$ and $x = 1$. We expect u to be like a Heaviside function, u' to be like a delta distribution, and u'' to be like a dipole distribution at $x = 0, 1$, which is what the extension of u'' implies.

In general, we can find the extended operator fairly easily. We note that for any functions u and v (i.e., without applying boundary constraints),

$$\langle Lu, v \rangle = J(u, v) + \langle u, L^*v \rangle,$$

where $J(u, v)$ is the boundary term found from integration by parts. Therefore

$$\langle \widehat{L}u, v \rangle = \langle Lu, v \rangle - J(u, v),$$

where v is in the domain of L^*, so that $\widehat{L}u$ is simply Lu modified by the addition of generalized functions at the boundary. For example, if we have separated boundary conditions, say $\alpha(x)u(x) + \beta(x)u'(x) = 0$ applied at $x = a$ or b, with $\beta(x) \neq 0$, then

$$\widehat{L}u = Lu - \frac{a_2(x_0)}{\beta(x_0)}(\alpha(x_0)u(x_0) + \beta(x_0)u'(x_0))\delta(x - x_0)\Big|_{x_0=a}^{x_0=b}.$$

The extension of Lu to $\widehat{L}u$ is unique as a distribution on the set of test functions also in the domain of L^*. Notice that L and $\widehat{L}$ agree on the domain of L (i.e., whenever $J(u, v) = 0$). The operator $\widehat{L}$ simply specifies what happens to functions not in the domain of L.

4.3.4 Inhomogeneous Boundary Data

If we wish to solve $Lu = f$ subject to homogeneous boundary data, we use the Green's function and the definition of the adjoint operator. If g^* is the adjoint Green's function, then

$$u(t) = \langle \delta(x - t), u \rangle = \langle L^* g^*, u \rangle = \langle g^*, Lu \rangle = \int_a^b g^*(x, t)f(x)dx.$$

Now, $g^*(x, t)$ is related to $g(x, t)$, the Green's function for L, by

$$
\begin{aligned}
g^*(\xi, \eta) &= \langle \delta(x - \xi), g^*(x, \eta) \rangle = \langle Lg(x, \xi), g^*(x, \eta) \rangle \\
&= \langle g(x, \xi), L^* g^*(x, \eta) \rangle = g(\eta, \xi),
\end{aligned}
$$

and, of course, the Green's function for a self-adjoint operator is symmetric. As a result

$$u(x) = \int_a^b g(x, t)f(t)dt$$

satisfies the differential equation $Lu = f$.

To solve a problem with inhomogeneous boundary data, we can still make use of the Green's function which satisfies homogeneous data. We use the extended definition of the operator and write

$$u(t) = \langle \delta(x - t), u \rangle = \langle L^* g^*, u \rangle = \langle g^*, \widehat{L}u \rangle.$$

For example, to solve $u'' = f$ subject to boundary conditions $u(0) = \alpha$, $u(1) = \beta$, we write

$$\widehat{L}u = f(x) + \alpha\delta'(x) - \beta\delta'(x - 1) = \hat{f}(x)$$

as the extended definition of the problem. It follows that

$$u(t) = \langle g^*, \hat{f} \rangle \quad = \quad \int_0^1 g(t,x)f(x)dx - \alpha \frac{\partial g(t,x)}{\partial x}\Big|_{x=0} + \beta \frac{\partial g(t,x)}{\partial x}\Big|_{x=1}$$

$$= \quad \int_0^1 g(t,x)f(x)dx + \alpha(1-t) + \beta t.$$

In practice one need not actually calculate the extended operator to use it. For example, if we wish to solve $Lu = f(x)$ subject to inhomogeneous boundary data it is easier to notice that

$$u(t) = \langle \delta(x-t), u \rangle = \langle L^* g^*, u \rangle = -J(u, g^*) + \langle g^*, Lu \rangle,$$

where g^* satisfies the homogeneous adjoint boundary conditions. As a result we find that

$$u(t) \quad = \quad \int_a^b g(t,x)f(x)dx$$

$$+ \quad \left[\frac{\partial}{\partial x}(a_2 g(t,x)) u(x) - a_2(x)g(t,x)u'(x) - a_1(x)u(x)g(t,x) \right]_{x=a}^{x=b}$$

For the special case of separated boundary conditions, this reduces to

$$u(t) \quad = \quad \int_a^b g(t,x)f(x)dx + \frac{B_2 u(b)u_1(t)}{B_2 u_1(b)} + \frac{B_1 u(a)u_2(t)}{B_1 u_2(a)},$$

where $u_1(x)$, $u_2(x)$ are particular solutions of the homogeneous differential equation, with $B_1 u_1(a) = 0$, $B_1 u_1(b) \neq 0$, $B_2 u_2(b) = 0$, $B_1 u_2(a) \neq 0$. This representation expresses the fact that the inhomogeneous boundary conditions are satisfied by adding to the solution an appropriate linear combination of solutions of the homogeneous differential equation. The important observation is that

$$u = \int_0^1 g(x,t)\hat{f}(t)dt$$

produces the correct solution, even with inhomogeneous boundary data, and so we need only concentrate on the problem with homogeneous conditions.

4.3.5 The Fredholm Alternative

The Fredholm alternative is so very important that it is worth restating here.

Theorem 4.3 (Fredholm Alternative)

For the differential operator $\hat{L}$ (in the extended sense)

1. The solution of $\widehat{L}u = \widehat{f}$ is unique if and only if $\widehat{L}u = 0$ has only the trivial solution $u = 0$.

2. The solution of $\widehat{L}u = \widehat{f}$ exists if and only if $\left\langle \widehat{f}, v \right\rangle = 0$ for all v for which $L^*v = 0$.

It follows that $\widehat{L}u = \widehat{f}$ can be solved for any $\widehat{f}$ if and only if $L^*v = 0$ implies $v = 0$.

The most direct proof of this theorem is to convert the differential equation to an integral equation for which the theorem is already known.

Example

Consider the differential operator $Lu = u'' + k^2 u$, $k \neq 0$, subject to homogeneous boundary conditions $u(0) = u(1) = 0$. By direct calculation we find that the Green's function is

$$g(x,t) = \begin{cases} \dfrac{\sin kx \sin k(t-1)}{k \sin k} & 0 \leq x < t \leq 1 \\ \dfrac{\sin kt \sin k(x-1)}{k \sin k} & 0 \leq t < x \leq 1 \end{cases}$$

provided $\sin k \neq 0$. The operator L is self adjoint and therefore the extension of L to $\widehat{L}$ satisfies $\langle \widehat{L}u, v \rangle = \langle u, Lv \rangle$ in the sense of distribution for all test functions v in the domain of L. As a result,

$$\widehat{L}u = \frac{d^2u}{dx^2} + k^2 u - u(1)\delta'(x-1) + u(0)\delta'(x).$$

To solve $u'' + k^2 u = f(x)$ with inhomogeneous boundary data $u(0) = \alpha$, $u(1) = \beta$ we find that $\widehat{f} = f - \beta\delta'(x-1) + \alpha\delta'(x)$ and

$$u = \int_0^1 g(x,t)\widehat{f}(t)dt = \int_0^1 g(x,t)f(t)dt - \alpha\frac{\partial g}{\partial t}\bigg|_{t=0} + \beta\frac{\partial g}{\partial t}\bigg|_{t=1}$$

$$= \int_0^1 g(x,t)f(t)dt - \alpha\frac{\sin k(x-1)}{\sin k} + \beta\frac{\sin kx}{\sin k}$$

provided $\sin k \neq 0$. If $\sin k = 0$, (or $k = m\pi$ for $m \neq 0$), there is a nontrivial homogeneous solution $v = \sin m\pi x$, so that a solution of $u'' + k^2 u = f(x)$, $u(0) = \alpha$, $u(1) = \beta$ exists if and only if

$$0 = \langle \widehat{f}, v \rangle = \langle \widehat{L}u, v \rangle = \langle u, L^*v \rangle,$$

that is,

$$0 = \int_0^1 (v'' + k^2 v)\, u\, dx \;=\; \int_0^1 \sin kx f(x) dx + k(\beta \cos k - \alpha)$$

$$= \int_0^1 \sin m\pi x f(x) dx - (\alpha + (-1)^m \beta)\, m\pi.$$

Probably the most direct way to find the solvability condition (not using extended operators explicitly) is to note that if $L^* v = 0$, then

$$0 = \langle u, L^* v \rangle = -J(u, v) + \langle Lu, v \rangle = -J(u, v) + \int_a^b f(x) v(x) dx.$$

For this example $v(x) = \sin kx$, and we find

$$0 = \int_0^1 f(x) \sin kx\, dx + \beta k \cos k - \alpha k$$

as stated before.

4.4 Least Squares Solutions and Modified Green's Functions

According to the Fredholm alternative, solutions of $\widehat{L}u = \hat{f}$ exist if and only if $\langle \hat{f}, v \rangle = 0$ for all v with $L^* v = 0$. If this condition fails, the best we can do is solve $\widehat{L}u = \hat{f}$ in a least squares sense.

Definition

The smallest least squares solution of $\widehat{L}u = \hat{f}$ satisfies

$$\widehat{L}u = \hat{f} - \sum_{i=1}^n \psi_i(x) \left\langle \hat{f}, \psi_i \right\rangle$$

and

$$\langle u, \phi_i \rangle = 0 \qquad i = 1, \ldots, n$$

where $\{\psi_i\}_{i=1}^n$ are a mutually orthonormal set of functions spanning the null space of L^*, $L^* \psi_i = 0$, and the functions $\{\phi_i\}_{i=1}^n$ span the null space of L, $L\phi_i = 0$.

This characterization of the least squares solution follows from the Fredholm alternative.

Example

Solve $u'' = \sin \pi x$, $u'(0) = \beta$, $u'(1) = \gamma$ in the least squares sense. (For this problem an exact solution exists if and only if $2/\pi - \gamma + \beta = 0$.)

We begin by formulating this problem in its extended sense. We define Lu to be the operator $Lu = u''$ with boundary conditions $u'(0) = u'(1) = 0$. On this domain the operator L is self adjoint.

Since $\langle \widehat{L}u, v \rangle = \langle u, Lv \rangle$ defines $\widehat{L}u$ in the generalized sense, we have that

$$\left\langle \widehat{L}u, v \right\rangle = v'u - u'v \Big|_0^1 + \langle u'', v \rangle = \langle u'', v \rangle + u'(0)v(0) - u'(1)v(1).$$

It follows that $\widehat{L}u = u'' + u'(0)\delta(x) - u'(1)\delta(x-1)$.

For this operator, $\hat{f} = \sin \pi x + \beta \delta(x) - \gamma \delta(x-1)$, and the least squares solution must satisfy

$$\widehat{L}u \equiv u'' + u'(0)\delta(x) - u'(1)\delta(x-1) = \hat{f} - \left\langle \hat{f}, 1 \right\rangle$$

$$= \sin \pi x - \frac{2}{\pi} + \gamma - \beta + \beta \delta(x) - \gamma \delta(x-1),$$

since $v = 1$ spans the null space of $Lv = 0$. We must therefore solve the boundary value problem

$$u'' = \sin \pi x - \frac{2}{\pi} + \gamma - \beta, \qquad u'(0) = \beta, \qquad u'(1) = \gamma$$

subject to the condition

$$\int_0^1 u \, dx = 0.$$

The unique solution of this problem is

$$u(x) = -\frac{1}{\pi^2} \sin \pi x - \left(\frac{2}{\pi} + \beta - \gamma \right) x^2 + \left(\beta + \frac{1}{\pi} \right) x + B$$

$$B = \frac{2}{\pi^3} + \frac{1}{3} \left(\frac{2}{\pi} + \beta - \gamma \right) - \frac{1}{2} \left(\frac{1}{\pi} + \beta \right).$$

There is a way to calculate this least squares solution that saves some computation. We realize that we must solve

$$u'' = \sin \pi x - K$$

$$u'(0) = \beta, \qquad u'(1) = \gamma$$

for some constant K. Rather than calculating K from the Fredholm alternative theorem, we let K remain an unknown parameter and choose it correctly later. For arbitrary K,

$$u = -\frac{1}{\pi^2}\sin \pi x - K\frac{x^2}{2} + Ax + B$$

$$u' = -\frac{1}{\pi}\cos \pi x - Kx + A.$$

We now use the boundary conditions $u'(0) = \beta$, $u'(1) = \gamma$ to determine that $A = \frac{1}{\pi} + \beta$, $K = \frac{2}{\pi} + \beta - \gamma$ so that

$$u = \frac{-1}{\pi^2}\sin \pi x - \left(\frac{2}{\pi} + \beta - \gamma\right)\frac{x^2}{2} + \left(\beta + \frac{1}{\pi}\right)x + B$$

and B is now chosen so that

$$\int_0^1 u\, dx = 0.$$

The solution obtained here is the same as before although we did not need to determine $\widehat{L}$ or directly calculate the projection of $\widehat{f}$ onto the range of $\widehat{L}$.

Now that we know how to find a least squares solution, it is natural to inquire about the possibility of finding a pseudo inverse operator for L. We know that not all differential operators are invertible, so Green's functions do not always exist. In fact, from Section 4.2, whenever the Wronskian $W(u_1, u_2) = u_1(x)u_2'(x) - u_2(x)u_1'(x)$ vanishes, the construction of the Green's function fails.

The significance of the Wronskian is that it is nonzero if and only if the functions u_1 and u_2 are linearly independent. If u_1 and u_2 satisfy the same second order differential equation, say

$$a_2(x)\frac{d^2 u_i}{dx^2} + a_1(x)\frac{du_i}{dx} + a_0(x)u_i = 0, \qquad i = 1,2$$

then $W = u_1 u_2' - u_2 u_1'$ satisfies

$$a_2\frac{dW}{dx} + a_1 W = 0$$

so that

$$W = W_0 \exp\left[-\int \frac{a_1(x)}{a_2(x)}dx\right].$$

As a result, if $W \neq 0$ for some x, then $W \neq 0$ for all x (assuming, of course, that

$$\int \frac{a_1(x)}{a_2(x)}dx$$

is bounded) and if $W = 0$ for some x, then $W = 0$ for all x. For Sturm-Liouville operators $a_2(x) = p(x)$, $a_1(x) = p'(x)$ so that $a_2(x)W(x) =$ constant.

For more general systems of differential equations, if u_i, $i = 1, 2, \ldots, n$ are vector solutions of

$$\frac{du}{dx} = A(x)u,$$

where $A(x)$ is an $n \times n$ matrix, then $W = \det[u_1, u_2, \ldots, u_n]$ satisfies the differential equation $\frac{dW}{dx} = \text{Trace}(A(x))W$ (called Abel's equation). As before, either $W = 0$ for all x or else $W \neq 0$ for all x.

Suppose that during the construction of a Green's function we find that $W = 0$. This means that $u_1(x)$ and $u_2(x)$ are linearly dependent so that $u_1(x)$ satisfies both boundary conditions. In other words, $u_1(x)$ is a nontrivial solution of $Lu = 0$ with boundary data $Bu = 0$. From the Fredholm alternative we expect difficulty. Since $\langle \delta(x - t), u_1(x) \rangle = u_1(t) \neq 0$ we do not expect a Green's function to exist, and the best we can hope for is a least squares solution of $Lu = f$.

We saw above that the least squares solution of $Lu = f$ is found by projecting f onto the range of L. If $\{\psi_i\}_{i=1}^{k}$ is a set of orthonormal functions which spans the null space of L^*, the projection of f onto the range of L is $f - \sum_{i=1}^{k} \langle f, \psi_i \rangle \psi_i$. For a second order differential operator, k is either one or two.

Applying this idea to Green's functions, we realize that since we cannot solve $Lg = \delta(x - t)$, the best we can do is to solve $Lg = \delta(x - t) - \omega(t) \sum_{i=1}^{k} \psi_i(x)\psi_i(t)$, where $\omega(x)$ is the weight function for the inner product. Of course, the function g is not uniquely determined by this differential equation, so we require $\langle g, \phi_i \rangle = 0$ where $\{\phi_i\}_{i=1}^{k}$ spans the null space of L. This function g is called the MODIFIED (or GENERALIZED) GREEN'S FUNCTION.

We take as our operational definition of the modified Green's function

1. $Lg = -\omega(t) \sum_{i=1}^{k} \psi_i(x)\psi_i(t)$ for $x \neq t$, where $\{\psi_i\}_{i=1}^{k}$ is an orthonormal basis for the null space of L^*.

2. $\frac{dg^k}{dx^k}$ is continuous at $x = t$ for $k = 0, 1, \ldots, n - 2$.

3. $\frac{dg^{n-1}}{dx^{n-1}}\Big|_{x=t^-}^{x=t^+} = \frac{1}{a_n(t)}$ (jump condition).

4. $\int_a^b g(x,t)\phi_i(x)\omega(x)dx = 0$, $i = 1, \ldots, k$ where $\{\phi_i\}_{i=1}^{k}$ forms a basis for the null space of L.

5. $g(x,t)$ satisfies the homogeneous boundary conditions, $Bg = 0$.

Example

The self-adjoint differential operator $Lu = \frac{d^2 u}{dx^2}$ with boundary conditions $u'(0) = u'(1) = 0$ has a null space spanned by $\phi_0 = 1$. The modified

Green's function must satisfy

$$u'' = \delta(x - t) - 1, \qquad u'(0) = u'(1) = 0$$

$$\int_0^1 u\, dx = 0$$

and is given by

$$g(x,t) = \begin{cases} -\frac{1}{3} + t - \left(\frac{x^2}{2} + \frac{t^2}{2}\right), & 0 \le x < t \le 1, \\ -\frac{1}{3} + x - \left(\frac{x^2}{2} + \frac{t^2}{2}\right), & 0 \le t < x \le 1. \end{cases}$$

The generalized Green's function can be found for very general differential operators. To see this in the special case of a one-dimensional null space and separated boundary conditions, suppose L is the self-adjoint operator $Lu = (pu')' + qu$ for $a \le x \le b$ subject to separated boundary conditions, $Bu = 0$, with weight function $\omega(x) = 1$. Suppose ϕ and ψ are linearly independent solutions of the homogeneous differential equation $Lu = 0$, and suppose that $B\phi = 0$ as well. For this equation, $p(x)(\phi\psi' - \psi\phi')$ is a constant which we can choose to be one. Furthermore, we choose ϕ to be normalized so that $\|\phi\| = 1$ with ϕ and ψ orthogonal.

The particular solution of $Lu = -\phi(x)\phi(t)$ is found using variation of parameters to be

$$u_p(x,t) = \phi(x)\phi(t)\int_a^x \phi(\tau)\psi(\tau)d\tau - \psi(x)\phi(t)\int_a^x \phi^2(\tau)d\tau.$$

We set

$$g(x,t) = \begin{cases} \alpha\phi(x) + \beta\psi(x) + u_p(x,t) & a \le x < t \le b, \\ \gamma\phi(x) + \delta\psi(x) + u_p(x,t) & a \le t < x \le b. \end{cases}$$

We must choose the constants α, β, γ, and δ so that g is continuous, satisfies the jump condition and the two boundary conditions. The function ϕ satisfies the boundary conditions $B_1\phi(a) = 0$ and $B_2\phi(b) = 0$, and $B_1\psi(a) \ne 0$ and $B_2\psi(b) \ne 0$. Since $B_1 u_p(a,t) = 0$, we must pick $\beta = 0$ so that $B_1 g(a,t) = 0$. It is easy to check that $u_p(b,t) = -\psi(b)\phi(t)$ and $\frac{\partial u_p(b,t)}{\partial x} = -\psi'(b)\phi(t)$ so that $B_2 u_p(b,t) = -\phi(t)B_2\psi(b)$. If we take $\delta = \phi(t)$, the boundary condition $B_2 g(b,t) = 0$ is satisfied.

For continuity, we require $\alpha = \gamma + \psi(t)$ and g becomes

$$g(x,t) = \begin{cases} \gamma\phi(x) + \psi(t)\phi(x) + u_p(x,t) & a \le x < t \le b \\ \gamma\phi(x) + \phi(t)\psi(x) + u_p(x,t) & a \le t < x \le b. \end{cases}$$

With g in this form, the jump condition holds automatically, and the constant γ is uniquely determined by the requirement that $\int_a^b g(x,t)\phi(x)dx = 0$.

Theorem 4.4

The modified Green's function for this self-adjoint operator satisfies the two properties:

1. $g(x,t) = g(t,x)$ (g is symmetric)
2. $u = \int_a^b g(x,t)f(t)dt$ satisfies the differential equation $Lu = f(x) - \phi(x)\int_a^b \phi(t)f(t)dt$ and $\int_a^b u(x)\phi(x)dx = 0$, that is, u is the smallest least squares solution of $Lu = f$.

Both of these properties are most easily verified by formal arguments. Since L is self adjoint,

$$\langle g(x,\xi), Lg(x,\eta)\rangle = \langle Lg(x,\xi), g(x,\eta)\rangle .$$

Now $Lg(x,t) = \delta(x-t) - \phi(x)\phi(t)$ so that

$$g(\eta,\xi) - \phi(\eta)\langle g(x,\xi),\phi(x)\rangle = g(\xi,\eta) - \phi(\xi)\langle g(x,\eta),\phi(x)\rangle ,$$

and since $\langle g(x,t),\phi(x)\rangle = 0$, symmetry follows. If $u = \int_a^b g(x,t)f(t)dt$ then $Lu = \langle Lg, f\rangle = f(x) - \phi(x)\langle \phi(t), f(t)\rangle$ as promised.

The extension of these statements to include a nonconstant weight function $\omega(x)$ is direct.

4.5 Eigenfunction Expansions for Differential Operators

An operator L is said to be positive definite if $\langle Lu, u\rangle > 0$ for all $u \neq 0$. If L is positive definite then the only solution of $Lu = 0$ is the trivial solution $u = 0$, since if there is a solution $u \neq 0$ with $Lu = 0$ then $\langle Lu, u\rangle = 0$ contradicts positive-definiteness. If L is positive definite, L has a Green's function. (We verified this for separated boundary conditions in Section 4.2.) Further, if L is self adjoint (with weight function $\omega(x) = 1$), then the Green's function $g(x,t)$ is symmetric $g(x,t) = g(t,x)$.

Definition

For the differential operator L (with boundary conditions specified), the pair $\{\phi, \lambda\}$ is called an eigenvalue, eigenfunction pair if $L\phi = \lambda\phi$, $\phi \neq 0$.

Examples of eigenvalues and eigenfunctions are numerous. For example, the operator $Lu = -\frac{d^2u}{dx^2}$ with boundary data $u(0) = u(1) = 0$ has an infinite set of eigenpairs $\{\phi_n, \lambda_n\} = \{\sin n\pi x, n^2\pi^2\}$.

The most important point of this chapter is now almost trivial to prove.

Theorem 4.5

The eigenfunctions of a self-adjoint, invertible second order differential operator form a complete set on $L^2[a, b]$.

Proof

Suppose L has a symmetric Green's function $g(x, t)$. The eigenvalue problem $L\phi = \lambda\phi$ is equivalent to the integral equation

$$\phi = \lambda \int_a^b g(x, t)\phi(t)dt = \lambda K\phi,$$

where K is a symmetric Hilbert-Schmidt operator. The eigenfunctions of K are the same as the eigenfunctions of L. However, from Chapter 3, we know that the eigenfunctions of K are complete over the range of K. The range of K is the domain of L, since $K = L^{-1}$ and the domain of L is the set of L^2 functions, with $Lu \in L^2$ which satisfy prescribed boundary conditions. The domain of L is dense in L^2, so the eigenfunctions of L are complete in $L^2[a, b]$.

The practical importance of this theorem is that for every self-adjoint second order differential equation, there is a natural coordinate system (i.e., a transform) with which to represent the inverse operator. This coordinate system is the set of mutually orthogonal eigenfunctions which we now know is complete, even though infinite. To solve $Lu = f$ using eigenfunctions, we express f in terms of the eigenfunctions of L,

$$f = \sum_{n=1}^{\infty} \alpha_n \phi_n, \qquad \alpha_n = \langle f, \phi_n \rangle$$

and seek a solution of the form $u = \sum_{n=1}^{\infty} \beta_n \phi_n$. Since $\{\phi_n\}$ are eigenfunctions, the equation $Lu = f$ transforms to the infinite set of separated algebraic equations

$$\beta_n \lambda_n = \alpha_n$$

which is always solvable and then $u = \sum_{n=1}^{\infty} \frac{\langle f, \phi_n \rangle \phi_n(x)}{\lambda_n}$ provided λ_n is not zero. Another way of expressing this is that, in terms of ϕ_n, the Green's function of L is

$$g(x, t) = \omega(t) \sum_{n=1}^{\infty} \frac{\phi_n(x)\phi_n(t)}{\lambda_n}.$$

Now would be a good time to review the commuting diagrams of transform theory shown in Chapter 1. What we have done here is to solve the equation $Lu = f$ by first representing u and f in the coordinate system of the eigenfunctions of L, solved the diagonalized system of equations in this coordinate representation, and then re-expressed the solution u in terms of these coordinates. It may happen that L has a complete set of orthogonal eigenfunctions, but some of the eigenvalues are zero. It follows that the modified Green's function is

$$g(x,t) = \omega(t) \sum_{\lambda_n \neq 0} \frac{\phi_n(x)\phi_n(t)}{\lambda_n}.$$

Notice that the eigenfunction expansion based on eigenfunctions of L will always satisfy homogeneous boundary conditions. If we want to use eigenfunctions to solve a problem with inhomogeneous data, a slightly different approach is necessary.

Suppose we wish to solve $Lu = f$ subject to inhomogeneous boundary conditions. If L is self adjoint we observe that

$$\langle \phi_n, f \rangle = \langle \phi_n, Lu \rangle = \langle L\phi_n, u \rangle - J(\phi_n, u) = \lambda_n \langle \phi_n, u \rangle - J(\phi_n, u)$$

so that $\langle \phi_n, u \rangle = \frac{1}{\lambda_n}(\langle \phi_n, f \rangle + J(\phi_n, u))$ is the algebraic equation which determines the Fourier coefficients of u, $\langle \phi_n, u \rangle$.

As a specific example, consider the problem $u'' = f(x)$ subject to inhomogeneous boundary conditions $u(0) = a$, $u'(1) = b$. The eigenfunctions for the operator $Lu = -u''$ with homogeneous boundary conditions $u(0) = 0$, $u'(1) = 0$ are the trigonometric functions

$$\phi_n(x) = \sin\left(\frac{2n-1}{2}\right)\pi x, \qquad \lambda_n = \left(\left(\frac{2n-1}{2}\right)\pi\right)^2$$

We calculate that

$$\int_0^1 \phi_n(x)f(x)dx = \int_0^1 \phi_n(x)u''dx = (\phi_n u' - \phi_n' u)\Big|_0^1 + \int_0^1 \phi_n'' u\, dx$$

so that

$$-\lambda_n \int_0^1 \phi_n u\, dx = \int_0^1 \phi_n(x)f(x)dx - b\phi_n(1) - a\phi_n'(0).$$

Since

$$u(x) = \sum_{n=1}^{\infty} a_n \phi_n(x), \qquad a_n = \frac{\langle \phi_n, u \rangle}{\|\phi_n\|^2}$$

we determine that

$$-\lambda_n a_n = 2\int_0^1 \sin\left(\frac{2n-1}{2}\pi x\right)f(x)dx + 2(-1)^n b - (2n-1)\pi a.$$

There is an apparent contradiction in the statement $u = \sum_{n=1}^{\infty} a_n \phi_n(x)$ satisfies $u(0) = a$, $u'(1) = b$, since if one evaluates the infinite sum at $x = 0$ or $x = 1$ one obtains $u = 0$. However, remember that convergence of this series is only in L^2. It can be shown that $\lim_{x \to 0+} u(x) = a$ and $\lim_{x \to 1-} u'(x) = b$, and that the function $u(x) = \sum_{n=1}^{\infty} a_n \phi_n(x)$ differs from a continuous function only at the two points $x = 0$ and $x = 1$. It is in this L^2 sense that u satisfies the inhomogeneous boundary conditions.

The most important group of eigenfunctions occur as eigenfunctions of Sturm-Liouville operators. Suppose $Lu = -(p(x)u')' + q(x)u$ where $p(x) > 0$ and $q(x) > 0$ except possibly on a set of measure zero of the interval $x \in [a, b]$. We can easily check that L is positive definite since

$$\langle Lu, u \rangle = -pu'u \Big|_a^b + \int_a^b \left(p(x)u'^2(x) + q(x)u^2(x) \right) dx$$

is positive for nonzero u, provided the boundary conditions are such that $pu'u \Big|_a^b \leq 0$. If the operator L is of the form $Lu = \frac{1}{\omega(x)}(-(p(x)u')' + q(x)u)$ then the appropriate inner product has weight function $\omega(x)$, with no other changes.

We can prove completeness of eigenfunctions for certain operators which are not positive definite. For example, suppose the Sturm-Liouville operator $Lu = -(p(x)u')' + q(x)u$, $p(x) > 0$, is not positive definite although $q(x)$ is bounded below. If $q(x) > -\lambda_0$, then $Lu = -(p(x)u')' + q(x)u + \lambda_0 u$ is a positive definite operator, and although the eigenvalues of L are all shifted, the eigenfunctions of L are unchanged and hence complete. For example, the operator $Lu = -u''$, with boundary conditions $u'(0) = u'(1) = 0$, is not positive definite, although the shifted operator $Lu = -u'' + u$ with the same boundary conditions is. As a result the eigenfunctions of this operator $\{\cos n\pi x\}_{n=0}^{\infty}$ are complete on $L^2[0, 1]$.

The remainder of this chapter gives a sample of the eigenfunctions that result from Sturm-Liouville operators.

4.5.1 Trigonometric Functions

The trigonometric functions are easily shown to be complete using this theory. The differential operator $Lu = -\frac{d^2u}{dx^2}$ with boundary conditions $u(0) = u(1) = 0$ has eigenfunctions $\{\sin n\pi x\}_{n=1}^{\infty}$ which are complete on $L^2[0, 1]$. Completeness of the full trigonometric series follows by examining the operator $Lu = -\frac{d^2u}{dx^2} + u$ subject to periodic conditions $u(0) = u(1)$ and $u'(0) = u'(1)$ for which the eigenfunctions are $\{\sin 2n\pi x\}_{n=1}^{\infty}$ and $\{\cos 2n\pi x\}_{n=0}^{\infty}$. Note that the eigenvalues with $n \neq 0$ have geometric multiplicity 2.

As we noted in Chapter 2, the completeness of Fourier series only guarantees convergence in the sense of L^2. Pointwise convergence is another matter entirely.

Theorem 4.6

If the function $f(x)$ is a periodic piecewise continuously differentiable function, its Fourier series is pointwise convergent to

$$\frac{f(x^+) + f(x^-)}{2}$$

where $f(x^+)$ and $f(x^-)$ are the right and left-hand limits, respectively of $f(x)$ at the point x.

To prove this result, (known as the Dirichlet-Jordan convergence theorem), we let

$$F_n(x) = 2 \int_0^1 f(\xi) \left(\frac{1}{2} + \sum_{k=1}^n \cos 2k\pi(\xi - x) \right) d\xi$$

be the nth partial sum of the Fourier series for the function $f(x)$ on $0 \le x \le 1$. From trigonometric identities, it follows that

$$F_n(x) = \int_0^1 f(\xi) \frac{\sin(2\pi(n+1/2)(\xi - x))}{\sin \pi(\xi - x)} d\xi$$

$$= \int_0^1 \frac{\pi f(\xi)(\xi - x)}{\sin \pi(\xi - x)} \cdot \frac{\sin(2\pi(n+1/2)(\xi - x))}{\pi(\xi - x)} d\xi.$$

Recall from the first section of this chapter that $S_k(x) = \frac{\sin kx}{\pi x}$ is a delta sequence when integrated against Lipschitz continuous functions. Since $g(x) = \frac{\pi x}{\sin \pi x}$ is a continuously differentiable function near $x = 0$ when we take $g(0) = 1$, it follows that

$$\lim_{n \to \infty} F_n(x) = \frac{f(x^+) + f(x^-)}{2}.$$

We cannot drop the requirement of differentiability with impunity, since if we try to use Fourier series on functions that are simply continuous, the results may be disastrous. In fact, one can show (cf. Rudin, Real and Complex Analysis) that there are continuous functions whose Fourier series fail to converge to $f(x)$ at any countable number of points. Another way of saying this is that the Fourier series of a continuous function is not necessarily continuous.

On the other hand, the requirement of differentiability can be weakened somewhat, as it can be shown (cf. G. A. Lion, A Simple Proof of the Dirichlet-Jordan Convergence Theorem, Am. Math. Monthly **93**, 281–282 (1986)) that the theorem still holds if $f(x)$ is a function with bounded variation. The variation of a function $f(x)$ on $[a, b]$ is defined as follows. Let

$P = \{x_0, x_1, \ldots, x_n\}$ be any ordered selection of points in the interval $[a, b]$, and let $\Delta f_i = f(x_i) - f(x_{i-1})$. The variation of f is given by

$$\text{Var}(f) = \text{l.u.b.} \sum_{i=1}^{n} |\Delta f_i|,$$

where the least upper bound is taken over all possible sets P. The function f has bounded variation if $\text{Var}(f) < \infty$. For example, the function $f(x) = x^2 \sin \frac{1}{x}$, $x > 0$ and $f(0) = 0$, has bounded variation on the interval $[0, 1]$, but $g(x) = x \sin \frac{1}{x}$, $x > 0$, $g(0) = 0$ does not have bounded variation. It can be shown that a function of bounded variation can be expressed as the sum of a monotone increasing and a monotone decreasing function. The proof of convergence of Fourier series for a function with bounded variation uses some ideas that are beyond the scope of this text (cf. Rudin, Principles of Mathematical Analysis, p. 180, Problems 15, 16, 17).

There are numerous other boundary conditions that lead to complete sets of trigonometric eigenfunctions. For example, with the boundary conditions $u(0) = 0$, $u(1) = \alpha u'(1)$, the operator $Lu = -u''$ has eigenvalues λ which satisfy the transcendental equation $\tan \lambda = \alpha \lambda$, for which there are an infinite number of positive roots. The eigenfunctions are $u_n(x) = \sin \lambda_n x$, which are again complete in $L^2[0, 1]$.

A more complicated example has the defining equation $u'' + \lambda u = 0$ with boundary conditions $u(0) = 0$, $u'(1) = \lambda u(1)$. To pose this problem as an operator equation we take

$$U = \begin{pmatrix} u(x) \\ \alpha \end{pmatrix}$$

where $\alpha \in \mathbb{R}$ and define

$$LU = L \begin{pmatrix} u \\ \alpha \end{pmatrix} = \begin{pmatrix} -u'' \\ u'(1) \end{pmatrix}$$

on the set with $u(0) = 0$, $u(1) = \alpha$. On this set the associated eigenvalue problem $LU = \lambda U$ is the same as the original problem. The appropriate inner product for this operator is $\langle U, V \rangle = \int_0^1 u(x)v(x)dx + \alpha\beta$ where

$$U = \begin{pmatrix} u(x) \\ \alpha \end{pmatrix}, \qquad V = \begin{pmatrix} v(x) \\ \beta \end{pmatrix},$$

and on the space with $u(0) = 0$, $u(1) = \alpha$, the operator L is self-adjoint. The operator L is positive semi-definite since

$$\langle LU, U \rangle = \int_0^1 u'^2 dx$$

is positive for any nonconstant function. Clearly, $LU = 0$ if $u(x) = 1$. The eigenfunctions of L are the functions

$$\phi = \begin{pmatrix} \sin \sqrt{\lambda} x \\ \sin \sqrt{\lambda} \end{pmatrix}$$

where the eigenvalues $\lambda_n = \mu_n^2$ satisfy the transcendental equation $\mu_n \tan \mu_n = 1$, for which there are an infinite number of positive solutions. The eigenfunctions are complete in $L^2 \times \mathbb{R}$ and the eigenfunction expansion takes the form

$$\begin{pmatrix} f(x) \\ \alpha \end{pmatrix} = \sum_{n=1}^{\infty} a_n \begin{pmatrix} \sin \sqrt{\lambda_n} x \\ \sin \sqrt{\lambda_n} \end{pmatrix},$$

where

$$a_n = \frac{\int_0^1 f(x) \sin \sqrt{\lambda_n} x \, dx + \alpha \sin \sqrt{\lambda_n}}{\int_0^1 \sin^2 \sqrt{\lambda_n} x \, dx + \sin^2 \sqrt{\lambda_n}}.$$

For example, if $f(x) = 0$, $\alpha = 1$ we find

$$a_n = \frac{2 \sin \mu_n}{1 + 3/2 \sin^2 \mu_n}$$

so that

$$2 \sum_{n=1}^{\infty} \frac{\sin \mu_n \sin \mu_n x}{1 + 3/2 \sin^2 \mu_n} = \begin{cases} 0 & x \neq 1 \\ 1 & x = 1, \end{cases}$$

4.5.2   Orthogonal Polynomials

All of the orthogonal polynomials listed in Chapter 2 are the eigenfunctions of some self-adjoint differential operator.

1. The most well known of the polynomials are the *Legendre* polynomials

$$P_n(x) = \frac{1}{2^n n!} \frac{d^n}{dx^n} (x^2 - 1)^n,$$

which are eigenfunctions of the differential operator

$$Lu = - \left((1 - x^2) u' \right)'$$

on the interval $x \in [-1, 1]$, $\lambda_n = n(n+1)$. No extra boundary conditions are needed as long as u is bounded on $-1 \leq x \leq 1$, since the function $p(x) = 1 - x^2$ vanishes at the endpoints of the interval. The operator L is not positive definite since $Lu = 0$ for $u = 1$. However the shifted operator $Lu = - \left((1 - x^2) u' \right)' + u$ is positive definite and completeness is assured.

2. *Associated Legendre* functions are eigenfunctions of the differential operator

$$Lu = -\left((1-x^2)u'\right)' + \left(\frac{m^2}{1-x^2}\right)u$$

on the interval $-1 \leq x \leq 1$. The operator L is positive definite and the eigenfunctions are

$$P_n^m(x) = \left(1-x^2\right)^{m/2} \frac{d^m P_n(x)}{dx^m}, \qquad n > m,$$

where $P_n(x)$ is the nth order Legendre polynomial. The eigenvalues are $\lambda_{n,m} = n(n+1)$.

3. *Chebyshev* polynomials $T_n(x) = \cos\left(n\cos^{-1}x\right)$ arise as eigenfunctions of the differential operator

$$Lu = -\left(1-x^2\right)^{1/2}\left(\left(1-x^2\right)^{1/2}u'\right)'$$

on $-1 \leq x \leq 1$ and are therefore complete in L^2 with respect to the inner product with weight function $w(x) = \left(1-x^2\right)^{-1/2}$. The eigenvalues are $\lambda_n = n^2$.

4. *Jacobi* polynomials are polynomial eigenfunctions of the operator

$$Lu = -\frac{1}{(1-x)^\alpha(1+x)^\beta}\left((1-x)^{\alpha+1}(1+x)^{\beta+1}u'\right)'$$

on the interval $-1 \leq x \leq 1$. The eigenvalues are $\lambda_n = n(n+\alpha+\beta+1)$ and the eigenfunctions are

$$P_n^{\alpha,\beta}(x) = \frac{(-1)^\alpha}{n!2^n}(1-x)^{-\alpha}(1+x)^{-\beta}\frac{d^n}{dx^n}\left[(1-x)^{\alpha+n}(1+x)^{\beta+n}\right].$$

For $\alpha = \beta = 0$, we recover the Legendre polynomials, and with $\alpha = \beta = -1/2$ we recover the Chebyshev polynomials, modulo a multiplicative scalar. *Gegenbauer* polynomials are those for which $\alpha = \beta$.

5. On the domain $0 < x < \infty$, *Laguerre* polynomials are the functions

$$L_n(x) = \sum_{k=0}^{n}\binom{n}{k}\frac{(-x)^k}{k!} = \frac{1}{n!}e^x\frac{d^n}{dx^n}\left(x^n e^{-x}\right)$$

which are orthogonal with respect to the inner product

$$\langle u,v\rangle = \int_0^\infty u(x)v(x)e^{-x}dx$$

and are eigenfunctions of the differential equation

$$xv'' + (1-x)v' + \lambda v = 0$$

with $\lambda_n = n$.

6. The *generalized Laguerre* functions are

$$L_n^{(k)}(x) = \frac{d^k L_n(x)}{dx^k}.$$

The eigenfunctions of the differential equation

$$xu'' + 2u' + \left(\lambda - \left(\frac{x}{4} + \frac{k^2 - 1}{4x}\right)\right) u = 0$$

are $u_n = e^{-x/2} x^{\frac{k-1}{2}} L_n^{(k)}(x)$, $\lambda_n = n - \frac{k-1}{2}$.

7. The SCHRÖDINGER EQUATION

$$u'' + (E - V(x)) u = 0$$

describes the density function for the location of an electron attracted by an energy potential $V(x)$. The eigenvalues E correspond to the energy levels of the electron, which, because the eigenvalues are discrete, correspond to the fact that energy levels are quantized. A simple example is $V(x) = x^2 - 1$. Setting $u = e^{-x^2/2} v$ we find

$$v'' - 2xv' + \lambda v = 0.$$

The eigenfunctions are $v = H_n(x)$, the *Hermite* polynomials, where

$$H_n(x) = (-1)^n e^{x^2/2} \frac{d^n}{dx^n} e^{-x^2/2}$$

$$\lambda_n = 2n.$$

Notice that because the Hermite, Laguerre and Associated Laguerre functions are defined on an infinite domain, our completeness theorem does not apply. Proof of completeness of these eigenfunctions will be deferred until later (Chapter 7).

4.5.3   Special Functions

Another important classical differential equation is BESSEL'S EQUATION $(xu')' - \frac{n^2}{x} u + \lambda x u = 0$. Solutions for this equation which are bounded at the origin are the nth order Bessel functions $J_n(\sqrt{\lambda} x)$. A boundary condition at $x = x_0$ such as $u(x_0) = 0$ (or $u'(x_0) = 0$), determines eigenvalues λ_k which satisfy the transcendental equation $J_n(\sqrt{\lambda_k}) = 0$ (or $J_n'(\sqrt{\lambda_k}) = 0$). The operator $Lu = -\frac{1}{x}(xu')' + \frac{n^2 u}{x^2}$ is positive definite and eigenfunctions $J_n(\sqrt{\lambda_k} x)$ are complete on the Hilbert space $L^2[0, x_0]$ with respect to the inner product $\langle u, v \rangle = \int_0^{x_0} u(x)v(x)x\,dx$. We will examine the properties of these special functions in more detail in Chapter 6.

4.5.4   Discretized Operators

One of the ways to solve boundary value problems numerically is to discretize the differential operator and then solve the resulting system of algebraic equations. As a simple example, consider the boundary value problem $u'' = f(x)$ subject to boundary conditions $u(0) = u(1) = 0$. If we introduce mesh points $x_j = j/N$, and set $u_j = u(x_j)$, $f_j = f(x_j)$ we can approximate the differential equation with the discrete system

$$u_{j+1} - 2u_j + u_{j-1} = h^2 f_j \qquad j = 1, 2, \ldots, N - 1$$

$$u_0 = u_N = 0.$$

This system of algebraic equations can be represented as a matrix equation with a tridiagonal matrix. The most direct (and fastest) way to solve this system of equations is to use Gaussian elimination specialized to tridiagonal matrices. However, it is valuable to see how transform methods work on this system of equations.

We consider the eigenvalue problem

$$\Delta u = \lambda u,$$

where Δu is the discretized operator

$$(\Delta u)_n = u_{n+1} - 2u_n + u_{n-1}$$

acting on the space of $\mathbb{R}^{N-1}$ vectors with $u_0 = u_N = 0$. The operator Δu is self adjoint with respect to the Euclidean inner product in $\mathbb{R}^{N-1}$ and therefore we expect to find a full set of mutually orthogonal eigenvectors.

Since Δu is the discretized version of u'', we guess that the eigenfunctions are also related through discretization. That is, we try a vector $\phi^{(k)}$ with entries

$$\phi_j^{(k)} = \sin \frac{\pi j k}{N} \qquad j = 1, \ldots, N - 1, k \text{ fixed}.$$

Notice that $\phi_0^{(k)} = \phi_N^{(k)} = 0$ and that

$$\left(\Delta \phi^{(k)} \right)_j = \phi_{j+1}^{(k)} - 2\phi_j^{(k)} + \phi_{j-1}^{(k)} = -4 \sin^2 \frac{\pi k}{2N} \cdot \phi_j^{(k)}.$$

In other words, the vector $\phi^{(k)}$ has eigenvalue $\lambda_k = -4 \sin^2 \frac{\pi k}{2N}$ and the eigenvalues λ_k are distinct for $k = 1, 2, \ldots, N-1$. As a result the $N-1$ eigenvectors $\phi^{(k)}$, $k = 1, 2 \ldots, N - 1$ are mutually orthogonal and diagonalize the operator Δu.

The solution of the difference equation using transforms is now clear. We suppose the solution u to be represented in terms of the eigenvectors $\phi^{(k)}$

$$u = \sum_{k=1}^{N-1} \alpha_k \phi^{(k)}$$

and represent f as

$$f = \sum_{k=1}^{N-1} \beta_k \phi^{(k)}.$$

Upon substitution into the difference equation, we find that we must take $\alpha_k = h^2 \beta_k / \lambda_k$.

The implementation of this solution can be made quite efficient. Notice that we have here applied the discrete Sine transform, which has a fast algorithm (Chapter 2). Thus, to implement this solution technique numerically, we first apply the fast Sine transform to the vector f to find the vector β. We divide the components β_k by λ_k to obtain α_k and then apply the inverse fast Sine transform to the vector α to find the solution vector u.

Further Reading

Two general references on the theory of distributions are

1. I. M. Gelfand and G. E. Shilov, *Generalized Functions*, Vol. I, Academic Press, New York, 1972,

and

2. A. H. Zemanian, *Distribution Theory and Transform Analysis*, McGraw-Hill, New York, 1965.

The properties of Green's functions are discussed in numerous books, including

3. R. Courant and D. Hilbert, *Methods of Mathematical Physics*, Vol. I, Interscience, New York, 1953.

4. I. Stakgold, *Boundary Value Problems of Mathematical Physics*, Macmillan, New York, 1968.

5. G. F. Roach, *Green's Functions: Introductory Theory with Applications*, Van Nostrand Reinhold, London, 1970.

6. B. Friedman, *Principles and Techniques of Applied Mathematics*, J. Wiley and Sons, New York, 1956.

This last book also uses the extended definition of an operator to deal with inhomogeneous boundary conditions.

Three references mentioned in the text in reference to convergence of Fourier series are

7. W. Rudin, *Principles of Mathematical Analysis*, McGraw-Hill, New York, 1964.

8. W. Rudin, *Real and Complex Analysis*, McGraw-Hill, 1974.

9. G. A. Lion, *A Simple Proof of the Dirichlet-Jordan Convergence Theorem*, Am. Math. Monthly **93**, 281–282 (1986).

Problems for Chapter 4

Section 4.1

1. Show that

$$\lim_{k \to \infty} \int_{-\infty}^{\infty} S_k(x)\phi(x)dx = \phi(0)$$

for any $\phi(x)$ which is continuously differentiable and bounded on the interval $-\infty < x < \infty$ where

a. $S_k(x) = \begin{cases} 1/2 + \sum_{j=1}^{k} \cos j\pi x, & |x| < 1 \\ 0, & x \geq 1 \end{cases}$

Remark: It is not actually necessary to assume that $\phi(x) \in C^1$. One only needs that $|\phi(x) - \phi(0)| \leq M|x|$ for x on some open interval containing the origin.

b. $S_k(x) = \begin{cases} -4k^2|x| + 2k & |x| \leq 1/2k \\ 0 & |x| \geq 1/2k \end{cases}$

c. $S_k(x) = \begin{cases} -k & |x| \leq 1/2k \\ 2k & 1/2k \leq |x| \leq 1/k \\ 0 & |x| > 1/k \end{cases}$

2. Show that

$$1 + 2\sum_{n=1}^{\infty} \cos 2n\pi x = \sum_{k=-\infty}^{\infty} \delta(x - k)$$

in the sense of distribution.

3. Prove the quotient rule for distributions. (If f is C^∞ and g is a distribution then $(g/f)' = \frac{g'f - f'g}{f^2}$ whenever $f \neq 0$.)

4. Show that a test function $\psi(x)$ is of the form $\psi = (x\phi)'$ where ϕ is a test function if and only if

$$\int_{-\infty}^{\infty} \psi(x)dx = 0 \qquad \text{and} \qquad \int_{0}^{\infty} \psi(x)dx = 0.$$

Hint: It is helpful to use that $\phi(x) - \phi(0) = x \int_0^1 \phi'(xt)dt$.

Solve the following differential equations in the sense of distribution.

5. $x^2 \frac{du}{dx} = 0$.

6. $\frac{d^2u}{dx^2} = \delta''(x)$.

7. $x\frac{du}{dx} = u$.

8. $\frac{d}{dx}(xu) = u$.

Express the following in terms of the usual δ-function.

9. $\delta\left(x^2 - a^2\right)$.

10. $H'\left(x^2 - a^2\right)$.

11. Find the derivative in the sense of distribution of the function

$$\chi(x) = \begin{cases} 1 & x \quad \text{irrational} \quad 0 < x < 1 \\ 0 & x \quad \text{elsewhere} \end{cases}$$

12. The convolution of regular functions is defined by

$$(f * g)(x) = \int_{-\infty}^{\infty} f(x - t)g(t)dt.$$

    a. Define the convolution of distributions.

    b. What is $\delta * \delta$?

13.   **a.** Find the weak formulation for the partial differential equation

$$\frac{\partial u}{\partial t} + c\frac{\partial u}{\partial x} = 0.$$

    Hint: Define the appropriate test functions in two variables x, t and define the correct rules for partial differentiation, integration by parts and change of variables.

    b. Show that $u = f(x - ct)$ is a generalized solution of $u_t + cu_x = 0$ if f is a generalized function.

Section 4.2

Construct Green's functions (if they exist) for the following equations.

1. $u'' = f(x)$, $u(0) = u'(1) = 0$.
2. $u'' = f(x)$, $u(0) = 0$, $u'(1) = u(1)$.
3. $u'' + \alpha^2 u = f(x)$, $u(0) = u(1)$, $u'(0) = u'(1)$.
   For what values of α does the Green's function fail to exist?
4. $u'' = f(x)$, $u(0) = 0$, $\int_0^1 u(x)dx = 0$.
5. $(xu')' = f(x)$, $u(1) = \alpha$, $\lim_{x \to 0} (xu'(x)) = 0$.
6. $u'' + \frac{3}{2x}u' - \frac{3}{2}\frac{u}{x^2} = f(x)$, $u(0) = 0$, $u'(1) = 0$.
7. Use a Green's function to solve $u'' - \alpha^2 u = f(x)$ on $L^2(-\infty, \infty)$.
8. The operator $Lu = u'' + 4u$ with boundary conditions $u'(0) = u'(\pi)$, $u(0) = u(\pi)$ has no Green's function. Why?

Convert the following differential equations to Fredholm integral equations of the form $u = \lambda \int k(x,\xi)u(\xi)d\xi + g(x)$.

9. $u'' + \lambda u = f(x)$, $u(0) = \alpha$, $u(1) = \beta$.
10. $u'' + \lambda u = f(x)$, $u(0) = u'(0)$, $u(1) = -u'(1)$.
11. $(xu')' - \frac{n^2 u}{x} + \lambda u = f(x)$, $u(0)$ finite, $u(1) = 0$.
12. $u'' - u + \lambda u = f(x)$, $u(\infty) = u(-\infty) = 0$.
13. $u'' + u = \lambda p(x)u$, $u(0) = 1$, $u'(0) = 0$.
14. $u'' + \frac{1}{x}u' + \lambda u = 0$, $u(0) = 1$, $u'(0) = 0$.
15. Suppose $A(t)$ is an $n \times n$ matrix, and suppose the matrix $X(t)$ consists of columns all of which satisfy the vector differential equation $\frac{dx}{dt} = A(t)x$ for $t \in [t_1, t_2]$, and $X(t)$ is invertible (the columns are linearly independent). Show that the boundary value problem

$$\frac{dx}{dt} = A(t)x + f(t)$$

$$Lx(t_1) + Rx(t_2) = \alpha,$$

where $f(t)$, α are vectors and L, R are $n \times n$ matrices, has a unique solution if and only if

$$M \equiv LX(t_1) + RX(t_2)$$

is a nonsingular matrix, and that, if this is the case, the solution can be written as

$$x(t) = X(t)M^{-1}\alpha + \int_{t_1}^{t_2} G(t,s)f(s)ds,$$

where

$$G(t, s) = \begin{cases} X(t)M^{-1}LX\,(t_1)\,X^{-1}(s) & s < t \\ -X(t)M^{-1}RX\,(t_2)\,X^{-1}(s) & s > t \end{cases}$$

(The matrix G is the Green's function for this problem.)

16. Suppose the matrix M in Problem 15 is singular. Find the modified Green's function for the problem as stated in Problem 15.

Section 4.3

Find the adjoint operator L^*, its domain, and the extended definition of L for the operators in Problems 1–3.

1. $Lu = u'' + a(x)u' + b(x)u$, $u(0) = u'(1)$, $u(1) = u'(0)$. (Assume $a(x)$ is continuously differentiable and $b(x)$ is continuous on the interval $[0, 1]$.)

2. $Lu = -\,(p(x)u')' + q(x)u$, $u(0) = u(1)$, $u'(0) = u'(1)$. (Assume $p(x)$ is continuously differentiable and $q(x)$ is continuous on the interval $[0, 1]$.)

3. $Lu = u'' + 4u' - 3u$, $u'(0) + 4u(0) = 0$, $u'(1) + 4u(1) = 0$.

4. a. Relate $g(x, \xi)$ to $g^*(x, \xi)$ when the inner product has a nonconstant weight function $w(x)$.

    b. Express the solution of $\widehat{L}u = \hat{f}$ in terms of the Green's function when the inner product has a nonconstant function $w(x)$.

5. Show that a Sturm-Liouville operator with separated boundary conditions is self adjoint.

6. Show that a Sturm-Liouville operator with periodic boundary conditions on $[a, b]$ is self adjoint if and only if $p(a) = p(b)$.

7. Find the adjoint operator for the matrix differential operator

$$Mx = \frac{dx}{dt} - A(t)x \qquad t_1 \le t \le t_2$$

with domain $Lx\,(t_1) + Rx\,(t_2) = 0$, where A, L, R are $n \times n$ matrices. Under what conditions is this a self-adjoint operator?

8. a. Is the operator $Lu = u''$ with boundary conditions $u(0) = u(1) = 0$ self adjoint with respect to the inner product for $H^1(0, 1)$?

    b. Can the domain of the operator $Lu = u''$ be defined so as to make it self-adjoint in $H^1(0, 1)$?

Find solvability conditions for the equations in Problems 9–12.

9. $u'' + u = f(x)$, $u(0) - u(2\pi) = \alpha$, $u'(0) - u'(2\pi) = \beta$.

10. $u'' = f(x)$, $u(0) - u(1) = \alpha$, $u'(0) - u'(1) = \beta$.

11. $u'' + \pi^2 u = f(x)$, $u(0) = \alpha$, $u'(1/2) = \beta$.

12. $(xu')' = f(x)$, $u'(1) = \alpha$, $\left. xu'(x) \right|_{x=0} = \beta$.

Section 4.4

Find modified Green's functions for the equations in Problems 1–6.

1. $u'' = f(x)$, $u(0) = u(1)$, $u'(0) = u'(1)$.

2. $u'' = f(x)$, $u(0) + u(1) = 0$, $u'(0) = u'(1)$.

3. $u'' + 4\pi^2 u = f(x)$, $u(0) = u(1)$, $u'(0) = u'(1)$.

4. $u'' + u = f(x)$, $u(0) = u(\pi) = 0$.

5. $u'' = f(x)$, $u(0) = 0$, $u(1) = u'(1)$.

6. $\left((1 - x^2)\, u' \right)' = f$, $x \in [-1, 1]$, u bounded.

Solve the problems 7–9 in the best possible (least squares) way.

7. $u'' = \sin^2 x$, $u'(0) = \alpha$, $u'(\pi) = \beta$.

8. $u'' + u = \sin^3 x$, $u(0) = 1$, $u(\pi) = 2$.

9. $u'' = x$, $u(0) = 0$, $u(1) = u'(1)$.

10. Suppose u_i, $i = 1, 2, \ldots, n$ are (column) vector solutions of the system of differential equation $du/dt = A(t)u$, where A is an $n \times n$ matrix. Define the Wronskian of $u_1, u_2, \ldots, u_n$ to be $W(u_1, u_2, \ldots, u_n) = \det[u_1, u_2, \ldots, u_n]$. Show that W satisfies $dW/dt = \text{trace}\,(A(t))\, W$.

Section 4.5

Use the appropriate eigenfunction expansion (if it exists) to represent the best solution of the following problems.

1. $u'' = f(x)$, $u'(0) = \alpha$, $u'(1) = \beta$.

2. $u'' + u = f(x)$, $u(0) = u(2\pi)$, $u'(0) = u'(2\pi)$.

3. $u'' = f(x)$, $u(0) = u(1)$, $u'(0) = u'(1)$.

4. $u'' = f(x)$, $u(0) + u(1) = 0$, $u'(0) = u'(1)$.

5. $u'' = \sin x$, $u'(0) = \alpha$, $u(\pi) = \beta$.

6. $\left((1 - x^2)\, u' \right)' = a + bx + cx^2$, $-1 < x < 1$, u bounded.

7. $xu'' + (1 - x)u' = a + bx + cx^2$, $0 < x < \infty$.

8. $u'' - x^2 u = (a + bx + cx^2)\, e^{-x^2/2}$, $x \in (-\infty, \infty)$, u bounded.

9. Expand $f(x) = \cos \pi x$ in terms of the eigenfunctions of $u'' + \lambda u = 0$ $u(-1) = u(1) = 0$.

10. Verify that $u = P_n^m(x)$ is an eigenfunction for the differential operator $Lu = \left((1-x^2)u'\right)' + \frac{m^2 u}{1-x^2}$ using that $v = P_n(x)$ satisfies Legendre's equation $\left((1-x^2)v'\right)' + n(n+1)v = 0$.

11. **a.** Find the eigenpairs associated with the difference operator $(\Delta u)_n = u_{n+1} - 2u_n + u_{n-1}$, $0 \le n \le N$, with conditions $u_0 = u_1$, $u_{N-1} = u_N$.

   b. Develop an algorithm for the fast implementation of the transform associated with these eigenfunctions.

12. **a.** Show that the eigenvalue problem

$$-u'' + 4\pi^2 \int_0^1 u(x)dx = \lambda u, \qquad 0 < x < 1,$$

$$u(0) = u(1), \qquad u'(0) = u'(1)$$

   has $\lambda = 4\pi^2$ as an eigenvalue of multiplicity 3.

   b. Find all eigenvalues and eigenfunctions for this problem. Do the eigenfunctions form a complete set?

13. Show that the variation of $f(x) = x^2 \sin \frac{1}{x}$ is bounded but that the variation of $g(x) = x \sin \frac{1}{x}$ is not bounded.

14. Use the tent finite element functions $S^h(1,0)$ and the Galerkin procedure to find a discretization of the differential equation $u'' = f(x)$, $u(0) = u(1) = 0$. How does this discretization compare with the finite difference approximation $u_{n+1} - 2u_n + u_{n-1} = h^2 f_n$?

5

CALCULUS OF VARIATIONS

As we saw in Chapter 1, the eigenvalues of a self-adjoint matrix A can be characterized by a variational principle. As a result, changes in A which produce monotone changes in $\langle Ax, x \rangle$ also produce monotone changes in the eigenvalues of A.

In this chapter we give a brief introduction to the CALCULUS OF VARIATIONS in order to show that a similar variational characterization applies to the eigenvalues of differential operators. In addition, in the process of deriving these properties we will learn how to solve other important variational (or extremal) problems.

5.1   Euler-Lagrange Equations

Recall from calculus that a necessary condition for x_0 to be a maximum or minimum of the function $f(x)$ is that $\nabla f(x_0) = 0$. A simple geometrical way to derive this is to observe that if $f(x_0)$ is the largest value taken on by $f(x)$, then for any $y \neq 0$, $\alpha \neq 0$, $f(x_0) > f(x_0 + \alpha y)$. Using Taylor's theorem we know that $f(x_0 + \alpha y) = f(x_0) + \alpha \nabla f(x_0) \cdot y +$ terms bounded by a constant times α^2, for all α small enough. Now, if $\nabla f(x_0) \neq 0$, there is a direction

y for which $\nabla f(x_0) \cdot y \neq 0$ and a change of sign of α changes the sign of $f(x_0) - f(x_0 + \alpha y)$, which is not permitted. In other words, if $\nabla f(x_0) \neq 0$, there is a direction in which to move which increases $f(x)$ and movement in the opposite direction decreases f, which contradicts that $f(x_0)$ was maximal (or minimal).

The problem of the calculus of variations is to maximize (or minimize) not functions, but FUNCTIONALS. For example, suppose $F(x, y, z)$ is a function of the three variables x, y, z and we wish to find a function $y(x)$ which maximizes $J(y) = \int_0^1 F(x, y(x), dy/dx)\, dx$ over the class of all piecewise C^1 functions $y(x) \in PC^1[0, 1]$ with $y(0) = y_0$ and $y(1) = y_1$, say. The set of piecewise C^1 functions $y(x)$ with appropriate boundary conditions is referred to as the class of admissible functions for this problem. Suppose $y_0(x)$ is the function that maximizes J and suppose $\eta(x)$ is any PC^1 function with $\eta(0) = \eta(1) = 0$. For any scalar α, $y_0 + \alpha\eta$ is admissible, and y_0 maximizes $J(y)$ only if $\frac{\partial J}{\partial \alpha}(y_0 + \alpha\eta)\,|_{\alpha=0} = 0$. The latter holds if and only if

$$\Delta J = \int_0^1 \left(\frac{\partial F}{\partial y}\eta + \frac{\partial F}{\partial y'}\eta'\right) dx = 0.$$

After integrating by parts, we require,

$$\Delta J = \int_0^1 \left[\frac{\partial F}{\partial y} - \frac{d}{dx}\left(\frac{\partial F}{\partial y'}\right)\right]\eta(x)dx = 0$$

for any fixed function $\eta(x)$, where we have used that $\eta(0) = \eta(1) = 0$ to eliminate the boundary contribution. Clearly, we must have

$$\frac{\partial F}{\partial y} - \frac{d}{dx}\frac{\partial F}{\partial y'} = 0.$$

This equation is called the EULER-LAGRANGE EQUATION for the functional $J(y)$.

The operator ΔJ, which corresponds to the derivative of the J with respect to the function y, is called the FRECHÉT DERIVATIVE of J. In general, the Frechét derivative of an operator $K(u)$ is defined as the linear operator L_u for which

$$K(u + h) = K(u) + L_u h + R(h)$$

where

$$\lim_{\|h\| \to 0} \frac{\|R(h)\|}{\|h\|} = 0.$$

Example

Find the arc with the shortest distance between the two points $(x, y) = (0, a)$ and $(x, y) = (1, b)$.

We suppose that the solution can be expressed as a function of one variable $y = y(x)$ and we seek the function $y = y(x)$ which minimizes

$$D = \int_0^1 \left(1 + \left(\frac{dy}{dx}\right)^2\right)^{1/2} dx, \qquad y(0) = a, \qquad y(1) = b.$$

For the functional D, the Euler-Lagrange equation is

$$\frac{d}{dx}\left(\frac{y'}{(1 + y'^2)^{1/2}}\right) = 0.$$

We easily solve this equation and find that we must have $y' = $ constant, and of course, this implies that $y(x) = a + (b - a)x$.

The solution of this problem was rather easy since the functional did not depend explicitly on $y(x)$. For the functional $J = \int_a^b F(x, y, y')dx$, the Euler-Lagrange equation simplifies in some special cases:

1. If $\partial F/\partial y \equiv 0$ the Euler-Lagrange equation becomes $\frac{\partial F(x,y')}{\partial y'} = $ constant so that $y'(x)$ is specified implicitly as some function of x, and $y(x)$ can be determined by quadrature.

2. If $\partial F/\partial y' \equiv 0$ the Euler-Lagrange equation is $\frac{\partial F(x,y)}{\partial y} = 0$ which specifies $y = y(x)$ implicitly.

3. If $\partial F/\partial x \equiv 0$, then

$$\frac{d}{dx}\left(F - y'\frac{\partial F}{\partial y'}\right) = \left(\frac{\partial F}{\partial y} - \frac{d}{dx}\frac{\partial F}{\partial y'}\right)y' = 0$$

so that $F - y'\frac{\partial F}{\partial y'} = $ constant is a first integral of the Euler-Lagrange equation.

Example

Find the path $y = y(x)$ which minimizes the time of descent for a sliding frictionless bead between $y(0) = y_0$ and $y(x_1) = 0$.

This is a special case (as was the first example) of FERMAT'S PROBLEM to find the path of minimal time of travel in a medium where the speed of travel is given at each point by $c(x, y)$. Since $\frac{ds}{dt} = c(x, y)$ where ds is incremental arclength, it follows that for any arc $y = y(x)$, the time of travel is

$$T = \int_0^{x_1} \frac{(1 + y'^2(x))^{1/2}}{c(x, y)} dx.$$

The ball is at rest at height y_0, and since the potential plus kinetic energies are a constant, the speed at height y is $c = \sqrt{2(y_0 - y)g}$, and the time of descent is

$$T = \int_0^{x_1} \left(\frac{1 + y'^2}{2(y_0 - y)g} \right)^{1/2} dx$$

$$= \int_0^{x_1} F(y, y') dx.$$

Since F does not depend explicitly on x, the Euler-Lagrange equation has the first integral

$$y' \frac{\partial F}{\partial y'} - F = \text{constant}$$

or

$$(2g(y_0 - y)(1 + y'^2))^{1/2} = \text{constant}.$$

From this first integral we can extract the first order separable differential equation

$$\frac{dy}{dx} = -\left(\frac{c_0^2 - (y_0 - y)}{y_0 - y} \right)^{1/2}.$$

with c_0 an arbitrary constant. The solution of this equation is found most readily by making the parametric substitution $y = y_0 + c_0^2(\cos t - 1)/2$ from which it follows that

$$\frac{dx}{dt} = \frac{1}{2}c_0^2(1 - \cos t)$$

or

$$x = \frac{1}{2}c_0^2(t - \sin t).$$

The constant c_0 must be chosen so that $y = 0$ at $x = x_1$. The solution curve is a cycloid.

This problem is a special case of the stratified medium problem where $c = c(y)$. Since there is no explicit dependence on x in a stratified medium, we know that a first integral is

$$\frac{1}{c(y)(1 + y'^2)^{1/2}} = K.$$

Since $dy/dx = \tan\theta$, where θ is the angle between the tangent to the curve $y(x)$ and the horizontal axis, it follows that

$$\frac{\cos\theta}{c(y)} = K$$

along the path of fastest travel. This relationship, called Snell's law, is fundamental in the study of wave propagation. From this follow the well-known equations for refraction of waves at an interface between two different media.

There are a number of generalizations of the Euler-Lagrange equation which we now consider.

5.1.1   Constrained Problems

Suppose we wish to minimize a functional

$$J = \int_a^b F(x, y, y')dx$$

subject to a constraint

$$K = \int_a^b G(x, y, y')dx = 0.$$

As in calculus, we use Lagrange multipliers, and extremize $J + \lambda K$. The resulting Euler-Lagrange equation is

$$\frac{\partial}{\partial y}(F + \lambda G) - \frac{d}{dx}\left[\frac{\partial}{\partial y'}(F + \lambda G)\right] = 0$$

which must be solved subject to $K = 0$.

Example   The Isoperimetric Problem

Find the curve $y(x)$, $0 \leq x \leq 1$, with $y(0) = y(1) = 0$ and fixed arclength that has maximal area $A = \int_0^1 y(x)dx$.

As a constrained problem we seek to maximize A subject to the constraint $\int_0^1 \left(1 + y'^2\right)^{1/2} dx - l = 0$. For this problem the Euler-Lagrange equation is

$$1 - \lambda\frac{d}{dx}\frac{y'}{\left(1 + y'^2\right)^{1/2}} = 0$$

from which it follows that

$$\frac{x - k_1}{\lambda} - \frac{y'}{\left(1 + y'^2\right)^{1/2}} = 0.$$

Perhaps the easiest way to solve this equation is in arclength coordinates. We set $x = x(s)$, $y = y(s)$ with $(dx/ds)^2 + (dy/ds)^2 = 1$ to obtain

$$\left(\frac{dx}{ds}\right)^2 = 1 - \left(\frac{x - k_1}{\lambda}\right)^2.$$

The solution of this equation is

$$x(s) = \lambda \sin\left(s/\lambda - \phi\right) + k_1$$

$$y(s) = \lambda \cos\left(s/\lambda - \phi\right) + k_2,$$

where the constants λ, ϕ, k_1, k_2 are arbitrary. We choose these constants to fit the data $x(0) = 0$, $x(l) = 1$, $y(0) = y(l) = 0$. It follows that

$$x(s) = \lambda\left(\sin\left(\frac{s - l/2}{\lambda}\right) + \sin\frac{l}{2\lambda}\right)$$

$$y(s) = \lambda\left(\cos\left(\frac{s - l/2}{\lambda}\right) - \cos\frac{l}{2\lambda}\right)$$

provided $2\lambda\sin\frac{l}{2\lambda} = 1$ has a solution with $\frac{l}{2\lambda} < \frac{\pi}{2}$. This solution which, unfortunately becomes invalid if $l > \pi/2$, is an arc of a circle.

Question: What is the solution of this problem when $l > \pi/2$?

5.1.2 Several Unknown Functions

Suppose we wish to extremize a functional with several unknown functions

$$J = \int_{x_0}^{x_1} F\left(x, y_1, y_1', y_2, y_2', \ldots, y_k, y_k'\right) dx.$$

Following our original argument we assume that we know the extremal functions y_1^*, y_2^*, ..., y_k^* and introduce the admissible functions $y_i = y_i^* + \alpha_i\eta_i$, $i = 1, 2, \ldots, k$ where α_i are independent parameters. Clearly, in order to extremize J we must require $\frac{\partial J}{\partial \alpha_i}\big|_{\alpha_i=0} = 0$ for $i = 1, 2, \ldots, k$, for every choice of fixed function η_i. As a result, we conclude that we must satisfy the system of equations

$$\frac{\partial F}{\partial y_i} - \frac{d}{dx}\frac{\partial F}{\partial y_i'} = 0, \qquad i = 1, 2, \ldots k,$$

which are the k Euler-Lagrange equations for this problem.

Example

Find the arc with shortest distance between two points.

Suppose $x(t)$, $y(t)$, $0 \le t \le 1$, is an arc between two specified points. The arclength of the curve is

$$L = \int_0^1 \left(\dot{x}^2 + \dot{y}^2\right)^{1/2} dt.$$

The corresponding Euler-Lagrange equations are

$$\frac{d}{dt}\left\{\frac{\dot{x}}{(\dot{x}^2 + \dot{y}^2)^{1/2}}\right\} = 0, \qquad \frac{d}{dt}\left\{\frac{\dot{y}}{(\dot{x}^2 + \dot{y}^2)^{1/2}}\right\} = 0$$

so that $\dot{x}$ and $\dot{y}$ are constants. The solution is, as expected, a straight line, although in this formulation we did not assume that the curve can be represented as a function of one variable.

This example is not particularly interesting because it yields a well-known result. A much more interesting problem is the following:

Example

Find the space-time curves which minimize distance when measured by the Schwarzschild metric

$$ds^2 = \left(c^2 - \frac{2GM}{r}\right) dt^2 - \frac{c^2 dr^2}{c^2 - \frac{2GM}{r}} - r^2 d\phi^2 .$$

This metric includes relativistic effects on the motion of a small body in a gravitational field produced by a much larger body. It is a special case of the Schwarzschild metric

$$ds^2 = \left(c^2 - \frac{2GM}{r}\right) dt^2 - \frac{c^2 dr^2}{c^2 - \frac{2GM}{r}} - r^2 d\phi^2 - r^2 \cos^2 \theta \, d\theta^2$$

with $\theta = 0$ (a planar orbit). The constants G and M are the gravitational constant and the mass of the large central body respectively, and c is the speed of light.

We look for geodesic arcs, that is, curves parameterized by a parameter p, say, that minimize

$$\int_{p_0}^{p_1} \frac{ds}{dp} dp,$$

where $t = t(p)$, $r = r(p)$, $\phi = \phi(p)$.

The Euler-Lagrange equations for this functional are

$$\frac{d}{dp}\left(\frac{e^\mu \dot{t}}{ds/dp}\right) = 0$$

$$\frac{d}{dp}\left(\frac{r^2 \dot{\phi}}{ds/dp}\right) = 0$$

$$\frac{1}{2}\left(\frac{d\mu}{dr}\left(e^\mu \dot{t}^2 + e^{-\mu}\dot{r}^2\right) - 2r\dot{\phi}^2\right)\frac{dp}{ds} + \frac{d}{dp}\left(\frac{e^{-\mu}\dot{r}}{ds/dp}\right) = 0,$$

where for notation we have taken $e^\mu = 1 - \frac{2GM}{c^2 r}$. It is convenient to choose the parameter p to be the arclength s so that $ds/dp = 1$. It follows that

$$e^\mu \frac{dt}{ds} = h_0$$

$$r^2 \frac{d\phi}{ds} = h_1$$

$$\frac{d^2 r}{ds^2} + \frac{c^2}{s}\mu' e^{2\mu}\left(\frac{dt}{ds}\right)^2 - \frac{1}{2}\mu'\left(\frac{dr}{ds}\right)^2 - re^\mu\left(\frac{d\phi}{ds}\right)^2 = 0$$

and

$$c^2 e^{\mu} \left(\frac{dt}{ds} \right)^2 - e^{-\mu} \left(\frac{dr}{ds} \right)^2 - r^2 \left(\frac{d\phi}{ds} \right)^2 = 1.$$

This last equation is a first integral of the system and therefore replaces the third equation of this set. Now we can eliminate $t(s)$ and obtain

$$h_0^2 - \left(\frac{dr}{ds} \right)^2 = \left(r^2 \left(\frac{d\phi}{ds} \right)^2 + 1 \right) \left(1 - \frac{2GM}{c^2 r} \right),$$

$$r^2 \frac{d\phi}{ds} = h_1,$$

which are the relativistic versions of the integrals of energy and angular momentum. We leave it as an exercise to the reader to show that with the change of variables $u = 1/r$ and one differentiation this system of equations becomes

$$\frac{d^2 u}{d\phi^2} + u = \frac{GM}{c^2 h_1^2} + \frac{3GM u^2}{c^2}.$$

We shall study this equation in Chapter 11 and show how it explains the advance of the perihelion of Mercury.

5.1.3   Higher Order Derivatives

Suppose we wish to extremize the functional

$$J = \int_{x_0}^{x_1} F \left(x, y, \frac{dy}{dx}, \frac{d^2 y}{dx^2}, \ldots, \frac{d^k y}{dx^k} \right) dx,$$

and suppose, to be specific, that $y, y', \ldots, \frac{d^k y}{dx^k}$ are specified at the two boundary points $x = x_0$ and $x = x_1$. We introduce an admissible function $y = y^* + \alpha \eta$ where y^* is the assumed extremal function with

$$\eta(x_0) = \eta(x_1) = \eta'(x_0) = \eta'(x_1) = \cdots = \frac{d^k \eta}{dx^k}(x_1) = 0.$$

Following the usual arguments, we find that we must require

$$F_y - \frac{d}{dx} F_{y'} + \frac{d^2}{dx^2} F_{y''} + \cdots + (-1)^k \frac{d^k}{dx^k} F_{y^{(k)}} = 0$$

which is the Euler-Lagrange equation for this problem.

5.1.4   Variable Endpoints

In the preceding discussion we have assumed that we are looking for a function $y(x)$ (or functions $y_i(x)$, $i = 1, 2, \ldots, k$) whose endpoint values are specified. If, however, the endpoint values are not specified a priori, the problem changes slightly.

Suppose we wish to find the function $y(x)$ which extremizes the functional

$$ J = \int_0^1 F(x, y, y') dx $$

but that $y(x)$ is not specified at $x = 0$ or $x = 1$. As before we suppose that $y_0(x)$ is the function that renders J an extremum and we let $y = y_0(x) + \alpha \eta$. Since J is extremal at $\alpha = 0$ it follows that we must have

$$ \Delta J = \int_0^1 \left(\frac{\partial F}{\partial y} \eta + \frac{\partial F}{\partial y'} \eta' \right) dx = 0 $$

for all choices of the function $\eta(x)$. When we integrate by parts we can no longer discard the boundary contribution and we find instead that

$$ \Delta J = \int_0^1 \left(\frac{\partial F}{\partial y} - \frac{d}{dx} \frac{\partial F}{\partial y'} \right) \eta \, dx + \frac{\partial F}{\partial y'} \eta \, \Big|_0^1 = 0 $$

for all choices of η. Since $\eta(x)$ for $0 < x < 1$ is independent of $\eta(0)$ and $\eta(1)$, we must require separately that

$$ \frac{\partial F}{\partial y} - \frac{d}{dx} \frac{\partial F}{\partial y'} = 0 $$

$$ \frac{\partial F}{\partial y'} \Big|_{x=0} = \frac{\partial F}{\partial y'} \Big|_{x=1} = 0. $$

The resulting boundary conditions are called the "free" or "natural" boundary conditions for this problem.

Example

Suppose a frictionless object slides downward along a curve $y = y(x)$ starting at $y = y_0$, $x = 0$. Find the curve for which the time to reach $x = x_1$ is the smallest.

This is a variant of an earlier problem, but here we do not specify at what height the object reaches the point $x = x_1$. The time of descent is

$$ T = \int_0^{x_1} \left(\frac{1 + y'^2}{2(y_0 - y)g} \right)^{1/2} dx, $$

and, as before, the Euler-Lagrange equation has a first integral

$$\left(2g\left(y_0 - y\right)\left(1 + y'^2\right)\right)^{1/2} = \text{constant}.$$

In addition we require $y(0) = y_0$ and the free boundary condition is

$$\frac{y'}{\left(1 + y'^2\right)^{1/2}} = 0$$

at $x = x_1$ so that $y'(x_1) = 0$.

Once again, the solution is a cycloid

$$y = y_0 + \frac{1}{2}c_0^2(\cos t - 1)$$

$$x = \frac{1}{2}c_0^2(t - \sin t),$$

where now $c_0^2 = \frac{2x_1}{\pi}$. It is an interesting exercise (which we leave to the reader) to show that the descent time for this problem is shorter than if we specify the height $y(x_1)$ at $x = x_1$. In fact the fastest cycloid is that one which drops a distance x_1/π in a length x_1.

5.1.5 Several Independent Variables

Suppose we wish to find a function y of more than one variable which extremizes the functional

$$J = \int_D F\left(x_1, x_2, \ldots, x_k, y, \frac{\partial y}{\partial x_1}, \frac{\partial y}{\partial x_2}, \ldots, \frac{\partial y}{\partial x_k}\right) dx_1 dx_2 \ldots dx_k,$$

where D is a k-dimensional region of interest.

Using the usual arguments, we require the first variation of J to vanish

$$\Delta J = \int_D \left(\frac{\partial F}{\partial y}\eta + \sum_{i=1}^{k} \frac{\partial F}{\partial z_i}\frac{\partial \eta}{\partial x_i}\right) dx_1 dx_2 \ldots dx_k = 0,$$

where $z_i = \frac{\partial y}{\partial x_i}$. Integration by parts in higher dimensions takes the form of the divergence theorem. For a vector valued function f

$$\int_D \text{div}(f)dV = \int_{\partial D} n \cdot f \, ds,$$

where n is the outward normal vector to the boundary ∂D. To use the divergence theorem, we observe that

$$\int_D \sum_{i=1}^k \frac{\partial}{\partial x_i} \left(\eta \frac{\partial F}{\partial z_i} \right) dV = \int_{\partial D} \eta \left(\frac{\partial F}{\partial z_i} \cdot n \right) ds$$

so that

$$\int_D \left(\sum_{i=1}^k \frac{\partial F}{\partial z_i} \frac{\partial \eta}{\partial x_i} \right) dV = \int_{\partial D} \eta \left(\frac{\partial F}{\partial z_i} \cdot n \right) ds$$

$$- \int_D \eta \sum_{i=1}^k \frac{\partial}{\partial x_i} \left(\frac{\partial F}{\partial z_i} \right) dV.$$

If we impose the condition that $\eta = 0$ on the boundary ∂D, we conclude that

$$\frac{\partial F}{\partial y} - \sum_{i=1}^k \frac{\partial}{\partial x_i} \left(\frac{\partial F}{\partial z_i} \right) = 0, \qquad z_i = \frac{\partial y}{\partial x_i},$$

which is the Euler-Lagrange equation for this problem.

It may be that the function y is not a priori specified on the boundary ∂D. In that case, we cannot assume that η is always zero on the boundary. Instead, we require

$$\frac{\partial F}{\partial z_i} \cdot n = 0, \qquad z_i = \frac{\partial y}{\partial x_i},$$

on that part of the boundary ∂D where y is not specified.

Throughout this discussion we have found that the Euler-Lagrange equations provide necessary conditions for an extremal function of a functional. To find sufficient conditions is a much more difficult problem. Recall from calculus that if a function f has $f'(x_0) = 0$ we cannot be sure we have a minimum or a maximum without checking higher derivatives of $f(x)$. An analogous statement holds for the calculus of functionals, namely, we are assured of an extremal value if the second variation of J is nonvanishing. Although this second variation can be written down, in practice, it is not often used, and other arguments are usually used to show that the derived function is indeed an extremal function. As these arguments are generally rather complicated, we will not discuss sufficient conditions here.

5.2   Hamilton's Principle

Many important equations of classical physics can be derived by invoking HAMILTON'S PRINCIPLE. Suppose the position of an object (or objects) can be described in terms of coordinates $q_1, q_2, \ldots, q_n$ and the velocity of the object is given by $\dot{q}_1, \dot{q}_2, \ldots, \dot{q}_n$, where $\dot{q}_i = \frac{dq_i}{dt}$. For this object, the kinetic energy is a quadratic form in $\dot{q}_i$,

$$T = \sum P_{ik}\left(q_1, q_2, \ldots, q_n, t\right) \dot{q}_i \dot{q}_k.$$

The potential energy U is a function of position given by $U = U\left(q_1, q_2, \ldots, q_n, t\right)$. If there are no frictional losses, i.e., if the system is conservative, then the equation of motion follows from

Hamilton's Principle

The trajectories of the object render

$$J = \int_{t_0}^{t_1} (T - U)dt$$

stationary. That is, the first variation ΔJ is zero.

The function $L = T - U$ is called the LAGRANGIAN for this system. Stationarity is equivalent to saying that the Euler-Lagrange equations hold for J, that is

$$\frac{d}{dt}\frac{\partial T}{\partial \dot{q}_i} - \frac{\partial (T - U)}{\partial q_i} = 0, \qquad i = 1, 2, \ldots, n.$$

If the object is at rest, $\dot{q}_i = 0$ so $T = 0$, and we must have $\frac{\partial U}{\partial q_i} = 0$, which determines the possible equilibria of the system.

There is an important subtlety to Hamilton's principle that needs explanation. We always suppose that we know the position of the object(s) at times $t = t_0$ and $t = t_1$. In coordinate space there are numerous trajectories connecting these two positions, but Hamilton's principle states that the path chosen by the physical system renders $\Delta J = 0$. Since we assume that the positions of the objects are known at time $t = t_0$ and $t = t_1$, there are no boundary conditions to compute.

If T and U are independent of t, we know that there is always a first integral. In particular,

$$\frac{d}{dt}\left(\sum_{i=1}^{n} \dot{q}_i \frac{\partial L}{\partial \dot{q}_i} - L\right) = \sum_{i=1}^{n} \dot{q}_i \left(\frac{d}{dt}\frac{\partial L}{\partial \dot{q}_i} - \frac{\partial L}{\partial q_i}\right) = 0.$$

The constant of the motion

$$H = \sum_{i=1}^{n} \dot{q}_i \frac{\partial L}{\partial \dot{q}_i} - L$$

is called the HAMILTONIAN of the system, and obviously, H, being constant, is conserved.

It is convenient to make the change of independent variables q_i, $\dot{q}_i \to \hat{q}_i$, p_i where $p_i = \frac{\partial L}{\partial \dot{q}_i}$, $\hat{q}_i = q_i$. To differentiate with respect to p_i and $\hat{q}_i$ we must apply the chain rule

$$\frac{\partial H}{\partial q_i} = \sum_{j=1}^{n} \left(\frac{\partial H}{\partial p_j} \frac{\partial p_j}{\partial q_i} + \frac{\partial H}{\partial \hat{q}_j} \frac{\partial \hat{q}_j}{\partial q_i} \right)$$

$$\frac{\partial H}{\partial \dot{q}_i} = \sum_{j=1}^{n} \left(\frac{\partial H}{\partial p_j} \frac{\partial p_j}{\partial \dot{q}_i} + \frac{\partial H}{\partial \hat{q}_j} \frac{\partial \hat{q}_j}{\partial \dot{q}_i} \right),$$

where $\frac{\partial \hat{q}_j}{\partial q_i} = \delta_{ij}$, $\frac{\partial \hat{q}_j}{\partial \dot{q}_i} = 0$. Since $H = \sum_{j=1}^{n} \dot{q}_j p_j - L$, it follows that

$$\frac{\partial H}{\partial \dot{q}_i} = \sum_{j=1}^{n} \frac{\partial H}{\partial p_j} \frac{\partial p_j}{\partial \dot{q}_i} = \sum_{j=1}^{n} \dot{q}_j \frac{\partial p_j}{\partial \dot{q}_i}.$$

or that

$$\frac{dq_i}{dt} = \frac{\partial H}{\partial p_i}.$$

(We assume that the matrix with entries $\partial p_j / \partial \dot{q}_i$ is nonsingular.) Next,

$$\frac{\partial H}{\partial \hat{q}_i} = -\sum_{j=1}^{n} \frac{\partial p_j}{\partial q_i} \frac{\partial H}{\partial p_j} + \frac{\partial H}{\partial q_i} = \frac{\partial}{\partial q_i} \left(H - \sum_{j=1}^{n} \dot{q}_j p_j \right)$$

$$= -\frac{\partial L}{\partial q_i} = -\frac{d}{dt} \frac{\partial L}{\partial \dot{q}_i} = -\dot{p}_i,$$

where we have used the Euler-Lagrange equations. The results are the familiar HAMILTON'S EQUATIONS

$$\frac{dq_i}{dt} = \frac{\partial H}{\partial p_i}, \qquad \frac{dp_i}{dt} = \frac{-\partial H}{\partial q_i}$$

which are the equations of motion for a conservative system, and the variables p_i, q_i are called canonical variables (we have now deleted the $\hat{\ }$).

Example   Swinging Pendulum

A pendulum hanging from a frictionless pivot point which makes an angle θ with the vertical has potential energy $U = mgl(1 - \cos\theta)$ and kinetic energy $T = \frac{ml^2\dot{\theta}^2}{2}$. The Lagrangian is

$$L = \frac{ml^2\dot{\theta}^2}{2} - mgl(1 - \cos\theta)$$

and the Euler-Lagrange equation is

$$l\ddot{\theta} + g\sin\theta = 0.$$

The Hamiltonian for this system is

$$H = \dot{\theta}\frac{\partial L}{\partial\dot{\theta}} - L = \frac{ml^2\dot{\theta}^2}{2} + mgl(1 - \cos\theta).$$

We introduce the coordinate $p = ml^2\dot{\theta}$ so that $H = \frac{p^2}{2ml^2} + mgl(1 - \cos\theta)$, and Hamilton's equations are

$$\frac{d\theta}{dt} = \frac{\partial H}{\partial p} = \frac{p}{ml^2}$$

$$\frac{dp}{dt} = -\frac{\partial H}{\partial\theta} = -mgl\sin\theta.$$

Example Vibrating String

The equation of motion for a vibrating string can be derived from Hamilton's principle. Suppose $u(x,t)$ is the displacement of a string from the horizontal. We assume that the potential energy is proportional to the change in arclength of the string

$$U = \mu\int_0^l\left((1 + u_x^2)^{1/2} - 1\right)dx$$

and the kinetic energy is $T = \frac{1}{2}\int_0^l \rho u_t^2 dx$ where μ is the appropriate constant of proportionality, called HOOKE'S CONSTANT, ρ is the density of the string, and l is its unstretched length. For this system, Hamilton's principle states that the motion renders the functional

$$J = \int_{t_0}^{t_1}\int_0^l\left[\frac{1}{2}\rho u_t^2 - \mu\left((1 + u_x^2)^{1/2} - 1\right)\right]dx\,dt$$

stationary. We suppose that the string is constrained to satisfy $u = 0$ at the ends $x = 0, l$. Since this is a functional in the two independent variables x, t, we find that the Euler-Lagrange equation is

$$\frac{\partial}{\partial t}\left(\rho\frac{\partial u}{\partial t}\right) - \frac{\partial}{\partial x}\frac{\mu u_x}{(1 + u_x^2)^{1/2}} = 0.$$

This equation is nonlinear and allows for the possibility of large extensions (within the range of validity of a linear Hooke's Law). If, however, we suppose that $|u_x| \ll 1$, so that u_x^2 can be ignored relative to 1, we find that

$$\frac{\partial}{\partial t}\left(\rho \frac{\partial u}{\partial t}\right) = \frac{\partial}{\partial x}\left(\mu \frac{\partial u}{\partial x}\right), \qquad 0 < x < l$$

which is the WAVE EQUATION.

Example   Vibrating Rod

The main difference between a rod and a string is that although both rod and string resist extensions, only the rod resists bending. The potential energy of a rod, therefore, should contain a term due to extension and a term due to bending. We assume the rod is thin and has only transverse displacement. If $u(x,t)$ is the transverse displacement of the rod at position x, the local bending is given by the curvature $\kappa = \frac{u_{xx}}{(1+u_{xx}^2)^{3/2}}$. The bending potential energy is proportional to κ^2 so that

$$U = \frac{\mu_1}{2}\int_0^l \frac{u_{xx}^2}{(1+u_{xx}^2)^3}dx + \mu_2\int_0^l \left((1+u_x^2)^{1/2} - 1\right) dx.$$

If we expect only small extensions $u_x^2 \ll 1$, and small bending $u_{xx}^2 \ll 1$, we can simplify the potential energy to

$$U = \int_0^l \left(\frac{\mu_1}{2}u_{xx}^2 + \frac{\mu_2}{2}u_x^2\right) dx.$$

If we do not make this smallness assumption, we will find a nonlinear equation of motion. An additional source of potential energy may come from external forcing $f(x,t)$, with contribution (Energy = force × distance)

$$U = \int_0^l fu\, dx.$$

Hamilton's principle states that the first variation of $J = \int_{t_0}^{t_1}(T - U)dt$ must vanish, that is

$$0 = \int_{t_0}^{t_1}\int_0^l (\rho u_t \eta_t - \mu_1 u_{xx}\eta_{xx} - \mu_2 u_x\eta_x - f\eta)\, dx\, dt$$

for all admissible functions $\eta(x,t)$. When we integrate this expression by parts, we obtain

$$\begin{aligned}
0 &= \int_{t_0}^{t_1}\int_0^l (-\rho u_{tt} - \mu_2 u_{xxxx} + \mu_1 u_{xx} - f)\eta\, dx\, dt \\
&\quad + \int_0^l \rho u_t\eta\Big|_{t_0}^{t_1} dx - \mu_1 \int_{t_0}^{t_1} (u_{xx}\eta_x - u_{xxx}\eta)\Big|_0^l dt
\end{aligned}$$

$$-\mu_2 \int_{t_0}^{t_1} \left(u_x \eta \Big|_0^l \right) dt.$$

The reason we display this detail is to show that there are several ways to determine boundary conditions for u. The Hamilton's principle implies that we should take $\eta = 0$ at $t = t_0$ and $t = t_1$. In addition, if we specify u a priori at $x = 0$ and $x = l$ then we must take $\eta = 0$ at $x = 0$ and $x = l$ as the condition of admissibility. If in addition we specify the slope u_x at $x = 0$ and $x = l$, we then require $\eta_x = 0$ at the endpoints for admissibility. However, if we choose to keep u_x unspecified, we are forced to take $u_{xx} = 0$ at the endpoints, so that the boundary terms vanish. With either of these choices, the boundary terms vanish and the result is the equation of motion

$$\rho u_{tt} = \mu_1 u_{xx} - \mu_2 u_{xxxx} + f$$

with boundary conditions for u specified at $x = 0, l$ and either u_x or $u_{xx} = 0$ specified at $x = 0, l$. If u_x is specified, the rod is said to be clamped and if $u_{xx} = 0$, it is said to be pinned.

Example   Vibrating Membrane

For a membrane, we assume that the potential energy is proportional to the change in surface area of the membrane due to stretching. Thus,

$$U = \mu \int_D \left(\left(1 + \left(\frac{\partial u}{\partial x} \right)^2 + \left(\frac{\partial u}{\partial y} \right)^2 \right)^{1/2} - 1 \right) dx\, dy$$

and

$$T = \frac{1}{2} \int_D \rho \left(\frac{\partial u}{\partial t} \right)^2 dx\, dy,$$

where D is the two-dimensional spatial domain of the membrane. For the Lagrangian $L = T - U$ the Euler-Lagrange equation is

$$\rho u_{tt} = \mu \left(\frac{\partial}{\partial x} \frac{u_x}{\left(1 + u_x^2 + u_y^2 \right)^{1/2}} + \frac{\partial}{\partial y} \frac{u_y}{\left(1 + u_x^2 + u_y^2 \right)^{1/2}} \right)$$

which simplifies to

$$\rho u_{tt} = \mu \left(u_{xx} + u_{yy} \right)$$

if we make the simplifying assumption that u_x^2 and u_y^2 are small compared with 1.

For the vibrating string and vibrating membrane, the ratio $c = (\mu/\rho)^{1/2}$ is the wavespeed. The wavespeed has been measured in many types of materials, such as lung tissue (30-45 m/s), air (300 m/s), water (1400 m/s), muscle tissue (150 m/s) and bone (3500 m/s).

5.3   Approximate Methods

The numerical solution of differential equations can be broadly classified as making use of one of two possible methods. In the first method we try to approximate the unknown function $u(x)$ by a large number of discrete values $u = u_j$ at $x = x_j$, $j = 0, 1, 2, \ldots, n$. We want the values u_j at x_j to satisfy the governing equation approximately and so we devise schemes by which to discretize the governing equation, which in the limit $n \to \infty$, look more and more like the original equation. The hope is that the approximate solution will get better and better as n, the order of the approximation gets larger.

This idea is the motivation for all finite difference methods (a vast subject worthy of much attention). For example, to solve $u' = u$ on the interval $x \in [0, 1]$ we might set $x_j = j/n$, $j = 0, 1, 2, \ldots, n$ and then approximate the differential equation $u' = u$ by the difference equations

$$u_{j+1} - u_j = \frac{1}{n} u_j,$$

or by

$$u_{j+1} - u_j = \frac{1}{n} u_{j+1},$$

or by

$$u_{j+1} - u_j = \frac{1}{2n} \left(u_{j+1} + u_j \right),$$

or by

$$u_{j+1} - u_{j-1} = \frac{2}{n} u_j,$$

or by a myriad of other possibilities.

The second way to approximate the solution of differential equations numerically is to look for, not specific values of the unknown function, but an approximate function representation that can be evaluated at any point. So, for example, if the functions $\{\phi_j\}_{j=1}^{\infty}$ form a basis for the relevant Hilbert space, we might try to find the best choice of constants, $\alpha_1, \alpha_2, \ldots, \alpha_n$, so that

$$u_n(x) = \sum_{j=1}^{n} \alpha_j \phi_j(x)$$

is the best possible approximation to the unknown function $u(x)$ using only n basis functions. One would hope that as n increases, the approximation gets not only better, but converges with some speed to the desired solution.

In Chapter 2 we addressed the problem of approximating a known function $u(x)$ by a linear combination of mutually orthogonal basis functions and found that, if we want to minimize the squared error $\|u(x) - u_n(x)\|^2$, (the norm is the natural norm for the Hilbert space), then the best we can do is to take

$$u_n(x) = \sum_{j=1}^{n} \alpha_j \phi_j(x), \qquad \alpha_j = \langle u(x), \phi_j(x) \rangle .$$

Here we have a slightly different problem since we do not know the function $u(x)$ a priori, but only an equation which defines (uniquely we hope) the function $u(x)$. Since we cannot directly approximate the function we try to minimize some other functional, which can be minimized only by the sought after function. So, for example, if we wish to solve $u' = u$ on $[0, 1]$ subject to $u(0) = a$, we might try to minimize the functional

$$F(u, u') = \int_0^1 (u' - u)^2 dx + (u(0) - a)^2 .$$

We see by inspection that the minimum of f is attained with $u' = u$, $u(0) = a$. We approximate $u(x)$ by minimizing F over the set of all approximating functions of the form

$$u_n(x) \;=\; \sum_{j=0}^{n} \alpha_j \phi_j(x)$$

where $\phi_j(x)$ are admissible functions for the variational problem. So, for example, if we want the best least squares quadratic approximation, we obtain

$$u_2(x) \;=\; \frac{a}{250} \left(249 + 216x + 210x^2\right)$$

$$=\; a \left(.996 + .864x + .84x^2\right) .$$

On the other hand, the best least squares quadratic approximation of the exact solution ae^x on the interval $[0, 1]$ is

$$v_2(x) \;=\; a \left(3(13e - 35) + 12(49 - 18e)x + 30(7e - 19)x^2\right)$$

$$=\; a \left(1.013 + .851x + .839x^2\right) .$$

Sketches of the differences $e^x - u_2(x)$ (solid curve) and $e^x - v_2(x)$ (dashed curve) are shown in Figure 5.1. Notice that although $v_2(x)$ provides a better estimate of e^x than does $u_2(x)$, neither approximation is noticeably bad, that is, the absolute error never exceeds .02 in either case.

We could also have tried to minimize the functional

$$F(u, u') = \int_0^1 (u' - u)^2 dx$$

subject to the admissibility condition $u(0) = a$. Naturally, this will give a different approximate solution than the one we already calculated.

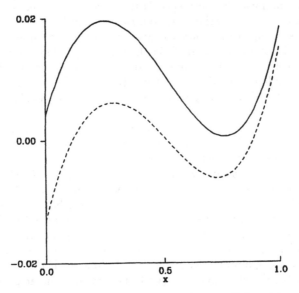

FIGURE 5.1 Error of the quadratic approximation to e^x found by minimizing a functional $F(u, u')$ (solid curve) and found by a least squares approximation of e^x (dashed curve).

In general, if we have a differential equation whose solution we wish to approximate, we try to find some functional whose minima (or extrema) are solutions of the differential equation. Then, we choose coefficients α_j so that

$$u_n(x) = \sum_{j=1}^{n} \alpha_j \phi_j(x)$$

renders the functional a minimum (or an extremum) over the set of all possible choices of α_j. The set $\{\phi_j\}_{j=1}^{\infty}$ must be an admissible set. If the set $\{\phi_j\}_{j=1}^{\infty}$ is also complete we can presumably make the approximation as good as we wish by taking n large. The method of minimizing a functional to approximate some unknown function is classically associated with the names of Rayleigh and Ritz.

The method of Galerkin finds approximate solutions using a different idea. If we wish to solve some differential equation, say $Lu = f$, we choose the coefficients α_j in

$$u_n(x) = \sum_{j=1}^{n} \alpha_j \phi_j(x)$$

in such a way that $\langle Lu - f, \phi_k \rangle = 0$ for $k = 1, 2, \ldots, n$. That is, we pick $Lu - f$ to be orthogonal to all the functions ϕ_k. Notice that if the trial functions

ϕ_k are test functions, this corresponds to approximately satisfying the weak formulation of the differential equation, rather than the strong formulation.

The interesting observation to make is that if the differential equation $Lu = f$ can be identified as an Euler-Lagrange equation, then the Galerkin procedure with the usual L^2 inner product is identical to the Rayleigh-Ritz procedure. Furthermore, if the trial functions $\{\phi_k\}_{k=1}^n$ are complete in the limit $n \to \infty$, the Galerkin approximation converges to a solution of $Lu = f$ as $n \to \infty$.

There are many good ways to choose basis functions. The orthogonal polynomials certainly work, since they are complete, although for these problems, there is no advantage to using an orthogonal basis, since the properties of orthogonality are generally not preserved when substituted into a functional involving derivatives. It is, however, helpful to choose basis functions which keep computation to a minimum. For this reason, the basis functions of choice for many applications are the finite element bases, discussed in Chapter 2. In particular, cubic spline approximations are quite popular in applications.

The Galerkin method also depends on the choice of Hilbert space. So, for example, approximate answers are much different if the space L^2 is used rather than the Sobolev space H^1 (with its associated inner product).

5.4   Eigenvalue Problems

The Sturm-Liouville eigenvalue problem first encountered in Chapter 4 can be formulated as a variational problem. Consider the functionals

$$D(\phi) \;=\; \int_0^1 \left(p(x)\phi'^2(x) + q(x)\phi^2(x) \right) dx$$

and

$$H(\phi) \;=\; \int_0^1 \omega(x)\phi^2(x)dx,$$

where the functions $p(x)$, $q(x)$, $\omega(x)$ are positive in the interval $0 < x < 1$. We propose to minimize $D(\phi)$ subject to the constraint $H(\phi) = 1$, which, using Lagrange multipliers, is equivalent to minimizing $D(\phi) - \lambda(H(\phi) - 1)$.

For this functional, the Euler-Lagrange equation is

$$(p\phi')' - q(x)\phi + \lambda\omega(x)\phi \;=\; 0$$

subject to any set of boundary conditions for which

$$p(x)\phi'\eta \left. \right|_0^1 \;=\; 0,$$

where $\eta(x)$ is any admissible variation function. Thus, for example, if $\phi'(0)$ or $\phi'(1)$ is specified, the function η must satisfy $\eta(0) = 0$ or $\eta(1) = 0$, respectively. On the other hand, if we do not specify η at $x = 0$ or $x = 1$, then the natural boundary conditions are $\phi'(0) = 0$ or $\phi'(1) = 0$. In either case, it is apparent that the minimizing function is an eigenfunction of the Sturm-Liouville problem $(pu')' - qu + \lambda w u = 0$ subject to appropriate boundary conditions. Furthermore, if the function ϕ is the minimizer of $D(\phi)$ with $H(\phi) = 1$ we see that $D(\phi) = \lambda$, the corresponding eigenvalue.

The functional $D(\phi)$, when minimized subject to $H(\phi) = 1$, yields the smallest eigenvalue of the Sturm-Liouville problem $(pu')' - qu + \lambda w u = 0$, at least with Dirichlet or Neumann boundary conditions. More general boundary conditions can be allowed if we consider the slightly modified functional

$$D(\phi) = \int_0^1 \left(p\phi'^2 + q\phi^2 \right) dx + \sigma(1)p(1)\phi^2(1) + \sigma(0)p(0)\phi^2(0)$$

subject to the constraint $H(\phi) = 1$. The Euler-Lagrange equation for $D(\phi) - \lambda \left(H(\phi) - 1 \right)$ is again

$$(p\phi')' - q\phi + \lambda w \phi = 0,$$

but now the natural boundary conditions are

$$p \left(\frac{\partial \phi}{\partial n} + \sigma \phi \right) \eta = 0 \qquad \text{at} \qquad x = 0 \qquad \text{and} \qquad x = 1,$$

where $\partial/\partial n$ represents the outward (normal) derivative at the boundary point. These considerations lead to the following

MINIMUM PRINCIPLE

1. If ϕ renders $D(\phi)$ minimal subject to the restriction $H(\phi) = 1$, then $(p\phi')' - q\phi + \lambda w \phi = 0$ and the minimal value is $D(\phi) = \lambda$.

2. Suppose the first k eigenfunctions $\phi_1, \phi_2, \ldots, \phi_k$ of $(p\phi')' - q\phi + \lambda w \phi = 0$ with $\lambda_1 \leq \lambda_2 \leq \cdots \leq \lambda_k$ are known. If ϕ minimizes $D(\phi)$ subject to the constraint $H(\phi) = 1$ and the k additional constraints $\int_0^1 \phi \phi_j w(x) dx = 0, j = 1, 2, \ldots, k$, then ϕ is the $(k+1)^{\text{st}}$ eigenfunction of $(p\phi')' - q\phi + \lambda w \phi = 0$ and $D(\phi) = \lambda_{k+1} \geq \lambda_k$ is the $(k + 1)^{\text{st}}$ eigenvalue.

This characterization of the kth eigenvalue and eigenfunction is inductive, depending on sequential knowledge of previous eigenpairs. A characterization which avoids reference to other eigenvalues is the

MINIMAX PRINCIPLE

Suppose the $k-1$ functions $v_1, v_2, \ldots, v_{k-1}$ are piecewise C^1 on $0 \leq x \leq 1$, and $d\{v_1, v_2, \ldots, v_{k-1}\}$ is the greatest lower bound of $D(\phi)$ subject to $H(\phi) = 1$ and $\int_0^1 \phi v_j \omega(x)dx = 0$, $j = 1, 2, \ldots, k-1$. Then λ_k is the largest value d assumes if $v_1, v_2, \ldots, v_{k-1}$ range over all possible sets of admissible functions.

Of course, ϕ and the functions v_i must be admissible functions, which simply means they must be piecewise C^1 on $0 \leq x \leq 1$ and, if we have Dirichlet conditions, they must also satisfy the boundary condition $\phi = 0$ at $x = 0$ or $x = 1$.

All of the above statements are proved in the same way as the analogous statements for matrices in Chapter 1, and we leave this as an exercise to the reader.

These variational characterizations of the eigenvalues of Sturm-Liouville equations lead to very useful comparisons of the eigenvalues of different problems. Before we make these observations, however, it is worthwhile to note one of the important physical interpretations of eigenvalues.

An eigenvalue of the Sturm-Liouville equation $(pu')' - qu' + \lambda \omega u = 0$ corresponds to a frequency of a mode of vibration of a string with equation of motion

$$\omega u_{tt} = (pu_x)_x - qu.$$

Since this equation is linear, its solution can be represented as a linear superposition of simpler solutions of the form $u = \phi(x)e^{i\kappa t}$. (We will discuss this important SUPERPOSITION PRINCIPLE in more detail in Chapter 8.) Any solution of this form must satisfy $(p\phi')' - q\phi + \kappa^2 \omega\phi = 0$ and we identify κ, the frequency of temporal vibration, with $\sqrt{\lambda}$, where λ is the eigenvalue of the Sturm-Liouville equation. The smallest eigenvalue corresponds to the fundamental frequency of vibration and larger eigenvalues correspond to overtones.

The minimax principle leads to several important observations for frequencies of vibration. We make use of two facts:

1. Strengthening admissibility conditions in a minimum problem cannot decrease the value of the minimum. Weakening conditions cannot increase the minimum.

2. If $D_1(\phi) < D_2(\phi)$ for all admissible ϕ then $\min D_1(\phi) \leq \min D_2(\phi)$.

Consequently for vibrating systems we conclude that

1. If $p(x)$ or $q(x)$ are increased or if ω is decreased, the fundamental tone and all overtones are increased. By increasing p or q the functional $D(\phi)$ increases for every admissible ϕ, while decreasing ω weakens the constraint $H(\phi) = 1$, by making all admissible ϕ larger.

2. For Dirichlet boundary conditions, the frequencies of vibration increase if the domain $0 \leq x \leq 1$ is shortened to a subdomain $0 < a \leq x \leq b < 1$. This follows since the set of admissible functions which are zero at $x = a$ and $x = b$ is a subset of those functions with $u = 0$ at $x = 0$ and $x = 1$.

3. Increasing σ at $x = 0$ or $x = 1$ increases the frequencies of vibration. The least restrictive boundary condition is $\phi' = 0$ with $\sigma = 0$. If $\sigma > 0$ at $x = 0$ or $x = 1$ then the functional $D(\phi)$ increases for every admissible ϕ. If the class of admissible ϕ is restricted so that $\phi = 0$ at a boundary, the minimum cannot decrease.

As a specific example, consider a piano or guitar string whose equation of motion is $\rho u_{tt} = \mu u_{xx}$ where ρ is the uniform density and μ is the uniform tension in the string. If the string has length l, we rescale space and write $\rho u_{tt} = \frac{\mu}{l^2} u_{xx}$ where $0 \leq x \leq 1$. The corresponding eigenvalue problem is

$$\frac{\mu}{l^2} \phi'' + \lambda \rho \phi \;=\; 0$$

$$\phi(0) = \phi(1) \;=\; 0.$$

The eigenvalues of this problem are

$$\lambda_n = \frac{n^2 \pi^2}{l^2} \frac{\mu}{\rho}$$

and the corresponding frequencies of vibration are

$$\kappa_n = \frac{n\pi}{l} \sqrt{\mu/\rho}.$$

It is now obvious (and it agrees with musical experience) that the fundamental frequency ($n = 1$) and all overtones are increased by shortening the string length l, by decreasing the string density per unit length ρ, or by tightening the string (adding more tension). Similarly, if we change the way the string is constrained at one end by letting it vibrate more freely, the frequencies of vibration decrease.

Further Reading

There are a number of good texts which give an overview of the calculus of variations. One of these is

1. I. M. Gelfand and S. V. Fomin, *Calculus of Variations*, Prentice-Hall, Englewood Cliffs, 1963.

A more extensive discussion of geodesics can be found in books on differential geometry, such as

2. J. J. Stoker, *Differential Geometry*, Wiley-Interscience, New York, 1969.

A derivation of the Schwarzschild metric and its use in finding relativistic space-time curves is given in

3. P. G. Bergmann, *Introduction to the Theory of Relativity*, Prentice-Hall, New York, 1942.

The importance of Hamilton's Principle to the study of mechanics is well documented in

4. H. Goldstein, *Classical Mechanics*, Addison-Wesley, Reading, 1950,

5. L. D. Landau and E. M. Lifshitz, *Mechanics*, Pergamon Press, Oxford, 1969.

Finally, a discussion of variational methods applied to finding eigenvalues can be found in

6. H. F. Weinberger, *Variational Methods for Eigenvalue Approximation*, SIAM, Philadelphia, 1974.

Problems for Chapter 5

Section 5.1

1. Find extremal curves for the following functionals

   a. $\int_a^b (\frac{y'^2}{x^3}) dx$

   b. $\int_a^b \left(y^2 - y'^2 - 2y \cos(x) \right) dx$

   c. $\int_a^b \left(y^2 + y'^2 - 2y e^x \right) dx$

2. Which curve minimizes

$$\int_0^1 \left(\frac{1}{2} y'^2 + y y' + y' + y \right) dx$$

   when $y(0)$ and $y(1)$ are not specified?

3. Show that the time of descent of a sliding frictionless bead is faster on a cycloid than on a path of constant slope. (If you are interested in an experimental verification that the cycloid gives fastest descent time, see F. M. Phelps III, F. M. Phelps IV, B. Zorn, and J. Gormley, An experimental study of the brachistochrome, Eur. J. Phys. **3** (1984) 1–4.)

4. What is the path that minimizes the time of descent for a rolling frictionless ball between $y(0) = y_0$ and $y(x_1) = 0$? Hint: The kinetic energy of a rolling ball is larger than that of a sliding bead of the same mass because of rotational energy. Find the increase in energy due to rotation and determine how this affects the moving velocity. Then determine how this affects the fastest path. Compare this path with that for the sliding bead. (Show that the kinetic energy of a rolling bead is 2/5 larger than that of the corresponding sliding bead.)

5. Suppose a curve $y = y(x)$, $3 \leq x \leq 5$, is used to generate a surface of revolution about the y-axis. Find the curve $y(x)$ giving the smallest surface area, with $y(3) = 3\ln 3$, $y(5) = 6\ln 3$.

6. Find the curve with fixed arclength $l > \pi/2$, passing through the points $(0,0)$ and $(1,0)$ which has maximal area between it and the x-axis, $0 \leq x \leq 1$.

7. Find the curve $y = y(x)$, $-a \leq x \leq a$ whose area $\int_{-a}^{a} y(x)dx = A$ is fixed, has minimal arclength, and $y(-a) = y(a) = 0$, $A \leq \pi a^2/2$.

8. Show that the shortest path on the surface of a sphere between two points on the sphere is a portion of a great circle.

9. Find extremals of the functional
$$\int_0^{\pi/2} \left(y'^2 + z'^2 + 2yz\right) dx$$

subject to $y(0) = 0$, $y(\pi/2) = 1$, $z(0) = 0$, $z(\pi/2) = 1$.

10. Find extremals of the functional
$$\int_0^1 \left(y''^2 - y^2 + x^2\right) dx$$

subject to boundary conditions $y(0) = 0$, $y'(0) = 1$, $y(\pi/2) = 0$, $y'(\pi/2) = 1$.

11. Show that the Euler-Lagrange equation of the functional
$$\int_{x_0}^{x_1} F(x, y, y', y'')dx$$

has the first integral $F_{y'} - \frac{d}{dx}F_{y''} = $ constant if $\partial F/\partial y \equiv 0$ and the first integral $F - y'\left(F_{y'} - \frac{d}{dx}F_{y''}\right) - y''F_{y''} = $ constant if $\partial F/\partial x \equiv 0$.

12. Find extremals of the functional
$$\int_0^1 \left(y'^2 + x^2\right) dx$$

subject to $y(0) = 0$, $y(1) = 0$, and $\int_0^1 y^2 dx = 1$.

Section 5.2

1. What are the natural boundary conditions for the vibrating string?

2. What are the natural boundary conditions for the vibrating rod?

3. Suppose an elastic string is attached horizontally between two smooth rings of mass m which are constrained to slide, without friction, in the vertical direction on pins located at $x = 0$, and $x = 1$. The rings are also attached to springs which are connected to the base of the pins. Find the equation of motion of this string-mass system.

4. A string of length l, density ρ, has a mass m attached at its midpoint. The ends of the string are held fixed. Find the equation of motion for transverse vibrations of the string.

5. A pendulum with length l and mass m at the end of a weightless rod is rotating about the vertical axis with angular frequency Ω. Show that, if θ is the angle the pendulum makes with the vertical axis, then

$$\ddot{\theta} + \left(g/l - \Omega^2 \cos \theta\right) \sin \theta = 0$$

is its equation of motion.

6. Find the Hamiltonian description for the vibrations of the nonlinear mass-spring system

$$m\ddot{x} + k_1 x + k_2 x^3 = 0.$$

7. A double pendulum consists of a weightless rod suspended from a frictionless pivot point. At the end of the rod is a weight and a frictionless joint from which is hung another weightless rod and a weight at its end. Find the equation of motion of this double pendulum.

8. The position of a rotating top can be described in terms of three angles: θ, the angle the top's axis of symmetry makes with a vertical line (the axis of gravity), ϕ, the angle of the projection of the top's axis of symmetry onto the plane on which it sits (orthogonal to the direction of gravity) and ψ, the angle of rotation of the top's own coordinate system about its axis of symmetry. In terms of these angles, the Lagrangian for a top is

$$L = \frac{I_1}{2}(\dot{\theta}^2 + \dot{\phi}^2 \sin^2 \theta) + \frac{I_2}{2}(\dot{\psi} + \dot{\phi} \cos \theta)^2 - mgl \cos \theta,$$

where I_1 is the moment of inertia about the top's axis of symmetry, and I_2 is the moment of inertia about the center of mass in the direction of the axis of symmetry, M is its mass and l the position from the bottom tip to the center of mass along the axis of symmetry. Find the equations of motion of the top and three integrals of the motion.

9. Find the equation of motion of a bead sliding without friction on one arc of a cycloid.
   Hint: Use arclength coordinates.

Section 5.3

1. Find the best quadratic approximation to $u' = u$ on $[0, 1]$ subject to $u(0) = a$ by minimizing the functional $F(u', u) = \int_0^1 (u' - u)^2 dx$ subject to the admissibility condition $u(0) = a$. Compare this approximation to those given in the text.

2. Find the best "two piece" piecewise linear approximation to $u' = u$ by minimizing $F(u', u) = \int_0^1 (u' - u)^2 dx$ subject to the admissibility condition $u(0) = a$.

3. Find the best cubic polynomial approximation to $u' = u$ on $[0, 1]$ subject to the admissibility condition $u(0) = a$ using the functional in problem 1. Discuss the improvement of this approximation over that found in Problem 1.

4. Find an approximate polynomial solution of the equation $u'' - u = 0$ on the interval $[0, 1]$ subject to admissibility conditions $u(0) = 1$, $u(1) = 0$ by minimizing the functional

$$F(u, u'') = \int_0^1 (u'' - u)^2 dx.$$

Compare the approximate solution with the exact solution.

5. Approximate the solution of the equation $u'' - u = 0$ on the interval $[0, 1]$ subject to admissibility conditions $u(0) = 1$, $u(1) = 0$ by extremizing the functional

$$F(u, u') = \int_0^1 (u^2 + u'^2) dx$$

over all quadratic polynomials. Compare the approximate solution with the exact solution and with the solution of Problem 4.

6. Find a cubic polynomial that approximates the solution of $u' = u$ with $u(0) = a$ using the Galerkin technique. Compare the result using the usual L^2 inner product with the result using a Sobolev H^1 inner product.

Section 5.4

1. Find an inequality relating the nth eigenvalues of $(p(x)u')' + \lambda w(x)u = 0$ with the following three sets of boundary conditions

   a. $u(0) = u(1) = 0$

   b. $u(0) = 0$, $\beta_0 u(1) + \beta_1 u'(1) = 0$, $\beta_0, \beta_1 > 0$

   c. $u(0) = 0$, $u'(1) = 0$

2. Bessel's equation $x(xu')' + (x^2 - \nu^2) u = 0$ has bounded solutions $J_\nu(x)$.

    a. If $J_n(\lambda_{n,k}) = 0$, and $\lambda_{n,k} < \lambda_{n,k+1}$ show that $\lambda_{n,k} \leq \lambda_{n+1,k}$ for all $n \geq 0$, $k \geq 1$.

    b. Define $\mu_{n,k}$ by $J'_n(\mu_{n,k}) = 0$, $0 < \mu_{n,k} < \mu_{n,k+1}$, $k \geq 1$. Show that $\lambda_{n,k} < \mu_{n,k}$.

3. Use a quadratic polynomial to estimate the smallest eigenvalue and associated eigenfunction of $u'' + \lambda u = 0$, $u(0) = u(1) = 0$.

4. Approximate the first and second eigenvalues and associated eigenfunctions of $u'' + \lambda u = 0$, $u(0) = u(1) = 0$, using cubic polynomials.

5. The equation of motion for a vibrating string is

$$\frac{\partial}{\partial t}\left(\rho\frac{\partial u}{\partial t}\right) = \frac{\partial}{\partial x}\left(\mu\frac{\partial u}{\partial x}\right),$$

where ρ is the density per unit length and μ is Hooke's constant. How will weakening the string affect the fundamental frequency?
How will a buildup of dirt on the string affect the fundamental frequency?

6. Formulate a variational characterization for the eigenvalues of the partial differential operator ∇^2, whose eigenfunctions satisfy $-\nabla^2\phi = \lambda\phi$, for $x \in D$ with $\phi = 0$ on ∂D, D some domain with piecewise C^2 boundary in n dimensions. What happens to the eigenvalue λ as D changes in size?

7. For many years train conductors would walk along their train and hit each of the wheels with a metal bar. They were listening for a lower than usual ringing tone. What would this indicate?

8. Invent a characterization of the eigenvalues of

$$\frac{\partial}{\partial x}\left(\mu\frac{\partial u}{\partial x}\right) + \frac{\partial}{\partial y}\left(\mu\frac{\partial u}{\partial y}\right) = -\lambda u$$

on a circular domain that explains your answer to Question 7. What boundary conditions should be applied?

6

COMPLEX VARIABLE THEORY

Until now in our development of Transform Theory, we have seen only one special kind of transform, namely transforms which correspond to discrete (possibly infinite) sums over discrete basis functions, usually the eigenfunctions of some linear operator. To extend the ideas of transform theory to other operators we must first become comfortable with functions defined on the complex plane. In this chapter we start from scratch, by discussing COMPLEX NUMBERS, COMPLEX VALUED FUNCTIONS, and COMPLEX CALCULUS, i.e. DIFFERENTIATION and INTEGRATION, of functions in the complex plane, as well as some of the important uses of complex variable theory to fluid motion and special functions.

6.1 Complex Valued Functions

The complex number z is defined by $z = x + iy$ where $i = \sqrt{-1}$ and where x and y are real numbers. The numbers x and y are called the REAL and IMAGINARY PARTS of z, respectively, denoted $x = \operatorname{Re} z$, $y = \operatorname{Im} z$. The COMPLEX CONJUGATE of z, denoted $\bar{z}$, is $\bar{z} = x - iy$. A convenient way

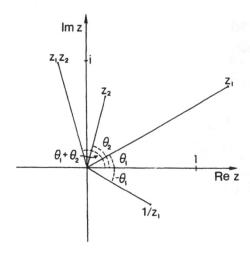

FIGURE 6.1 Argand diagram.

to visualize complex numbers is in the two-dimensional plane, called the AR-
GAND DIAGRAM with ordinate Re z and abscissa Im z (see Fig. 6.1). In the
Argand diagram it is clear that $z = r(\cos\theta + i\sin\theta)$ where $r = (x^2 + y^2)^{1/2}$
and $\theta = \tan^{-1}(y/x)$. The angle θ is called the argument of z, $\theta = \arg(z)$ and
$r = |z|$ is called the modulus of z, $|z|^2 = z\bar{z}$. An amazing fact, called the Euler
identity, is that $\cos\theta + i\sin\theta = e^{i\theta}$. To see this, recall from calculus the Taylor
series expansions for $\cos\theta$, $\sin\theta$, and e^θ. Using these expansions, we see that

$$
\cos\theta + i\sin\theta = \sum_{n=0}^{\infty}(-1)^n\frac{\theta^{2n}}{2n!} + i\sum_{n=0}^{\infty}(-1^n)\frac{\theta^{2n+1}}{(2n+1)!}
$$

$$
= \sum_{n=0}^{\infty}i^{2n}\frac{\theta^{2n}}{2n!} + \sum_{n=0}^{\infty}i^{2n+1}\frac{\theta^{2n+1}}{(2n+1)!}
$$

$$
= \sum_{n=0}^{\infty}\frac{(i\theta)^n}{n!} = e^{i\theta},
$$

as predicted. (Here we have used the power series expansion

$$
e^z = \sum_{n=0}^{\infty}\frac{z^n}{n!}
$$

as the definition of e^z for complex z.) The expression $z = re^{i\theta}$ gives geometrical
insight into the meaning of multiplication and division in the complex plane.
For example, if $z_1 = r_1 e^{i\theta_1}$ and $z_2 = r_2 e^{i\theta_2}$ it follows that $z_1 z_2 = r_1 r_2 e^{i(\theta_1+\theta_2)}$.

In particular

$$iz_1 = r_1 e^{i(\theta_1 + \pi/2)}$$

$$-z_1 = r_1 e^{i(\theta_1 + \pi)}$$

and

$$-iz_1 = r_1 e^{i(\theta_1 - \pi/2)} = r_1 e^{i(\theta + 3\pi/2)}.$$

Clearly the Argand diagram is 2π periodic in θ.

To define complex valued functions of z it is sufficient to write $f(z) = f(x + iy) = u(x, y) + iv(x, y)$ or $f(z) = r(x, y)e^{i\theta(x,y)}$. Easy examples are the powers of z, $f(z) = z^k$ with k integer, in which case $f(z) = z^k = (re^{i\theta})^k = r^k e^{ik\theta} = r^k \cos k\theta + ir^k \sin k\theta$, which defines the function z^k uniquely. Another useful function is $f(z) = e^z = e^{x+iy} = e^x(\cos y + i \sin y)$ which defines e^z uniquely.

We define the natural logarithm $\ln z$ as the inverse function of e^z, that is $\ln z = \omega$ if and only if $z = e^\omega$, but immediately we encounter a problem. Since the Argand diagram is periodic in θ, the function e^z is a periodic function of z, that is, $e^z = e^{z+2\pi i}$ for all z. As a result, if we know a value ω_0 for which $z = e^{\omega_0}$ then also $z = e^{\omega_0 + 2\pi in}$ so that $\ln z = \omega_0 + 2\pi in$ for all integers n. In other words, the logarithm function is not uniquely defined. We see the dilemma more clearly when we write

$$\ln z = \ln \left(re^{i\theta} \right) = \ln r + \ln e^{i\theta} = \ln r + i\theta.$$

Since a fixed z can correspond to the infinite number of choices of θ of the form $\theta_0 + 2\pi n$, the logarithm function is not uniquely defined.

The obvious remedy for this problem is to require θ to lie in a specific interval, say, $0 \le \theta < 2\pi$. We could equally well require θ to be in the interval $2\pi \le \theta < 4\pi$, etc. and this would give us a different definition of the function $\ln z$. For example, in the first definition $\ln i = \frac{\pi i}{2}$ and in the second $\ln i = \frac{5\pi i}{2}$. Any valid region of definition for θ is called a BRANCH of the function, while the branch that is most commonly used is called the PRINCIPAL BRANCH for that function. However, the definition of the principal branch is not universally agreed upon. For example, many texts use $-\pi \le \theta < \pi$ for the principal branch of $\ln z$, but we shall see an example shortly where the interval $0 \le \theta < 2\pi$ is more useful. In any case, the definition of a function is not complete until its branch structure is clarified.

Because of our definition of the principal branch of $\ln z$, if we want to evaluate $\ln z$ along a curve in the Argand plane keeping θ continuous, we are never permitted to enclose the origin since doing so would lead to a nonunique evaluation of $\ln z$. For example, if $z = e^{i\theta}$ for $\theta > 0$ then $\ln z = i\theta$. If we now move counterclockwise through an angle 2π to $z = e^{i(2\pi + \theta)}$, which is the same value of z, we find $\ln z = i(\theta + 2\pi)$ which is not the same as at first.

Clearly, the problem is that we traversed the line $\theta = 2\pi$. This line, which is forbidden to cross, is called a BRANCH CUT, because it is a cut between different branches of definition of the function $\ln z$. The point $z = 0$ is called a BRANCH POINT, being the point at which a branch cut ends. Just as the definition of the principal branch is arbitrary, the location of the branch cut is arbitrary, as long as all calculations are made consistent with the definition once it is made.

Branch cuts are a necessary part of the definition of many functions. For $f(z) = z^{1/2} = r^{1/2}e^{i\theta/2}$ we see that there is nonuniqueness unless a restriction is placed on θ. If we restrict $0 \le \theta < 2\pi$ we get a different value of $z^{1/2}$ than if we take $2\pi \le \theta < 4\pi$ even though $e^{i\theta} = e^{i(\theta+2\pi)}$. In this book we usually take $0 \le \theta < 2\pi$ to be the principal branch of $z^{1/2}$. For example, on the principal branch,

$$\sqrt{i} = \frac{\sqrt{2}}{2} + i\frac{\sqrt{2}}{2},$$

whereas on the second branch

$$\sqrt{i} = -\left(\frac{\sqrt{2}}{2} + i\frac{\sqrt{2}}{2}\right).$$

In this example, choosing the principal branch of definition for $z^{1/2}$ is the same as taking $\sqrt{x}$ to be positive, rather than negative, for positive real numbers x. Also notice that for the principal branch, $z^{1/2}$ lies in the upper half plane with non-negative imaginary part, while on the nonprincipal branch $z^{1/2}$ lies in the lower half of the complex plane.

Example

We can illustrate the freedom of choice in assigning branch cuts by considering the function $f(z) = (z^2 - 1)^{1/2}$. This function has branch points at $z = \pm 1$, but there are a number of ways to choose the branch cuts. One could take two separate cuts, along $z = 1+\rho$, $\rho > 0$ and along $z = -1-\rho$, ρ real and $\rho > 0$, or one could have one branch cut which joins the two branch points along $z = \rho$ with $|\rho| \le 1$. These different choices of branch cut structure give distinct definitions for the function $f(z)$.

Example

Suppose $f(z) = (z^2 - 1)^{1/2}$ is defined so that $f(i) = i\sqrt{2}$. Evaluate $f(-i)$. The natural temptation is to take $(-i)^2 = (i)^2$ so that $f(-i) = i\sqrt{2}$, but this may be incorrect. To evaluate $f(z)$ correctly we write $(z-1) = \rho_1 e^{i\theta_1}$ and $(z + 1) = \rho_2 e^{i\theta_2}$ where ρ_1, θ_1 are the distance and angle from

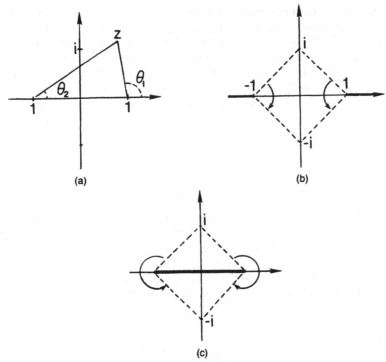

(a)

(b)

(c)

FIGURE 6.2   a) Definition of the angles θ_1 and θ_2; and b), c) Two possible ways to define the branch cuts for the function $f(z) = \left(z^2 - 1\right)^{1/2} = (\rho_1\rho_2)^{1/2}e^{(\theta_1 + \theta_2)/2}$.

1 to z while ρ_2, θ_2 are the distance and angle from -1 to z. At $z = i$, we take $\rho_1 = \rho_2 = \sqrt{2}$ and $\theta_1 = 3\pi/4$, $\theta_2 = \pi/4$ and find that $f(i) = (\rho_1\rho_2)^{1/2} e^{i(\theta_1 + \theta_2)/2} = i\sqrt{2}$ as required. To evaluate $f(z)$ at $z = -i$ in a way consistent with the first branch cut structure (Fig. 6.2) we are only permitted to increase θ_1 to $5\pi/4$ while we must decrease θ_2 to $-\pi/4$, so that $f(i) = i\sqrt{2}$. On the other hand, with the second branch cut structure, we are allowed to move from $z = i$ to $z = -i$ by decreasing θ_1 to $-3\pi/4$ and simultaneously decreasing θ_2 to $-\pi/4$, or we could increase θ_1 to $5\pi/4$ and θ_2 to $7\pi/4$, and if the branch cut structure is valid, get the same answer. Indeed, using either of these processes, we find that, for the second branch cut structure, $f(-i) = -i\sqrt{2}$, which is not the same as for the first branch cut definition.

Example

The structure of $g(z) = (1 + \sqrt{z})^{1/2}$ illustrates other complications that can occur in the branch cut selection. Suppose we take the branch cut for the square root function $\sqrt{z}$ to be along $\theta = \pi$. We can then define $\sqrt{z}$ by $\sqrt{z} = \rho^{1/2} e^{i\theta/2}$ or by $\sqrt{z} = -\rho^{1/2} e^{i\theta/2}$ where $z = \rho e^{i\theta}$ with $-\pi < \theta \le \pi$. With the first definition $\sqrt{z}$ lies in the right half plane $\mathrm{Re}\,\sqrt{z} \ge 0$ and with the second $\sqrt{z}$ lies in the left half plane $\mathrm{Re}\,\sqrt{z} \le 0$. Therefore, $w = 1 + \sqrt{z}$ has $\mathrm{Re}\,(1 + \sqrt{z}) \ge 1$ in the first definition and $\mathrm{Re}\,(1 + \sqrt{z}) \le 1$ in the second. Now to evaluate $g(z)$ we must take a second square root. There is no problem doing this if $\mathrm{Re}(1 + \sqrt{z}) \ge 1$. If we write $w = 1 + \sqrt{z} = r e^{i\phi}$ we note that $r \ge 1$ and $-\pi/2 < \phi < \pi/2$ and we take

$$g_1(z) = r^{1/2} e^{i\phi/2}$$

or

$$g_2(z) = -r^{1/2} e^{i\phi/2}$$

The complex numbers $g_1(z)$ lie to the right of the hyperbolic arc $\mathrm{Re}\,g = \sqrt{1 + (\mathrm{Im}\,g)^2}$ and the complex numbers $g_2(z)$ lie to the left of the hyperbolic arc $\mathrm{Re}\,g = -\sqrt{1 + (\mathrm{Im}\,g)^2}$ in the complex g plane. If, however, we start with $\mathrm{Re}\,(1 + \sqrt{z}) \le 1$, there is a problem in taking the square root since the origin lies in this region. Once again the curve $\theta = \pi$ is a branch cut of $\sqrt{w}$ in the w complex plane. If $w = r e^{i\phi}$ can take

$$g_3(z) = r^{1/2} e^{i\phi/2}$$

or

$$g_4(z) = -r^{1/2} e^{i\phi/2}$$

and the complex numbers given by g_3 satisfy

$$0 \le \mathrm{Re}\,g \le \sqrt{1 + (\mathrm{Im}\,g)^2}$$

and the complex numbers given by g_4 lie in the region

$$-\sqrt{1 + (\mathrm{Im}\,g)^2} \le \mathrm{Re}\,g \le 0.$$

We see that there are four different ways to define $g(z)$, and the four ways taken together cover the whole g complex plane. However, for any one of the definitions, there is a one to one relationship between the z plane and the ith region of the g complex plane. The four different definitions of g_i correspond to four branches of the function $\sqrt{1 + \sqrt{z}}$, and the boundaries of the four regions in the g complex plane are the images under this map of the branch cuts in the z plane. It follows that the functions $g_1(z)$ and $g_2(z)$ have a branch cut on the $\theta = \pi$ axis in the z plane. The functions $g_3(z)$ and $g_4(z)$ have this branch cut and they have another cut as well. The line $\theta = 0$ with $\mathrm{Re}\,z \ge 1$ must also be a branch cut for these two functions since under the map $w(z) = 1 + \sqrt{z}$ with $\sqrt{z}$ defined as $\sqrt{z} = -\rho^{1/2} e^{i\theta/2}$, the curve $\theta = 0$, $\mathrm{Re}\,z \ge 1$ is mapped to the cut $\theta = \pi$ in the w complex plane, and this in turn is mapped to the imaginary axis in the g complex plane by the final square root operation.

6.2 The Calculus of Complex Functions; Differentiation and Integration

6.2.1 Differentiation—Analytic Functions

Once we have complex functions that are well defined we would like to differentiate and integrate them. We take our definition of the derivative of f exactly as with real variable functions, namely,

$$f'(z) = \lim_{\xi \to z} \frac{f(\xi) - f(z)}{\xi - z}.$$

A function which is continuously differentiable is said to be ANALYTIC. This innocuous looking definition is actually much more restrictive than for real variables, since by $\lim_{\xi \to z}$ we mean that ξ can approach z along any crazy path in the complex plane. No such freedom is available with real variables.

There are many reasonable looking functions that are not differentiable by this definition. For example, to differentiate $f(z) = z\bar{z}$ we evaluate

$$\frac{f(\xi) - f(z)}{\xi - z} = \bar{\xi} + z\left(\frac{\bar{\xi} - \bar{z}}{\xi - z}\right).$$

If we take $\xi = z + \rho e^{i\theta}$ and let $\rho \to 0$ we see that

$$\lim_{\rho \to 0} \frac{f\left(z + \rho e^{i\theta}\right) - f(z)}{\rho e^{i\theta}} = \bar{z} + z e^{-2i\theta}$$

which is not independent of the direction of approach θ.

If a function is to be differentiable, the limits taken along the direction of the real and imaginary axes must certainly be the same. Taking $f(z) = u(x,y) + iv(x,y)$,

$$\frac{f(z + \Delta x) - f(z)}{\Delta x} = \frac{u(x + \Delta x, y) + iv(x + \Delta x, y) - u(x,y) - iv(x,y)}{\Delta x}$$

$$\to \frac{\partial u}{\partial x} + i\frac{\partial v}{\partial x},$$

whereas

$$\frac{f(z + i\Delta y) - f(z)}{i\Delta y} \to -i\frac{\partial u}{\partial y} + \frac{\partial v}{\partial y}.$$

These two expressions are equal if and only if

$$\frac{\partial u}{\partial x} = \frac{\partial v}{\partial y} \quad \text{and} \quad \frac{\partial v}{\partial x} = -\frac{\partial u}{\partial y}.$$

We have just proven the first half of the following theorem.

Theorem 6.1

The function $f(z) = u(x, y) + iv(x, y)$ is continuously differentiable in some region D if and only if $\partial u/\partial x$, $\partial u/\partial y$, $\partial v/\partial x$, $\partial v/\partial y$ exist, are continuous, and satisfy the conditions

$$\frac{\partial u}{\partial x} = \frac{\partial v}{\partial y}, \qquad \frac{\partial v}{\partial x} = -\frac{\partial u}{\partial y}.$$

These conditions are called the CAUCHY-RIEMANN conditions.

The second half of the proof is not substantially harder. Suppose the derivatives $\partial u/\partial x$, $\partial u/\partial y$, $\partial v/\partial x$, $\partial v/\partial y$ exist and are continuous in a neighborhood of some point $z_0 = (x_0, y_0)$. Then, from real function theory

$$
\begin{aligned}
u(x, y) &= u(x_0, y_0) + u_x(x_0, y_0)(x - x_0) + u_y(x_0, y_0)(y - y_0) \\
&\quad + R_1(\Delta x, \Delta y) \\
v(x, y) &= v(x_0, y_0) + v_x(x_0, y_0)(x - x_0) + v_y(x_0, y_0)(y - y_0) \\
&\quad + R_2(\Delta x, \Delta y),
\end{aligned}
$$

where the remainder terms R_1, R_2 satisfy

$$\lim_{\substack{\Delta x \to 0 \\ \Delta y \to 0}} \frac{R_i}{(\Delta x^2 + \Delta y^2)^{1/2}} = 0.$$

Now

$$
\begin{aligned}
f(z_0 + \Delta z) - f(z_0) &= (u_x(x_0, y_0)\Delta x + u_y(x_0, y_0)\Delta y + R_1(\Delta x, \Delta y)) \\
&\quad + i(v_x(x_0, y_0)\Delta x + v_y(x_0, y_0)\Delta y + R_2(\Delta x, \Delta y)).
\end{aligned}
$$

If the Cauchy-Riemann conditions hold, then $u_x = v_y$ and $v_x = -u_y$ at x_0, y_0 and

$$f(z_0 + \Delta z) - f(z_0) = u_x(\Delta x + i\Delta y) + iv_x(\Delta x + i\Delta y) + R_1 + iR_2$$

so that

$$\lim_{\Delta z \to 0} \frac{f(z_0 + \Delta z) - f(z_0)}{\Delta z} = u_x(x_0, y_0) + iv_x(x_0, y_0) = f'(z_0),$$

independent of how the limit $\Delta z \to 0$ is taken.

Products of analytic functions are analytic, and quotients of analytic functions are analytic unless the denominator function is zero. Rules for differentiation are the same as the product rule and quotient rule for real variables, that is, if $f(z) = g(z)h(z)$ then $f'(z) = g'(z)h(z) + h'(z)g(z)$, provided $g(z)$ and $h(z)$ are individually analytic. Also, if $f(z)$ is analytic, then $(\frac{1}{f(z)})' = -\frac{f'(z)}{f^2(z)}$, provided $f(z) \neq 0$.

The proof of these facts is direct. For example, if $f(z)$ is analytic, then

$$\frac{1}{\Delta z}\left(\frac{1}{f(z+\Delta z)} - \frac{1}{f(z)}\right) = -\left(\frac{f(z+\Delta z) - f(z)}{\Delta z}\right)\frac{1}{f(z)f(z+\Delta z)}$$

$$\longrightarrow \frac{-f'(z)}{f^2(z)}$$

as $\Delta z \to 0$.

Other rules of differentiation from real variable calculus are valid for complex valued functions. If $f(z)$ and $g(z)$ are analytic, then their composition is analytic and the chain rule

$$(f(g(z))' = f'(g(z))g'(z)$$

holds.

When they exist, inverse functions of analytic functions are analytic. Suppose $w = g(z)$ is analytic. For any Δz define $w + \Delta w = g(z + \Delta z)$ for z fixed. Then

$$\frac{g^{-1}(w+\Delta w) - g^{-1}(w)}{\Delta w} = \frac{\Delta z}{\Delta w} = \frac{\Delta z}{g(z+\Delta z) - g(z)} \to \frac{1}{g'(z)}$$

as Δz, hence Δw, approaches 0, which makes sense provided $g'(z) \neq 0$.

In spite of our earlier warning of caution, there are many analytic functions. All powers of z and polynomials of z are analytic, and rational functions of z are analytic, except at points where their denominators vanish. Differentiation for powers of z is the same as in real variables, namely, $\frac{d}{dz}(z^k) = kz^{k-1}$. Roots of z are analytic except at branch points. The transcendental functions and trigonometric functions are analytic as well. To see that e^z is analytic, use that

$$e^z = \sum_{n=0}^{\infty}\frac{z^n}{n!},$$

so that

$$\frac{e^{z+\Delta z}}{\Delta z} - e^z = e^z\left(\frac{e^{\Delta z} - 1}{\Delta z}\right) \to e^z \quad \text{as} \quad \Delta z \to 0.$$

Since e^z is analytic and $(e^z)' = e^z$, it follows that $\ln z$, e^{iz}, $\tanh z$, etc. are analytic functions.

6.2.2   Integration

Integration in the complex plane is defined in terms of real line integrals of the complex function $f = u + iv$. If C is any curve in the complex plane we define the line integral

$$\int_C f(z)dz = \int_C u\,dx - v\,dy + i\int_C v\,dx + u\,dy,$$

where the integrals $\int_C u\,dx$, $\int_C u\,dy$, $\int_C v\,dx$, $\int_C v\,dy$ are real line integrals in the usual calculus sense (i.e., Riemann integrals).

Typically, evaluation of line integrals is a tedious task, because the integrals depend nontrivially on the actual path C traversed. However, if f is analytic, a remarkable, and extremely important, fact results.

Theorem 6.2

If $f(z)$ is analytic in any domain R, then

$$\int_C f(z)dz = 0$$

for any closed curve C whose interior lies entirely in R.

Proof

Recall from calculus, that if $\phi(x, y)$ and $\psi(x, y)$ are continuously differentiable in some region R, for any closed curve whose interior is inside R,

$$\int_C \phi\,dx + \psi\,dy = \int_R \left(\frac{\partial \psi}{\partial x} - \frac{\partial \phi}{\partial y}\right) dx\,dy,$$

where R represents the interior of the closed curve C. This identity, (the analogue of integration by parts), is called Green's Theorem, and can be applied to the integral of a complex function f to find

$$\int_C f\,dz = \int_C u\,dx - v\,dy + i\int_C v\,dx + u\,dy$$

$$= -\int_R (u_y + v_x)\,dx\,dy + i\int_R (u_x - v_y)\,dx\,dy.$$

If f is analytic, we apply the Cauchy-Riemann conditions, and see that $\int_C f\,dz = 0$.

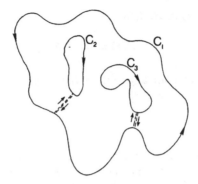

FIGURE 6.3   Contour of integration for a multiply connected domain.

The above statement presumes that the region R is simply connected. Using a simple trick, we can show that it is also true for multiply connected regions. Suppose the region R has closed boundary curves $C_1, C_2, \ldots, C_n$, where C_1 is the exterior boundary and $C_2, \ldots, C_n$ are interior boundaries. If f is analytic on R,

$$I = \int_{C_1} f\, dz + \int_{C_2} f\, dz + \ldots + \int_{C_n} f\, dz = 0$$

provided the sense of integration is chosen correctly. As a matter of convention we always define the integral $\int_{C_i} f\, dz$ so that as we traverse the arc C_i in the forward direction, the interior of the region R lies to the left of C_i. This direction of traversal is counterclockwise for exterior boundaries and clockwise for interior boundaries (a description which is rapidly becoming antiquated with the advent of digital watches).

The way to prove this statement is to connect all the arcs $C_1, C_2, \ldots, C_n$ by new arcs, so that by traversing the connecting arcs in both directions, we traverse all of these arcs as one closed arc (see Fig. 6.3). Since the contribution to the integral I is unchanged by the dual traversal of the connecting arcs, the integral I is the same as the integral of f around a closed curve, on the interior of which f is analytic, so I is zero.

One consequence of this theorem is that the integrals of f around closed curves $\int_{C_1} f\, dz$ and $\int_{C_2} f\, dz$ are equal if the arc C_1 can be deformed into the arc C_2 and if $f(z)$ is analytic on the region lying between the arcs C_1 and C_2.

A second observation is the analogue of the FUNDAMENTAL THEOREM OF CALCULUS.

Theorem 6.3

If $f(z)$ is analytic, and $\int_C f(z)dz = 0$ for all closed arcs C in the interior of R, then for z, z_0 in R, define

$$F(z) = \int_{z_0}^{z} f(\xi) d\xi$$

as the line integral along any path in R connecting z and z_0. Then, $F(z)$ is analytic, and $F'(z) = f(z)$.

Proof

The analyticity of $f(z)$ is used to establish that $\int_{z_0}^{z} f(\xi) d\xi$ is independent of path and therefore depends only on the endpoints z_0 and z. Hence for fixed z_0, $\int_{z_0}^{z} f(\xi) d\xi$ is indeed a function of z only.

To find the derivative of $F(z)$ note that

$$\frac{F(\xi) - F(z)}{\xi - z} = \frac{\int_{z_0}^{\xi} f(\eta) d\eta - \int_{z_0}^{z} f(\eta) d\eta}{\xi - z} = \frac{\int_{z}^{\xi} f(\eta) d\eta}{\xi - z} \to f(z)$$

as $\xi \to z$. In the above statement we have made strong use of independence of path in order to write

$$\int_{z_0}^{\xi} f(\eta) d\eta - \int_{z_0}^{z} f(\eta) d\eta = \int_{\xi}^{z} f(\eta) d\eta.$$

Example

For any closed curve C containing the origin, $\int_C z^n dz = 0$ for any integer n except $n = -1$, in which case $\int_C \frac{dz}{z} = 2\pi i$. If the closed curve C does not contain the origin, then $\int_C z^n dz = 0$, for all n. Obviously, for any closed curve $\int_C z^n dz = 0$ if $n \geq 0$, since positive powers of z are analytic everywhere. The powers z^{-n} are analytic except at the origin so that $\int_C z^{-n} dz = 0$ if C does not contain the origin. If C contains the origin it can be deformed into the unit circle $|z| = 1$ without changing the value of the integral, whereby

$$\int_C z^{-n} dz = \int_{|z|=1} z^{-n} dz = i \int_0^{2\pi} e^{(1-n)i\theta} d\theta = 2\pi i \delta_{n1}.$$

6.2.3   Cauchy Integral Formula

This rather simple example provides the basis for essentially all integration in the complex plane. This is because of the remarkable fact known as the CAUCHY INTEGRAL FORMULA.

Theorem 6.4

Let C be a simple, nonintersecting, closed curve traversed counterclockwise. Suppose $f(z)$ is analytic everywhere inside C. For any point z inside C

$$\int_C \frac{f(\xi)d\xi}{\xi - z} = 2\pi i f(z).$$

Proof

If C contains the point z in its interior we can deform the curve C into a small circle C' so that $|f(z) - f(\xi)| < \epsilon$ for every ξ on the circle C' for any fixed small number $\epsilon > 0$. Then

$$
\begin{aligned}
\int_C \frac{f(\xi)d\xi}{\xi - z} &= \int_{C'} \frac{f(\xi)d\xi}{\xi - z} \\
&= f(z)\int_{C'}\frac{d\xi}{\xi - z} + \int_{C'}\frac{f(\xi) - f(z)}{\xi - z}d\xi \\
&= 2\pi i f(z) + i\int_0^{2\pi}\left(f\left(z + re^{i\theta}\right) - f(z)\right)d\theta,
\end{aligned}
$$

where $\left|f\left(z + re^{i\theta}\right) - f(z)\right| < \epsilon$. Notice that $\int_{C'}\frac{d\xi}{\xi - z} = 2\pi i$ from the previous example. Since ϵ is arbitrary, the conclusion follows.

The Cauchy Integral Formula can be interpreted in two ways. First, as an integration formula, we see how to easily evaluate an integral around a closed contour with an integrand of the form $f(z)/(z - z_0)$ with z_0 a fixed point. Alternately, we see that if $f(z)$ is known on some closed contour and if $f(z)$ is analytic inside the contour, we can determine $f(z)$ uniquely at every point inside the contour as a weighted average of $f(z)$ on the contour.

Consequences of the Cauchy Integral Formula are numerous. For example, if $f(z)$ is analytic inside C, we know that $f'(z)$ exists and is continuous inside C, and

$$
\begin{aligned}
\frac{f(\eta) - f(z)}{\eta - z} &= \frac{1}{2\pi i}\int_C \left(\frac{f(\xi)}{\xi - \eta} - \frac{f(\xi)}{\xi - z}\right)\frac{d\xi}{\eta - z} \\
&= \frac{1}{2\pi i}\int_C \frac{f(\xi)}{(\xi - \eta)(\xi - z)}d\xi \to \frac{1}{2\pi i}\int_C \frac{f(\xi)}{(\xi - z)^2}d\xi
\end{aligned}
$$

as $\eta \to z$. In other words,

$$f'(z) = \frac{1}{2\pi i}\int_C \frac{f(\xi)d\xi}{(\xi - z)^2}.$$

This formula can be interpreted in two ways, namely, as a representation of the derivative $f'(z)$ in terms of an integral of the analytic function $f(z)$, or as an integration formula. For example, from this formula it follows that $\int_C \frac{dz}{z^2} = 0$ for any closed curve C containing the origin, since to identify this integral with the above we take $f(z) = 1$ so that $f'(z) = 0$. Notice also, that if $f(z)$ is analytic on the entire complex plane and bounded with $|f(z)| \le M$, then taking C to be a circle of radius R about z,

$$|f'(z)| \le \frac{M}{R}$$

for all $R > 0$. Hence, $f(z)$ is a constant (this result is known as Liouville's theorem).

To extend the Cauchy Integral Formula to higher order derivatives, we use induction. If

$$g(z) = \frac{(n-1)!}{2\pi i} \int_C \frac{f(\xi)d\xi}{(\xi - z)^n},$$

then

$$\frac{g(z) - g(\eta)}{z - \eta} = \frac{(n-1)!}{2\pi i} \int_C \frac{(\xi - \eta)^n - (\xi - z)^n}{(\xi - z)^n(\xi - \eta)^n} \frac{f(\xi)d\xi}{(z - \eta)}.$$

We use that

$$x^n - y^n = (x - y)\sum_{k=0}^{n-1} x^{n-k-1}y^k,$$

so that

$$\frac{g(z) - g(\eta)}{z - \eta} \to \frac{n!}{2\pi i} \int_C \frac{f(\xi)d\xi}{(\xi - z)^{n+1}} = g'(z).$$

In other words, the function $g(z)$ is continuously differentiable, hence analytic, for z inside C. Thus, if $f(z)$ is analytic inside C, the nth derivative of f is given by

$$f^{(n)}(z) = \frac{n!}{2\pi i} \int_C \frac{f(\xi)d\xi}{(\xi - z)^{n+1}}.$$

Further consequences of the Cauchy Integral Formula are the MEAN VALUE THEOREMS. The Cauchy Integral Formula expresses the fact that $f(z)$ is a weighted average of $f(\xi)$ on a curve surrounding the point z. If the curve C is a circle, $\xi = z + re^{i\theta}$, then the Cauchy Integral Formula says that

$$f(z) = \frac{1}{2\pi} \int_0^{2\pi} f\left(z + re^{i\theta}\right) d\theta,$$

that is, $f(z)$ is precisely the average value of f on a circle of radius r about z. Similarly, since this is true independent of r, we conclude that

$$f(z) = \frac{1}{\pi R^2} \int_0^{2\pi} \int_0^R f\left(z + re^{i\theta}\right) r\,dr\,d\theta$$

and again, $f(z)$ is the average value of f on the interior of any circle of radius R centered about z.

This last expression can be used to prove the MAXIMUM MODULUS THEOREM.

Theorem 6.5

If $f(z)$ is analytic on the interior of a region, then $|f(z)|$ cannot attain a local maximum on the interior of the region.

To prove this, suppose z is an interior point of a region on which $|f(z)| \geq |f(\xi)|$ for all ξ in the region. We surround the point z with a circle of radius R centered at $\xi = z$, lying entirely within the region. Then

$$|f(z)| \leq \frac{1}{\pi R^2} \int_A |f(\xi)| \, dA$$

which can hold if and only if $|f(z)| = |f(\xi)|$ for all ξ in the disc. Finally, if $|f(\xi)|$ is constant everywhere inside a circle, $f(\xi)$ is constant (use the Cauchy-Riemann conditions). Thus, if $|f(z)|$ attains its maximum on the interior of some region, it does so only because $f(z)$ is exactly a constant.

There are two important consequences of the maximum modulus theorem that we will use later.

Theorem 6.6 (Phragmén-Lindelof)

If $|f(iy)| \leq M$ and if $|f(z)| \leq Ke^{|z|^\beta}$ for $\beta < 1$ and $\operatorname{Re} z \geq 0$ then $|f(z)| \leq M$ for $\operatorname{Re} z \geq 0$.

The second result is a consequence of the Phragmén-Lindelof theorem.

Theorem 6.7

If $f(z)$ is bounded on the real axis and if $|f(z)| \leq Ae^{\sigma|z|}$, then $|f(x+iy)| \leq me^{\sigma|y|}$.

To prove the Phragmén-Lindelof theorem we set $F(z) = f(z)e^{-\epsilon z^\gamma}$ with $\beta < \gamma < 1$. Then, on a circle of radius r

$$|F(z)| \leq K \exp\left(r^\beta - \epsilon r^\gamma \cos \gamma \theta\right)$$

which for r sufficiently large and $-\pi/2 \leq \theta \leq \pi/2$ is arbitrarily small and hence bounded by M. From the maximum modulus theorem, $|F(z)| \leq M$ everywhere in $\operatorname{Re} z \geq 0$. We let $\varepsilon \to 0$ to obtain that $|f(z)| \leq M$ as well.

The proof of the second fact we leave to the interested reader (see exercises 6.2.13,14).

We can use these observations about the Cauchy Integral Formula to solve an important partial differential equation, specifically Laplace's equation on the unit circle with specified data on the boundary of the circle. Notice that if $f(z) = u(x, y) + iv(x, y)$ is analytic, then

$$\frac{\partial u}{\partial x} = \frac{\partial v}{\partial y} \quad \text{and} \quad \frac{\partial v}{\partial x} = -\frac{\partial u}{\partial y} \quad \text{so that} \quad \frac{\partial^2 u}{\partial x^2} + \frac{\partial^2 u}{\partial y^2} = 0.$$

That is, the real (and imaginary) part of an analytic function is a solution of Laplace's equation.

From the Cauchy Integral Formula,

$$f(z) = \frac{1}{2\pi i} \int_C \frac{f(\eta) d\eta}{\eta - z},$$

which, if C is chosen to be the unit circle $\eta = e^{i\theta}$ becomes

$$f(z) = \frac{1}{2\pi} \int_0^{2\pi} \frac{f(\eta)\eta \, d\theta}{\eta - z}.$$

Since z is inside the unit circle, $1/\bar{z}$ lies outside the unit circle and

$$\frac{1}{2\pi} \int_0^{2\pi} \frac{f(\eta)\eta \, d\theta}{\eta - 1/\bar{z}} = 0.$$

Subtracting these two expressions we obtain (since $\eta = 1/\bar{\eta}$)

$$f(z) \;=\; \frac{1}{2\pi} \int_0^{2\pi} f(\eta) \left(\frac{\eta}{\eta - z} - \frac{1/\bar{\eta}}{1/\bar{\eta} - 1/\bar{z}} \right) d\theta$$

or

$$f(z) \;=\; \frac{1}{2\pi} \int_0^{2\pi} f(\eta) \frac{1 - |z|^2}{|\eta - z|^2} d\theta.$$

We take the real part of $f(z)$ to find

$$u(r, \phi) = \frac{1}{2\pi} \int_0^{2\pi} u_0(\theta) \frac{1 - r^2}{1 - 2r \cos(\phi - \theta) + r^2} d\theta$$

which solves Laplace's equation subject to $u = u_0(\phi)$ on the boundary of the unit circle. The kernel of this integral operator is called the POISSON KERNEL.

6.2.4   Taylor and Laurent Series

We can use the Cauchy Integral Formula to deduce the properties of TAYLOR SERIES in the complex plane. Suppose $f(z)$ is analytic in the interior of the closed arc C. We know that

$$f(z) = \frac{1}{2\pi i} \int_C \frac{f(\xi)d\xi}{\xi - z}.$$

If z_0 is any point inside C, we re-express this formula in terms of $z - z_0$, as

$$f(z) = \frac{1}{2\pi i} \int_C \frac{f(\xi)}{\xi - z_0} \left(\frac{\xi - z_0}{(\xi - z_0) - (z - z_0)} \right) d\xi$$

$$= \frac{1}{2\pi i} \int_C \frac{f(\xi)}{\xi - z_0} \frac{d\xi}{\left(1 - \frac{z - z_0}{\xi - z_0} \right)}.$$

We use the fact that the geometric series

$$\sum_{n=0}^{\infty} z^n = \frac{1}{1-z}$$

converges provided $|z| < 1$, which is proven in the usual way. (Show that

$$S_k = \sum_{n=0}^{k} z^n = \frac{1 - z^{k+1}}{1 - z} \rightarrow \frac{1}{1-z}$$

provided $|z| < 1$.) Then

$$f(z) = \frac{1}{2\pi i} \int_C \sum_{n=0}^{\infty} \left(\frac{z - z_0}{\xi - z_0} \right)^n \frac{f(\xi)}{\xi - z_0} d\xi$$

$$= \sum_{n=0}^{\infty} a_n (z - z_0)^n$$

where

$$a_n = \frac{1}{2\pi i} \int_C \frac{f(\xi)}{(\xi - z_0)^{n+1}} d\xi = \frac{f^{(n)}(z_0)}{n!}$$

provided $|z - z_0| < |\xi - z_0|$ for every ξ on the arc C. To determine the region of convergence of this formula we let C be the largest possible circle about z_0 inside of which $f(z)$ is analytic. If $|z - z_0|$ is smaller than the radius of this largest possible circle, we have uniform convergence, hence validity, of this series, called the TAYLOR SERIES EXPANSION of f about z_0.

It is often useful to find a power series expansion for a function $f(z)$ about a point z_0 at which $f(z)$ is not analytic. As a next best alternative to analyticity on the interior of some region, we might have that $f(z)$ is analytic on an annular region about z_0, say $r_1 \leq |z - z_0| \leq r_2$. For z in this region we can write

$$f(z) = \frac{1}{2\pi i} \int_{C_2} \frac{f(\xi)d\xi}{\xi - z} - \frac{1}{2\pi i} \int_{C_1} \frac{f(\xi)d\xi}{\xi - z},$$

where C_2 and C_1 are circles of radius r_2 and r_1, respectively, about z_0. The reason we can write $f(z)$ in this way is because we can connect C_1 and C_2 into one closed arc by traversing a connecting arc twice (once in each direction) and apply Cauchy's Integral Formula to the new, single closed arc with z in its interior. Expanding about z_0 we find

$$
\begin{aligned}
f(z) &= \frac{1}{2\pi i} \int_{C_2} \frac{f(\xi)d\xi}{\xi - z_0} \left(1 - \frac{z - z_0}{\xi - z_0}\right)^{-1} d\xi \\
&+ \frac{1}{2\pi i} \int_{C_1} \frac{f(\xi)d\xi}{z - z_0} \left(1 - \frac{\xi - z_0}{z - z_0}\right)^{-1} d\xi \\
&= \sum_{n=0}^{\infty} (z - z_0)^n \frac{1}{2\pi i} \int_{C_2} \frac{f(\xi)}{(\xi - z_0)^{n+1}} d\xi \\
&+ \sum_{n=0}^{\infty} (z - z_0)^{-(n+1)} \frac{1}{2\pi i} \int_{C_1} f(\xi)(\xi - z_0)^n \, d\xi.
\end{aligned}
$$

Since f is analytic in the annulus between C_1 and C_2, we can deform the paths C_1 and C_2 into one path to write

$$f(z) = \sum_{n=-\infty}^{\infty} a_n (z - z_0)^n, \qquad a_n = \frac{1}{2\pi i} \int_C \frac{f(\xi)d\xi}{(\xi - z_0)^{n+1}},$$

where C is any closed curve lying entirely in the annulus of analyticity. The region of convergence of this power series is the largest possible annulus about z_0 containing C on which f is everywhere analytic.

It is tempting to write $a_n = \frac{f^{(n)}(z_0)}{n!}$, since the formula for a_n looks to be the same as with Taylor series. However, since f need not be analytic at z_0, this identification is incorrect. On the other hand, if f is analytic everywhere inside the circle C, then indeed $a_n = \frac{f^{(n)}(z_0)}{n!}$ for $n \geq 0$ and $a_n = 0$ for $n < 0$ and this power series reduces exactly to Taylor series.

Notice also that for $n = -1$ we have that

$$\frac{1}{2\pi i} \int_C f(z)dz = a_{-1},$$

which expresses the fact that the integral of the function f around a closed path is the $n = -1$ coefficient of this power series. If $z = z_0$ is the only singularity of $f(z)$ inside C, the term a_{-1} is called the RESIDUE of $f(z)$ at z_0 and this identity is called the RESIDUE THEOREM. Obviously, the integral of an analytic function around a closed path is zero since $a_{-1} = 0$.

This power series, called the LAURENT SERIES, is a generalization of the Taylor series. Although there can only be one Taylor series for f about a point z_0, there may be numerous Laurent series. For example, the function $f(z) = \frac{1}{(z-1)(z+2)}$ has three possible power series expansions about $z = 0$. These are

$$f(z) = \frac{1}{3} \sum_{n=0}^{\infty} \left(-1 + \left(\frac{-1}{2} \right)^{n+1} \right) z^n, \qquad |z| < 1$$

which is a Taylor series, and

$$f(z) = \frac{1}{3} \sum_{n=-\infty}^{-1} z^n - \frac{1}{6} \sum_{n=0}^{\infty} (-z/2)^n, \qquad 1 < |z| < 2$$

and

$$f(z) = \frac{1}{6} \sum_{n=1}^{\infty} \left(\frac{1}{2^{n-1}} + (-1)^n \right) \left(\frac{2}{z} \right)^n, \qquad |z| > 2$$

which are Laurent series.

A function that is analytic in the entire complex plane is called an EN-TIRE function. The usual way for the analyticity of a function f to fail is for f to have ISOLATED SINGULARITIES. If f is analytic on the disc $0 < |z - z_0| < r$ but not at the point $z = z_0$, then z_0 is said to be an isolated singularity of f. If f is analytic and single valued (no branch cuts) on the disc $0 < |z - z_0| < r$, we can write

$$f(z) = \sum_{n=-\infty}^{\infty} a_n (z - z_0)^n, \qquad a_n = \frac{1}{2\pi i} \int_C \frac{f(\xi) d\xi}{(\xi - z_0)^{n+1}},$$

where C is any contour in the annulus $0 < |z - z_0| < r$. The singularities of f are classified by the principal part of f,

$$\text{principal part } (f) = \sum_{n=-\infty}^{-1} a_n (z - z_0)^n$$

which are the negative terms of the Laurent series. There are three different possibilities:

1. If $f(z)$ is bounded near $z = z_0$, then $|f(z)| < M$ for $0 < |z - z_0| < r_0$. Taking C to be a circle of radius r, we find that $a_n < \frac{M}{r^n}$ for any n, so that $a_n = 0$ for $n < 0$. Thus, the principal part of f is zero. The point

z_0 is said to be a REMOVABLE SINGULARITY, since if we simply redefine $f(z_0) = a_0$, then $f(z)$ is analytic and $f(z) = \sum_{n=0}^{\infty} a_n (z - z_0)^n$ throughout the disc $0 \le |z - z_0| < r$.

2. If the principal part of $f(z)$ has a finite number of terms, and $(z - z_0)^k f(z)$ is analytic for all $k \ge n$, then z_0 is said to be a POLE OF ORDER n.

3. If the principal part of $f(z)$ has an infinite number of terms, z_0 is said to be an ESSENTIAL SINGULARITY. As an example, the function

$$e^{1/z} = \sum_{n=0}^{\infty} \frac{1}{n!} \frac{1}{z^n}$$

has an essential singularity at $z = 0$.

If $f(z)$ has an essential singularity at z_0, then in every neighborhood of z_0, $f(z)$ comes arbitrarily close to every complex number, because if not, there is a number ω for which $\frac{1}{f(z)-\omega}$ is bounded and analytic. Hence,

$$\frac{1}{f(z) - \omega} = \sum_{n=0}^{\infty} b_n (z - z_0)^n$$

in some nontrivial neighborhood of z_0. If $b_0 \neq 0$, then $\lim_{z \to z_0} f(z) = \omega + \frac{1}{b_0}$ is finite and contradicts that z_0 is an essential singularity of $f(z_0)$. If $b_0 = 0$, either there is a first nonvanishing b_k in which case f has a pole of order k, or else the b_n are all zero in which case $f = \omega$, again a contradiction.

The Taylor series has an important interpretation. If $f(z)$ is analytic and its values are known on some open (possibly small) region, its Taylor series determines $f(z)$ uniquely on the interior of a (probably larger) region centered about the point z. For any point z_1 about which the function $f(z)$ is known, a new Taylor series can be defined with the result that the largest circle of definition is most likely different than the original region of definition. In fact, by continually changing the center of the expansion, the region of definition of $f(z)$ can be continually enlarged. Unless the boundary of definition of $f(z)$ (i.e., the boundary of the union of all circles for which Taylor series can be given) is densely packed with singularities of f, the domain of definition of f can be extended to cover the entire z plane, excluding, of course, points where f has singularities. Thus, if $f(z)$ is known on some small region, it can be extended analytically and uniquely to a much larger region, possibly the entire z plane. The new values are called the ANALYTIC CONTINUATION of $f(z)$.

The practical implication of this is that if two different analytic expressions agree on some open neighborhood, they are representations of exactly the same analytic function.

Example

Evaluate $f(z) = \sum_{n=0}^{\infty} z^n$ at $z = 2i$.

Strictly speaking, $f(z)$ cannot be evaluated at $z = 2i$ since this representation of $f(z)$ makes sense only within the region of convergence $|z| < 1$. However, if $|z| < 1$, we know that

$$f(z) = \frac{1}{1-z}.$$

This representation of $f(z)$ is valid for all $z \neq 1$ and so, since it is analytic and agrees with $f(z)$ on the original domain of definition, it represents the unique extension of $f(z)$ to the entire z plane. We conclude that $f(2i) = \frac{1}{1-2i}$.

Example

The integral

$$I(a) = \int_0^{\pi} \frac{x \sin x \, dx}{a^2 - 2a \cos x + 1}$$

can be evaluated via contour integral techniques (see Problem 6.4.8) to be $I(a) = \frac{2\pi}{a} \ln(1 + a)$ if $|a| < 1$ and $I(a) = \frac{2\pi}{a} \ln\left(1 + \frac{1}{a}\right)$ if $|a| > 1$, and is analytic as a function of a provided $|a| < 1$ or $|a| > 1$. If $|a| = 1$, however, the integral $I(a)$ does not exist. In fact, if $a = e^{i\phi}$, the denominator $a^2 - 2a \cos x + 1$ vanishes whenever $\cos x = \cos \phi$. Thus the circle $|a| = 1$ is a closed arc of singularities through which analytic expressions cannot be continued.

6.3   Fluid Flow and Conformal Mappings

6.3.1   Laplace's Equation

One of the most interesting classical applications of complex variable theory is to the flow of two-dimensional INVISCID FLUIDS. This subject matter is of limited usefulness for the study of real, viscous, three-dimensional flows, and in a modern context, most fluid flow problems require the use of heavy machinery (i.e., large scale computations). However, this idealized subject remains interesting because of its esthetic appeal and for the conclusions which are easily obtained. We begin by deriving the relevant equations.

Suppose $u(x, y)$ is a vector field in $\mathbb{R}^2$ representing the steady velocity of a fluid. We assume that u is independent of the third coordinate. For any region V with boundary ∂V the total flow of fluid across the boundary is

$$F = \int_{\partial V} \rho u \cdot n \, dA,$$

where ρ is the density of the material, n is a unit vector normal to the boundary, and dA represents the element of surface area. If fluid is neither created nor destroyed inside V, fluid is conserved, so $F = 0$. From the divergence theorem

$$0 = \int_{\partial V} \rho u \cdot n \, dA = \int_V \text{div}(\rho u) dV,$$

where the second integral is an integral over the area (or volume) of V. Since V is an arbitrary volume, it must be that $\text{div}(\rho u) = 0$ everywhere. This law, which is a basic conservation law, simplifies to

$$\nabla \cdot u = 0$$

if ρ is independent of x, y, i.e., if the fluid is incompressible.

We introduce a scalar function $\phi(x, y)$, called a POTENTIAL FUNCTION, whose gradient is u,

$$u = \nabla \phi$$

or

$$u_1 = \frac{\partial \phi}{\partial x}, \quad u_2 = \frac{\partial \phi}{\partial y} \text{ where } u = (u_1, u_2).$$

For consistency we must have

$$\frac{\partial^2 \phi}{\partial y \partial x} = \frac{\partial^2 \phi}{\partial x \partial y}$$

or

$$\frac{\partial u_1}{\partial y} = \frac{\partial u_2}{\partial x}.$$

If ϕ is a potential function, then

$$\phi(x, y) = \int_{x_0, y_0}^{x, y} u \cdot ds,$$

a line integral from some fixed, but arbitrary, point x_0, y_0 to x, y along some path connecting the two points. Since we want this to uniquely define ϕ (up to an additive constant) we require that the line integral around any closed path be zero so that the function $\phi(x, y)$ depends only on the endpoints and not on the actual path of integration. If this is to be true, we must have

$$0 = \int_{\partial V} u_1 dx + u_2 dy = \int \int_V \left(\frac{\partial u_2}{\partial x} - \frac{\partial u_1}{\partial y} \right) dx \, dy,$$

where $u = \left(\begin{smallmatrix} u_1 \\ u_2 \end{smallmatrix} \right)$ and V is any closed region with boundary ∂V. This integral vanishes identically if and only if

$$\frac{\partial u_2}{\partial x} = \frac{\partial u_1}{\partial y},$$

which we already specified for consistency. If we think of u as a vector field in $\mathbb{R}^3$ dependent on the independent variables x, y, z with $u_3 = $ constant and u_1, u_2 independent of z, then $\frac{\partial u_2}{\partial x} = \frac{\partial u_1}{\partial y}$ is equivalent to requiring that $\operatorname{curl} u = \nabla \times u = 0$. Thus, to represent u in terms of a potential function we require that the flow be IRROTATIONAL, or curl free.

If, as hypothesized, the flow is irrotational, then the conservation equation $\operatorname{div} u = \frac{\partial u_1}{\partial x} + \frac{\partial u_2}{\partial y} = 0$ becomes

$$\frac{\partial^2 \phi}{\partial x^2} + \frac{\partial^2 \phi}{\partial y^2} = 0,$$

which is LAPLACE'S EQUATION. The function ϕ is the potential function for a steady, incompressible, inviscid, irrotational flow in two-dimensions. (This equation was derived for flows in an incompressible fluid, but it can also be derived for flows in a compressible fluid, such as air, for subsonic speeds. Laplace's equation is not valid at supersonic speeds.)

If $f(z) = \phi(x, y) + i\psi(x, y)$ is an analytic function, then both $\phi(x, y)$ and $\psi(x, y)$ satisfy Laplace's equation. In addition, the reverse statement is true, namely, if $\phi(x, y)$ satisfies Laplace's equation, there is a function $\psi(x, y)$, unique up to an additive constant, for which $\phi(x, y) + i\psi(x, y)$ is analytic as a function of $z = x + iy$. In fact, if we set $\psi = \int_C \phi_x \, dy - \phi_y \, dx$ where C is any path from a fixed point x_0, y_0, to x, y, then ψ is independent of the path of integration C, since $\phi_{xx} + \phi_{yy} = 0$, and $\phi_x = \psi_y$, $\phi_y = -\psi_x$, which are the Cauchy-Riemann conditions.

The physical interpretation of $\psi(x, y)$ is interesting. Since $\phi_x = \psi_y$ and $\psi_x = -\phi_y$, the gradients of ϕ and ψ are orthogonal $\nabla \phi \cdot \nabla \psi = 0$. Thus, the level surfaces of ϕ and ψ are orthogonal, so that the level surfaces of ψ are tangent to $\nabla \phi$, the velocity vector field. The level surfaces of ψ are called STREAMLINES of the flow.

To study fluid flow in this context we need consider only analytic functions. Some very simple analytic functions give interesting flows.

Examples

1. Suppose $f = Az = A(x + iy)$, then $\phi = Ax$, $\psi = Ay$ so that streamlines are the lines $y = $ constant. The corresponding flow is uniform steady flow parallel to the x-axis with velocity A.

2. The function $f = A \ln z = A \ln r + Ai\theta$ where $z = re^{i\theta}$ has $\phi = A \ln r$, $\psi = A\theta$. Streamlines are $\theta = $ constant representing flow outward from the origin, which is a simple SOURCE if $A > 0$ or a SINK if $A < 0$ (Fig. 6.4a).

3. The function $f = iA \ln z = -A\theta + iA \ln r$, $\phi = A\theta$, $\psi = A \ln r$ has streamlines $r = $ constant so that the flow is uniform clockwise rotation

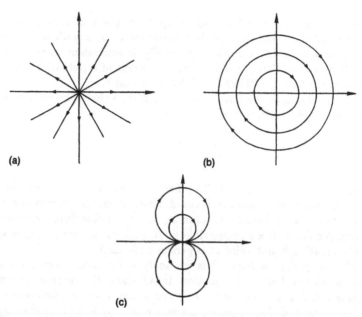

FIGURE 6.4 Streamlines corresponding to the analytic functions
a) $f(z) = -A \ln z$, b) $f(z) = iA \ln z$, c) $f(z) = a/z$.

(a VORTEX) about the origin (Fig. 6.4b). It is important to note that even though this flow is rotating, it is still an irrotational flow, except at the origin, where analyticity fails.

4. $f = A/z$, $\phi = \frac{Ax}{x^2+y^2}$, $\psi = \frac{-Ay}{x^2+y^2}$, has streamlines $x^2 + (y-\alpha)^2 = \alpha^2$ which is a family of circles centered on the y-axis, all passing through the origin. This flow is that of a DIPOLE (Fig. 6.4c).

A standard problem is to determine the flow past a circular cylinder. Consider $f(z) = V(z + \frac{a^2}{z})$ and notice that on the circle $|z| = a$, $(\bar{z} = a^2/z)$

$$f(z) = V(z + \bar{z}) = 2Vx$$

so that $\phi = 2Vx$ and $\psi = 0$ on $|z| = a$. In other words, the circle $|z| = a$ is a streamline of this flow. In fact, since

$$\psi = V\left(1 - \frac{1}{x^2 + y^2}\right)y,$$

the streamlines are parallel to the x-axis a long way from the origin, and they flow around a circle of radius a.

STAGNATION POINTS are those at which the flow is quiescent, i.e., $\nabla \phi = 0$. Since $f = \phi + i\psi$, it follows from the chain rule (make the two-dimensional change of variables $z = x + iy$, $\bar{z} = x - iy$) and analyticity that

$$\frac{df}{dz} = \frac{1}{2}(\phi_x - i\phi_y) + \frac{i}{2}(\psi_x - i\psi_y) = \phi_x - i\phi_y,$$

and

$$\frac{d\bar{f}}{d\bar{z}} = \phi_x + i\phi_y.$$

The expression

$$\frac{df}{dz} = \phi_x - i\phi_y$$

is called the COMPLEX VELOCITY of the flow, and a point is a stagnation point if and only if $\frac{df}{dz} = 0$. For the flow

$$f = V\left(z + \frac{a^2}{z}\right),$$

$$\frac{df}{dz} = V\left(1 - \frac{a^2}{z^2}\right) = 0.$$

if and only if $z^2 = a^2$, that is, if and only if $x = \pm a$, $y = 0$.

If $f(z)$ is any flow with all singularities outside the circle $|z| > a$, then the function

$$w(z) = f(z) + \bar{f}\left(\frac{a^2}{z}\right)$$

corresponds to the perturbation of $f(z)$ due to flow about a circle of radius a. The function $\bar{f}(z)$ is defined as $\bar{f}(z) = \overline{f(\bar{z})} = \phi(x, -y) - i\psi(x, -y)$. For example, if $f(z) = z$, then $\bar{f}(z) = z$, or if $f(z) = e^{iz}$ then $\bar{f}(z) = e^{-iz}$. Notice that with this definition, if $f(z)$ is analytic, then $\bar{f}(z)$ is also an analytic function of z.

To understand the flow corresponding to $w(z)$, note that if all the singularities of $f(z)$ lie outside the circle $|z| = a$, the singularities of $w(z)$ lying outside $|z| = a$ are the same as for $f(z)$. Furthermore, on the circle $|z| = a$, $\bar{z} = a^2/z$ and

$$w(z) = f(z) + \bar{f}(\bar{z})$$

is real so that $\psi = 0$. In other words, the circle $|z| = a$ is a streamline of the flow $w(z)$.

Suppose we have a fluid flow $w(z)$ for which the circle $|z| = a$ is a streamline. We want to determine what forces are exerted on the circle by the flow. The force F on the circle has horizontal component F_x and vertical component F_y corresponding to drag and lift, respectively. The important THEOREM of BLASIUS relates the lift and drag on an object to a contour integral around the object.

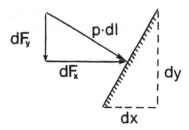

FIGURE 6.5 Forces due to pressure on a wall segment of length dl.

Theorem 6.8

If w is an analytic function corresponding to the flow about an object, then

$$F_x - iF_y = \frac{i\rho}{2} \int_C \left(\frac{dw}{dz}\right)^2 dz,$$

where ρ is the density of the fluid, and C is the contour given by the boundary of the object and integration is counter-clockwise. (Of course, if w is analytic, the contour can always be deformed to make integration more convenient.)

Since the fluid is assumed inviscid, the only forces exerted on the boundary of an object are the normal forces of fluid pressure. Suppose a small arc of boundary of length dl can be represented by the incremental vector (dx, dy). Since force is pressure times length, the increment of force generated by the pressure p is

$$d(F_x - iF_y) = -p(dy + i\,dx)$$

which is normal to the object's boundary (see Fig. 6.5).

We know from Newton's law, that momentum of the fluid must also be balanced. If $V(t)$ is a volume of fluid at time t, then Newton's law takes the form

$$\frac{d}{dt} \int_{V(t)} \rho u\, dV = \int_{V(t)} \rho f\, dV + \int_{\partial V} T\, dS,$$

where the vector f represents body forces, i.e., forces acting on the body as a whole such as gravity, and the vector T represents stress or surface forces, i.e., forces on this body of fluid due to contact with neighbors. Surface forces are of two kinds. There are pressures which act only normal to the surface and shear forces which act tangentially to the surface and are responsible for viscosity in a fluid.

If we assume that mass is conserved, we can rewrite Newton's law as

$$\int_{V(t)} \left(\rho \frac{Du}{Dt} - \rho f - \nabla \cdot T\right) dV = 0,$$

where the differential operator $\frac{Du}{Dt} = \frac{\partial u}{\partial t} + u \cdot \nabla u$ is called the MATERIAL or SUBSTANTIAL derivative of u and is the total derivative of u with respect to t. (For any quantity c moving with velocity u, the total derivative is

$$\frac{dc}{dt} = \frac{\partial c}{\partial t} + \frac{\partial c}{\partial x} \cdot \frac{dx}{dt} = \frac{\partial c}{\partial t} + u \cdot \nabla c.)$$

Of course, the volume $V(t)$ is arbitrary so we conclude that

$$\rho \frac{Du}{Dt} = \rho f + \nabla \cdot T$$

which is Cauchy's equation for the balance of linear momentum.

We do not need to use Cauchy's equation in its full generality. Since we have assumed the fluid is inviscid, the stress tensor T has only normal components of pressure, so $\nabla \cdot T = -\nabla p$. Furthermore, we have assumed the motion is steady and irrotational so that $\frac{\partial u}{\partial t} = 0$, and $\nabla \times u = 0$. If we have only conservative body forces f (such as gravity) we can express f in terms of a potential $\rho f = -\nabla \Phi$. Combining these assumptions it follows (after some vector calculus) that

$$\frac{1}{2}|u|^2 + p/\rho + \Phi = \text{constant},$$

which is called BERNOULLI'S LAW. For the problems that we discuss here the effects of the gravitational potential Φ will be ignored.

From Bernoulli's law, $p = p_0 - \frac{1}{2}\rho|u|^2$ along streamlines, where the vector u is the velocity vector, $|u|$ is the magnitude of the velocity, i.e., speed. Thus

$$d(F_x - iF_y) \quad = \quad \frac{1}{2}\rho|u|^2(dy + i\,dx) = \frac{i}{2}\rho|u|^2 d\bar{z}$$

along streamlines. The complex velocity vector is dw/dz so that $|u|^2 = \frac{dw}{dz}\frac{d\bar{w}}{d\bar{z}}$, and

$$d(F_x - iF_y) \quad = \quad \frac{i\rho}{2}\frac{dw}{dz}d\bar{w}.$$

On a streamline $d\bar{w} = d\phi - id\psi = d\phi + id\psi = \frac{dw}{dz}dz$ since $d\psi = 0$, so we can integrate around a closed streamline to obtain the conclusion of our theorem

$$F_x - iF_y \quad = \quad \frac{i\rho}{2}\int_C \left(\frac{dw}{dz}\right)^2 dz.$$

What kinds of fluid flows produce lift? None of the flows already discussed produce lift. For the flow

$$w(z) = V\left(z + a^2/z\right)$$

we can directly calculate that $F_x = F_y = 0$. In order to produce lift, a flow must have CIRCULATION. If ϕ is the potential function for a flow, the circulation around any closed path is defined as

$$\Gamma = \int_C \frac{d\phi}{ds} ds,$$

where ds is an element of arc length and C is a closed contour. A simple flow with circulation is the VORTEX FLOW $f(z) = iA \ln z$ where $\phi = A\theta$ so that for any closed path surrounding the origin, $\Gamma = -2\pi A$.

If $w(z)$ is a flow which has a streamline at $|z| = a$, then $w(z) + iA \ln z$ will also have a streamline at $|z| = a$, but this perturbed flow will now have circulation. Consider, for example, the flow $w(z) = V(z + a^2/z) + iA \ln z$. For this flow we calculate that

$$F_x - iF_y = -2\pi i \rho V A$$

so that $F_x = 0$ (there is no drag), and the lift is $F_y = 2\pi \rho V A$. If V is complex, of course, there is drag as well.

The three factors that determine the lift generated by a circular cylinder are the density of the fluid, the forward velocity, and the speed of rotation. We can now understand, at least qualitatively, what makes a baseball curve. Because of viscosity, the spinning of the ball entrains air and produces circulation superimposed on a uniform velocity field. A ball with topspin will drop, and with sidespin will curve away from the side with excess forward velocity due to spin. However, a ball with more velocity curves more, and a ball thrown in the Rocky Mountains curves less than one thrown at sea level or in air with high humidity. Slicing and hooking would be less of a problem at a golf course on the moon.

The lift produced by an object is independent of its shape. If $w' = V + A/z +$ (terms that vanish faster than $1/z$ as $z \to \infty$), and if $w(z)$ has no singularities outside of the object, the lift is still

$$F_y = 2\pi \rho V A.$$

Thus, in a uniform wind, the circulation is the only factor that determines lift.

6.3.2 Conformal Mappings

Airplane wings are generally not made from circular cylinders, but to find the fluid flow past a noncircular object is, in general, a difficult problem. However, using the chain rule, we recognize that the composition of two analytic functions is analytic. If we know the desired solution for one region, and we know how to map (analytically) the original region into a second region, the map composed with the original solution provides the desired solution for the second region.

To illustrate this point, consider two complex planes with complex variables ζ and z. The function $z = f(\zeta)$ can be viewed as a map of the ζ plane to the z plane. Until now we have viewed $f(\zeta)$ as a complex function of the complex variable ζ. Now we want to think of $f(\zeta)$ as a map that takes the complex number ζ and maps it to a new point on the complex plane z at $z = f(\zeta)$. (If $f(\zeta)$ is multivalued we confine our attention to some single valued branch of f.) For example, the map $z = a\zeta$ is a map which expands (or contracts) and rotates objects from the ζ plane onto the z plane. The circle $|\zeta| = 1$ in the ζ plane becomes the circle $|z| = |a|$ in the z plane. If $w = g(z)$ represents some flow about the unit circle, then $w = g(\zeta/a)$ represents the flow about the circle $|\zeta| = |a|$ in the ζ plane, rotated by the angle $\arg a$.

A map between two complex planes $z = f(\zeta)$ is said to be CONFORMAL (i.e., a mapping with form preserved) at any point ζ_0 for which $f'(\zeta_0) \neq 0$. If $f'(\zeta_0) = 0$ then the map is not invertible at ζ_0 and ζ_0 is a branch point of the inverse map. One feature of a conformal map is that it preserves form. In particular, it rotates all infinitesimal line elements through the angle $\arg f'(\zeta_0)$ and stretches line elements by the factor $|f'(\zeta_0)|$. Any infinitesimal object at ζ_0 transforms into a similar rotated figure at $z_0 = f(\zeta_0)$ with area scaled by the factor $|f'(\zeta_0)|^2$. This statement becomes apparent by looking at the Taylor series for z in a neighborhood of ζ_0. For $z = z_0 + \delta z$ we observe that

$$\delta z = f'(\zeta_0) \, \delta \zeta$$

to leading order in $\delta \zeta$. Therefore, $\arg \delta z = \arg \delta \zeta + \arg f'(\zeta_0)$ to leading order, and $|\delta z| = |f'(\zeta_0)| \, |\delta \zeta|$ from which the above statements follow.

The central problem of conformal mappings is to determine when a given domain D in the z plane can be mapped conformally onto a given domain D' in the ζ plane. The main theorem in this direction is for simply connected domains, and is known as the Riemann Mapping Theorem.

Theorem 6.9

Any two simply connected domains, each having more than one boundary point, can be mapped conformally one onto the other. For a simply connected domain D in the z plane with more than one boundary point, there is a three parameter family of transformations mapping D onto the circular domain $|\zeta| < 1$. A transformation $\zeta = F(z)$ is uniquely specified by any three independent conditions.

Even though Riemann's Theorem guarantees that the desired transformation exists, it provides no constructive technique to actually obtain it. To gain experience in how to build specific transformations, we will examine a number of examples.

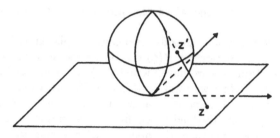

FIGURE 6.6 The stereographic projection.

The work horse of conformal mappings is the BILINEAR TRANSFOR-MATION

$$\zeta = \frac{az + b}{cz + d}$$

with inverse

$$z = \frac{-d\zeta + b}{c\zeta - a}$$

provided $ad - bc \neq 0$. The special case of a linear map $\zeta = az + b$ is easily seen to translate objects by b, to rotate them by $\arg a$ and scale them by $|a|$. Thus we have uniform rotation, linear scaling and uniform translation. The second special case $\zeta = 1/z$ corresponds to reflection through the unit circle followed by reflection in the real axis. This is apparent by noting that if $z = re^{i\theta}$ then $\zeta = \frac{1}{r}e^{-i\theta}$. The map $\zeta = 1/z$ is not defined at $z = 0$. However as $z \to 0$, $\zeta \to \infty$ so it is conventional to describe the image of $z = 0$ as $\zeta = \infty$. As there is no single point called "infinity," we make sense of this identification by using the *stereographic projection* of a plane onto a sphere. To define the stereographic projection we place a unit sphere on the z plane in contact with the origin at its south pole (Fig. 6.6). Connect any point on the z plane to the north pole with a straight line. The point z on the sphere cut by the line is then the image of the z point. Of course as $z \to \infty$ in any direction, the image point z approaches the north pole (a single point) which we identify as the "point" at infinity.

The linear map transforms circles into circles and the map $\zeta = 1/z$ also transforms circles into circles, so that any composition of these maps also transforms circles into circles. (A straight line is viewed as a degenerate circle through infinity.) The composition of these two maps is bilinear and any bilinear map can be viewed as a composition of maps of these two forms. Thus a bilinear map transforms circles into circles. At first glance it may appear that a bilinear map has four arbitrary constants. However, one of the constants can always be taken to be one leaving only three free constants.

Some standard examples of bilinear transformations are the following:

Example

The map $\zeta = \frac{z-i}{z+i}$ transforms the real axis onto the unit circle, the upper half plane onto the interior of the unit circle and the lower half plane onto the exterior of the unit circle. To see this, notice that $|\zeta| = 1$ is equivalent to $|z - i| = |z + i|$ which is the real axis in the z plane.

Example

The map $\zeta = \frac{z+1/4}{z+4}$ maps the family of nonconcentric circles $|z + 1/4| = \kappa|z + 4|$ into the family of concentric circles $|\zeta| = \kappa$ in the ζ plane. To be convinced of this one must verify that the equation $|z - a| = \kappa|z - b|$ represents a family of nonconcentric circles (see Problem 6.3.4).

We make use of this map to solve the following problem:

Example

Solve Laplace's equation $\nabla^2 \phi = 0$ between the two nonconcentric circles $|z| = 1$ and $|z - 1| = 5/2$, subject to the boundary conditions $\phi = a$ on $|z| = 1$ and $\phi = b$ on $|z - 1| = 5/2$.

We seek a bilinear transformation that maps the two nonconcentric bounding circles into two concentric circles. To do this we first imbed the circles $|z| = 1$ and $|z - 1| = 5/2$ into a family of circles of the form $|z - \alpha| = \kappa|z - \beta|$ with α, β fixed.

Any circle $|z - \delta| = \gamma$ can be written in the form $|z - \alpha| = \kappa|z - \beta|$ provided $\alpha = \delta + r_1 e^{i\theta}$ and $\beta = \delta + r_2 e^{i\theta}$ for some fixed θ, and $r_1 r_2 = \gamma$ (see Problem 6.3.4). To put two nonconcentric circles into this same form, α and β must lie on the line that joins the centers of the circles, in this case, the real axis. To pick α and β uniquely we must also satisfy the conditions $\alpha\beta = 1$ and $(1 - \alpha)(1 - \beta) = (5/2)^2$. Thus $\alpha = -1/4$, $\beta = -4$, and the circles $|z| = 1$ and $|z - 1| = 5/2$ can be written as $|z + 1/4| = \kappa|z + 4|$ where $\kappa = 1/4$ and $1/2$, respectively. We now see that the map $\zeta = \frac{z+1/4}{z+4}$ takes the circles $|z| = 1$ onto $|\zeta| = 1/4$ and the circle $|z - 1| = 5/2$ onto $|\zeta| = 1/2$ (see Fig. 6.7).

The solution of Laplace's equation in the ζ plane can now be written down by inspection. The function

$$f(\zeta) = (2b - a) + (b - a)\frac{\ln \zeta}{\ln 2}$$

is analytic and satisfies the boundary conditions $\mathrm{Re}\, f(\zeta) = a$ on $|\zeta| = 1/4$ and $\mathrm{Re}\, f(\zeta) = b$ on $|\zeta| = 1/2$. Composing this with the bilinear transformation we obtain

$$\phi = \mathrm{Re}\, f\left(\frac{z + 1/4}{z + 4}\right) = 2b - a + \frac{b - a}{\ln 2}\ln\left|\frac{z + 1/4}{z + 4}\right|$$

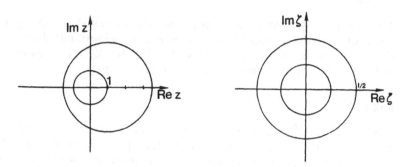

FIGURE 6.7   The circles $|z| = 1$ and $|z - 1| = 5/2$ in the z plane and in the ζ plane where $\zeta = (z + 1/4)/(z + 4)$.

as the desired solution of Laplace's equation.

Example

Find the flow past the two unit circles centered at $z = ie^{i\theta}$ and $z = -ie^{i\theta}$, with velocity at infinity parallel to the x-axis having amplitude U and circulation γ.

We first rotate the z plane by the angle θ, by setting $t = ze^{-i\theta}$ and next apply the bilinear map $\omega = \frac{it-2}{t}$ which maps the circles $|t - i| = 1$ and $|t + i| = 1$ onto the straight lines $\operatorname{Im}\omega = 2$ and $\operatorname{Im}\omega = 0$, respectively. The exterior of the circles is mapped onto the strip between the two lines. (The easiest way to see this is to look at the map $\omega = \frac{it-2}{t} = i - 2/t$ as a shift of the map $\omega = -2/t$ and to check that the circles $|t - i| = 1$, $|t + i| = 1$ map onto lines with $\operatorname{Im}\omega = $ constant.) Next we apply the transformation $s = e^{\pi\omega/2}$ which maps the strip onto the upper half plane. With this transformation, the line $\operatorname{Im}\omega = 0$ maps onto the positive real s axis and the line $\operatorname{Im}\omega = 2$ maps onto the negative real s axis. Finally we apply the bilinear transformation

$$\zeta = \frac{s + i}{s - i}$$

which takes the real s axis onto the unit circle $|\zeta| = 1$ and the upper half s plane to the exterior of the circle $|\zeta| = 1$. We have mapped the circle $|t - i| = 1$ onto the upper semicircle $|\zeta| = 1$, $\operatorname{Im}\zeta > 0$ while $|t + i| = 1$ has been mapped to $|\zeta| = 1$ with $\operatorname{Im}\zeta < 0$ (see Fig. 6.8). Composing these maps we obtain

$$\zeta = -\coth\left(\frac{\pi e^{-i\theta}}{2z}\right)$$

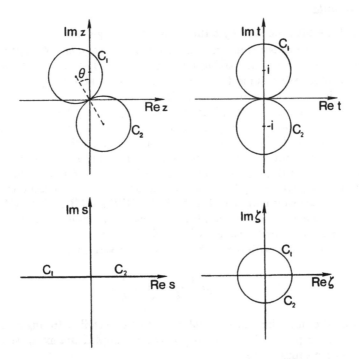

FIGURE 6.8 Images of the two unit circles centered at $z = ie^{i\theta}$ and at $z = -ie^{i\theta}$ under the transformations, $t = ze^{-i\theta}$, $w = i - 2/t$ (not shown), $s = e^{\pi w/2}$, and $\zeta = (s+i)/(s-i)$.

which maps the original domain onto the exterior of the unit circle.

The complex potential we seek must satisfy $\lim_{z\to\infty} f'(z) = U$. To find the condition at infinity in the ζ plane, note that if $f(z) = g(\zeta(z))$ then $g'(\zeta) = \frac{f'(z)}{\zeta'(z)}$ and

$$\lim_{\zeta\to\infty} g'(\zeta) = \lim_{z\to\infty} \frac{f'(z)}{\zeta'(z)} = -\frac{U\pi e^{-i\theta}}{2}.$$

We are now able to write that

$$g(\zeta) = -\frac{\pi U}{2}\left(e^{-i\theta}\zeta + \frac{e^{i\theta}}{\zeta}\right) - \frac{i\gamma}{2\pi}\ln\zeta$$

from which it follows that the complex potential for the flow we seek is

$$f(z) = \frac{\pi U}{2}\left(e^{-i\theta}\coth\frac{\pi e^{-i\theta}}{2z} + e^{i\theta}\tanh\frac{\pi e^{-i\theta}}{2z}\right) - \frac{i\gamma}{2\pi}\ln\coth\frac{\pi e^{-i\theta}}{2z}.$$

Example

Find the lift generated by a thin airfoil with angle of attack α.

We take as our thin airfoil the straight line segment of length $2a$, $z = re^{-i\alpha}$, $-a \leq r \leq a$, making an angle α, $-\pi/2 < \alpha < \pi/2$, with the horizontal. We assume there is a uniform flow at infinity and suppose there is some (as yet undetermined) circulation γ.

We first rotate the z plane by angle α by setting $t = e^{i\alpha}z$ so that the airfoil becomes the line segment in the t-plane on the real axis with $-a \leq \mathrm{Re}\, t \leq a$. We look for a transformation that maps the unit circle in the ζ plane onto our "airfoil" in the t plane. The bilinear transformation $t = \frac{a}{2}(\zeta + \frac{1}{\zeta})$ has this property since on $\zeta = e^{i\theta}$, $t = a\cos\theta$ is a double cover of the line segment $-a \leq \mathrm{Re}\, t \leq a$. Note also that $t = \infty$ corresponds to $\zeta = \infty$. This transformation, called the JOUKOWSKI TRANSFORMATION is quite useful in airflow problems, because it provides a transformation between circles and a variety of "airfoil" shapes.

To continue with our problem, the composite map

$$z = \frac{e^{-i\alpha}a}{2}\left(\zeta + \frac{1}{\zeta}\right)$$

takes the unit circle in the ζ plane onto the airfoil with angle of attack α in the z plane. If $f(z) = g(\zeta(z))$ is the complex potential we seek, as $\zeta \to \infty$ we must have

$$\lim_{\zeta \to \infty} g'(\zeta) = \lim_{z \to \infty} \frac{f'(z)}{\zeta'(z)} = \frac{Ue^{-i\alpha}a}{2}.$$

The complex potential for the flow about the circle $|\zeta| = 1$ satisfying this condition at infinity is

$$g(\zeta) = \frac{aU}{2}\left(e^{-i\alpha}\zeta + \frac{e^{i\alpha}}{\zeta}\right) - \frac{i\gamma}{2\pi}\ln\zeta.$$

From the inverse map

$$\zeta = \frac{1}{a}\left(e^{i\alpha}z + \left(e^{2i\alpha}z^2 - a^2\right)^{1/2}\right),$$

we find (up to an unimportant additive constant)

$$\begin{aligned}
f(z) &= Ue^{i\alpha}\left(z\cos\alpha - i\sin\alpha\left(z^2 - e^{-2i\alpha}a^2\right)^{1/2}\right) \\
&\quad - \frac{i\gamma}{2\pi}\ln\left(z + \left(z^2 - a^2e^{-2i\alpha}\right)^{1/2}\right).
\end{aligned}$$

We calculate the complex velocity as

$$f'(z) = Ue^{i\alpha}\cos\alpha - i\frac{Ue^{i\alpha}(\sin\alpha)z + \gamma/2\pi}{(z^2 - a^2e^{-2i\alpha})^{1/2}}$$

for arbitrary γ. Unfortunately, there are singularities at the ends of the airfoil $z = \pm ae^{-i\alpha}$. For physical reasons, we require that the velocity be finite at the trailing end of the airfoil. This requirement, called the KUTTA CONDITION, is equivalent to requiring that the flow at the trailing end leaves the airfoil tangent to the airfoil. We can satisfy this condition by setting $\gamma = -2\pi aU\sin\alpha$ in which case the complex velocity is

$$f'(z) = Ue^{i\alpha}\left(\cos\alpha - i\sin\alpha\left(\frac{z - ae^{-i\alpha}}{z + ae^{-i\alpha}}\right)^{1/2}\right)$$

and the corresponding lift generated by the circulation is

$$F_y = 2\pi\rho aU^2\sin\alpha.$$

For α near $\pi/2$ this formula is counterintuitive and is certainly wrong since the Kutta condition fails in real flows with nearly vertical plates. The Kutta condition works satisfactorily only for α small enough.

Use of the Joukowski transformation makes it possible to find the flow patterns and lift for a variety of airfoil shapes. One finds different shapes by examining the image of a circle of radius r centered at $z_0 = x_0 + iy_0$ in the complex plane under the Joukowski transformation $z = \zeta + L^2/\zeta$. By applying the Kutta condition to these flows with velocity U at infinity and angle of attack α one obtains the circulation κ and lift for the given airfoil. Some examples are shown in Figs. 6.9, with the parameters U, α, x_0, y_0, r, L, and κ indicated.

6.3.3 Free Boundary Problems— The Hodograph Transformation

Suppose a flow having velocity U at infinity ($\mathrm{Re}\, z \to -\infty$) flows past a flat vertical plate on the imaginary axis $-a \le \mathrm{Im}\, z \le a$, $\mathrm{Re}\, z = 0$. Behind the plate ($\mathrm{Re}\, z > 0$) a cavity of unknown shape is formed, the boundary of which is a streamline of the flow. We wish to determine this cavitating flow, and in particular, find the shape of the cavity.

Why do we expect a cavity to form? Quite simply, from Bernoulli's law $\frac{1}{2}|u|^2 + p/\rho = \text{constant}$, p must decrease as the speed $|u|$ increases. If p falls below atmospheric pressure p_* (or the vapor pressure of the fluid), the fluid must boil, or cavitate, since pressures below p_* are not possible in the fluid. Thus, near corners where speeds become very large we expect cavities to form.

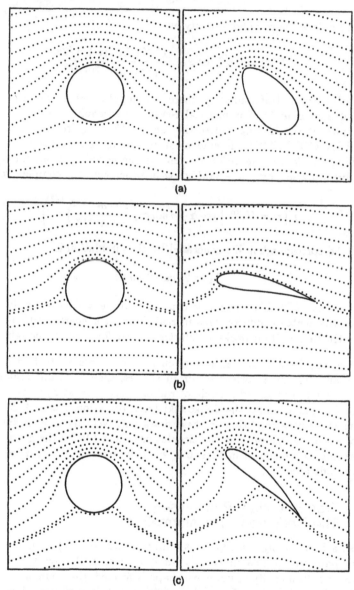

FIGURE 6.9 Flow with circulation κ and velocity at infinity U from an angle of attack α around a circle of radius r centered at x_0 and y_0 before and after application of the Joukowski transformation $z = \zeta + L^2/\zeta$. a) $U = 0.244$, $\kappa = 0.148$, $\alpha = 50.9$ degrees, $x_0 = 0.088$, $y_0 = 0.0$, $r^2 = 0.25$, and $L^2 = 0.078$, b) $U = 1.00$, $\kappa = 0.357$, $\alpha = 12.0$ degrees, $x_0 = -0.056$, $y_0 = 0.05$, $r^2 = 0.25$, and $L^2 = 0.188$, c) $U = 1.00$, $\kappa = 0.761$, $\alpha = 41.74$ degrees, $x_0 = 0.056$, $y_0 = 0.05$, $r^2 = 0.25$, and $L^2 = 0.188$.

(In the thin airfoil problem we did not have a cavitating flow, only because we did not apply Bernoulli's law at the leading corner.)

A similar problem involves the flow from the nozzle of a jet. Suppose a flow from $\text{Re}\, z = -\infty$ hits a flat plate occupying the imaginary axis except for an opening at $\text{Re}\, z = 0, -a \leq \text{Im}\, z \leq a$. The boundary of the flow from this nozzle is of course a streamline, which is unknown.

We want to use conformal mappings to determine the solution of these free boundary problems. The solution $w(z) = \phi + i\psi$ of these problems must satisfy the following conditions:

1. $\psi = 0$ on the free boundary of the fluid domain, since the boundary is defined as a streamline,

2. $w'(z) \to U > 0$ as $z \to \infty$,

3. $|w'(z)| = U$ on the free (undetermined) boundary. This follows from Bernoulli's law $p = p_0 - \frac{1}{2}\rho|u|^2$ along streamlines without gravitational effects. Since the boundary is a streamline, and in the cavity, pressure is everywhere atmospheric, the speed of the fluid is constant on the boundary. However, this velocity is U as $z \to \infty$ so $|u|^2 = U^2$ along the streamline forming the cavity.

4. The flow is symmetric about the real z axis so that $w(z) = \overline{w(\bar{z})}$. As a result $\psi = 0$ on the real axis with $\text{Re}\, z \leq 0$.

The idea to solve this problem is to view the desired solution $w(z) = \phi + i\psi$ as a mapping, called the HODOGRAPH, of the complex z plane to the complex w plane, and also to view the equation $t = \frac{1}{w'(z)}$ as a map from the z plane to the t plane. The goal is to find the map between the w plane and t plane $t = h(w)$. Once $h(w)$ is found, then $t = dz/dw = h(w)$ so

$$z = \int_0^w h(w)\,dw$$

is the inverse of the hodograph map $z = z(w)$.

To carry out this procedure for the cavity problem, we view the unknown function $w = w(z)$ as a map and note that whatever else it does, it must map the wall and unknown free stream to the positive half line $\text{Im}\, w = \psi = 0, \phi > 0$ and the center line of the flow for $\text{Re}\, z < 0$ must map to the negative half line $\text{Im}\, w = \psi = 0, \phi < 0$. Thus, $w(z) = \phi + i\psi$ is a map of the flow domain to the w plane and the boundary has collapsed to a slit on the positive $\text{Im}\, w = 0$ axis. The corners of the plate correspond to $\phi = \mu$, where μ is yet to be determined. Notice that along the streamline $\psi = 0$, the fluid is moving from left to right so that ϕ is increasing along the streamline.

The second map $t = \frac{1}{w'(z)}$ must satisfy

$$|t| = \frac{1}{|w'(z)|} = \frac{1}{U}$$

on the free stream. Thus, the free stream is mapped onto an arc of the circle $|t| = 1/U$ in the t plane.

Along the flat vertical plate, $\psi = 0$, so $\psi_y = \phi_x = 0$ with the result that $\omega'(z) = \phi_x - i\phi_y$ is imaginary on the plate, and $t = \frac{1}{\omega'(z)}$ maps the plate to a section of the imaginary t axis. We expect that $\phi_y > 0$ (the fluid flows upwards) on the plate above the x axis and $\phi_y < 0$ on the plate below the x axis, so the upper half of the flow maps to the upper half of the t domain. Notice finally that the point $z = 0$ is a point where the streamlines have different directions so it must be that $\omega'(z) = 0$ at $z = 0$, hence $z = 0$ maps to $t = \infty$ under the map $t = \frac{1}{\omega'(z)}$. We conclude that $t = \frac{1}{\omega'(z)}$ maps the boundary of the cavity onto the arc of the circle $|t| = 1/U$ with $\operatorname{Re} t > 0$ and maps the plate onto the imaginary axis $|\operatorname{Im} t| > 1/U$ (Fig. 6.10). Since $\phi_x \geq 0$, the flow domain is in the right half t domain ($\operatorname{Re} t > 0$). Bernoulli's law implies that the flow domain has $|t| \geq 1/U$.

Using only simple observations we have now mapped the physical z plane to the ω plane. We do not actually know the function $\omega(z)$ but we know how the boundary arcs map. Similarly we know how the flow domain and boundaries map to the t plane under $t = \frac{1}{\omega'(z)}$ even though we do not, as yet, know $\omega'(z)$.

The problem is to see if we can complete the cycle and map the ω plane to the t plane. We first map the ω plane to the $\operatorname{Im}\zeta < 0$ half plane using $\zeta = \omega^{-1/2}$. Note that the $\operatorname{Re}\omega > 0$ axis maps to the $\operatorname{Re}\zeta$ axis. The part of the $\operatorname{Re}\omega$ axis corresponding to the plate $0 < \operatorname{Re}\omega < \mu$, maps to $|\operatorname{Re}\zeta| > \mu^{-1/2}$ and $\mu < \operatorname{Re}\omega < \infty$, corresponding to the free stream, maps to $|\operatorname{Re}\zeta| < \mu^{-1/2}$. Now, the Joukowski transformation

$$\zeta = \frac{1}{2i\mu^{1/2}}\left(Ut - \frac{1}{Ut}\right)$$

maps the $\operatorname{Im}\zeta < 0$ half plane onto the t plane, the slit $-\mu^{-1/2} \leq \operatorname{Re}\zeta \leq \mu^{-1/2}$ is mapped to the circular arc $|t| = \frac{1}{U}$, and the rest of the real ζ axis is mapped to the $|\operatorname{Im} t| > 1/U$ axis. It follows that

$$t = \frac{1}{U\sqrt{\omega}}\left(\sqrt{\omega - \mu} + i\sqrt{\mu}\right)$$

is a map of the ω plane to the t plane for which there is the correct correspondence between flow domains and boundaries, and therefore,

$$\frac{dz}{d\omega} = \frac{1}{U\sqrt{\omega}}\left(\sqrt{\omega - \mu} + i\sqrt{\mu}\right).$$

This can be integrated to yield the inverse map

$$z = \frac{i}{U}\left[\frac{\mu\pi}{2} - \mu\cos^{-1}\left(\sqrt{\frac{\omega}{\mu}}\right) + (\omega(\mu - \omega))^{1/2} + 2\sqrt{\mu\omega}\right].$$

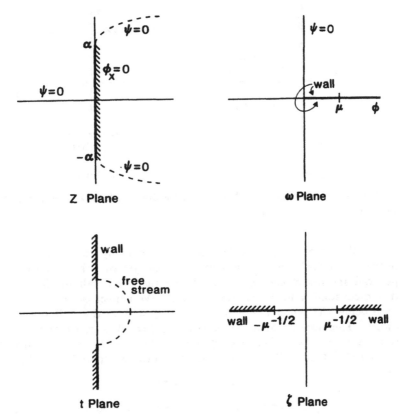

FIGURE 6.10 Image of the flow boundaries for the flow past a flat plate in the z plane, the $w(z) = \phi + i\psi$ plane, the $t = 1/w'(z)$ plane, and the $\zeta = \left(Ut + (Ut)^{-1}\right)/\left(2i\mu^{1/2}\right)$ plane.

The constant of integration is chosen so that $z(0) = 0$. To determine μ we use that $w = \mu$ must correspond to the corners of the flow at $z = \pm ia$ so that

$$\mu = \frac{2Ua}{\pi + 4}.$$

The free stream corresponds to $w = \phi$ (since $\psi = 0$) for $\phi > \mu$ and we obtain

$$x = \frac{1}{U}\left[-\left(\phi(\phi - \mu)\right)^{1/2} + \mu \ln\left(\sqrt{\phi/\mu} + \sqrt{\phi/\mu - 1}\right)\right],$$

$$y = a + \frac{2}{U}\left(\sqrt{\mu\phi} - \mu\right),$$

as its parametric representation.

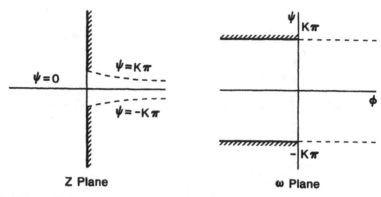

FIGURE 6.11   Image of the flow boundaries for the flow from a jet in the z plane and the $\omega(z) = \phi + i\psi$ plane.

The solution of the jet nozzle problem is found by noting its similarity with the cavity problem we just solved. For the jet we take $\psi = k\pi$ on the upper wall and free stream and $\psi = -k\pi$ on the lower wall and free stream, with the constant k to be determined later. We expect $\phi_y < 0$ above and $\phi_y > 0$ below the center line $\psi = 0$, while $\phi_x = 0$ on the wall and $\phi_x > 0$ elsewhere. Therefore the unknown function $\omega(z) = \phi + i\psi$ maps the flow region onto the strip in the ω plane $-k\pi \leq \psi \leq k\pi$ (Fig. 6.11). Next, we map the strip $-k\pi \leq \psi \leq k\pi$ in the ω plane onto the entire complex ξ plane by using

$$\omega = k \ln \xi - k\pi i,$$

with the boundary $\psi = k\pi$ mapped to the line $\xi = \rho e^{2\pi i}$ and $\psi = -k\pi$ mapped to $\xi = \rho, \rho \geq 0$. This ξ plane has the same structure as the hodograph plane for the previous flat plate problem, and it is easy to check that the $t = \frac{1}{\omega'(z)}$ planes are similar for the two problems. It is now simply a matter of calculating the details of the map from the ξ to the t plane and then calculating the inverse hodograph $z = z(\omega)$. We leave these details as an exercise for the reader as the steps are nearly the same as above.

6.4   Contour Integration

The properties of analytic functions allow us to evaluate many real integrals that cannot be evaluated by other techniques, and there are many integrals whose integration becomes much easier than with other methods. The main tool we use is the Cauchy Integral Formula which is implemented by manipulating contours into the most agreeable form.

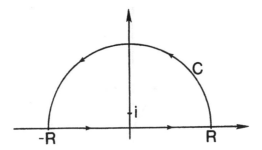

FIGURE 6.12   Semi-circular contour of integration.

Our first example is to evaluate $\int_0^\infty \frac{dx}{1+x^2}$. This integral can be evaluated using real variable techniques (if you happen to remember the correct trick, or have a table of integrals nearby). To use complex variable techniques we consider the integral

$$I = \frac{1}{2} \int_C \frac{dz}{1+z^2},$$

where the contour C consists of the real axis with $-R \le x \le R$ and the large circular arc $z = Re^{i\theta}$ (Fig. 6.12). The integrand has poles at $z = \pm i$ and can be written as

$$\frac{1}{z^2+1} = \frac{1}{(z-i)(z+i)}.$$

Since $\frac{1}{z+i}$ is analytic inside C, we can use the Cauchy Integral Formula to find that $I = \frac{2\pi i}{2(z+i)} \big|_{z=i} = \frac{\pi}{2}$.

On the other hand I is a line integral with two different contours. If we write I in terms of these two line integrals we find

$$I = \frac{1}{2} \int_{-R}^{R} \frac{dx}{1+x^2} + \frac{1}{2} \int_0^\pi \frac{iRe^{i\theta}\, d\theta}{1+R^2 e^{2i\theta}}.$$

For R sufficiently large

$$\left| \frac{iRe^{i\theta}}{1+R^2 e^{2i\theta}} \right| \le \frac{R}{R^2 - 1},$$

and taking the limit $R \to \infty$, we find

$$I = \frac{1}{2} \int_{-\infty}^{\infty} \frac{dx}{1+x^2} = \int_0^\infty \frac{dx}{1+x^2} = \pi/2.$$

It is a bit more difficult to evaluate

$$I = \int_0^\infty \frac{dx}{1+x^3}$$

using real variable methods. To use complex variable techniques, we consider the integral

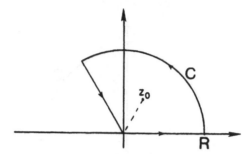

FIGURE 6.13 Sectorial contour of integration.

$$I = \int_C \frac{dz}{1+z^3}$$

over some contour which traverses the positive real axis and a large radius arc. The trick is to figure out how to close the arc C.

Suppose C consists of two spokelike arcs with $z = re^{i\phi}$, $\phi = 0$ or θ, on $0 \le r \le R$, and a circular arc $z = Re^{i\phi}$, $0 \le \phi \le \theta$ (Fig. 6.13). For this choice of C,

$$I = \int_0^R \frac{dr}{1+r^3} + \int_0^\theta \frac{Rie^{i\phi}d\phi}{1+R^3e^{3i\phi}} + \int_R^0 \frac{e^{i\theta}dr}{1+r^3e^{3i\theta}}.$$

Since

$$\left| \frac{iRe^{i\phi}}{1+R^3e^{3i\phi}} \right| \le \frac{R}{R^3-1}$$

for large R, we can take the limit as R goes to infinity and obtain

$$I = \int_0^\infty \frac{dr}{1+r^3} - \int_0^\infty \frac{e^{i\theta}dr}{1+r^3e^{3i\theta}}.$$

If we pick $\theta = 2\pi/3$ (we could also take $\theta = 4\pi/3$) we find that

$$I = \left(1 - e^{2\pi i/3}\right) \int_0^\infty \frac{dx}{1+x^3}.$$

Now that we know how I relates to the original real integral, we evaluate it using the Cauchy Integral Formula. Specifically, since $z_0 = e^{i\pi/3}$ is the only root of $z^3 + 1 = 0$ inside the contour C, we know that

$$I = \int_C \frac{dz}{1+z^3} = 2\pi i f\left(e^{i\pi/3}\right),$$

where $\frac{1}{1+z^3} = \frac{f(z)}{z-e^{i\pi/3}}$. It follows that $I = \frac{2\pi i}{3e^{2\pi i/3}}$ so that

$$\int_0^\infty \frac{dx}{1+x^3} = \frac{2\pi}{3\sqrt{3}}.$$

We used a simple, but important, trick to evaluate $f\left(e^{i\pi/3}\right)$. Notice that we do not need to know $f(z)$ but only $f(z_0)$, so we do not actually need to factor $z^3 + 1$. Instead, we realize, using Taylor's theorem, that for any analytic function $g(z)$, if $g(z_0) = 0$, then

$$g(z) = g'(z_0)(z - z_0) + \text{terms that vanish like } (z - z_0)^2.$$

Thus,

$$\lim_{z \to z_0} \frac{g(z)}{z - z_0} = g'(z_0)$$

and for this example

$$f(z_0) = \lim_{z \to z_0} \frac{z - e^{i\pi/3}}{1 + z^3} = \left.\frac{1}{3z^2}\right|_{z=z_0}.$$

(This is no more than an application of L'Hôpital's rule.)

This observation can be stated in a more general way. If $f(z) = p(z)/q(z)$ is analytic in a neighborhood of $z = z_0$ except at $z = z_0$ where $q(z_0) = 0$, $q'(z_0) \neq 0$ (i.e., z_0 is a simple pole) then

$$\int_C f(z)dz = 2\pi i \frac{p(z_0)}{q'(z_0)}$$

for any small circle C surrounding the point z_0.

The problem of finding a contour on which to evaluate the integral

$$I = \int_0^\infty \frac{dx}{x^2 + 3x + 2}$$

is complicated by the fact that the polynomial $z^2 + 3z + 2$ is not invariant under certain rotations in the complex plane as were the polynomials $z^2 + 1$ and $z^3 + 1$. As a result, the methods used on the previous problems will not work. Instead we try the contour integral

$$I = \int_C \frac{\ln z\, dz}{z^2 + 3z + 2},$$

where the contour C is shown in Figure 6.14. The contour C consists of the four arcs $z = x$ for $\varepsilon \leq x \leq R$, $z = Re^{i\theta}$ for $0 \leq \theta < 2\pi$, $z = xe^{2\pi i}$ for $\varepsilon < x < R$, and $z = \varepsilon e^{i\theta}$ for $0 \leq \theta < 2\pi$. It is important to note that the function $\ln z$ has a branch cut along the real axis and therefore the arcs $z = x$ and $z = xe^{2\pi i}$ must be viewed as different, lying on opposite sides of the branch cut. On these four arcs we find

$$I = \int_\varepsilon^R \frac{\ln x\, dx}{x^2 + 3x + 2} + \int_0^{2\pi} \frac{(\ln R + i\theta)Rie^{i\theta}\,d\theta}{R^2 e^{2i\theta} + 3Re^{i\theta} + 2}$$

$$+ \int_R^\varepsilon \frac{(\ln x + 2\pi i)dx}{x^2 + 3x + 2} + \int_{2\pi}^0 \frac{(\ln \varepsilon + i\theta)\varepsilon i e^{i\theta}\,d\theta}{\varepsilon^2 e^{2i\theta} + 3\varepsilon e^{i\theta} + 2}.$$

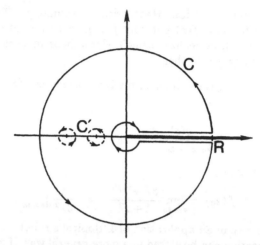

FIGURE 6.14 Contour of integration avoiding the logarithmic branch cut.

In the limit $R \to \infty$ and $\varepsilon \to 0$ the contributions from the two circular arcs vanish and we have

$$I = -2\pi i \int_0^\infty \frac{dx}{x^2 + 3x + 2},$$

which is proportional to the real integral we desire to evaluate. Now we can evaluate I directly via the Cauchy Integral Formula. We observe that $x^2 + 3x + 2 = (x+2)(x+1)$ and we collapse the contour C onto a contour consisting of two small circles surrounding $z = -2$ and $z = -1$. Using these two contours we have

$$I = 2\pi i \left(\frac{\ln z}{z+2} \bigg|_{z=-1} + \frac{\ln z}{z+1} \bigg|_{z=-2} \right) = -2\pi i \ln 2.$$

It follows that

$$\int_0^\infty \frac{dx}{x^2 + 3x + 2} = \ln 2.$$

Of course, this integral can also be evaluated using partial fractions to get the same answer, but the importance of this example is that it illustrates the value of branch cuts when evaluating contour integrals.

A slightly more complicated example which has a branch cut is

$$I = \int_0^\infty \frac{x^\alpha dx}{(1 + x)^2},$$

where $|\operatorname{Re} \alpha| < 1$. We consider the integral

$$I = \int_C \frac{z^\alpha dz}{(1+z)^2}$$

on the same contour (Fig. 6.14) as in the previous example, namely a large circle with an indentation along the branch cut on the positive real axis. We use the Cauchy Integral Formula to evaluate the contour integral, but now, since the point $z = -1$ is a pole of order 2 we have that

$$I = 2\pi i \frac{d}{dz} \left(z^\alpha \right) \Big|_{z=e^{\pi i}} = -2\pi i \alpha e^{\pi i \alpha}.$$

Evaluating the line integrals, we learn that the large and small circular arcs do not contribute since $|\operatorname{Re}\alpha| < 1$, and

$$I = \int_0^\infty \frac{x^\alpha dx}{(1+x)^2} + e^{2\pi i \alpha} \int_\infty^0 \frac{x^\alpha dx}{(1+x)^2}$$

or

$$\int_0^\infty \frac{x^\alpha dx}{(1+x)^2} = \frac{\pi \alpha}{\sin \pi \alpha}, \qquad |\operatorname{Re}\alpha| < 1.$$

The idea of using contour integration to evaluate real integrals is clear. We first try to identify a complex integral and pick a closed contour for which one portion of the integral relates to the real integral, and the remaining (often circular) arcs have negligible contribution. After we have identified the appropriate integral and contour, we use analyticity to deform the path and Cauchy's Integral Formula to evaluate the integral.

In the previous examples we were able to close the contours because the integrands disappeared sufficiently rapidly in the limit $R \to \infty$. There is another class of integrals whose contours can be closed by large circular arcs. These are specified by Jordan's Lemma.

Theorem 6.10 (Jordan's Lemma)

Let $I(R) = \int_{\Gamma(R)} e^{iaz} g(z)dz$ where Γ is the arc $z = Re^{i\theta}$, $0 \leq \theta \leq \pi$, and $a > 0$ is real. If $|g(Re^{i\theta})| \leq G(R) \to 0$ as $R \to \infty$, $0 \leq \theta \leq \pi$, then $\lim_{R\to\infty} I(R) = 0$.

In other words, for an integral of the form $\int e^{iaz} g(z)dz$ the contour can be closed without changing the value of the integral using a large circular arc above the $\operatorname{Re} z$ axis if $a > 0$ or below the $\operatorname{Re} z$ axis if $a < 0$.

To prove Jordan's Lemma, observe that along the arc Γ we have

$$I = \int_\Gamma e^{iaz} g(z)dz = \int_0^\pi ie^{iaR\cos\theta} e^{-aR\sin\theta} g\left(Re^{i\theta}\right) Re^{i\theta} d\theta.$$

so that

$$|I| \leq \int_0^\pi e^{-aR\sin\theta} G(R) R\, d\theta = 2\int_0^{\pi/2} e^{-aR\sin\theta} G(R) R\, d\theta.$$

On the interval $0 \leq \theta \leq \pi/2$, $\sin\theta \geq 2\theta/\pi$ so that

$$|I| \leq 2G(R) \int_0^{\pi/2} \exp\left(\frac{-2aR\theta}{\pi}\right) R\, d\theta = \frac{\pi}{a} G(R) \left(1 - e^{-aR}\right)$$

and $|I| \to 0$ as $R \to \infty$.

To illustrate the usefulness of the Jordan Lemma, we evaluate

$$\int_0^\infty \frac{\cos ax\, dx}{1+x^2}, \qquad \text{for} \qquad a > 0.$$

Consider the integral

$$I = \int_C \frac{e^{iaz}dz}{1+z^2}, \qquad \text{for} \qquad a > 0$$

over the contour C which consists of the real axis $z = x$ for $-R \leq x \leq R$, and a large circular arc in the upper half plane (Fig. 6.12 again). We are able to close the contour in the upper half plane because of Jordan's Lemma. As a result

$$I = \int_{-\infty}^\infty \frac{e^{iax}dx}{1+x^2} = \int_{-\infty}^\infty \frac{\cos ax\, dx}{1+x^2}.$$

Now we evaluate I using the Cauchy Integral Formula. The only root of $z^2 + 1 = 0$ in the upper half plane is at $z = i$. Therefore,

$$I = 2\pi i \frac{e^{iaz}}{z+i}\bigg|_{z=i} = \pi e^{-a},$$

and

$$\int_0^\infty \frac{\cos ax\, dx}{1+x^2} = \frac{\pi}{2} e^{-a} \qquad \text{for} \qquad a > 0.$$

Sometimes the method we have outlined appears to fail, but still gives useful information. For example, to evaluate the integrals $\int_0^\infty \cos\left(ax^2\right) dx$ or $\int_0^\infty \sin\left(ax^2\right) dx$ we try

$$I = \int_C e^{iaz^2} dz.$$

However, since e^{iaz^2} is analytic, $I = 0$ around any closed contour and it appears that the trial fails. However, using the proof of Jordan's Lemma we can show that the large circular arcs $z = Re^{i\theta}$ with $0 \leq \theta \leq \pi/2$ or $\pi \leq \theta \leq 3\pi/2$ will not contribute to the integral. (To apply Jordan's Lemma directly first make the change of variables $t = z^2$.) We take the wedge-shaped contour $z = x$, $0 \leq x \leq R$, $z = Re^{i\theta}$, $0 \leq \theta \leq \pi/4$ and $z = xe^{i\pi/4}$, $0 \leq x \leq R$, let $R \to \infty$ and find

$$0 = \int_0^\infty e^{iax^2}\,dx - e^{i\pi/4}\int_0^\infty e^{-ax^2}\,dx.$$

In other words, although we have not been able to evaluate integral using this contour, we have found a change of variables, which is quite useful since the new integral converges much more rapidly than the original one. This is certainly valuable if numerical integration becomes necessary, although in this case we can evaluate the new integral exactly. In fact, for $a > 0$

$$\left(\int_0^\infty e^{-ax^2}\,dx\right)^2 = \int_0^\infty\int_0^\infty e^{-ax^2}e^{-ay^2}\,dx\,dy = \int_0^{\pi/2}\int_0^\infty e^{-ar^2}r\,dr\,d\theta = \frac{\pi}{4a}.$$

It follows that

$$\int_0^\infty e^{iax^2}\,dx = \frac{e^{i\pi/4}}{2}\left(\frac{\pi}{|a|}\right)^{1/2}$$

and

$$\int_0^\infty \cos ax^2\,dx = \int_0^\infty \sin ax^2\,dx = \frac{1}{2}\left(\frac{\pi}{2|a|}\right)^{1/2}$$

It is a fact of analytic function theory that if $f(z)$ is oscillatory along one path, there is always another contour on which it is decaying. We will make strong use of this fact in Chapter 10.

As a note for future reference, it follows from this last calculation that

$$\int_0^\infty \frac{e^{i\mu t}}{\sqrt{\mu}}\,d\mu = \left(\frac{\pi}{2|t|}\right)^{1/2}\left(1 + i\frac{t}{|t|}\right)$$

for $t \neq 0$.

Another example of the use of Jordan's Lemma is found in the evaluation of $\int_0^\infty \frac{\sin ax}{x}\,dx$, $a > 0$. We first observe that

$$I = \int_0^\infty \frac{\sin ax}{x}\,dx = \frac{1}{2}\int_\Gamma \frac{(e^{iaz} - e^{-iaz})}{2iz}\,dz,$$

where the path Γ is precisely the x-axis. Unfortunately, as it stands, we cannot close the contour either above or below the Re z axis. To close the contour, we must separate the integral I into two integrals, but then there is a singularity at $z = 0$ on the path of integration, which is also not permissible. To avoid this trouble we first deform the path Γ into Γ', which has a slight indentation below the x-axis near the origin $z = 0$. This change of contour does not change the value of I. Now we can split the integrals into

$$I = \frac{1}{4i}\int_{\Gamma'}\frac{e^{iaz}}{z}\,dz - \frac{1}{4i}\int_{\Gamma'}\frac{e^{-iaz}}{z}\,dz$$

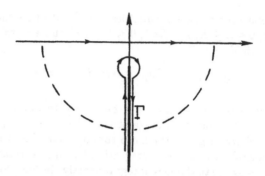

FIGURE 6.15 Keyhole contour of integration avoiding the square root branch cut.

and by Jordan's Lemma, the first of these integrals can be closed above the $\mathrm{Re}\,z$ axis and the second can be closed below. Since the only singularity is at $z = 0$, the second integral is zero and the first is

$$ I = \frac{2\pi i}{4i} e^{iaz} \Big|_{z=0} = \pi/2 $$

provided $a > 0$.

Our last example of the usefulness of Jordan's Lemma is the evaluation of the integral

$$ I = \int_{-\infty}^{\infty} e^{iaz}(z + i\alpha)^{-1/2} dz $$

which arises as an Inverse Fourier Transform (Chapter 7). The integrand has a branch cut along the negative imaginary axis below the branch point $z = -i\alpha$ and is analytic in the upper half plane. If $a > 0$ we can close the contour above the real z axis where the integrand is analytic, and obtain $I = 0$ for $a > 0$. For $a < 0$ we can close the contour in the lower half plane except for the fact that to do so requires crossing a branch cut, which is not permitted. Instead we close the contour in the lower half plane with a keyhole-shaped notch along the negative imaginary axis which avoids crossing the branch cut. For this closed contour (see Fig. 6.15), the integral is zero, or, since the large circular arc does not contribute

$$ I = \int_{-\infty}^{\infty} e^{iaz}(z + i\alpha)^{-1/2} dz = \int_{\Gamma} e^{iaz}(z + i\alpha)^{-1/2} dz, $$

where Γ is a "key hole" around the branch cuts on the negative imaginary axis. We use this contour to provide a change of variables. We first deform Γ into the contour with

$$ z = -i\alpha + xe^{-i\pi/2} \qquad \text{for} \qquad \varepsilon \leq x < \infty, $$

$$z = -i\alpha + \varepsilon e^{i\theta} \qquad \text{for} \qquad -\pi/2 \le \theta \le 3\pi/2,$$

and

$$z = -i\alpha + x e^{3\pi i/2} \qquad \text{for} \qquad \varepsilon \le x < \infty.$$

Thus,

$$-I = \int_{-\Gamma} e^{iaz}(z + i\alpha)^{-1/2} dz$$

$$= e^{a\alpha} \int_{\infty}^{\varepsilon} \frac{e^{ax} e^{-\pi i/4} dx}{\sqrt{x}} + \varepsilon^{1/2} i e^{a\alpha} \int_{-\pi/2}^{3\pi/2} e^{i\left(a\varepsilon e^{i\theta} + \theta/2\right)} d\theta$$

$$+ e^{a\alpha} \int_{\varepsilon}^{\infty} \frac{e^{ax} e^{3\pi i/4}}{\sqrt{x}} dx.$$

Taking the limit $\varepsilon \to 0$ we find

$$I = 2 e^{a\alpha} e^{-\pi i/4} \int_0^{\infty} \frac{e^{ax}}{\sqrt{x}} dx, \qquad \text{where} \qquad a < 0.$$

After making the change of variables $x = s^2$, we obtain

$$I = 4 e^{a\alpha} e^{-\pi i/4} \int_0^{\infty} e^{as^2} ds = 2 e^{a\alpha} e^{-\pi i/4} \left(\frac{\pi}{|a|}\right)^{1/2}.$$

The Jordan Lemma provides one way to choose contours. To evaluate the integral

$$I = \int_0^{2\pi} \frac{d\theta}{a + b\cos\theta}, \qquad a > b > 0$$

another method works quite nicely. We make the change of variables $z = e^{i\theta}$ so that $\cos\theta = \frac{1}{2}\left(z + \frac{1}{z}\right)$, $d\theta = \frac{dz}{iz}$ and

$$I = \frac{2}{ib} \int_C \frac{dz}{z^2 + \frac{2a}{b}z + 1},$$

where the contour C is the unit circle. The integrand is analytic except at the roots of $z^2 + \frac{2a}{b}z + 1 = 0$. Since $a > b > 0$, this equation has two real roots, one inside and one outside the unit circle. The one singularity inside the contour is at $z_+ = \frac{-a}{b} + \left((a/b)^2 - 1\right)^{1/2}$. Using the Cauchy Integral Formula, we learn that

$$I = 2\pi i \left(\frac{2}{ib}\right) \cdot \frac{1}{2z + \frac{2a}{b}}\bigg|_{z_+} = \frac{2\pi}{\sqrt{a^2 - b^2}}.$$

The same change of variables can be used to evaluate

$$I = \int_0^\alpha \frac{d\theta}{a + b\cos\theta} = \frac{2}{ib}\int_1^{e^{i\alpha}} \frac{dz}{z^2 + \frac{2a}{b}z + 1}.$$

Since $z^2 + \frac{2a}{b}z + 1 = 0$ has roots at $z_\pm = \frac{-a}{b} \pm \left((a/b)^2 - 1\right)^{1/2}$ we can write

$$I = \frac{2}{ib}\left(\frac{1}{z_+ - z_-}\right)\int_1^{e^{i\alpha}} \left(\frac{1}{z - z_+} - \frac{1}{z - z_-}\right) dz$$

which can be integrated directly to

$$I = \frac{2}{ib}\frac{1}{z_+ - z_-}\ln\left(\frac{z - z_+}{z - z_-}\right)\Big|_1^{e^{i\alpha}}.$$

For any complex number w, $\ln w = \ln(|w|) + i\arg(w)$, and since this integral is real we must have

$$I = \frac{2}{b}\frac{1}{z_+ - z_-}\arg\left(\frac{e^{i\alpha} - z_+}{e^{i\alpha} - z_-}\right).$$

With some more computation, one can show that this reduces to

$$I = \frac{1}{\sqrt{a^2 - b^2}}\tan^{-1}\left(\frac{\sqrt{a^2 - b^2}\sin\alpha}{b + a\cos\alpha}\right).$$

One does not always use circular arcs. For example, rectangular arcs are often useful when the integrand is periodic. To evaluate the integral

$$\int_{-\infty}^\infty \frac{x\,dx}{\sinh x - ia}, \qquad a \neq 0 \qquad \text{real},$$

we propose to evaluate the contour integral

$$I = \int_C \frac{z\,dz}{\sinh z - ia}$$

over the contour C which is the boundary of the rectangular region $-R \leq \operatorname{Re} z \leq R$ and $0 \leq \operatorname{Im} z \leq 2\pi$ (Fig. 6.16).

The integrand is analytic except at the roots of $\sinh z = ia$. Since $\sinh z = \sinh(x + iy) = \sinh x \cos y + i \sin y \cosh x$, $\sinh z$ is 2π-periodic in $y = \operatorname{Im} z$. If $|a| < 1$, the roots of $\sinh z = ia$ are located on the imaginary axis at $z = iy_0 + 2n\pi i$ and $z = (2n + 1)\pi i - iy_0$, $n = 0, \pm 1, \pm 2, \ldots$, where y_0 is the smallest (in absolute value) root of $\sin y = a$. If $|a| > 1$ the roots of $\sinh z = ia$ are located at $z = x_0 + (\pi/2 + 2n\pi)i$ and $z = -x_0 + (2n\pi + \pi/2)i$, where x_0 satisfies $\cosh x_0 = a$. In other words, if z_0 is a root of $\sinh z = ia$, so also is $z = z_0 + 2n\pi i$ and $z = -z_0 + (2n + 1)\pi i$, $n = 0, \pm 1, \pm 2, \ldots$, and there are no

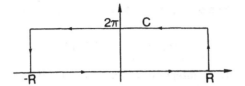

FIGURE 6.16 Rectangular contour of integration.

more roots. There are exactly two roots inside the contour of integration, so we can evaluate the integral using residues as

$$I \;=\; 2\pi i \left\{ \left(\frac{z_0}{\cosh z_0} \right) + \frac{\pi i - z_0}{\cosh (\pi i - z_0)} \right\} = 2\pi i \left(\frac{2z_0 - \pi i}{\cosh z_0} \right).$$

Evaluating the line integral around the contour we find

$$I \;=\; \int_{-R}^{R} \frac{x \, dx}{\sinh x - ia} + \int_{0}^{2\pi} \frac{i(R + iy) dy}{\sinh(R + iy) - ia}$$

$$+ \int_{R}^{-R} \frac{(x + 2\pi i) dx}{\sinh x - ia} + \int_{2\pi}^{0} \frac{i(-R + iy) dy}{\sinh(-R + iy) - ia}.$$

In the limit $R \to \infty$, the integrals on the vertical boundaries vanish and

$$I \;=\; -2\pi i \int_{-\infty}^{\infty} \frac{dx}{\sinh x - ia} = 2\pi i \left(\frac{2z_0 - \pi i}{\cosh z_0} \right).$$

Unfortunately, although we have correctly evaluated the integral I, it is not the integral we originally wanted. To do this we try instead a different integral

$$I \;=\; \int_{C} \frac{z(z - 2\pi i)}{\sinh z - ia} dz,$$

where the contour C is chosen to be rectangular as before. The calculations are similar and we successfully determine that

$$\int_{-\infty}^{\infty} \frac{x \, dx}{\sinh x - ia} = \frac{\pi}{2} \frac{\pi + 2iz_0}{\cosh z_0}.$$

6.5 Special Functions

6.5.1 The Gamma Function

The GAMMA FUNCTION is a complex valued function which for positive integers agrees with the factorial function, but is defined on the entire complex plane. The most common representation of the gamma function is (due to Euler)

$$\Gamma(z) = \int_0^\infty e^{-t} t^{z-1} dt.$$

If z is not an integer, the integrand has a branch point at the origin, a branch cut along the $\mathrm{Re}\, z < 0$ axis, and the integral does not converge unless $\mathrm{Re}\, z > 0$. However, the definition of $\Gamma(z)$ can be continued analytically into the left half z plane.

The first thing to note, from integration by parts, is that

$$\Gamma(z) \;=\; \int_0^\infty e^{-t} t^{z-1} dt = -e^{-t} t^{z-1} \big|_0^\infty + (z-1) \int_0^\infty e^{-t} t^{z-2} dt$$

$$=\; (z-1)\Gamma(z-1) \qquad \text{provided} \qquad \mathrm{Re}\, z > 1.$$

Second, we observe that the integral

$$\Gamma'(z) \;=\; \int_0^\infty e^{-t} t^{z-1} \ln t \, dt$$

makes sense whenever $\mathrm{Re}\, z > 0$, so that $\Gamma(z)$ is analytic for $\mathrm{Re}\, z > 0$.

We can evaluate $\Gamma(z)$ for a few special values of z,

$$\Gamma(1) = \int_0^\infty e^{-t} dt = 1$$

so that $\Gamma(n) = (n-1)!$, and

$$\Gamma\left(\frac{1}{2}\right) = \int_0^\infty e^{-t} t^{-1/2} dt = 2 \int_0^\infty e^{-s^2} ds = \sqrt{\pi}.$$

One of the important identities relating the gamma function to things a bit more familiar is

$$\Gamma(z+1)\Gamma(1-z) = \frac{\pi z}{\sin \pi z}.$$

We prove this identity by considering

$$\Gamma(z+1)\Gamma(1-z) \;=\; \int_0^\infty e^{-t} t^z dt \int_0^\infty e^{-s} s^{-z} ds$$

$$=\; \lim_{R\to\infty} \int_0^R \int_0^R e^{-(s+t)} t^z s^{-z} ds \, dt$$

provided $|\mathrm{Re}\, z| < 1$. We view the square $0 \le s, t \le R$, as two triangles with vertices $(0,0)$, $(0,R)$, $(R,0)$ and $(0,R)$, (R,R), $(R,0)$ and treat the two integrals over these triangles separately. On the first triangle, we set $s+t = x$, $t = y$ to obtain

$$T_1 = \lim_{R \to \infty} \int_0^R e^{-x} \int_0^x y^z (x-y)^{-z} \, dy \, dx$$

and setting $y = x\eta$ we find

$$T_1 = \int_0^1 \eta^z (1-\eta)^{-z} \, d\eta.$$

The final change of variables $\eta = \frac{u}{1+u}$ yields

$$T_1 = \int_0^\infty \frac{u^z}{(1+u)^2} \, du = \frac{\pi z}{\sin \pi z}.$$

Recall that we were able to evaluate this last integral using contour integral methods in the previous section of this chapter.

It remains to show that the second triangle contributes nothing to the integral. Indeed on the triangle with corners at $(R,0)$, (R,R), $(0,R)$

$$|T_2| = \left| \int \int e^{-(s+t)} t^z s^{-z} \, ds \, dt \right|$$

$$\leq e^{-R} \int_0^R \int_0^R t^\xi s^{-\xi} \, ds \, dt$$

$$= e^{-R} \frac{R^2}{(1-\xi^2)} \to 0$$

as $R \to \infty$, where $\xi = \text{Re}\, z$, provided $|\xi| < 1$.

Now we know that

$$\Gamma(z+1)\Gamma(1-z) = \frac{\pi z}{\sin \pi z}$$

for $|\text{Re}\, z| < 1$, but since $\Gamma(z)$ and $\sin \pi z$ are analytic for $\text{Re}\, z > 0$, we can define $\Gamma(z)$ for $\text{Re}\, z < 0$ using analytic continuation, by

$$\Gamma(-z) = \frac{-\pi(1+z)}{\Gamma(z+2)\sin \pi z}$$

which is valid for all z with $\text{Re}\, z > 0$ as long as z is not equal to a positive integer.

Another important identity involves the product of gamma functions $\Gamma(p)\Gamma(q)$ with $\text{Re}\, p, q > 0$. We calculate that

$$\Gamma(p)\Gamma(q) = \int_0^\infty e^{-t} t^{p-1} \, dt \int_0^\infty e^{-s} s^{q-1} \, ds$$

$$= \int_0^\infty e^{-t} t^{p-1} \left(t^q \int_0^\infty e^{-tx} x^{q-1} \, dx \right) dt$$

$$= \int_0^\infty x^{q-1} \left(\int_0^\infty e^{-t(x+1)} t^{p+q-1} dt \right) dx$$

$$= \int_0^\infty \frac{x^{q-1}}{(1+x)^{p+q}} dx \int_0^\infty e^{-s} s^{p+q-1} ds$$

$$= \Gamma(p+q) \int_0^\infty \frac{x^{q-1}}{(1+x)^{p+q}} dx.$$

The integral

$$B(p,q) = \int_0^\infty \frac{x^{q-1}}{(1+x)^{p+q}} dx$$

is commonly called the BETA FUNCTION, and we see that

$$B(p,q) = \frac{\Gamma(p)\Gamma(q)}{\Gamma(p+q)}.$$

Other representations of the beta function are found by making appropriate changes of variables, so that

$$B(p,q) = \int_0^\infty \frac{x^{q-1}}{(1+x)^{p+q}} dx$$

$$= \int_0^1 t^{p-1}(1-t)^{q-1} dt$$

$$= 2 \int_0^{\pi/2} (\sin \theta)^{2p-1} (\cos \theta)^{2q-1} d\theta.$$

These last representations are useful for solving Abel's integral equation (Chapter 3) and its generalizations (Problem 6.5.4).

6.5.2 Bessel Functions

It is often the case that differential equations with real independent variables can be extended into the complex plane and there the power of analytic function theory can be used to gain useful information. As an example of this idea we study BESSEL'S EQUATION and later LEGENDRE'S EQUATION. The Bessel's equation results from separation of variables of Laplace's equation in polar coordinates (see Chapter 8), and is

$$z^2 \frac{d^2 y}{dz^2} + z \frac{dy}{dz} + \left(z^2 - \nu^2 \right) y = 0,$$

where the function $y = y(z)$ is complex valued, and, of course, derivatives are defined in the sense of complex differentiation. Anywhere except $z = 0$ in the complex plane that this equation can be satisfied, the solution is analytic, since if y is continuous then its derivatives are defined and continuous. The only point in the complex plane where solutions may fail to be analytic is at $z = 0$, since there, even if we know y, we cannot determine its derivatives from the differential equation. This observation gives a quick proof that $u = e^{az}$ is analytic everywhere, since it satisfies the differential equation $du/dz = au$.

Since we expect solutions of the equation to be analytic in the complex plane, a natural way to find these solutions is via power series. Actually, for most practical purposes, power series representations are not particularly useful, and other solution representations are usually preferable. For example, to determine qualitative features of the solutions, integral representations and asymptotic methods (Chapter 10) are preferable by far. We view a power series solution as but a crude but necessary way to solve a differential equation.

If we seek a solution of the form

$$y = z^p \sum_{n=0}^{\infty} a_n z^n$$

and substitute it into Bessel's differential equation, we find that we must have

$$a_0 \left(p^2 - \nu^2\right) + a_1 \left((p+1)^2 - \nu^2\right) z + \sum_{n=2}^{\infty} \left(a_n \left((n+p)^2 - \nu^2\right) + a_{n-2}\right) z^n = 0.$$

A power series is identically zero if all its coefficients are identically zero. Therefore, we require $p^2 = \nu^2$ and

$$a_n = -\frac{a_{n-2}}{(n+p)^2 - \nu^2}, \qquad a_1 = 0, \qquad a_0 = 1.$$

Other choices are possible, but give nothing new. For example, the choice $a_0 = 0$, $a_1 = 1$ gives precisely the same end result, and since the equation is linear, all other solutions are linear combinations of these two, so that by taking $a_0 = 1$, $a_1 = 0$ there is no loss of generality.

We must pick p to satisfy $p^2 = \nu^2$ so that $p = \pm\nu$. (The equation $p^2 - \nu^2 = 0$ is called the INDICIAL EQUATION.) If we take $p = \nu$ we find (taking $m = 2n$)

$$y = J_\nu(z) = \left(\frac{z}{2}\right)^\nu \sum_{m=0}^{\infty} \frac{(-1)^m}{m!\,\Gamma(m+\nu+1)} \left(\frac{z}{2}\right)^{2m},$$

which is called the BESSEL FUNCTION of the FIRST KIND, and is an analytic function of z for $\text{Re}\, z > 0$.

The power series for $J_\nu(z)$ and $J_{-\nu}(z)$ are convergent for $0 < |z| < \infty$ and if ν is an integer, the origin is not a branch point and they are analytic in the entire complex plane. The function $(z/2)^{-\nu} J_\nu(z)$ is analytic for all z, $|z| < \infty$. If ν is not an integer $J_\nu(z)$ and $J_{-\nu}(z)$ are linearly independent. However, if

ν is an integer $\nu = n$, then $J_n(z)$ and $J_{-n}(z)$ are linearly dependent. We can see this by noting that $\Gamma(k)$ is unbounded for every nonpositive integer $k \leq 0$, so that $1/\Gamma(m - n + 1) = 0$ when $m - n + 1 \leq 0$. From the power series representation we see that

$$J_{-n}(z) = \left(\frac{z}{2}\right)^{-n} \sum_{m=n}^{\infty} \frac{(-1)^m}{m!\Gamma(m - n + 1)} \left(\frac{z}{2}\right)^{2m}$$

$$= \sum_{k=0}^{\infty} \frac{(-1)^{k+n}(z/2)^{2k+n}}{k!\Gamma(k + n + 1)} = (-1)^n J_n(z)$$

For integer ν we need to find a second linearly independent solution. To do so, we observe that $J_\nu(z)$ is an analytic function of ν. We differentiate the governing equation with respect to ν and determine that

$$z^2 \left(\frac{\partial y}{\partial \nu}\right)'' + z \left(\frac{\partial y}{\partial \nu}\right)' + (z^2 - \nu^2) \left(\frac{\partial y}{\partial \nu}\right) - 2\nu y = 0.$$

Since $J_{-n}(z) = (-1)^n J_n(z)$, it follows that

$$\left(\frac{\partial J_\nu}{\partial \nu} - (-1)^n \frac{\partial J_{-\nu}}{\partial \nu}\right)\bigg|_{\nu=n}$$

is a solution of Bessel's equation for $\nu = n$. Motivated by this and L'Hôpital's rule, we define

$$Y_\nu(z) = \frac{J_\nu(z) \cos \nu\pi - J_{-\nu}(z)}{\sin \nu\pi}.$$

We leave as an exercise for the reader the verification of the Wronskians $W(J_\nu(z), J_{-\nu}(z)) = -2\frac{\sin \nu\pi}{\pi z}$ and $W(J_\nu(z), Y_\nu(z)) = \frac{2}{\pi z}$.

One can now calculate directly the power series representation of $Y_n(z)$. Probably the most important result of this calculation is that it shows $Y_n(z)$ to have a combination of a logarithmic branch point and nth order pole at the origin. For example, $Y_0(z) \sim \frac{2}{\pi} \ln z$ for small z. $Y_n(z)$ is analytic everywhere else in the z plane.

In Chapter 10 we will show that the behavior of $J_\nu(z)$ for large z is

$$J_\nu(z) \sim \left(\frac{2}{\pi z}\right)^{1/2} \cos\left(z - \frac{\nu\pi}{2} - \frac{\pi}{4}\right)$$

It follows from its definition that

$$Y_\nu(z) \sim \left(\frac{2}{\pi z}\right)^{1/2} \sin\left(z - \frac{\nu\pi}{2} - \frac{\pi}{4}\right)$$

for large z.

All solutions of Bessel's equation are linear combinations of $J_\nu(z)$, and $Y_\nu(z)$. However, in different applications, other linear combinations are useful and have been given special names. For example, the HANKEL FUNCTIONS of FIRST and SECOND kind, defined by

$$H_\nu^{(1)}(z) = J_\nu(z) + iY_\nu(z)$$
$$H_\nu^{(2)}(z) = J_\nu(z) - iY_\nu(z)$$

(also called Bessel functions of the third kind) are useful because,

$$J_\nu \pm iY_\nu \sim \left(\frac{2}{\pi z}\right)^{1/2} e^{\pm i(z - \nu\pi/2 - \pi/4)}.$$

That is, $H_\nu^{(1)}(z)$ decays exponentially in the upper half complex plane and $H_\nu^{(2)}(z)$ decays exponentially in the lower half plane.

The change of independent variables $z = it$ transforms the Bessel differential equation into

$$t^2\omega'' + t\omega' - \left(t^2 + \nu^2\right)\omega = 0.$$

For nonintegral ν, $J_\nu(it)$ and $J_{-\nu}(it)$ are linearly independent solutions. However, it is usual to write the solutions as

$$I_\nu(z) = e^{-i\nu\pi/2} J_\nu(iz)$$

and call $I_\nu(z)$ the MODIFIED BESSEL FUNCTION of FIRST KIND. One can easily verify from the power series representation of $J_\nu(z)$ that $I_\nu(z)$ is real for real positive z and from the asymptotic representation of $J_\nu(z)$ that for large z, $I_\nu(z)$ grows exponentially

$$I_\nu(z) \sim \frac{e^z}{(2\pi z)^{1/2}}.$$

For integral ν another linearly independent solution (similar to $Y_\nu(z)$) is needed. The choice

$$K_\nu(z) = \frac{1}{2}\pi i e^{i\nu\pi/2} H_\nu^{(1)}(iz)$$
$$= -\frac{1}{2}\pi i e^{-i\nu\pi/2} H_\nu^{(2)}(-iz)$$

is natural because, like $I_\nu(z)$, $K_\nu(z)$ is real for real positive z. Its behavior for large z is given by $K_\nu(z) \sim (\pi/2z)^{1/2} e^{-z}$. The function $K_\nu(z)$ is called the MODIFIED BESSEL FUNCTION of SECOND KIND.

It helps to realize that these functions are related to each other in much the way that $\sin z$, $\cos z$, e^{iz}, e^{-iz}, e^z, and e^{-z} are related. That is, two linearly independent solutions of $\frac{d^2u}{dz^2} + u = 0$ are $\sin z$ and $\cos z$. Of course, an equally valid set is e^{iz}, e^{-iz}. If we make the change of variables $z = it$ the differential equation becomes $\frac{d^2u}{dt^2} - u = 0$ and the solutions are e^t, e^{-t} or $\cosh t$, $\sinh t$. The relationship between Hankel functions and Bessel functions is analogous to Euler's formula $e^{\pm iz} = \cos z \pm i \sin z$.

As mentioned earlier, power series representations do not provide much qualitative information about Bessel functions and other representations are needed. To this end we use the recurrence formulae

$$\frac{d}{dz}\left(z^{-\nu}J_\nu(z)\right) = -z^{-\nu}J_{\nu+1}(z)$$

$$\frac{d}{dz}\left(z^{\nu}J_\nu(z)\right) = z^{\nu}J_{\nu-1}(z)$$

which can be verified from the power series representations or directly from the differential equation. (The functions $Y_\nu(z)$, $H_\nu^{(1)}(z)$, and $H_\nu^{(2)}(z)$ satisfy recurrence formulae of the same form.) We define the GENERATING FUNCTION

$$\phi(z,t) = \sum_{n=-\infty}^{\infty} J_n(z)t^n$$

and calculate that

$$\frac{\partial \phi}{\partial z} = \sum_{n=-\infty}^{\infty} J_n'(z)t^n = \frac{1}{2}\sum_{n=-\infty}^{\infty}(J_{n-1}(z) - J_{n+1}(z))\,t^n = \frac{1}{2}\left(t - \frac{1}{t}\right)\phi.$$

The solution of this differential equation is

$$\phi(z,t) = \phi_0(t)e^{\frac{1}{2}(t-\frac{1}{t})z}.$$

At $z = 0$, we have $\phi(0,t) = \phi_0(t) = \sum_{n=-\infty}^{\infty} J_n(0)t^n = 1$ since $J_0(0) = 1$, and $J_n(0) = 0$ for $n \neq 0$, so the generating function is

$$\phi(z,t) = \sum_{n=-\infty}^{\infty} J_n(z)t^n = e^{\frac{1}{2}(t-\frac{1}{t})z}.$$

The important observation to make is that here we have a Laurent series representation of the function $e^{\frac{1}{2}(t-\frac{1}{t})z}$ for each fixed z, and its coefficients are

$$J_n(z) = \frac{1}{2\pi i}\int_C e^{\frac{1}{2}(t-\frac{1}{t})z}t^{-(n+1)}dt$$

for any contour C surrounding the origin. If we take the specific contour $t = e^{i\theta}$ (the unit circle) we obtain

$$J_n(z) \;=\; \frac{1}{2\pi} \int_0^{2\pi} e^{i(z\sin\theta - n\theta)}\,d\theta$$

or from symmetry

$$J_n(z) \;=\; \frac{1}{\pi} \int_0^{\pi} \cos(z\sin\theta - n\theta)\,d\theta.$$

We will find specific use for these formulae in later chapters.

The most important fractional order Bessel functions are those of order $\nu = n + 1/2$, with n an integer. For $n = 0$ we make the change of variables $y = u/\sqrt{z}$ and find that Bessel's equation becomes

$$u'' + u = 0.$$

The solutions of this equation are $\sin z$ and $\cos z$ so that

$$J_{1/2}(z) \;=\; \frac{A}{\sqrt{z}}\sin z, \qquad J_{-1/2}(z) = \frac{A}{\sqrt{z}}\cos z.$$

The constant A can be determined from the power series representation to be $A = (2/\pi)^{1/2}$. From the recurrence formulae

$$J_{n+1/2}(z) \;=\; \frac{n - 1/2}{z} J_{n-1/2}(z) - J'_{n-1/2}(z),$$

so that

$$J_{3/2}(z) \;=\; \left(\frac{2}{\pi z}\right)^{1/2} \left(\frac{1}{z}\sin z - \cos z\right),$$

and, of course, $J_{n+1/2}(z)$ is in general a combination of $\sin z$ and $\cos z$. The functions $J_{n+1/2}(z)$ occur naturally in the study of the Laplacian in spherical coordinates and will be discussed further in that context in Chapter 8.

6.5.3   Legendre Functions

Other special functions which are used frequently are the Legendre polynomials and the associated Legendre functions. The Legendre equation

$$\left(1 - z^2\right) y'' - 2zy' + n(n+1)y \;=\; 0$$

and the associated Legendre equation

$$\left(1 - z^2\right) y'' - 2zy' + \left(\nu(\nu+1) - \frac{m^2}{1 - z^2}\right) y \;=\; 0$$

arise from separation of variables of Laplace's equation in spherical coordinates (see Chapter 8).

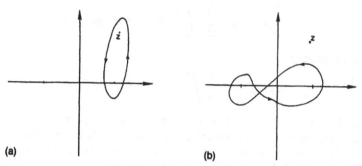

FIGURE 6.17 Contours of integration for Legendre functions a) of the first kind, b) of the second kind.

We define the Legendre polynomial $P_n(z)$ by Rodrigue's formula

$$P_n(z) = \frac{1}{2^n n!} \frac{d^n}{dz^n} \left[(z^2 - 1)^n\right].$$

Certainly $P_n(z)$ is a polynomial of degree n and it is easy to check that this formula agrees with those used in earlier chapters. We use the Cauchy Integral Formula to write

$$P_n(z) = \frac{1}{2\pi i} \int_C \frac{(t^2 - 1)^n}{2^n (t - z)^{n+1}} dt,$$

where the contour C encircles the point z and is traversed in the counter-clockwise direction.

We can use this representation to show that Legendre's equation is satisfied by $P_n(z)$. Substituting directly into the equation we find that

$$(1 - z^2)\, P_n''(z) - 2z P_n'(z) + n(n + 1)P_n(z) = \frac{n + 1}{2\pi i} \int_C \frac{d}{dt} \left[\frac{(t^2 - 1)^{n+1}}{(t - z)^{n+2}}\right] dt.$$

Whenever the function $(t^2 - 1)^{n+1} / (t - z)^{n+2}$ returns to its original value after one traversal of C, this expression is zero, and the Legendre equation is satisfied. This observation leads to a generalization of $P_n(z)$ to nonintegral n. In fact, the function

$$\frac{1}{2\pi i} \int_C \frac{(t^2 - 1)^\nu}{2^\nu (t - z)^{\nu+1}} dt$$

satisfies the Legendre equation whenever the function $(t^2 - 1)^{\nu+1} / (t - z)^{\nu+2}$ returns to its original value after one traversal of C. One such choice for C (Fig. 6.17a) is to enclose the point $t = z$ and the point $t = 1$ but not the point

$t = -1$, with the resulting definition

$$P_\nu(z) = \frac{1}{2\pi i} \int_C \frac{(t^2 - 1)^\nu}{2^\nu (t - z)^{\nu+1}} dt$$

called the LEGENDRE FUNCTION of the FIRST KIND.

A different choice for the contour C (Fig. 6.17b) is to enclose $t = 1$ and $t = -1$ but not $t = z$ and to go around the points $t = 1$ and $t = -1$ in opposite directions, that is, in a figure eight. This leads to the definition of LEGENDRE FUNCTIONS of the SECOND KIND

$$Q_\nu(z) = \frac{1}{2^{\nu+1}} \int_{-1}^1 \frac{(1 - t^2)^\nu \, dt}{(z - t)^{\nu+1}}.$$

We can construct a generating function for $P_n(z)$. We define

$$\phi(z, s) \quad = \quad \sum_{n=0}^\infty P_n(z) s^n$$

and use the integral representation of $P_n(z)$ to write

$$\phi(z, s) \quad = \quad \frac{1}{2\pi i} \int_C \sum_{n=0}^\infty \frac{(t^2 - 1)^n s^n}{2^n (t - z)^{n+1}} dt$$

$$= \quad \frac{1}{2\pi i} \int_C \frac{1}{(t - z)} \left(1 - \frac{s(t^2 - 1)}{2(t - z)} \right)^{-1} dt$$

$$= \quad \frac{1}{2\pi i} \int_C \frac{dt}{t - z - (s/2)(t^2 - 1)} \quad \text{provided} \quad \left| \frac{s(t^2 - 1)}{2(t - z)} \right| < 1$$

on the contour C. Notice that although the denominator has two roots, for small s one root is quite close to $t = z$ and the other is very large and therefore outside the contour. From the Cauchy Integral Formula we obtain that

$$\phi(z, s) \quad = \quad (1 - 2zs + s^2)^{-1/2}$$

valid for small s. Once again we recognize Taylor's theorem and write that

$$P_n(z) = \frac{1}{2\pi i} \int_C \frac{ds}{s^{n+1} (1 - 2zs + s^2)^{1/2}}$$

for any contour C enclosing the origin, but not enclosing the singularities at $s = z \pm (1 - z^2)^{1/2}$.

Once the Legendre functions are known the Associated Legendre functions follow quickly. It is a direct calculation to show that solutions of the associated Legendre's equation are

$$P_\nu^m(z) \;=\; \left(1-z^2\right)^{m/2} \frac{d^m}{dz^m}\left(P_\nu(z)\right)$$

$$Q_\nu^m(z) \;=\; \left(1-z^2\right)^{m/2} \frac{d^m}{dz^m}\left(Q_\nu(z)\right)$$

for integer values of m, and noninteger values of m rarely occur in applications.

6.5.4   Sinc Functions

The SINC FUNCTIONS (or Whittaker Cardinal functions) are interpolating functions defined by

$$C(f)(x) \;=\; \sum_{j=-\infty}^{\infty} f(jh)S_j(x) \qquad h>0$$

where

$$S_0(x) \;=\; \frac{\sin(\pi x/h)}{\pi x/h}$$

$$S_k(x) \;=\; S_0(x-kh).$$

The Sinc functions have a number of remarkable properties that make them an excellent choice for a transform or for approximation schemes such as Galerkin approximation of integral equations and integral transforms.

A most remarkable fact of Sinc functions is that

$$C(f)(z) = f(z)$$

for all z in the complex plane, for all $f(z)$ with $f(x)$ square integrable on $-\infty < x < \infty$ and $|f(z)| \le Ae^{\pi|z|/h}$. Furthermore, on this set of functions (which we call $B(h)$), the expression $C(f)(z)$ is actually a Fourier series representation of $f(z)$ in terms of the mutually orthogonal function $S_k(x)$. The Fourier coefficient of $S_k(x)$ is trivial to calculate, being $f(kh)$.

To prove these (and other) facts, we need a few preliminary results. First, if $f(z)$ is entire with $|f(z)| \le e^{\pi|z|/h}$ and if $f(z)$ vanishes at the equidistant points $x_k = kh$, $k = 0, \pm 1, \pm 2, \ldots$, then

$$f(z) \;=\; C\sin \pi z/h.$$

This follows by examining the function

$$g(z) \;=\; \frac{f(z)}{\sin \pi z/h}$$

and noting that $g(z)$ is bounded on the boundary of the square $|x| \le (k+1/2)h$, $|y| \le (k+1/2)h$. Since $g(z)$ is entire (the points $z = kh$ are regular points), the maximum modulus theorem implies that $|g(z)|$ is bounded and hence $g(z) = C$, as required.

Now we conclude that any two functions that are both in $B(h)$ and which agree at the points $z_k = kh$, $k = 0, \pm 1, \pm 2, \ldots$ are in fact exactly the same. It follows that $S_0(z)$ is the only function in $B(h)$ that satisfies $g(0) = 1$, $g(kh) = 0$. Furthermore, if $C(f)(z)$ is in $B(h)$ (which it is), then $f(z) = C(f)(z)$. (We leave verification of this fact for the interested reader.)

Now we observe that $S_j(x)$ and $S_k(x)$ are mutually orthogonal. In fact (Problem 6.5.27)

$$\int_{-\infty}^{\infty} S_j(x)S_k(x)dx = h\delta_{jk}.$$

As a result, the Whittaker Cardinal function $C(f)(x)$ is a Fourier series of the form

$$f(z) = \sum_{k=-\infty}^{\infty} a_k S_k(z)$$

where

$$f(kh) = a_k = \frac{1}{h}\int_{-\infty}^{\infty} f(x)S_k(x)dx$$

are the Fourier coefficients. Since this Fourier series is exact, Parseval's equality holds, implying that

$$\int_{-\infty}^{\infty} |f(x)|^2 \, dx = h \sum_{k=-\infty}^{\infty} |f(kh)|^2 .$$

Finally, using that

$$\int_{-\infty}^{\infty} \frac{\sin t}{t} \frac{\sin(t-z)}{t-z} dt = \frac{\pi \sin z}{z}$$

(Problem 6.4.22), it is not difficult to verify that

$$f(z) = \frac{1}{h}\int_{-\infty}^{\infty} f(t)S_0(t-z)dt.$$

In other words, on the space of functions $B(h)$, the function $S_0(t)$ acts like a delta function. We already know from Chapter 4 that $S_0(t)$ forms a delta sequence as h goes to zero. However, for a function in $B(h)$, the evaluation of this integral is the same for all $h \le h_1$, since $B(h)$ is a subset of $B(h_1)$ for all $h \le h_1$.

Some of the important uses of Sinc functions are to give approximate representations of functions, for example

$$C_n(f)(x) = \sum_{k=-n}^{n} f(kh)S_k(x)$$

and in their ability to approximate functions that are not entire. For example, if $f(z)$ is analytic in the strip $|\operatorname{Im} z| < d$, if

$$\int_{-d}^{d} |f(x+iy)|\, dy \to 0$$

as $|x| \to \infty$, and if

$$N = \lim_{y \to d^-} \int_{-\infty}^{\infty} (|f(x+iy)| + |f(x-iy)|)\, dx < \infty$$

then the error between the function of f and its Sinc interpolant denoted $E(f) = f - C(f)$, is bounded. It can be shown that

$$\|E(f)\|_\infty \;\leq\; \frac{1}{2\pi d}\frac{N}{\sinh(\pi d/h)}$$

and

$$\|E(f)\|_2 \;\leq\; \frac{1}{2\sqrt{\pi d}}\frac{N}{\sinh(\pi d/h)}.$$

That is, the L^2 and L^∞ norms of the error are bounded and become exponentially small as h goes to zero or as d, the width of the strip of analyticity becomes larger.

As an example of how to use Sinc functions, suppose we wish to solve approximately the Fredholm integral equation on the infinite line

$$u(x) + \int_{-\infty}^{\infty} k(x,y)u(y)dy = f(x),$$

where $\int_{-\infty}^{\infty} k^2(x,y)dx\, dy < \infty$. We project the integral onto a finite Sinc expansion by writing

$$u_n(x) \;+\; \sum_{j=-n}^{n}\left(\int_{-\infty}^{\infty} k(jh,y)u_n(y)dy\right) S_j(x)$$

$$=\; \sum_{j=-n}^{n} f(jh)S_j(x).$$

The approximate solution $u_n(x)$ will itself be a Sinc function

$$u_n(x) \;=\; \sum_{j=-n}^{n} \alpha_j S_j(x).$$

The only difficulty with this approximation is the evaluation of the integrals

$$\int_{-\infty}^{\infty} k(jh,y)S_l(y)dy.$$

Notice, however, that these are the Fourier coefficients of $k(jh,y)$ so that

$$\frac{1}{h}\int_{-\infty}^{\infty} k(jh,y)S_l(y)dy = k(jh,lh)$$

whenever $k(jh,y)$ is in $B(h)$. If this is not the case, then this is still a good approximation if $k(jh,y)$ is analytic in a strip.

This approximation reduces the integral equation to the matrix equation

$$\alpha_j + h \sum_{l=-n}^{n} k(jh,lh)\alpha_l = f_j$$

or

$$(I + h\tilde{K})\alpha = f,$$

where the entries of the matrix $\tilde{K}$ are $k(jh,lh)$, and this system can be readily solved numerically.

In Chapter 7 we will examine some further properties of Sinc functions.

Further Reading

Two good books on complex variable theory (there are many) are

1. E. T. Copson, *Theory of Functions of a Complex Variable*, Oxford at the Clarendon Press, 1935.

2. G. F. Carrier, M. Krook, and C. E. Pearson, *Functions of a Complex Variable*, McGraw-Hill, New York, 1966.

The latter of these is an excellent source of very challenging problems. The three volume set

3. P. Henrici, *Applied and Computational Complex Analysis*, Wiley and Sons, 1986

is an excellent source of information on how to use complex variable theory in applied problems.

The problems of fluid dynamics are beautifully discussed in the classic work

4. L. M. Milne-Thomson, *Theoretical Aerodynamics*, MacMillan, London, 1966.

A most interesting picture book which shows what actual fluids look like is

5. M. Van Dyke, *An Album of Fluid Motion*, The Parabolic Press, Stanford, 1982.

An overview of Sinc functions and a good list of references is in the review article

6. F. Stenger, "Numerical Methods Based on Whittaker Cardinal, or Sinc Functions," SIAM Rev. *23*, 165–224, (1981).

Problems for Chapter 6

Section 6.1

1. **a.** The function $f(z) = (z(z-1)(z-4))^{1/2}$ is defined to have branch cuts given by $z = \rho e^{-i\pi/2}$, $z = 1 + \rho e^{i\pi/2}$ and $z = 4 + \rho e^{-i\pi/2}$, $0 \le \rho < \infty$, and with $f(2) = -2i$. Evaluate $f(-3)$, $f(1/2)$, and $f(5)$.

   b. Determine the branch cut structure of $f(z) = \ln\left(5 + \left(\frac{z+1}{z-1}\right)^{1/2}\right)$. Make a plot in the f complex plane of the different branches of this function.

2. What is wrong with the following "proof"?

$$e^z = e^{2\pi i(z/2\pi i)} = 1^{z/2\pi i} = 1 \qquad \text{for all } z.$$

3. Find all values of z for which

   a. $\sin z = 2$

   b. $\sin z = i$

   c. $\tan^2 z = -1$

4. Compute all values of i^i, $\ln(1+i)^{\pi i}$, and arctanh 1.

5. Suppose we define $\sqrt{z}$ to be either $\sqrt{z} = \rho^{1/2}e^{i\theta/2}$ or $\sqrt{z} = -\rho^{1/2}e^{i\theta/2}$ where θ is restricted by $0 \le \theta < 2\pi$. Find the four ways the function $g(z) = \sqrt{1 + \sqrt{z}}$ can be defined and find the image of these four definitions in the complex g plane. Where are the branch cuts in the z plane?

6. Is it always true that $\ln z_1 z_2 = \ln z_1 + \ln z_2$? If so, prove the result and if not, provide a counterexample.

7. Show that the function $w(z) = \frac{1}{2}\left(z + \frac{1}{z}\right)$ is a two to one function (every value w is attained twice). Find the branch cut structure for the inverse function $z = z(w)$. What are the two regions of the z plane corresponding to separate branches of the inverse function?

Section 6.2

1. The function $f(z)$ is analytic in the entire z plane including $z = \infty$ except at $z = i/2$, where it has a simple pole, and at $z = 2$ where it has a pole of order 2. It is known that

$$\int_{|z|=1} f(z)dz = 2\pi i$$

$$\int_{|z|=3} f(z)dz = 0$$

$$\int_{|z|=3} f(z)(z-2)dz = 0$$

Find $f(z)$ (unique up to an arbitrary additive constant).

2. For

$$f(z) = \frac{(z-16)^{1/3}(z-20)^{1/2}}{z-1}$$

the vertical lines $z = 16 + iy$, $z = 20 - iy$, $y \geq 0$ are branch cuts and $f(24) = \frac{4}{23}$. Evaluate $\int_C f(z)dz$ for the contour $C : |z-2| = 2$.

3. Evaluate the line integral $\int_C z^{-1/3}dz$ on the path C traversing from $z = 1 + i$ to $z = 1 - i$ along the following composite path

$$C_1 : 2y^2 - x^2 = 1 \text{ from } (x,y) = 1,1 \text{ to } (x,y) = \left(-2, \sqrt{5/2}\right),$$

$$C_2 : x = -2 \text{ from } y = \sqrt{5/2} \text{ to } y = -\sqrt{5/2},$$

$$C_3 : 2y^2 - x^2 = 1 \text{ from } (x,y) = \left(-2, -\sqrt{5/2}\right) \text{ to } (x,y) = (1, -1).$$

The positive x-axis is a branch cut.

4. Verify, without direct differentiation, that

$$u(x,y) = (x^2 + y^2)^{1/4} \cos\left(\frac{1}{2}\tan^{-1} y/x\right)$$

is a solution of Laplace's equation $\frac{\partial^2 u}{\partial x^2} + \frac{\partial^2 u}{\partial y^2} = 0$.

5. In what annular regions centered at $z = -i$ can Laurent expansions for the following functions be obtained?

    a. $(z^2 - 1)^{1/2}$

    b. $(z^2 + 1)^{1/2}$

    c. $\ln\left(\frac{z+1}{z-1}\right)$

    d. $(z(z^2 - 1))^{1/2}$

    e. $(z(z - i)(z^2 - 1))^{1/2}$

Evaluate the following integrals

6. $\int_{|z|=1/2} \frac{z+1}{z^2+z+1} dz.$

7. $\int_{|z|=1/2} \exp\left[z^2 \ln(1+z)\right] dz.$

8. $\int_{|z|=1/2} \arcsin z \, dz.$

9. $\int_{|z|=1} \frac{\sin z}{2z+i} dz.$

10. $\int_{|z|=1} \frac{\ln(z+2)}{z+2} dz.$

11. $\int_{|z|=1} \cot z \, dz.$

12. Show that, if $f(z)$ is analytic and nonzero on the interior of some region, then $|f(z)|$ cannot attain a local minimum on the interior of the domain.

13. Prove a version of the Phragmén-Lindelof theorem valid on sectors: If $f(z)$ is analytic on the sector $|\arg z| \leq \pi/2\alpha, |f(z)| \leq M$ on the boundary of the sector, and $|f(z)| < K \, e^{|z|^\beta}$, then either $|f(z)| \leq M$ or $\beta \geq \alpha$.

14. Prove that if $F(z)$ is analytic on the entire complex plane, bounded on the real axis and $|F(z)| \leq Ae^{\sigma|z|}$, then $|F(x + iy)| \leq me^{\sigma|y|}$.
Hint: The function $G(z) = F(z)e^{i(\sigma+\epsilon)z}$ satisfies $|G(iy)| \leq A$ and $|F(x)| \leq \sup_{-\infty<x<\infty} |G(x)|$. Apply problem 6.2.13.

15. Prove the Fundamental Theorem of Algebra: An nth order polynomial $P_n(z)$ has n roots (counting multiplicity) in the complex plane.
Hint: If $P_n(z)$ has no roots then $1/P_n(z)$ is bounded. Apply Liouville's Theorem.

Section 6.3

1. Verify that the map $\zeta = \frac{1}{z}$ transforms circles into circles.

2. Verify that the composition of two bilinear maps is bilinear.

3. Verify that a bilinear map takes circles into circles.

4.   **a.** Show that $|z - \alpha| = k|z - \beta|$ is a circle.

    b. Show that $|z - \gamma| = R$ can be expressed as $|z - \alpha| = k|z - \beta|$ by taking $\alpha = \gamma + r_1 e^{i\phi}, \beta = \gamma + r_2 e^{i\phi}, r_1 r_2 = R^2$. What is k?

5. Find the bounded solution of Laplace's equation subject to

$$\phi = a \quad \text{on} \quad |z| = 1, \qquad \text{Im } z \geq 0$$

$$\phi = b \quad \text{on} \quad |z - i| = \sqrt{2}, \quad \text{Im } z \leq 0.$$

Hint: Explore the maps $w = \frac{z-1}{z+1}$ and $\xi = \ln w - i\pi/2$.

6. Solve Laplace's equation on the interior of the domain with boundary $|z| = 1$, $\text{Im } z \geq 0$, and $|z - i| = \sqrt{2}$, $\text{Im } z \leq 0$, subject to $\frac{\partial \phi}{\partial n} = 0$ on $|z| = 1$, $\text{Im } z \geq 0$, and $\frac{\partial \phi}{\partial n} = b$ on $|z - i| = \sqrt{2}$, $\text{Im } z \leq 0$. Hint: $\frac{\partial \phi}{\partial n} = $ constant implies $\psi = $ constant; use the transformation of problem 5.

7. Find the image of the unit circle $|\xi| = 1$ under the transformation

$$z = \int_0^\xi \left(1 - \eta^4\right)^{-1/2} d\eta.$$

8. Find the image of the region $|\text{Re } z| \leq k$, $\text{Im } z > 0$ under the transformation $\xi = \sin \frac{\pi z}{2k}$.

9. Describe the flow corresponding to the function $f(z) = U_0 z^\beta$ if $1/2 < \beta < 1$ or if $\beta \geq 1$.

10. Calculate the drag on a thin airfoil of length $2a$ with angle of attack α. Where is the stagnation point of the flow?

11. Calculate the lift for a thin airfoil in a flow that is not horizontal to the real axis, but rather has velocity field $f'(z) \sim U e^{i\beta}$ as $z \to \infty$. Interpret your results in terms of the lift for a climbing or descending aircraft.

12. Find a noncavitating symmetric flow past a flat vertical plate. How does this compare with the thin airfoil solution?

13. Find the parametric representation of the free stream from a jet nozzle. Show that the width of the jet at $x = \infty$ is $d = \frac{\pi L}{\pi+2}$ where $L = 2a$ is the width of the orifice.
    Hint: Show that

$$\frac{dz}{dw} = \frac{1}{U} \left(e^{-w/2k} + \left(1 + e^{-w/k}\right)^{1/2} \right).$$

Section 6.4

Evaluate the following integrals

1. $\int_{-\infty}^{\infty} \frac{dx}{ax^2 + bx + c}$, $b^2 - 4ac < 0$.

2. $\int_0^\infty \frac{x \sin x}{a^2 + x^2} dx$.

3. $\int_0^\infty \frac{dx}{1 + x^k}$, $k \geq 2$.

4. $\int_0^\infty \frac{dx}{(x+1)x^p}$, $0 < p < 1$.

5. $\int_1^\infty \frac{x}{(x^2+4)(x^2-1)^{1/2}}dx$.

6. $\int_0^\infty \frac{\sqrt{x}}{x^2+1}dx$.

7. $\int_{-\infty}^\infty \frac{e^{ikx}}{a+ik}dk$.

8. $\int_0^\pi \frac{x\sin x}{1-2a\cos x+a^2}dx$, $a > 0$. Hint: Consider $\int_C \frac{z\,dz}{a-e^{iz}}$ on the rectangular contour with corners at $z = \pm\pi$ and $z = \pm\pi - iR$, let $R \to \infty$.

9. $\int_{-\infty}^\infty \frac{e^{i\omega x}}{\cosh x}dx$, ω real.

10. $\int_{-\infty}^\infty \frac{\sec x\,dx}{x^2+1}$.

11. $\int_0^\infty \frac{dx}{2+x+x^3}$.

12. $\int_0^{2\pi} \ln(a + b\cos\theta)d\theta$, $a > b > 0$.

13. $\int_0^\infty \frac{\sin \alpha x}{\sinh \pi x}dx$. Hint: Use the answer to problem 9.

14. $\int_0^\pi \ln(\sin x)dx$.

15. $\int_0^\infty \frac{\ln(1+x^2)}{1+x^2}dx$.

16. Show that
$$\int_0^\infty \frac{\cos\alpha x\,dx}{x+\beta} = \int_0^\infty \frac{xe^{-\alpha\beta x}}{x^2+1}dx, \quad \alpha,\beta > 0.$$

17. Suppose $P_n(z)$ is an nth order polynomial $n \geq 2$ with simple roots $\{z_k\}_{k=1}^n$. Show that
$$\sum_{k=1}^n \frac{1}{P_n'(z_k)} = 0.$$

Hint: Evaluate $\int_C \frac{dz}{P_n(z)}$ on some contour C.

18. Evaluate
$$\frac{\Gamma(\alpha)}{2\pi i} \int_{c-i\infty}^{c+i\infty} \frac{e^{\beta t}}{(t-z)^\alpha}dt,$$

$0 < \mathrm{Re}\,\alpha < 1$, β real, where the path of integration is to the right of all singularities.

19. Show that
$$2rz \int_z^r \frac{T_n(\rho/r)T_n(\rho/z)d\rho}{\rho\,(r^2-\rho^2)^{1/2}\,(\rho^2-z^2)^{1/2}} = \pi$$

for n even, where
$$T_n(x) = \cos\left(n\cos^{-1}x\right)$$

is the nth Chebyshev polynomial.
Hint: The change of variables $\rho^2 = r^2\cos^2\theta + z^2\sin^2\theta$ converts this to an integral for which complex variable techniques work nicely.

20. Evaluate $\int_0^\infty t^{-1/2} e^{i\mu t} dt$ for all real μ.

21. Evaluate $\int_0^\infty t^{1/2} e^{i\mu t} dt$ for $\mathrm{Im}\,\mu > 0$.

22. Show that $\int_{-\infty}^\infty \frac{\sin x}{x} \cdot \frac{\sin(x-x_0)}{x-x_0} dx = \frac{\pi \sin x_0}{x_0}$.

23. Show that for any entire function $f(z)$ with $|f(z)| \le e^{k|z|}$, $f(x)$ square integrable,

$$g(\omega) = \int_{-\infty}^\infty e^{i\omega x} f(x) dx = 0$$

for $|\omega| > k$.

24. Evaluate $\int_{-\infty}^\infty \frac{e^{iax}}{(z+i\alpha)^\beta} dz$ for $0 < \beta < 1$.

25. Evaluate $\int_{-\infty}^\infty \frac{e^{iax} dz}{\sqrt{z+i\alpha} + \sqrt{z+i\beta}}$, $\alpha < \beta < 0$.

Hint: Use that $\frac{1}{\sqrt{z+i\alpha} + \sqrt{z+i\beta}} = \frac{\sqrt{z+i\alpha} - \sqrt{z+i\beta}}{i(\alpha-\beta)}$, but be careful to use Jordan's lemma correctly.

Section 6.5

1. Verify that

$$\Gamma(z) = \frac{1}{2i \sin \pi z} \int_C e^t t^{z-1} dt,$$

where the contour C goes from $t = -\infty - i\varepsilon$ to $t = -\infty + i\varepsilon$, $\varepsilon > 0$ small, around the branch point at $t = 0$, agrees with the gamma function $\Gamma(z) = \int_0^\infty e^{-t} z^{z-1} dt$ for $\mathrm{Re}\,z > 0$.

2. Verify the binomial theorem

$$(1+z)^\alpha = \sum_{n=0}^\infty \frac{\Gamma(\alpha+1)}{\Gamma(\alpha-n+1)\Gamma(n+1)} z^n.$$

For what values of α is this a correct formula?

3. **a.** Show that

$$\frac{\pi}{\cosh \pi \alpha} = \Gamma\left(\frac{1}{2} - i\alpha\right) \Gamma\left(\frac{1}{2} + i\alpha\right)$$

and

$$\frac{\pi \alpha}{\sinh \pi \alpha} = \Gamma(1 - i\alpha)\Gamma(1 + i\alpha)$$

b. Show that $\pi \alpha \coth \pi \alpha = K_+(\alpha)K_-(\alpha)$ where

$$K_+(\alpha) = K_-(-\alpha) = \sqrt{\pi} \frac{\Gamma(1 - i\alpha)}{\Gamma(\frac{1}{2} - i\alpha)}.$$

4. Solve the generalization of Abel's integral equation

$$T(x) = \int_0^x \frac{f(y)dy}{(x-y)^\alpha}$$

for $f(x)$ in terms of $T(x)$, where $0 < \alpha < 1$. Hint: Consider the expression $\int_0^\eta \frac{T(x)dx}{(\eta-x)^{1-\alpha}}$.

Evaluate the following Wronskians

5. $W\left(J_\nu(z), J_{-\nu}(z)\right)$.

6. $W\left(J_\nu(z), Y_\nu(z)\right)$.

7. $W\left(J_\nu(z), H_\nu^{(1)}(z)\right)$.

8. $W\left(H_\nu^{(1)}(z), H_\nu^{(2)}(z)\right)$.

9. Verify that $Y_0(z) \sim \frac{2}{\pi}\ln z$ for small z. Find the small z behavior for $Y_n(z)$.

10. Show that

$$\cos(z\sin\theta) = J_0(z) + 2\sum_{n=1}^\infty J_{2n}(z)\cos 2n\theta$$

and

$$\sin(z\sin\theta) = 2\sum_{n=0}^\infty J_{2n+1}(z)\sin(2n+1)\theta.$$

11. Prove that

 a. $J_0(z) + 2\sum_{n=1}^\infty J_{2n}(z) = 1$

 b. $J_n(y+z) = \sum_{m=-\infty}^\infty J_m(y)J_{n-m}(z)$.

12. Show that

$$J_{-1/2}(z) = \left(\frac{2}{\pi z}\right)^{1/2}\cos z$$

$$J_{3/2}(z) = \left(\frac{2}{\pi z}\right)^{1/2}\left(\frac{\sin z}{z} - \cos z\right)$$

$$J_{-3/2}(z) = -\left(\frac{2}{\pi z}\right)^{1/2}\left(\frac{\cos z}{z} + \sin z\right).$$

13. Find the generating function $\phi(z,t) = \sum_{n=-\infty}^\infty I_n(z)t^n$.

14. Show that $J_0(z) = J_0(-z)$ and $H_0^{(1)}(z) - H_0^{(1)}(-z) = 2J_0(z)$ for real valued z.

15. Show that

$$\int_0^\infty \frac{xJ_0(\alpha x)dx}{x^2 + k^2} = K_0(\alpha k).$$

Hint: Evaluate $\frac{1}{2\pi i}\int \frac{zH_0^{(1)}(\alpha z)dz}{z^2+k^2}$ on a closed semicircular arc in the upper half complex z plane.

16. Use the generating function for Legendre polynomials to evaluate $\int_{-1}^1 P_n^2(t)dt$.

17. Show that $P_\nu(z) = \frac{1}{2\pi}\int_{-\pi}^\pi \left(z + (z^2 - 1)^{1/2}\cos\theta\right)^{-\nu-1} d\theta$ for $|\arg z| < \pi/2$.

18. Show that $Q_\nu(z) = \int_0^\infty \left(z + (z^2 - 1)^{1/2}\cosh\theta\right)^{-\nu-1} d\theta$.

19. Verify that

$$Q_0(z) = \frac{1}{2}\ln\frac{z+1}{z-1}$$

$$Q_1(z) = \frac{1}{2}z\ln\frac{z+1}{z-1}.$$

20. Show that $P_n(z) = \frac{1}{\pi}\int_0^\pi \frac{d\theta}{\left(z+(z^2-1)^{1/2}\cos\theta\right)^{n+1}}$.

21. Show that $P_n(\cos\phi) = \frac{2}{\pi}\int_0^\phi \frac{\cos(n+1/2)\theta\, d\theta}{\sqrt{2\cos\theta-2\cos\phi}}$.

22. What is the Green's function for the equation

$$u'' + \frac{1}{x}u' + \left(k^2 - \frac{16}{x^2}\right)u = f(x), \quad x \in (0,\alpha), \quad u(0)\text{ bounded}, \quad u(\alpha) = 0.$$

Under what conditions, if any, does the Green's function fail to exist?

23. Show that a generating function for the Hermite polynomials is

$$\sum_{n=0}^\infty \frac{H_n(x)t^n}{n!} = e^{tx - t^2/2}.$$

24. Show that a generating function for the generalized Laguerre functions is

$$\sum_{n=0}^\infty L_n^\alpha(x)t^n = \frac{e^{-xt/(1-t)}}{(1-t)^{\alpha+1}}.$$

Remark: The generalized Laguerre polynomials $L_n^\alpha(x)$ are orthogonal with respect to weight function $w(x) = x^\alpha e^{-x}$.

25. Show that a generating function for Chebyshev polynomials is

$$\sum_{n=0}^{\infty} T_n(x)t^n = \frac{1-xt}{(1-2xt+t^2)}.$$

26. A pendulum swings from $\theta = \pi/2$ to $\theta = -\pi/2$. Express the period of oscillation in terms of the Beta function.

27. Show that $\int_{-\infty}^{\infty} S_j(x)S_k(x)dx = h\delta_{jk}$ where $S_j(x) = S_0(x-jh)$, $S_0(x) = \frac{\sin y}{y}$, $y = x\pi/h$.

7

TRANSFORM AND SPECTRAL THEORY

7.1 Spectrum of an Operator

In Chapter 1 we learned that a natural basis for $\mathbb{R}^n$ is the set of eigenvectors of a self-adjoint matrix, and that relative to this basis the matrix operator is diagonal. Similarly, in Chapters 3 and 4 we learned that certain integral and differential operators have the same property, namely, that their eigenvectors form a complete, orthonormal basis for the appropriate space, and that relative to this basis, the operator is diagonal. Unfortunately, this property of eigenvalue-eigenvector pairs is not true in general, namely, the eigenvectors of an operator need not form a basis for the space and, in fact, there need not be any eigenvectors at all.

The representation of functions in the Hilbert space in terms of eigenfunctions, or something like eigenfunctions, of an operator L is called the SPECTRAL REPRESENTATION for L. The point of this chapter is to show that the spectral representation of L depends on the behavior of the inverse of the operator $L - \lambda$, and to determine how to find the appropriate spectral representation.

Suppose L is a linear operator on a Hilbert space. There are four possibilities for the behavior of the operator $(L - \lambda)^{-1}$.

1. If $(L - \lambda)^{-1}$ exists and is bounded, λ is said to belong to the RESOLVENT SET for L. From the Fredholm alternative (Chapter 3) we know that if $L - \lambda$ has closed range, then $(L - \lambda)^{-1}$ exists and is unique if and only if both $(L - \lambda)u = 0$ and $(L^* - \bar{\lambda})v = 0$ have only the trivial solutions $u = v = 0$.

2. If there is a nontrivial function $u \neq 0$ in the Hilbert space for which $(L - \lambda)u = 0$, λ is said to belong to the POINT or DISCRETE SPECTRUM of L. The value λ is called an eigenvalue and u is the eigenfunction.

3. If the closure of the range of an operator $L - \lambda$ is the whole space, but the inverse $(L - \lambda)^{-1}$ is not bounded, λ is said to belong to the CONTINUOUS SPECTRUM of L.

4. Finally, if the range of $L - \lambda$ is not dense in the Hilbert space, (i.e., the closure of the range is a proper subset of the Hilbert space), but λ is not in the point spectrum of L, then λ is said to belong to the RESIDUAL SPECTRUM of L.

There is a nice way to organize this information by asking three questions:

1. Is $L - \lambda$ a one to one mapping? If no, then λ is in the discrete spectrum. If yes,

2. Is the range of $L - \lambda$ dense in H? If no, then λ is in the residual spectrum. If yes,

3. Is the inverse of $L - \lambda$ on its range a continuous operator? If no, λ is in the continuous spectrum. If yes, λ is in the resolvent set.

We make the following observation: The value λ is in the residual spectrum of L only if $\bar{\lambda}$ is in the discrete spectrum of L^*. To prove this, recall from the Fredholm Alternative theorem, that $H = \bar{R}(L) + N(L^*)$, that is, any vector in the Hilbert space H is an orthogonal linear combination of an element in $\bar{R}(L)$, the closure of the range of L, and $N(L^*)$, the null space of L^*. Thus the range of $L - \lambda$ (or its closure, if L is not closed) is a proper subset of H if and only if the null space of $L^* - \bar{\lambda}$ is not empty, so that there is a vector $v \neq 0$ with $L^*v = \bar{\lambda}v$, i.e., $\bar{\lambda}$ belongs to the point spectrum of L^*. Consequently, if L is a self-adjoint operator, it has no residual spectrum.

In previous discussions, the point spectrum of an operator L was adequate to determine a basis for the Hilbert space, but it is possible that the continuous spectrum may also play a role. Just what this role is will be illustrated in the next sections.

7.2   Fourier Transforms

7.2.1   Transform Pairs

There is an interesting relationship between Green's functions, eigenvalue expansions and contour integration in the complex plane. Suppose we have an operator L with a complete set of orthonormal eigenfunctions $\{\phi_k\}$ where $L\phi_k = \lambda_k \phi_k$. For every u in the Hilbert space, $u = \sum_{k=1}^{\infty} \alpha_k \phi_k$ where $\alpha_k = \langle u, \phi_k \rangle$. For this u, $Lu = \sum_{k=1}^{\infty} \alpha_k \lambda_k \phi_k$, if Lu makes sense. Furthermore, $L^2 u = L(Lu) = \sum_{k=1}^{\infty} \alpha_k \lambda_k^2 \phi_k$ and obviously, $L^{(n)} u = \sum_{k=1}^{\infty} \alpha_k \lambda_k^n \phi_k$, again, provided these all make sense. If $f(x)$ is a polynomial, one can define the operator $f(L)$ by $f(L)u = \sum_{k=1}^{\infty} \alpha_k f(\lambda_k) \phi_k$ and finally, if f is an analytic function, the same definition makes good sense, namely

$$f(L)u = \sum_{k=1}^{\infty} \alpha_k f(\lambda_k) \phi_k.$$

If we wish to invert the operator $L - \lambda$ for some complex number λ, we can identify $f(x) = \frac{1}{x - \lambda}$ and the operator $(L - \lambda)^{-1} = f(L)$ is, by our previous definition,

$$(L - \lambda)^{-1} u = \sum_{k=1}^{\infty} \frac{\alpha_k \phi_k}{\lambda_k - \lambda}.$$

Of course, we already know that, if L is a differential operator, the inverse operator $(L - \lambda)^{-1}$ is represented by the Green's function, say

$$(L - \lambda)^{-1} u = \int G(x, \xi; \lambda) u(\xi) d\xi$$

where λ is treated as a parameter, and integration is taken over the appropriate spatial domain. Notice that this implies that

$$-\int G(x, \xi; \lambda) u(\xi) d\xi = \sum_{k=1}^{\infty} \frac{\alpha_k \phi_k}{\lambda - \lambda_k}.$$

If both of these expressions are defined for λ in the complex plane, we can integrate counterclockwise with respect to λ around a large circle, and let R, the radius of the circle, go to infinity. The result is

$$\int_{C_\infty} \left(\int G(x, \xi; \lambda) u(\xi) d\xi \right) = -\int_{C_\infty} \sum_{k=1}^{\infty} \frac{\alpha_k \phi_k}{\lambda - \lambda_k} d\lambda$$

$$= -2\pi i \sum_{k=1}^{\infty} \alpha_k \phi_k = -2\pi i u(x),$$

because of the residue theorem. Apparently,

$$u(x) = \frac{-1}{2\pi i} \int_{C_\infty} \left(\int G(x,\xi;\lambda) u(\xi) d\xi \right) d\lambda.$$

Here we have an interesting statement. Namely, there is some function which when integrated against reasonable functions $u(\xi)$ reproduces $u(x)$. We know from the theory of distributions that this cannot occur for any honest function, but it must be that

$$-\frac{1}{2\pi i} \int_{C_\infty} G(x,\xi;\lambda) d\lambda = \delta(x-\xi)$$

interpreted in the sense of distribution.

This rather curious formula has a most powerful interpretation, namely it shows how to get representations of the "δ-function" by knowing only the Green's function for the differential operator of interest. As we shall see, this generates directly the transform pair appropriate to the operator L.

To see how this formula works to generate transform pairs, we examine the familiar operator $Lu = -u''$ with boundary conditions $u(0) = u(1) = 0$. We calculate the Green's function for $(L - \lambda)u$ to be

$$G(x,\xi;\lambda) = \begin{cases} \dfrac{\sin \sqrt{\lambda} x \; \sin \sqrt{\lambda}(1-\xi)}{\sqrt{\lambda} \sin \sqrt{\lambda}} & , \quad 0 \le x < \xi \le 1, \\[3mm] \dfrac{\sin \sqrt{\lambda} \xi \; \sin \sqrt{\lambda}(1-x)}{\sqrt{\lambda} \sin \sqrt{\lambda}} & , \quad 0 \le \xi < x \le 1. \end{cases}$$

To evaluate $\int_{C_\infty} G(x,\xi;\lambda) d\lambda$, observe that there are simple poles at $\lambda = n^2\pi^2$ for $n = 1, 2, \ldots$. The point $\lambda = 0$ is not a pole or branch point since for small λ,

$$G(x,\xi;\lambda) \sim \begin{cases} x(1-\xi) & 0 \le x < \xi \le 1 \\[2mm] \xi(1-x) & 0 \le \xi < x \le 1. \end{cases}$$

Applying the residue theorem, we obtain that

$$\delta(x-\xi) = -\frac{1}{2\pi i} \int_{C_\infty} G(x,\xi;\lambda) d\lambda = 2 \sum_{k=1}^{\infty} \sin k\pi x \sin k\pi \xi.$$

This representation of the delta function is equivalent to the sine Fourier series since for any smooth f with compact support in the interval $[0,1]$,

$$f(x) = \int_0^1 \delta(x-\xi) f(\xi) d\xi$$

$$= \sum_{k=1}^{\infty} \sin k\pi x \left(2 \int_0^1 f(\xi) \sin k\pi\xi\, d\xi \right)$$

$$= \sum_{k=1}^{\infty} \alpha_k \sin k\pi x$$

where $\alpha_k = 2 \int_0^1 f(\xi) \sin k\pi\xi\, d\xi$.

It is a straightforward calculation to extend the representation of the delta function to Sturm-Liouville operators

$$Lu - \lambda u = - \left(\frac{1}{\omega}(pu')' + qu + \lambda u \right).$$

We assume that the boundary conditions correspond to a self-adjoint operator, and that the eigenfunctions satisfy

$$L\phi_k - \lambda_k \phi_k = 0$$

where $Lu = -\frac{1}{\omega}(pu')' - qu$, and are complete and orthonormal with respect to the inner product with weight function $\omega > 0$. Since the functions ϕ_k are complete, we can represent the Green's function as $G(x,\xi;\lambda) = \sum_{k=1}^{\infty} \alpha_k(\xi)\phi_k(x)$, where $\alpha_k(\xi)$ are the Fourier coefficients $\alpha_k = \langle \phi_k, G \rangle$. Since G is defined by

$$LG - \lambda G = \delta(x - \xi)$$

then

$$\alpha_k(\xi) = \langle \phi_k, G \rangle = \frac{\phi_k(\xi)\omega(\xi)}{\lambda_k - \lambda}$$

so that

$$G(x,\xi;\lambda) = \sum_{k=1}^{\infty} \frac{\phi_k(\xi)\phi_k(x)\omega(\xi)}{\lambda_k - \lambda}.$$

It follows that

$$\frac{1}{2\pi i} \int_{C_\infty} G(x,\xi;\lambda)d\lambda = - \sum_{k=1}^{\infty} \phi_k(\xi)\phi_k(x)\omega(\xi)$$

where the contour integration is taken in the counterclockwise direction around the infinitely large circle C_∞. Formally, we know that

$$\delta(x - \xi) = \sum_{k=1}^{\infty} \phi_k(x)\phi_k(\xi)\omega(\xi)$$

so that,

$$\frac{1}{2\pi i} \int_{C_\infty} G(x,\xi;\lambda)d\lambda = -\delta(x - \xi),$$

which is the formula we sought.

As an illustration of how to use this formula, consider the equation

$$-\frac{1}{x}(xG')' - \lambda G = \delta(x - \xi)$$

with boundary conditions $G'(0) = 0$, $G(1) = 0$. Recalling that the solutions of the homogeneous equation are the Bessel functions $J_0(\sqrt{\lambda}x)$ and $Y_0(\sqrt{\lambda}x)$, we can write down that

$$G(x, \xi; \lambda) = \begin{cases} \dfrac{-J_0(\sqrt{\lambda}x)V(\sqrt{\lambda}\xi)}{\sqrt{\lambda}W\left(J_0(\sqrt{\lambda}\xi), V(\sqrt{\lambda}\xi)\right)} & 0 \le x < \xi \le 1 \\[3mm] \dfrac{-J_0(\sqrt{\lambda}\xi)V(\sqrt{\lambda}x)}{\sqrt{\lambda}W\left(J_0(\sqrt{\lambda}\xi), V(\sqrt{\lambda}\xi)\right)} & 0 \le \xi < x \le 1 \end{cases}$$

where

$$V(\sqrt{\lambda}x) = Y_0(\sqrt{\lambda})J_0(\sqrt{\lambda}x) - Y_0(\sqrt{\lambda}x)J_0(\sqrt{\lambda}),$$

and $W(u, v) = uv' - vu'$ is the Wronskian of its two arguments.

We must first calculate the Wronskian

$$W\left(J_0(\sqrt{\lambda}z), V(\sqrt{\lambda}z)\right) = -J_0(\sqrt{\lambda})W\left(J_0(\sqrt{\lambda}z), Y_0(\sqrt{\lambda}z)\right)$$

Recall from the definition of $Y_\nu(z)$ (Chapter 6) that

$$Y_\nu(z) = \frac{J_\nu(z)\cos\nu\pi - J_{-\nu}(z)}{\sin\nu\pi}$$

so that $W\left(J_\nu(z), Y_\nu(z)\right) = \frac{1}{\sin\nu\pi}W(J_{-\nu}, J_\nu)$. We know that for any two solutions u, v of Bessel's equation, $W(u, v) = k/z$. To calculate the correct value of k, we note that near $z = 0$, $J_\nu(z) \sim \left(\frac{z}{2}\right)^\nu \frac{1}{\Gamma(\nu+1)}$ so that

$$W\left(J_{-\nu}(z), J_\nu(z)\right) = \frac{2\nu}{z\Gamma(1 + \nu)\Gamma(1 - \nu)} = \frac{2}{\pi z}\sin\pi\nu$$

using an identity for the gamma function (Chapter 6).

It follows that

$$G(x, \xi; \lambda) = \begin{cases} \dfrac{\pi J_0(\sqrt{\lambda}x)V(\sqrt{\lambda}\xi)\xi}{2J_0(\sqrt{\lambda})}, & 0 \le x < \xi \le 1, \\[3mm] \dfrac{\pi J_0(\sqrt{\lambda}\xi)V(\sqrt{\lambda}x)\xi}{2J_0(\sqrt{\lambda})}, & 0 \le \xi < x \le 1. \end{cases}$$

To calculate the integral $\frac{1}{2\pi i}\int_{C_\infty} G(x, \xi; \lambda)d\lambda$, we use that there are an infinite number of positive real roots of $J_0(\mu_k) = 0$, so that for $\lambda = \mu_k^2$ the

Green's function has simple poles. To evaluate the residue at these poles, we note that $W\left(J_\nu(z), Y_\nu(z)\right) = \frac{2}{\pi z}$, and if $\lambda = \mu_k^2 = \lambda_k$, then $J_0'(\sqrt{\lambda_k})Y_0(\sqrt{\lambda_k}) = \frac{-2}{\pi\sqrt{\lambda_k}}$, and $V(\sqrt{\lambda}x) = Y_0(\sqrt{\lambda_k})J_0(\sqrt{\lambda}x)$. It follows that

$$\delta(x - \xi) = 2\sum_{k=1}^{\infty} \frac{J_0\left(\sqrt{\lambda_k}\xi\right) J_0\left(\sqrt{\lambda_k}x\right) x}{\left[J_0'\left(\sqrt{\lambda_k}\right)\right]^2}.$$

This representation of the delta function induces the transform based on Bessel functions

$$f(x) = \sum_{k=1}^{\infty} \alpha_k J_0\left(\mu_k x\right),$$

$$\alpha_k = \frac{2}{\left[J_0'\left(\mu_k\right)\right]^2} \int_0^1 f(x)x J_0\left(\mu_k x\right) dx.$$

The beauty of the formula relating the contour integral of the Green's function and the delta function is that it gives us an algorithm to generate many different transforms, even those for which the usual eigenfunction expansion techniques do not work. Of course, if we use this formula as an algorithm, it behooves us to come up with an independent proof of the validity of the derived transform pair.

As an example of a transform pair for an operator that does not have eigenfunctions, consider the Green's function

$$-G'' - \lambda G = \delta(x - \xi)$$

with $G(0) = 0$ and G in $L^2[0,\infty)$. If λ is not a positive real number, we calculate directly that

$$G(x,\xi;\lambda) = \begin{cases} \dfrac{\sin\sqrt{\lambda}x e^{i\sqrt{\lambda}\xi}}{\sqrt{\lambda}}, & 0 \le x < \xi < \infty, \\[2ex] \dfrac{\sin\sqrt{\lambda}\xi e^{i\sqrt{\lambda}x}}{\sqrt{\lambda}}, & 0 \le \xi < x < \infty. \end{cases}$$

Notice that $e^{i\sqrt{\lambda}x}$ is exponentially decaying as $x \to \infty$ if $\text{Im}(\sqrt{\lambda}) > 0$. We want to evaluate $\frac{1}{2\pi i}\int_{C_\infty} G(x,\xi;\lambda)d\lambda$ around the large circle C_∞ in the counterclockwise direction. However, there is a branch point at $\lambda = 0$ and therefore, a branch cut along the positive real axis. Since $G(x,\xi;\lambda)$ is analytic in λ everywhere else, we can deform C_∞ down to a contour C_2 which traverses in from $\text{Re}(\lambda) = \infty$ along the branch cut with $\text{Im}(\lambda)$ slightly positive and then out to $\text{Re}(\lambda) = \infty$ along the branch cut with $\text{Im}(\lambda)$ slightly negative. We make the change of variables $\lambda = \nu^2$ so that the branch cut is "unfolded" and our path of integration becomes the real axis from $\operatorname{Re}\mu = \infty$ to $\operatorname{Re}\mu = -\infty$. Notice that $\frac{\sin\sqrt{\lambda}x}{\sqrt{\lambda}}$ and $\cos\sqrt{\lambda}x$ are entire functions, and the integral of an entire function around any closed path is zero. As a result,

$$G(x,\xi;\lambda) = i\frac{\sin\sqrt{\lambda}x\sin\sqrt{\lambda}\xi}{\sqrt{\lambda}} + \text{an entire function}$$

for all x,ξ. Thus we evaluate

$$\frac{1}{2\pi i}\int_{C_\infty} G(x,\xi;\lambda)d\lambda = \frac{1}{2\pi}\int_{C_2}\frac{\sin\sqrt{\lambda}x\sin\sqrt{\lambda}\xi d\lambda}{\sqrt{\lambda}}$$

$$= \frac{-1}{\pi}\int_{-\infty}^{\infty}\sin\mu x\sin\mu\xi d\mu = -\delta(x-\xi),$$

so that

$$\delta(x-\xi) = \frac{2}{\pi}\int_0^\infty \sin\mu x\sin\mu\xi d\mu.$$

This representation is equivalent to the transform pair

$$\phi(x) = \frac{2}{\pi}\int_0^\infty \hat\phi(\mu)\sin\mu x d\mu$$

$$\hat\phi(\mu) = \int_0^\infty \phi(x)\sin\mu x dx,$$

and is called the FOURIER SINE INTEGRAL TRANSFORM.

We can verify the validity of the Fourier Sine transform directly for any function ϕ which is continuously differentiable and has compact support. Consider first the integral

$$\int_{-\infty}^{\infty}\left(\int_{-\infty}^{\infty}e^{i\mu x}d\mu\right)\phi(x)dx = \lim_{R\to\infty}\int_{-\infty}^{\infty}\left(\int_{-R}^{R}e^{i\mu x}d\mu\right)\phi(x)dx$$

$$= 2\pi\lim_{R\to\infty}\int_{-\infty}^{\infty}\phi(x)\frac{\sin Rx}{\pi x}dx.$$

Recall from Chapter 4 that $S_k(x) = \frac{\sin kx}{\pi x}$ is a delta sequence for Lipschitz continuous functions. As a result,

$$\int_{-\infty}^{\infty}\left(\int_{-\infty}^{\infty}e^{i\mu x}d\mu\right)\phi(x)\,dx = 2\pi\phi(0)$$

so that

$$\frac{1}{2\pi}\int_{-\infty}^{\infty}e^{i\mu x}d\mu = \delta(x)$$

is a representation of the delta function. It follows that

$$\frac{1}{\pi i}\int_{-\infty}^{\infty}\sin(\mu x)e^{i\mu\xi}d\mu = \frac{-1}{2\pi}\int_{-\infty}^{\infty}\left(e^{i\mu(x+\xi)}-e^{i\mu(\xi-x)}\right)d\mu$$

$$= \delta(x-\xi)-\delta(x+\xi).$$

Since the domain of the differential operator was defined to be $L^2[0,\infty)$ with $u(0)=0$, test functions $\phi(x)$ have support on $[0,\infty)$ and hence are identically zero for $x<0$. For these test functions $\delta(x+\xi)\equiv 0$, and the validity of the transform follows.

We leave it as an exercise for the reader to verify that the FOURIER-COSINE INTEGRAL TRANSFORM

$$F(\mu) = \int_0^\infty f(x)\cos\mu x\,dx$$

$$f(x) = \frac{2}{\pi}\int_0^\infty F(\mu)\cos\mu x\,d\mu$$

arises from the delta function representation for $Lu=-u''$ with $u'(0)=0$, u in $L^2[0,\infty)$.

The FOURIER INTEGRAL TRANSFORM is one of the most important transform pairs. It can be derived from the operator $Lu=-u''$ defined on $L^2(-\infty,\infty)$ using the Green's function G defined by $-G''-\lambda G=\delta(x-\xi)$. In Chapter 4 we found that this Green's function is given by

$$G(x,\xi;\lambda) = \frac{-e^{i\sqrt{\lambda}|x-\xi|}}{2i\sqrt{\lambda}}.$$

We evaluate

$$-\frac{1}{2\pi i}\int_{C_\infty}G(x,\xi;\lambda)d\lambda = \frac{1}{2\pi}\int_{-\infty}^\infty e^{i\mu|x-\xi|}d\mu = \delta(x-\xi)$$

which agrees with our earlier calculation. This representation leads to the Fourier Transform Theorem.

Theorem 7.1

If $f(t)$ is piecewise continuously differentiable and is in $L^1(-\infty,\infty)$, then

$$f(x) = \frac{1}{2\pi}\int_{-\infty}^\infty e^{-i\mu x}\hat{f}(\mu)d\mu$$

$$\hat{f}(\mu) = \int_{-\infty}^\infty e^{i\mu x}f(x)dx.$$

The function $\hat{f}(\mu)$ is called the Fourier transform of $f(x)$. We want $f(x)$ to be in $L^1(-\infty, \infty)$ since then we are assured of the existence of the transform $\hat{f}(\mu)$ for real valued arguments μ. However, it is not hard to see that this classical theorem is not nearly general enough to cover all of the possible transformable functions.

Consider, for example, the function $g(t) = t^{-1/2}H(t)$. This function is singular at the origin and is not integrable on the entire real line. Nonetheless, we can define the Fourier transform of $g(t)$ by

$$\hat{g}(\mu) = \int_{-\infty}^{\infty} g(t)e^{i\mu t}dt = \int_0^{\infty} e^{i\mu t}t^{-1/2}dt.$$

Using contour integral methods (in particular, Jordan's Lemma) this integral can be evaluated as

$$\hat{g}(\mu) = \begin{cases} \left(e^{i\pi/4}\right)\left(\frac{\pi}{\mu}\right)^{1/2}, & \mu > 0 \\[2mm] \left(e^{-i\pi/4}\right)\left(\frac{\pi}{|\mu|}\right)^{1/2}, & \mu < 0 \end{cases}$$

or, more compactly as

$$\hat{g}(\mu) = \left(\frac{\pi}{2|\mu|}\right)^{1/2}\left(1 + \frac{i\mu}{|\mu|}\right), \quad \mu \neq 0.$$

We can find the inverse transform of $\hat{g}(\mu)$ by direct computation of the integral

$$\frac{1}{2\pi}\int_{-\infty}^{\infty}\hat{g}(\mu)e^{-i\mu t}d\mu$$

$$= \frac{1}{2\sqrt{\pi}}\left[e^{-i\pi/4}\int_{-\infty}^0 \frac{e^{-i\mu t}}{\sqrt{|\mu|}}d\mu + e^{i\pi/4}\int_0^{\infty}\frac{e^{-i\mu t}}{\sqrt{|\mu|}}d\mu\right]$$

$$= \frac{1}{2\sqrt{2|t|}}\left\{e^{-i\pi/4}\left(1 + \frac{it}{|t|}\right) + e^{i\pi/4}\left(1 - \frac{it}{|t|}\right)\right\} = \frac{1}{\sqrt{|t|}}H(t).$$

In other words, the Fourier Integral Theorem gives a correct result even though the function $g(t)$ is not a piecewise continuously differentiable L^1 function.

If we are willing to generalize the Fourier Transform formulae slightly, the Fourier Transform can be applied to an even larger class of functions. Consider, for example, the function $f(t) = t^{1/2}H(t)$. We define the transform of $f(t)$

$$\hat{f}(\mu) = \int_{-\infty}^{\infty} e^{i\mu t}f(t)dt$$

in the complex μ-plane, with $\operatorname{Im}\mu > 0$. (Up until now we have taken μ to be a real number). If we denote $\mu = \rho e^{i\theta}$ where $0 < \theta < \pi$, then, using Jordan's lemma, we are permitted to make the change of variables $t = re^{i(\pi/2-\theta)}$ with the result that

$$\hat{f}(\mu) = \frac{\sqrt{\pi}e^{\frac{3\pi i}{4}}}{2\mu^{3/2}} \quad , \quad \text{provided } \operatorname{Im}\mu > 0.$$

However, this expression is an analytic function in the entire complex plane, excluding, of course, the branch point at $\mu = 0$, so we take the transform to be

$$\hat{f}(\mu) = \frac{\sqrt{\pi}e^{\frac{3\pi i}{4}}}{2\mu^{3/2}}$$

defined by analytic continuation, everywhere in the complex plane except at $\mu = 0$.

To find the inverse transform of $\hat{f}(\mu)$, observe that for any real number $\alpha > 0$, if $g(t) = e^{-\alpha t}f(t)$ is a transformable function, then the transforms $\hat{g}(\mu)$ and $\hat{f}(\mu)$ are related by $\hat{g}(\mu) = \hat{f}(\mu + i\alpha)$. If the Fourier Transform Theorem applies to $\hat{g}(\mu)$, then

$$g(t) = \frac{1}{2\pi}\int_{-\infty}^{\infty} e^{-i\mu t}\hat{g}(\mu)d\mu$$

so that

$$\begin{aligned}
f(t) &= \frac{1}{2\pi}\int_{-\infty}^{\infty} e^{-i(\mu+i\alpha)t}\hat{f}(\mu + i\alpha)d\mu \\
&= \frac{1}{2\pi}\int_{-\infty+i\alpha}^{+\infty+i\alpha} e^{-i\mu t}\hat{f}(\mu)d\mu = \frac{1}{2\pi}\int_C e^{-i\mu t}\hat{f}(\mu)d\mu
\end{aligned}$$

where the contour C is any contour in the complex plane which parallels and lies above the real axis. The moral is that the Fourier Transform still works provided the contour of integration for the inversion lies above the singularity of $\hat{f}(\mu)$ at $\mu = 0$.

A similar story is true for the function $h(t) = t^{1/3}e^{i\beta t}H(-t)$, β real. Since $h(t)$ is nonzero for $t < 0$, we can define the Fourier Transform $\hat{h}(\mu)$ provided $\operatorname{Im}\mu < 0$, in which case

$$\hat{h}(\mu) = \int_{-\infty}^{0} e^{i(\mu+\beta)t}t^{1/3}dt = \frac{e^{\frac{2\pi i}{3}}}{(\mu+\beta)^{4/3}}\Gamma(4/3).$$

Obviously, this expression is analytic except at the branch point $\mu = -\beta$, so, using analytic continuation, we are able to define $\hat{h}(\mu)$ on the entire complex plane.

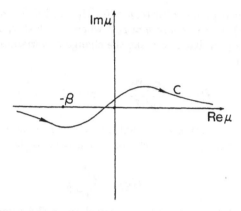

FIGURE 7.1 Contour of integration for the inverse Fourier Transform for $k(t) = t^{1/2}H(t) + t^{1/3}e^{i\beta t}H(-t)$.

Now, suppose we let $g(t) = e^{-\gamma t}h(t)$ for some real $\gamma < 0$. The transforms $\hat{g}(\mu)$, $\hat{h}(\mu)$ are related by $\hat{g}(\mu) = \hat{h}(\mu + i\gamma)$. It follows that

$$h(t) = \frac{1}{2\pi}\int_{-\infty}^{\infty} e^{-i(\mu+i\gamma)t}\hat{h}(\mu + i\gamma)d\mu = \frac{1}{2\pi}\int_{C} e^{-i\mu t}\hat{h}(\mu)d\mu$$

on any contour C which parallels and lies below the real axis. Of course, since $\hat{h}(\mu)$ is analytic, we can deform the contour C into a number of equivalent paths.

As a final example, consider the function $k(t) = t^{1/2}H(t) + t^{1/3}e^{i\beta t}H(-t)$ which is the sum of the previous two functions. The obvious choice for the transform is the sum $\hat{k}(\mu) = \hat{f}(\mu) + \hat{h}(\mu)$ provided $\mu \neq 0, -\beta$, although there is no way to write down a single integral transform that yields this result.

The definition of the inverse is also clear, although again, its derivation is indirect. It is apparent that the inversion formula

$$k(t) = \frac{1}{2\pi}\int_{C} e^{i\mu t}\hat{k}(\mu)d\mu$$

holds provided the contour C is chosen parallel to the real axis for large $|\mu|$, but it must pass above the singularity at $\mu = 0$ and below the singularity at $\mu = -\beta$ (see Figure 7.1).

There is a relationship between a function $f(x)$ and its Fourier Transform $F(\mu)$ analogous to Parseval's identity for Fourier series. The identity is

$$\int_{-\infty}^{\infty} |f(x)|^2 dx = \frac{1}{2\pi}\int_{-\infty}^{\infty} |F(\mu)|^2 d\mu,$$

and is known as PLANCHEREL'S EQUATION. This identity (which is easily verified for functions in L^2 by direct calculation) leads to an interesting "uncertainty principle."

Theorem 7.2

Suppose $F(\mu)$ is the Fourier transform of $f(x)$. If

$$E_x = \frac{\int_{-\infty}^{\infty} x^2 |f(x)|^2 dx}{\int_{-\infty}^{\infty} |f(x)|^2 dx}$$

and

$$E_\mu = \frac{\int_{-\infty}^{\infty} \mu^2 |F(\mu)|^2 d\mu}{\int_{-\infty}^{\infty} |F(\mu)|^2 d\mu}$$

exist, then $E_x E_\mu \geq \frac{1}{4}$.

The interpretation of this inequality as an uncertainty principle stems from the observation that E_x and E_μ correspond to second moments of the functions $|f(x)|^2$ and $|F(\mu)|^2$, respectively, and therefore measure the spread or bandwidth of the respective signals. Thus, if a function $f(x)$ is very narrowly concentrated, its transform must be broadly spread. We cannot simultaneously concentrate a function and its spectral frequency.

The proof of this inequality uses the Schwarz inequality and Plancherel's identity. By the Schwarz inequality

$$\left| \int_{-\infty}^{\infty} x\, f(x) f'(x) dx \right|^2 \leq \int_{-\infty}^{\infty} |x\, f(x)|^2 dx \int_{-\infty}^{\infty} |f'(x)|^2 dx.$$

Integrating by parts, we learn that

$$\int_{-\infty}^{\infty} x\, f(x) f'(x) dx = \lim_{\xi \to \infty} \frac{x}{2} f^2(x) \Big|_{-\xi}^{\xi} - \frac{1}{2} \int_{-\infty}^{\infty} f^2(x) dx$$

and we assume that $x\, f^2(x)$ goes to zero as $x \to \pm\infty$. If $F(\mu)$ is the transform of $f(x)$, then $-i\mu F(\mu)$ is the transform of $f'(x)$. So our original inequality becomes (use Plancherel's identity)

$$\frac{1}{4} \int_{-\infty}^{\infty} |f(x)|^2 dx \cdot \int_{-\infty}^{\infty} |F(\mu)|^2 d\mu \leq \int_{-\infty}^{\infty} |x f(x)|^2 dx \int_{-\infty}^{\infty} |\mu F(\mu)|^2 d\mu$$

from which it follows that $E_x E_\mu \geq \frac{1}{4}$.

No discussion of Fourier Transforms is complete without mention of the Convolution Theorem

Theorem 7.3

If the functions $f(x)$, $g(x)$ have Fourier Transforms $F(\mu)$, $G(\mu)$, respectively, then the Fourier transform of

$$\int_{-\infty}^{\infty} g(\xi)f(x-\xi)d\xi \quad \text{is} \quad F(\mu)G(\mu).$$

Similarly, the inverse Fourier transform of

$$\frac{1}{2\pi}\int_{-\infty}^{\infty} F(s)G(\mu-s)ds \quad \text{is} \quad f(x)g(x).$$

Proof

We calculate the inverse Fourier Transform of $F(\mu)G(\mu)$ to be

$$\frac{1}{2\pi}\int_{-\infty}^{\infty} F(\mu)G(\mu)e^{-i\mu x}d\mu$$

$$= \frac{1}{2\pi}\int_{-\infty}^{\infty} F(\mu)e^{-i\mu x}\left(\int_{-\infty}^{\infty} g(\xi)e^{i\mu\xi}d\xi\right)d\mu$$

$$= \frac{1}{2\pi}\int_{-\infty}^{\infty} g(\xi)\left(\int_{-\infty}^{\infty} F(\mu)e^{-i\mu(x-\xi)}d\mu\right)d\xi$$

$$= \int_{-\infty}^{\infty} g(\xi)f(x-\xi)d\xi.$$

Similarly, the Fourier transform of $f(x)g(x)$ is

$$\int_{-\infty}^{\infty} f(x)g(x)e^{i\mu x}dx = \int_{-\infty}^{\infty} g(x)e^{i\mu x}\left(\frac{1}{2\pi}\int_{-\infty}^{\infty} F(\omega)e^{-i\omega x}d\omega\right)dx$$

$$= \int_{-\infty}^{\infty} F(\omega)\left(\int_{-\infty}^{\infty} g(x)e^{i(\mu-\omega)x}dx\right)d\omega$$

$$= \frac{1}{2\pi}\int_{-\infty}^{\infty} F(\omega)G(\mu-\omega)d\omega.$$

Some texts do not have the factor $\frac{1}{2\pi}$ for this latter result because their definition of the Fourier transform is

$$F(\mu) = \frac{1}{\sqrt{2\pi}}\int_{-\infty}^{\infty} f(x)e^{i\mu x}dx$$

$$f(x) = \frac{1}{\sqrt{2\pi}}\int_{-\infty}^{\infty} F(\mu)e^{-i\mu x}d\mu.$$

One motivation for the latter definition is that with it the improper eigenfunctions $e^{ikx}/\sqrt{2\pi}$ are normalized. How one chooses to include the scale factor of $1/2\pi$ is a matter of taste rather than substance, although as we see here, it does influence the statement of the convolution theorem, and care must be taken to get this factor correct.

All of the transforms introduced in Chapter 2 have convolution theorems (see, for example, problems 2.2.12, 13, 20). The usefulness of the convolution theorem is illustrated by the following examples.

Example

Solve $-u'' + a^2 u = f(x)$ where $f(x) \in L^2(-\infty, \infty)$.

We already know how to solve this problem using Green's functions and the techniques introduced in Chapter 4. However, we can also use Fourier Transforms to solve this problem rather easily.

We let $U(\mu)$ and $F(\mu)$ be the Fourier transforms of u and f, respectively. We transform the differential equation by multiplying by $e^{i\mu x}$ and integrating over $-\infty < x < \infty$. After integrating by parts we find

$$(\mu^2 + a^2)U = F$$

so that $U = \frac{F}{\mu^2 + a^2}$. Since $U(\mu)$ is the product of two transforms, we invoke the convolution theorem. The inverse transform of

$$G(\mu) = \frac{1}{\mu^2 + a^2} \text{ is } g(x) = \frac{e^{-a|x|}}{2a}.$$

It follows that

$$u(x) = \int_{-\infty}^{\infty} \frac{e^{-a|x-t|}}{2a} f(t)dt$$

solves the differential equation.

Example

Solve the integral equation

$$\int_{-\infty}^{\infty} k(x-y)u(y)dy - \lambda u(x) = f(x).$$

This equation is a Fredholm integral equation with a convolution kernel. It is natural to take the Fourier Transform of this equation. We define $U(\mu)$, $K(\mu)$ and $F(\mu)$ to be the Fourier transforms of $u(x)$, $k(x)$ and $f(x)$, respectively. The transform of the integral equation is

$$(K(\mu) - \lambda)U = F$$

so that

$$u(x) = \frac{1}{2\pi} \int_{-\infty}^{\infty} \frac{F(\mu)e^{-i\mu x}}{K(\mu) - \lambda} d\mu.$$

As a specific example, if we take $k(x) = \frac{1}{2}\frac{x}{|x|}$ and $\lambda = -1$, then since $K(\mu) = \frac{1}{i\mu}$,

$$u(x) = \frac{1}{2\pi} \int_{-\infty}^{\infty} \frac{i\mu}{1 + i\mu} F(\mu)e^{-i\mu x} d\mu.$$

Applying the convolution theorem once again (Problem 7.2.8), we learn that

$$u(x) = \int_{-\infty}^{x} e^{y-x} f'(y) dy.$$

The Fourier Transform theorem is an important example of how spectral theory works when the operator has no discrete spectrum. What we see from the transform is that functions are represented as continuous sums (that is, integrals), of basis functions, but in this case the basis functions $e^{\pm i\sqrt{\lambda}x}$ are not eigenfunctions since they are not square integrable. The functions

$$e^{\pm i\sqrt{\lambda}x}$$

are called IMPROPER EIGENFUNCTIONS and the values λ with $\text{Re}\,\lambda > 0$, $\text{Im}\,\lambda = 0$, are called IMPROPER EIGENVALUES.

In fact, the values λ, with $\text{Re}\,\lambda > 0$, $\text{Im}\,\lambda = 0$ are continuous spectrum for the operator $Lu = -u''$ with $-\infty < x < \infty$. To see this we must show that if λ is real and positive the inverse of $Lu - \lambda u = -u'' - \lambda u$ exists but is unbounded. It is certainly apparent from the Green's function

$$g(x, \xi; \lambda) = \frac{-e^{i\sqrt{\lambda}|x-\xi|}}{2i\sqrt{\lambda}}$$

that the inverse exists, at least for $\lambda \neq 0$. To show that the inverse is unbounded we will find a sequence of square integrable functions v_n and solutions of $(L - \lambda)u_n = v_n$ for which $\|u_n\|/\|v_n\|$ is unbounded as $n \to \infty$.

If L has an improper eigenfunction $u(x)$ we can always construct such sequences. We take

$$u_n(x) = \begin{cases} u(x) & -a_n < x < a_n \\ f_n(x) & a_n < |x| < b_n \\ 0 & |x| > b_n \end{cases}$$

with $\lim_{n\to\infty} a_n = \infty$, and $f_n(x)$ chosen to be any C^2 function defined on $a_n \leq |x| \leq b_n$ with $f_n(\pm a_n) = u(\pm a_n)$, $f_n(\pm b_n) = 0$ with $v_n = (L - \lambda)f_n$ satisfying $\int_{a_n}^{b_n} |v_n|^2 dx < K$ for all n. If this can be done, then λ is in the continuous spectrum.

To see how this works in a specific example, consider the operator $Lu = -d^2u/dx^2$ on the infinite interval $-\infty < x < \infty$. An improper eigenfunction is $u(x) = \cos\sqrt{\lambda}x$ so we take (for $\lambda \neq 0$)

$$u_n(x) = \begin{cases} \cos\sqrt{\lambda}x & |x| < \frac{2n\pi}{\sqrt{\lambda}} \\ f\left(|x| - \frac{2n\pi}{\sqrt{\lambda}}\right) & 0 < |x| - \frac{2n\pi}{\sqrt{\lambda}} < 1 \\ 0 & |x| > 1 + \frac{2n\pi}{\sqrt{\lambda}} \end{cases}$$

where $f(x)$ is any C^2 function defined for $0 \leq x \leq 1$ with the property that $f(0) = 1$, $f'(0) = f'(1) = f(1) = 0$. Certainly, many such functions $f(x)$ can be constructed. The important observation is that $v_n = Lu_n - \lambda u_n$ has norm

$$\begin{aligned} \|v_n\|^2 &= \int_{-\infty}^{\infty} v_n^2(x)dx \\ &= 2\int_0^1 \left(-f''(x) - \lambda f(x)\right)^2 dx \end{aligned}$$

independent of n. On the other hand,

$$\|u_n\|^2 \geq \frac{2n\pi}{\sqrt{\lambda}},$$

so that $\|u_n\|^2/\|v_n\|^2$ is unbounded as $n \to \infty$. A similar argument can be used when $\lambda = 0$. In either case, we have that if λ is real, with $\lambda \geq 0$, then λ is in the continuous spectrum of the operator.

We now recognize that the Fourier Transform Theorem is a spectral representation whereby functions in L^2 can be represented as integrals over all points of the continuous spectrum of the operator $Lu = -u''$ with basis functions the properly normalized improper eigenfunctions. For the Fourier Transform the improper eigenfunctions have multiplicity two whereas for the Sine or Cosine Transforms the multiplicity is one.

7.2.2 Completeness of Hermite and Laguerre Polynomials

The Fourier Transform can be used to prove an important fact about the Hermite and Laguerre polynomials. In Chapter 2 we proved that orthogonal polynomials form a complete set whenever they are defined on a space with finite domain. The proof does not, however, apply to the Hermite and Laguerre polynomials.

We want to show that the Hermite polynomials form a complete set of functions in the space of all functions $f(x)$ with $\int_{-\infty}^{\infty} e^{-x^2/2} f^2(x)dx < \infty$. To do this we demonstrate that if

$$\int_{-\infty}^{\infty} e^{-x^2/2} x^n f(x) dx = 0 \quad \text{for} \quad n = 0, 1, 2, \ldots$$

then $f(x)$ is the zero function and this is sufficient to guarantee completeness (see Chapter 2).

The function

$$F(z) = \int_{-\infty}^{\infty} e^{izx} e^{-x^2/2} f(x) dx$$

is the Fourier transform of the function $e^{-x^2/2} f(x)$. The function $F(z)$ is an analytic function in the complex z plane and, in fact, since

$$|F(z)| \leq \int_{-\infty}^{\infty} e^{(-\text{Im } z)x - x^2/4} \left| e^{-x^2/4} f(x) \right| dx$$

$$\leq \left[\int_{-\infty}^{\infty} e^{2(-\text{Im } z)x - x^2/2} dx \int_{-\infty}^{\infty} e^{-x^2/2} f^2(x) dx \right]^{1/2} < \infty,$$

$F(z)$ has no singularities for bounded z. The function $F(z)$ is therefore an entire function. We can represent $F(z)$ in terms of its Taylor series

$$F(z) = \sum_{n=0}^{\infty} a_n z^n$$

and since $F(z)$ is entire we know that the radius of convergence is infinite. Furthermore, by assumption,

$$a_n = \frac{1}{n!} F^{(n)}(0) = \frac{i^n}{n!} \int_{-\infty}^{\infty} x^n e^{-x^2/2} f(x) dx = 0,$$

so that $F(z) = 0$. The Fourier transform has a unique inverse so we conclude that $f(x) = 0$.

The proof that Laguerre polynomials are complete uses the same ideas. We consider the set of all real functions $f(x)$ with $\int_0^{\infty} x^{\alpha} e^{-x} f^2(x) dx < \infty$, $\alpha > -1$ and suppose that

$$\int_0^{\infty} x^{\alpha} e^{-x} x^n f(x) dx = 0, \quad n = 0, 1, 2, \ldots .$$

We make the change of variables $x = t^2/2$ and write this as

$$\int_0^{\infty} t^{2n} e^{-t^2/2} g(t) dt = 0 \text{ for } n = 0, 1, 2, \ldots$$

where $g(t) = t^{2\alpha+1} f(t^2/2)$. Since $g(t)$ is only defined for $t > 0$, we extend its definition by setting $g(-t) = g(t)$. It follows that

$$\int_{-\infty}^{\infty} t^{2n} e^{-t^2/2} g(t) dt = 0$$

for $n = 0, 1, 2, \ldots$. Of course, since $g(t)$ is even, we know that

$$\int_{-\infty}^{\infty} t^n e^{-t^2/2} g(t) dt = 0$$

for all n odd, hence for all n. Using our previous result, $g(t)$ is the zero function, and therefore, so also is $f(x)$. As a result, the Laguerre polynomials form a complete set on this space.

7.2.3   Sinc Functions

Suppose $f(z)$ is a function whose Fourier transform $F(\mu)$ has $F(\mu) = 0$ for $|\mu| > \pi/h$. The function $f(z)$ is said to be BAND LIMITED since its component frequencies lie in the restricted band of width $|\mu| \leq \pi/h$. (The Wiener-Paley theorem states that any entire function which is square integrable on the real line and exponentially bounded is band limited).

The transform $F(\mu)$ is a square integrable function representable on the interval $|\mu| \leq \pi/h$ as a Fourier series

$$F(\mu) = \sum_{k=-\infty}^{\infty} a_k e^{ikh\mu} \quad , \quad |\mu| \leq \pi/h$$

where

$$a_k = \frac{h}{2\pi} \int_{-\pi/h}^{\pi/h} F(\mu) e^{-ikh\mu} d\mu.$$

Now we take the inverse Fourier transform of $F(\mu)$ to find

$$f(t) \quad = \quad \frac{1}{2\pi} \int_{-\pi/h}^{\pi/h} F(\mu) e^{-i\mu t} d\mu$$

$$= \quad \frac{1}{2\pi} \int_{-\pi/h}^{\pi/h} \sum_{k=-\infty}^{\infty} a_k e^{i(kh-t)\mu} d\mu$$

$$= \quad \frac{1}{h} \sum_{k=-\infty}^{\infty} a_k S_0(t - kh)$$

where

$$S_0(t) = \frac{\sin \pi t/h}{\pi t/h}.$$

This last identity results from interchanging the order of integration and summation and evaluating the integral. Setting $t = nh$ in this identity, we find that

$$a_n = hf(nh)$$

so that

$$f(t) = \sum_{k=-\infty}^{\infty} f(kh)S_0(t - kh)$$

and

$$f(kh) = \frac{1}{2\pi} \int_{-\pi/h}^{\pi/h} F(\mu)e^{-ikh\mu}d\mu$$

$$F(\mu) = h \sum_{k=-\infty}^{\infty} f(kh)e^{ikh\mu} \quad \text{for} \quad |\mu| < \pi/h.$$

The implications of this calculation are profound. We learn that a band limited function can be exactly represented as a Whittaker cardinal function and one only needs to know the values of the function at the equally spaced mesh points $x = kh$.

The human auditory system is band limited since the frequencies we can hear are limited. Thus, audible signals can be reconstructed very accurately from knowledge of the signal at equally space times. This observation is the basis of digital sound recording (whose fidelity far exceeds that of conventional analogue recording) and the recent emergence of compact discs and players.

To make a compact disc, an audio signal is sampled at evenly timed intervals, and the amplitude of the signal is digitized as a binary number. These numbers are encoded into a compact disc as a series of small holes or absence of small holes (a zero or one) which can be read by a laser beam. (A compact disc has room for 10^{10} such holes.) A compact disc player reconstructs the audio signal approximately by using a finite, rather than infinite, sum of Sinc functions.

7.3   Laplace, Mellin and Hankel Transforms

There are a number of important integral transforms which can be derived in much the same way as the Fourier Transform. Arguably the most important of these is the LAPLACE TRANSFORM.

Suppose we have a function $f(t)$ defined only for $t > 0$. If we do not care about $t < 0$, it does no harm to set $f(t) = 0$ for $t < 0$. For this function the

Fourier Transform theorem gives that

$$F(\mu) = \int_0^\infty e^{i\mu t} f(t)dt \quad , \quad f(t) = \frac{1}{2\pi} \int_{-\infty}^\infty e^{-i\mu t} F(\mu)d\mu.$$

If $F(\mu)$ exists for real valued μ, it is analytic in the upper half complex plane $\text{Im }\mu > 0$. We make the change of variables $\mu = is$, $s > 0$ and find

$$F(is) = \int_0^\infty e^{-st} f(t)dt \quad , \quad f(t) = \frac{i}{2\pi} \int_{-i\infty}^{i\infty} e^{st} F(is)ds.$$

The quantity $\hat{f}(s) = F(is)$ is the Laplace transform of $f(t)$ and the Laplace transform formulae read

$$\hat{f}(s) = \int_0^\infty e^{-st} f(t)dt \quad , \quad f(t) = \frac{i}{2\pi} \int_C e^{st} \hat{f}(s)ds,$$

where the contour C is vertical and to the right of all singularities of $\hat{f}(s)$ in the complex plane.

Although this inversion formula is perfectly valid, it is somewhat cumbersome in daily use. The way these formulae are usually used is to make a big table of transforms, and then when a function presents itself, to look up the transform or its inverse in the table.

There are two ways to derive the MELLIN TRANSFORM. In the first we use the fundamental relationship

$$\delta(x - \xi) = -\frac{1}{2\pi i} \int_C G(x,\xi;\lambda)d\lambda$$

for the operator $Lu - \lambda u = -x(xu')' - \lambda u$ on the interval $0 < x < \infty$ subject to the condition $u(0) = 0$.

It is a straightforward matter to find the Green's function of this operator. The fundamental solutions of $Lu - \lambda u = 0$ are

$$u = x^{\pm i\sqrt{\lambda}}.$$

If we take $\sqrt{\lambda}$ to always have nonnegative imaginary part (i.e., define $\sqrt{\lambda}$ as the principal square root), then $x^{i\sqrt{\lambda}}$ is bounded for large positive x and $\lim_{x\to 0} x^{-i\sqrt{\lambda}} = 0$. It follows that

$$G(x,\xi;\lambda) = \begin{cases} \frac{i}{2\sqrt{\lambda}\xi}(\xi/x)^{i\sqrt{\lambda}} & 0 < x < \xi < \infty \\[2mm] \frac{i}{2\sqrt{\lambda}\xi}(x/\xi)^{i\sqrt{\lambda}} & 0 < \xi < x < \infty \end{cases} \quad ,$$

provided λ is not a positive real number.

Now we observe that for ξ, x fixed

$$G(x, \xi; \lambda) = \frac{i}{4\sqrt{\lambda\xi}} \left(\left(\frac{x}{\xi}\right)^{i\sqrt{\lambda}} + \left(\frac{\xi}{x}\right)^{i\sqrt{\lambda}} \right) + \text{an entire function,}$$

and we calculate that

$$\frac{1}{2\pi i} \int_C G(x, \xi; \lambda) d\lambda = -\frac{1}{2\pi} \int_{-\infty}^{\infty} \frac{1}{\xi} (\xi/x)^{i\mu} d\mu$$

so that

$$\delta(x - \xi) = \frac{1}{2\pi} \int_{-\infty}^{\infty} \frac{1}{\xi} (\xi/x)^{i\mu} d\mu,$$

or after the change of variable $\mu = -is$,

$$\delta(x - \xi) = \frac{1}{2\pi i} \int_{-i\infty}^{i\infty} \frac{1}{\xi} \left(\frac{\xi}{x}\right)^s ds.$$

We use this to define the Mellin Transform pair $f(x)$, $F(s)$ as

$$F(s) = \int_0^{\infty} x^{s-1} f(x) dx \quad , \quad f(x) = \frac{1}{2\pi i} \int_{-i\infty}^{i\infty} x^{-s} F(s) ds.$$

An alternate derivation of this formula is by a direct change of variables from the Fourier Transform. To motivate that a change of variables should work, notice that the operation $x\frac{d}{dx}$ becomes $\frac{d}{dt}$ under the change of variables $x = e^t$. Thus the operator $Lu - \lambda u = -x(xu')' - \lambda u$ for $x \in [0, \infty)$ becomes

$$Lu - \lambda u = -\frac{d^2 u}{dt^2} - \lambda u$$

for $t \in (-\infty, \infty)$. This suggests that the same change of variables relates the Mellin Transform to the Fourier Transform. Indeed, if we start with the Mellin Transform pair and make the change of variables $x = e^t$, $s = i\mu$ we find that

$$F(i\mu) = \int_{-\infty}^{\infty} e^{it\mu} f(e^t) dt$$

$$f(e^t) = \frac{1}{2\pi} \int_{-\infty}^{\infty} e^{-it\mu} F(i\mu) d\mu.$$

This is the Fourier Transform pair for the function $g(t) = f(e^t)$ for $-\infty < t < \infty$. In other words, if $F(s)$ is the Mellin Transform of $f(x)$ then $F(i\mu)$ is the Fourier Transform of $f(e^t)$.

The third important transform pair is the HANKEL TRANSFORM. This transform is derived in the same way as the Mellin Transform, using as the basic operator the Bessel's operator

$$Lu - \lambda u = -\frac{1}{x}(xu')' - \lambda u \quad , \quad 0 < x < \infty$$

with

$$\lim_{x \to 0} u' = 0.$$

The Green's function for this operator is

$$G(x, \xi; \lambda) = \begin{cases} -\frac{\pi i \xi}{2} J_0(\sqrt{\lambda}x) H_0^{(1)}(\sqrt{\lambda}\xi) , & 0 \leq x < \xi < \infty, \\ -\frac{\pi i \xi}{2} J_0(\sqrt{\lambda}\xi) H_0^{(1)}(\sqrt{\lambda}x) , & 0 \leq \xi < x < \infty, \end{cases}$$

provided λ is not on the positive, real axis, where $J_0(x)$ is the zeroth order Bessel function of the first kind, and $H_0^{(1)}(x)$ is the zeroth order Hankel function of the first kind behaving like

$$H_0^{(1)}(x) \sim \sqrt{\frac{2}{\pi}x} \, e^{i(x - \pi/4)} \quad \text{as} \quad x \to +\infty.$$

(Use that $W(J_\nu(z), H_\nu^{(1)}(z)) = \frac{2i}{\pi z}$.) Applying the standard formula

$$-\frac{1}{2\pi i} \int_C G(x, \xi; \lambda) d\lambda = \delta(x - \xi)$$

and using that $J_0(-x) = J_0(x)$, $H_0^{(1)}(x) - H_0^{(1)}(-x) = 2J_0(x)$ (Problem 6.5.14) we find that

$$\delta(x - \xi) = x \int_0^\infty \mu J_0(\mu x) J_0(\mu \xi) d\mu$$

or, in terms of transform pairs,

$$F(\mu) = \int_0^\infty x J_0(\mu x) f(x) dx \quad , \quad f(x) = \int_0^\infty \mu J_0(\mu x) F(\mu) d\mu.$$

7.4 Z Transforms

Up until now our discussion of transform theory has included only consideration of differential operators. Another important class of linear operators are difference operators. For example, the difference operator

$$(\Delta u)_n = u_{n+1} - 2u_n + u_{n-1}, \quad -\infty < n < \infty$$

is an operator on $\ell^2(-\infty,\infty)$, (the set of all real doubly infinite sequences $u = \{u_n\}$ with $\sum_{n=-\infty}^{\infty} u_n^2 < \infty$). The inner product for this space is $\langle u, v \rangle = \sum_{n=-\infty}^{\infty} u_n v_n$.

Of course we already know plenty about difference operators on finite dimensional spaces because they are equivalent to matrix operators and we have studied spectral theory for matrices in some detail.

One natural occurrence of difference operators is from finite difference approximations of differential operators, whereby one replaces the differential operator $Lu = \frac{du}{dx}$ by a difference operator $(\Delta u)_n = \frac{u_{n+1}-u_n}{h}$, say. Difference operators also occur naturally in the study of lattices. We shall see an example of a nonlinear lattice (the Toda Lattice) in Chapter 9.

Suppose Δu is some difference operator on a sequence space. The first question to ask is what we mean by a Green's function, or inverse operator, and the obvious answer is that for each fixed j we define a vector $g^j = \{g_i^j\}$ by the equation

$$\left(\Delta g^j\right)_i = \delta_{ij}$$

where δ_{ij} is the Kronecker delta, and of course, we want g_i^j to live in the appropriate Hilbert space. It follows that if g_i^j is so defined, then the solution of $(\Delta u)_n = f_n$ is $u = \{u_n\}$ where

$$u_n = \sum_j g_n^j f_j$$

since $(\Delta u)_n = \sum_j (\Delta g^j)_n f_j = \sum_j \delta_{nj} f_j = f_n$.

If we follow our previous outline for developing transforms, we must next find the inverse for the operator $-(\Delta u)_n - \lambda u_n$. For the specific operator $(\Delta u)_n = u_{n+1} - 2u_n + u_{n-1}$ acting on $\ell^2(-\infty,\infty)$, the solution of $(\Delta u)_n + \lambda u_n = 0$ has the form

$$u_n = z^n$$

provided $\lambda + z + \frac{1}{z} - 2 = 0$. The equation $\lambda = 2 - z - \frac{1}{z}$ is a mapping of the z-complex plane to the λ-complex plane that double covers the λ-plane (a Joukowski transformation). Under this map the unit circle $|z| = 1$ goes to the real line segment $\lambda = 2(1 - \cos\theta)$, $0 \leq \theta \leq 2\pi$, which is the real axis $0 \leq \operatorname{Re}\lambda \leq 4$. The interior of the unit circle maps to the entire λ-plane as does the exterior of the unit circle. Hence, for every complex λ not on the real line between 0 and 4, there is a root, say ξ, of the characteristic equation $\lambda = 2 - z - \frac{1}{z}$ with $|\xi| < 1$.

Now we define the inverse function $G_n^j(\lambda)$ to satisfy

$$(\Delta G^j)_n + \lambda G_n^j = -\delta_{nj}.$$

It is not difficult to verify that

$$G_n^j(\lambda) = \frac{-\xi^{|n-j|}}{\xi - \frac{1}{\xi}}$$

is in $\ell^2(-\infty, \infty)$, where $\lambda = 2 - \xi - 1/\xi$ and $|\xi| < 1$. As is our custom, we integrate $G_n^j(\lambda)$ in the complex λ plane counterclockwise around a large circle. The expression

$$I = \frac{-1}{2\pi i} \int_C G_n^j(\lambda) d\lambda$$

should, if all goes well, be the Kronecker delta, $I = \delta_{nj}$. In fact

$$I = \frac{1}{2\pi i} \int_C \frac{\xi^{|n-j|}}{\xi - \frac{1}{\xi}} d\lambda = \frac{1}{2\pi i} \int_{C'} \xi^{|n-j|-1} d\xi = \delta_{nj}$$

since $d\lambda = \left(\frac{1}{\xi^2} - 1\right) d\xi$. In other words, we find the representation of the Kronecker delta to be

$$\delta_{nj} = \frac{1}{2\pi i} \int_{C'} z^{n-j-1} dz$$

which we already know is correct. From this we define a transform pair, called the z-transform, for sequences in ℓ^2, by

$$U(z) = \sum_{n=-\infty}^{\infty} u_n z^n$$

and

$$u_n = \frac{1}{2\pi i} \int_{C'} U(z) z^{-n-1} dz.$$

The countour C' must enclose the origin and be traversed in the counterclockwise direction. For z restricted to the unit circle, the z-transform becomes

$$V(\theta) = \sum_{n=-\infty}^{\infty} u_n e^{in\theta} \quad , \quad V(\theta) = U(e^{i\theta})$$

$$u_n = \frac{1}{2\pi} \int_0^{2\pi} V(\theta) e^{-in\theta} d\theta$$

which is exactly the complex version of the Fourier series representation of $V(\theta)$. In other word, the z-transform for $\ell^2(-\infty, \infty)$ sequences is exactly the inverse of the Fourier series transform of 2π periodic functions.

Observe that this transform arises from the continuous spectrum of the operator Δu. As we saw earlier, all solutions of $\Delta u + \lambda u = 0$ are of the form $u = z^n$ where z is a root of $\lambda + z + \frac{1}{z} - 2 = 0$. Since the two roots of this equation satisfy $z_1 z_2 = 1$, there are no solutions in ℓ^2, i.e., the operator has no discrete spectrum. If $z = e^{i\theta}$, then $u = z^n$ is a bounded, but not ℓ^2, solution, so we have an improper eigenfunction when $\lambda = 2 - 2\cos\theta$. If $z = e^{i\theta}$, we construct a sequence of sequences $u^k = u_n^k$ defined by $u_n^k = e^{in\theta}$ for $|n| < k$, $u_n^k = 0$ for $n \geq k$. Then $(\Delta u^k + \lambda u^k)_j = 0$ for $|j| \neq k, k-1$, and $\|\Delta u^k + \lambda u^k\|$ is bounded independent of k. However, $\|u^k\| = 2k - 1$ which is unbounded as $k \to \infty$, so that $\lambda = 2 - 2\cos\theta$ is continuous spectrum.

7.5   Scattering Theory

So far we have seen two types of transform pairs, the discrete transform, typified by Fourier series, and the continuous transform, typified by the Fourier integral. A third type of transform that is important in applications is actually not new, but a mixture of the previous two.

A differential operator which gives rise to the mixed case is the SCHRÖDINGER OPERATOR

$$Lu = -u'' + q(x)u \quad , \quad x \in (-\infty, \infty),$$

where $\int_{-\infty}^{\infty} |q(x)| dx < \infty$.

There are many ways to motivate this operator, two of which we mention here. The first comes from classical mechanics. Consider a vibrating inhomogeneous string whose transverse vibrations are governed by

$$\rho(x)u_{tt} = (E_0 u_x)_x$$

where u is the transverse displacement of the string, $\rho(x)$ is the nonuniform density, and E_0 is the lateral tension in the string (assumed constant). If we define a new independent variable

$$\xi = \int_0^x \frac{dx}{c(x)} \quad , \quad c = \left(\frac{E_0}{\rho}\right)^{1/2},$$

the equation transforms into

$$\rho c u_{tt} = (\rho c u_\xi)_\xi.$$

We now set $\phi = \sqrt{\rho c}\, u$ and find $\phi_{tt} = \phi_{\xi\xi} - q(\xi)\phi$ where $q(\xi) = n_{\xi\xi}/\eta$, $\eta = \sqrt{\rho c}$, $\xi = \int_0^x \frac{dx}{c(x)}$. Obviously, the right-hand side of this equation involves the Schrödinger operator.

The second motivation for this equation (and the origin of its name) is from quantum physics. The idea occured to Schrödinger that elementary particles (electrons, for example) should not be viewed as having an exact position but rather should be viewed as being spread out with a position described in terms of a density. We call $\psi(x)$ in $L^2(-\infty, \infty)$ a wave function for the particle if $\psi\bar\psi$ is a density, that is, if $\int_{-\infty}^{\infty} |\psi|^2 dx = 1$. Schrödinger concluded that, in steady state, ψ should satisfy

$$-\frac{\hbar^2}{2m}\frac{d^2\psi}{dx^2} + V(x)\psi = E\psi$$

where E is the total energy of the particle, $\hbar$ is Planck's constant divided by 2π, $2\pi\hbar = 6.546 \times 10^{-27}$ erg secs, m is the mass of the particle, and $V(x)$ is the potential energy field for the particle at position x. If we set $\lambda = \frac{2mE}{\hbar^2}$ and $q(x) = \frac{2mV}{\hbar^2}$, we obtain

$$\psi'' + (\lambda - q(x))\psi = 0$$

which is again Schrödinger's equation.

The properties of this equation are easy to discuss, although explicit solutions are nearly impossible to obtain. Since $q(x)$ is assumed to be absolutely integrable, $\int_{-\infty}^{\infty} |q(x)| dx < \infty$, we use variation of parameters to find integral equations for the fundamental solutions of this equation. For example, solutions of $u'' + \lambda u = f$ can be written as

$$u = \frac{1}{2ik} \int_{x_0}^{x} e^{ik(x-y)} f(y) dy - \frac{1}{2ik} \int_{x_1}^{x} e^{-ik(x-y)} f(y) dy \quad , \quad \lambda = k^2$$

where x_0 and x_1 are arbitrary constants. We take $f(x) = q(x)u(x)$ and obtain

$$u_1(x, k) = e^{ikx} - \frac{1}{k} \int_{x}^{\infty} \sin k(x - y) q(y) u_1(y, k) dy$$

and

$$u_2(x, k) = e^{-ikx} + \frac{1}{k} \int_{-\infty}^{x} \sin k(x - y) q(y) u_2(y, k) dy$$

as two solutions of the Schrödinger equation. Notice that $u_1(x, k)$ behaves like e^{ikx} as $x \to \infty$ and $u_2(x, k)$ behaves like e^{-ikx} as $x \to -\infty$.

These statements do not prove that the functions $u_1(x, k), u_2(x, k)$ exist, but only that if they do exist, they can be represented in this particular form. To actually prove that they exist, we use a variant of the contraction mapping principle used in Chapter 3 to solve integral equations.

To prove that $u_1(x, k)$ exists, we make the change of variables $v(x) = e^{-ikx} u_1(x, k)$ and find that $v(x)$ must satisfy

$$v(x) = 1 - \frac{1}{2ik} \int_{x}^{\infty} \left(1 - e^{-2ik(x-y)}\right) q(y) v(y) dy.$$

The Neumann iterates for this problem are

$$v_1(x) = 1$$
$$v_{n+1}(x) = 1 - \frac{1}{2ik} \int_{x}^{\infty} \left(1 - e^{-2ik(x-y)}\right) q(y) v_n(y) dy.$$

Clearly, $v_1(x)$ is bounded for all x. Suppose $v_n(x)$ is bounded for all x. If k is restricted to be real or to lie in the upper half k-plane, then $v_{n+1}(x)$ is also bounded, provided the function

$$Q(x) = \int_{x}^{\infty} |q(\xi)| d\xi$$

is bounded for all x. In fact,

$$|v_{n+1}(x)| \leq 1 + \frac{1}{|k|} \int_x^\infty |q(\xi)| d\xi \cdot \sup_x |v_n(x)|.$$

Now, we examine the difference

$$v_2(x) - v_1(x) = -\frac{1}{2ik} \int_x^\infty \left(1 - e^{-2ik(x-y)}\right) q(y) dy$$

and observe that $|v_2(x) - v_1(x)| \leq \frac{Q(x)}{|k|}$. Furthermore,

$$v_{n+1}(x) - v_n(x) = -\frac{1}{2ik} \int_x^\infty \left(1 - e^{-2ik(x-y)}\right) q(y) \left(v_n(y) - v_{n-1}(y)\right) dy.$$

Suppose $|v_n(x) - v_{n-1}(x)| \leq \frac{1}{(n-1)!} \left(\frac{Q(x)}{|k|}\right)^{n-1}$. Then

$$|v_{n+1}(x) - v_n(x)| \leq \frac{1}{(n-1)!|k|} \int_x^\infty |q(\xi)| \left(\frac{Q(\xi)}{|k|}\right)^{n-1} d\xi$$

$$= \frac{-1}{(n-1)!|k|^n} \int_x^\infty (Q(\xi))^{n-1} \frac{dQ(\xi)}{d\xi} d\xi$$

$$= \frac{1}{n!} \left(\frac{Q(x)}{|k|}\right)^n ,$$

which verifies the inductive statement. It follows that

$$|v_n(x)| \leq |v_1(x)| + |v_2(x) - v_1(x)| + \cdots + |v_n(x) - v_{n-1}(x)|$$

$$\leq \sum_{j=0}^{n-1} \frac{1}{j!} \left(\frac{Q(x)}{|k|}\right)^j \leq \exp\left(\frac{Q(x)}{|k|}\right).$$

It is now easy to show that $v_n(x)$ converges to a limit, say $v(x)$, that $v(x)$ satisfies the integral equation and that $\lim_{x \to \infty} v(x) = 1$, provided $\text{Im } k \geq 0$, $k \neq 0$. It follows that $u_1(x, k) = e^{ikx}v(x)$ satisfies the Schrödinger equation, that $u_1(x, k) \sim e^{ikx}$ for large $x > 0$ and that $u_1(x, k)$ is analytic in the upper half k-plane. A similar argument shows that $u_2(x, k)$ satisfies the Schrödinger equation, behaves like e^{-ikx} for large $x < 0$ and is analytic in the upper half k-plane.

The choices of e^{ikx} and e^{-ikx} as the asymptotic behavior of $u_1(x, k)$ and $u_2(x, k)$ are motivated by both mathematical and physical reasoning. For k in the upper half of the complex plane (which it is since $\lambda = k^2$ and we take k to the principal square root of λ), e^{ikx} is square integrable for large positive

x and e^{-ikx} is square integrable for large negative x. Physicists refer to e^{ikx} as "outgoing at $x = +\infty$" and e^{-ikx} is "outgoing at $x = -\infty$" whereas e^{ikx} is "incoming at $x = -\infty$" and e^{-ikx} is "incoming at $x = \infty$". The reasoning behind this nomenclature is taken from an examination of the wave equation $u_{tt} = u_{xx}$. The general solution of this equation is

$$u = f(x - t) + g(x + t).$$

The expression $f(x - t)$ represents a right moving object and $g(x + t)$ is left moving. If we apply the Fourier Transform (in time) to the wave equation we obtain $U_{xx} + \omega^2 U = 0$ where $U(x, \omega)$ is the Fourier time transform of $u(x, t)$. Clearly the general solution is $U(x, \omega) = A(\omega)e^{i\omega x} + B(\omega)e^{-i\omega x}$ and, via the inverse Fourier transform,

$$
\begin{aligned}
u(x, t) &= \frac{1}{2\pi} \int_{-\infty}^{\infty} e^{-i\omega t} \left[A(\omega)e^{i\omega x} + B(\omega)e^{-i\omega x} \right] d\omega \\
&= \frac{1}{2\pi} \int_{-\infty}^{\infty} \left[A(\omega)e^{i\omega(x - t)} + B(\omega)e^{-i\omega(x + t)} \right] d\omega.
\end{aligned}
$$

Comparing the two representations for the solution of the wave equation, we see that solutions with spatial dependence $e^{i\omega x}$ correspond to right moving waves, and those with spatial dependence $e^{-i\omega x}$ correspond to left moving waves, from which physicists obtain their nomenclature. We will henceforth refer to $e^{i\omega x}$ as right moving, and similarly $e^{-i\omega x}$ as left moving.

Given this nomenclature, we see that the two solutions of Schrödinger's equation $u_1(x, k)$ and $u_2(x, k)$ are right moving as $x \to \infty$ and left moving as $x \to -\infty$, respectively. Notice also that $u_1(x, -k)$ and $u_2(x, -k)$ are solutions of the Schrödinger equation as well.

We know that any second order differential equation can have at most two linearly independent solutions, and that any solution can be represented as a linear combination of a linearly independent pair. We can readily show that $u_1(x, k)$ and $u_1(x, -k)$ are linearly independent, as are $u_2(x, k)$ and $u_2(x, -k)$. In fact, the Wronskian of any two solutions must be a constant, independent of x. Thus, as $x \to +\infty$ we calculate

$$W\left(u_1(x, k), u_1(x, -k)\right) = W\left(e^{ikx}, e^{-ikx}\right) = -2ik$$

and similarly

$$W\left(u_2(x, k), u_2(x, -k)\right) = 2ik.$$

Since we know two linearly independent pairs, we can express solutions u_1 and u_2 as

$$
\begin{aligned}
u_2(x, k) &= c_{11}(k)u_1(x, k) + c_{12}(k)u_1(x, -k), \\
u_1(x, k) &= c_{21}(k)u_2(x, -k) + c_{22}(k)u_2(x, k).
\end{aligned}
$$

These expressions contain much more than just mathematical information. Since $u_1(x, k)$ and $u_1(x, -k)$ are outgoing and incoming respectively as $x \to \infty$, the first of these expresses in mathematical language the result of a "scattering" experiment. That is, $c_{12}(k)$ is the amplitude of an incoming wave from $x \to +\infty$, $c_{11}(k)$ is the amplitude of the wave reflected back to $x \to +\infty$ and 1, the coefficient of $u_2(x, k)$, is the wave transmitted through to $x \to -\infty$. When $c_{12}(k) \neq 0$, it is convenient to normalize the incoming incident wave to have amplitude 1, and denote

$$T_r = \frac{1}{c_{12}(k)} \quad , \quad R_r = \frac{c_{11}(k)}{c_{12}(k)}$$

as the TRANSMISSION and REFLECTION COEFFICIENTS for waves incoming from the right. Similarly,

$$T_L = \frac{1}{c_{21}(k)}, \quad R_L = \frac{c_{22}(k)}{c_{21}(k)}$$

are the transmission and reflection coefficients, respectively, for waves incoming from the left.

There are some important consistency relationships between the coefficients $c_{ij}(k)$. The definitions of $c_{ij}(k)$ hold if and only if

$$\begin{aligned}
c_{11}(k)c_{21}(k) + c_{21}(k)c_{22}(-k) &= 0, \\
c_{11}(k)c_{22}(k) + c_{12}(k)c_{21}(-k) &= 1, \\
c_{11}(k)c_{22}(k) + c_{21}(k)c_{12}(-k) &= 1, \\
c_{11}(-k)c_{12}(k) + c_{22}(k)c_{12}(k) &= 0.
\end{aligned}$$

It follows from these identities that $c_{12}(k)c_{21}(-k) = c_{12}(-k)c_{21}(k)$ and that $c_{11}(k)c_{22}(k) = c_{11}(-k)c_{22}(-k)$, and using these identities one can readily show that $c_{12}(k) = c_{21}(k)$ and $c_{11}(k) = -c_{22}(-k)$. Furthermore, we can calculate the Wronskians

$$W\left(u_1(x, k), u_2(x, k)\right) = c_{12}(k)W\left(u_1(x, k), u_1(x_1, -k)\right) = -2ikc_{12}(k)$$

and

$$W\left(u_1(x, k), u_2(x, -k)\right) = c_{22}(k)W\left(u_2(x, k), u_2(x, -k)\right) = 2ikc_{22}(k).$$

A zero of $c_{12}(k)$ with $\operatorname{Im} k > 0$, if it exists, has an important physical interpretation. If $c_{12}(k_0) = 0$, then $W\left(u_1(x, k_0), u_2(x, k_0)\right) = 0$ so that $u_2(x, k_0) = \alpha u_1(x, k_0)$ for some nonzero constant α. Since $u_1(x, k_0) \sim e^{ik_0 x}$ as $x \to +\infty$ and $u_2(x, k_0) \sim e^{-ik_0 x}$ as $x \to -\infty$, we see that $u_2(x, k_0)$ is a square integrable solution of the Schrödinger equation. Square integrable solutions, if they exist, are called BOUND STATES and are analogous to eigenfunctions of the Sturm-Liouville problems. The values k_0^2 are eigenvalues of the problem.

The title bound state refers to the fact that these wave functions are "bound" to the scattering center by the potential.

We shall show that the roots of $c_{12}(k) = 0$ in the upper half plane occur only on the imaginary axis and are always simple (recall that a self-adjoint operator has only real eigenvalues). The Schrödinger equation at $k = k_0$ is $u'' + k_0^2 u = q(x)u$ so that

$$\bar{u}u'' - u\bar{u}'' + \left(k_0^2 - \bar{k}_0^2\right)u\bar{u} = 0.$$

Integrating, we find that

$$\int_{-\infty}^{\infty}(\bar{u}u'' - u\bar{u}'')dx = (\bar{u}u' - u\bar{u}')\Big|_{-\infty}^{\infty} = \left(\bar{k}_0^2 - k_0^2\right)\int_{-\infty}^{\infty}u\bar{u}dx.$$

Since $c_{12}(k_0) = 0$, u and $\bar{u}$ have exponentially decaying tails as $x \to \pm\infty$,

$$\left(\bar{k}_0^2 - k_0^2\right)\int_{-\infty}^{\infty}|u|^2 dx = 0,$$

implying that $k_0^2 = \bar{k}_0^2$ or that k_0^2 is real. The only roots of $c_{12}(k)$ that correspond to bound states are those with $\operatorname{Im} k_0 > 0$. A root with $\operatorname{Im} k_0 < 0$ is said to lie on the unphysical sheet, that is the non-principal branch of the $\sqrt{\lambda}$-plane.

We now show that a root of $c_{12}(k)$ at $k = k_0$ is simple. We know that

$$c_{12}(k) = \frac{-1}{2ik}W\left(u_1(x, k), u_2(x, k)\right),$$

so that

$$\frac{d}{dk}c_{12}(k)\Big|_{k=k_0} = \frac{-1}{2ik_0}\left(W(\dot{u}_1, u_2) + W(u_1, \dot{u}_2)\right)$$

where $\dot{u}_i \equiv \frac{d}{dk}u_i\Big|_{k=k_0}$, $i = 1, 2$. At $k = k_0$,

$$u_2 = c_{11}u_1 \quad \text{and} \quad u_1 = c_{22}u_2$$

so that

$$\dot{c}_{12}(k_0) = \frac{-1}{2ik_0}\left(c_{11}W(\dot{u}_1, u_1) + c_{22}W(u_2, \dot{u}_2)\right).$$

(We have assumed here that $c_{11} \neq 0$, which in rare cases is not true). Now $u_1'' + k^2 u_1 = qu_1$ so that

$$\frac{d}{dx}\left(u_1'(x, k)u_1(x, k_0) - u_1'(x, k_0)u_1(x, k)\right) + (k^2 - k_0^2)u_1(x, k)u_1(x, k_0) = 0.$$

Differentiating this last equation with respect to k and setting $k = k_0$ we obtain

$$\frac{d}{dx}W(u_1, \dot{u}_1) + 2k_0 u_1^2(x, k_0) = 0$$

or

$$W(u_1, \dot{u}_1) = 2k_0 \int_x^\infty u_1^2(x, k_0)dx,$$

since $W(u_1, \dot{u}_1) \to 0$ as $x \to +\infty$. In a similar way

$$W(u_2, \dot{u}_2) = -2k_0 \int_{-\infty}^x u_2^2(x, k_0)dx.$$

From this we calculate that

$$
\begin{aligned}
\dot{c}_{12}(k_0) &= \frac{-1}{2ik_0}\left(c_{11}(k_0)W(\dot{u}_1, u_1) + c_{22}(k_0)W(u_2, \dot{u}_2)\right) \\[2mm]
&= -ic_{11}(k_0)\int_x^\infty u_1^2(x, k_0)dx - ic_{22}(k_0)\int_{-\infty}^x u_2^2(x, k_0)dx \\[2mm]
&= -i\int_{-\infty}^\infty u_1(x, k_0)u_2(x, k_0)dx = -ic_{11}(k_0)\int_{-\infty}^\infty u_1^2(x, k_0)dx \neq 0
\end{aligned}
$$

if $c_{11}(k_0) \neq 0$, so that the pole at k_0 is simple.

Some examples of scattering problems should be helpful.

Example   Scattering by an inhomogeneous string

Suppose we have an infinitely long string with two different compositions. The wave equation is

$$u_{tt} = c^2(x)u_{xx}$$

where $c(x) = c_1$ for $x < 0$ and $c(x) = c_2$ for $x > 0$. To determine the vibrations of this string, we set $u(x, t) = U(x)e^{-i\omega t}$ and find that

$$U_{xx} + k^2 U = 0$$

where $k(x) = \omega/c(x)$. If we have waves incident from $x = -\infty$, then the solution is

$$
U = \begin{cases}
e^{ik_1 x} + Re^{-ik_1 x} & x < 0 \\[2mm]
Te^{ik_2 x} & x > 0
\end{cases}
$$

where $k_i = \omega/c_i$. Requiring the solution $U(x)$ to be continuously differentiable at the origin, we obtain that

$$R = \frac{c_2 - c_1}{c_2 + c_1}$$

$$T = \frac{2c_2}{c_2 + c_1}.$$

Observe that if $c_1 = c_2$ there is no reflection and full transmission, however, in the limit $c_1 \gg c_2$ there is no transmission and full reflection.

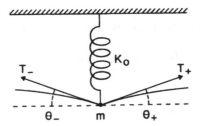

FIGURE 7.2 Forces acting on the mass-spring-string system.

Example Scattering by an oscillator

An infinite homogeneous string is attached at $x = 0$ to a mass-spring oscillator (Figure 7.2). We want to determine the reflection and transmission coefficients. If we take $u(x, t) = U(x)e^{-i\omega t}$ and have waves incident from $x = -\infty$, the solution is

$$U(x) = \begin{cases} e^{ikx} + Re^{-ikx} & x < 0 \\ Te^{ikx} & x > 0 \end{cases}$$

where $k = \omega/c$. The only question is how to match the solution at $x = 0$.

The equation of motion of the mass at the origin comes from Newton's law. The forces on the mass are the restoring force of the spring and the vertical components of tension in the string. If we let u_0 be the displacement of the mass from equilibrium, then

$$m\frac{d^2 u_0}{dt^2} = -K_0 u_0 + T_+ \sin\theta_+ + T_- \sin\theta_-$$

where T_+, T_- are the tensions and θ_+, θ_- are the angles with the horizontal for $x = 0^+$, $x = 0^-$, respectively (see Figure 7.2). The constant K_0 is the Hooke's constant for the string. Since vibrations are assumed to be transverse, horizontal forces must balance, so that $T_+ \cos\theta_+ = T_- \cos\theta_- = T_0$, a constant. As a consequence

$$m\frac{d^2 u_0}{dt^2} + K_0 u_0 = T_0 \left(\frac{\partial u}{\partial x}\Big|_{x=0+} - \frac{\partial u}{\partial x}\Big|_{x=0-} \right).$$

With the assumed temporal behavior $u_0(t) = U_0 e^{-i\omega t}$, we require

$$(-m\omega^2 + K_0)\, U_0 = T_0 \left(U_x\Big|_{x=0+} - U_x\Big|_{x=0-} \right).$$

where $c^2 = T_0/\rho$, ρ being the density of the string. A straightforward calculation reveals that the transmission and reflection coefficients are

$$T = U_0 = \frac{2i\gamma k}{k^2 - k_0^2 + 2i\gamma k}$$

$$R = \frac{k_0^2 - k^2}{k^2 - k_0^2 + 2i\gamma k} \quad , \quad k_0^2 = \frac{K_0}{mc^2} \quad , \quad \gamma = \rho/m.$$

To understand the physical implications of this solution, we substitute into the equation of motion for u_0,

$$m\frac{d^2 u_0}{dt^2} + K_0 u_0 = T_0 \left(U_x \Big|_{x=0^+} - U_x \Big|_{x=0^-} \right) e^{-i\omega t}$$

$$= 2mc\gamma(U_0 - 1)i\omega e^{-i\omega t} = -2mc\gamma \left(\frac{du_0}{dt} + i\omega e^{-i\omega t} \right).$$

Written slightly differently, we have

$$m\frac{d^2 u_0}{dt^2} + 2mc\gamma\frac{du_0}{dt} + K_0 u = -2mc\gamma i\omega e^{-i\omega t}$$

which is the equation of motion of a forced, damped oscillator. Consequently, the effect of the incoming waves on the string is to force and also to damp the motion of the mass-spring system, resulting in a forced, periodic vibration of the mass that is out of phase from the incoming oscillation.

Example Scattering by a delta function potential

We consider the Schrödinger equation $u'' + (k^2 - q(x)) u = 0$ with the potential function $q(x) = A\delta(x)$. For this problem we take

$$u_2(x, k) = \begin{cases} e^{-ikx} & x < 0 \\ \alpha e^{ikx} + \beta e^{-ikx} & x > 0 \end{cases}.$$

Notice that $\alpha = c_{11}(k)$ and $\beta = c_{12}(k)$. To calculate α and β, we require that u be continuous at $x = 0$ and satisfy the jump condition

$$u'\Big|_{x=0^-}^{x=0^+} = Au\Big|_{x=0} .$$

It follows that $c_{11} = \frac{A}{2ik}$, $c_{12} = \frac{2ik-A}{2ik}$. The reflection and transmission coefficients are

$$T_r = \frac{2ik}{2ik - A} \quad , \quad R_r = \frac{A}{2ik - A}.$$

The important observation for this example is to note that $c_{12}(k) = 0$ for $k = -iA/2$. If $A < 0$, this corresponds to a bound state solution of the form

$$u(x) = e^{-|Ax|/2}.$$

If $A > 0$, the root of $c_{12}(k) = 0$ occurs in the lower half k plane and there is no bound state.

Example   Square well potential

We consider the Schrödinger equation $u'' + \left(k^2 - q(x)\right) u = 0$ with $q(x) = -A$ for $0 < x < \alpha$ and $q(x) = 0$ elsewhere. We take the solution $u_2(x, k)$ to be

$$u_2(x, k) = \begin{cases} e^{-ikx} & x < 0 \\[2mm] ae^{ik_1x} + be^{-ik_1x} & 0 < x < \alpha \\[2mm] c_{11}e^{ikx} + c_{12}e^{-ikx} & x > \alpha \end{cases}$$

where $k_1^2 = k^2 + A$. We require $u_2(x, k)$ to be continuously differentiable at $x = 0, \alpha$ and so calculate

$$c_{11}(k) = \frac{i\left(k_1^2 - k^2\right)}{2kk_1} e^{-ik\alpha} \sin k_1\alpha,$$

$$c_{12}(k) = e^{ik\alpha} \left(\cos k_1\alpha - i\frac{k^2 + k_1^2}{2kk_1} \sin k_1\alpha \right).$$

To determine if there are bound states, we try to solve $c_{12}(k) = 0$ for $k = i\mu$, μ real. We look for real roots of

$$\frac{\tan\left(\alpha(A - \mu^2)^{1/2}\right)}{(A - \mu^2)^{1/2}} = \frac{2\mu}{A - 2\mu^2}.$$

With A positive there is always at least one root of this equation. For $0 < \alpha\sqrt{A} < \pi$ there is exactly one real root and as αA increases, the number of roots increases. In fact, for $k\pi < \alpha\sqrt{A} < (k+1)\pi$ there are $k+1$ roots. Sketches of the two curves $f_1 = \tan\left(\alpha(A-\mu^2)^{1/2}\right)/(A-\mu^2)^{1/2}$ (solid curve) and $f_2 = 2\mu/(A-2\mu^2)$ (dashed curve) are shown in Figure 7.3 with $A = 4$, $\alpha = 2.4$.

If we set $A\alpha = B$ and let α approach zero while keeping B constant, we find that

$$c_{11}(k) = \frac{-B}{2ik} \qquad c_{12}(k) = \frac{2ik + B}{2ik}$$

which is exactly what we expect, since if $A\alpha = B$, then $q(x)$ is a delta sequence $q(x) \longrightarrow -B\delta(x)$.

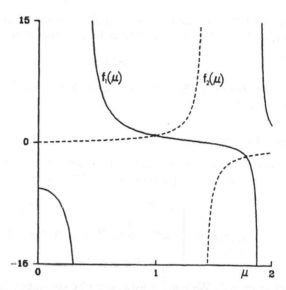

FIGURE 7.3   Plot of functions $f_1(\mu)$ (solid curve) and $f_2(\mu)$ (dashed curve). The intersections of these curves determine the bound states for the square well potential of depth $-A$ and width α. Here $A = 4.0$, and $\alpha = 2.4$.

Example   The n-barrier potential

The previous example is a special case of the problem where $q(x)$ is piecewise constant in $n + 1$ pieces. We suppose that there are points $-\infty < x_1 < x_2 < \cdots < x_n < \infty$, that $q(x) = 0$ for $x < x_1$ and $x > x_n$ and that $q(x) = q_i$ for $x_i < x < x_{i+1}$, $i = 1, 2, \ldots, n - 1$. To construct a solution to the Schrödinger equation we take

$$u(x, k) = A_j e^{ik_j x} + B_j e^{-ik_j x} \quad , \quad k_j^2 = k^2 - q_j$$

on the jth interval $x_j < x < x_{j+1}$. (Set $x_0 = -\infty$ and $x_{n+1} = +\infty$). We relate the coefficients A_j, B_j to A_{j+1}, B_{j+1} by requiring $u(x, k)$ to be continuously differentiable at $x = x_{j+1}$. It follows that

$$\begin{pmatrix} A_{j-1} \\ B_{j-1} \end{pmatrix} = S_{j-1} \begin{pmatrix} A_j \\ B_j \end{pmatrix}$$

where the connection matrix S_{j-1} is given by

$$S_{j-1} = \begin{pmatrix} \frac{k_{j-1}+k_j}{2k_{j-1}} e^{i(k_j - k_{j-1})x_j} & \frac{k_{j-1}-k_j}{2k_{j-1}} e^{-i(k_j + k_{j-1})x_j} \\ \frac{k_{j-1}-k_j}{2k_{j-1}} e^{i(k_j + k_{j-1})x_j} & \frac{k_{j-1}+k_j}{2k_{j-1}} e^{i(k_{j-1} - k_j)x_j} \end{pmatrix} .$$

We multiply all these together to find that

$$\begin{pmatrix} A_0 \\ B_0 \end{pmatrix} = (S_0 S_1 \cdots S_n) \begin{pmatrix} A_{n+1} \\ B_{n+1} \end{pmatrix}$$

If we take $A_0 = 0$, $B_0 = 1$, then $A_{n+1} = c_{11}(k)$ and $B_{n+1} = c_{12}(k)$ whereas if we take $A_{n+1} = 1$, $B_{n+1} = 0$ we find $A_0 = c_{21}(k)$ and $B_0 = c_{22}(k)$. It follows that the connection matrix

$$S = S_0 S_1 \cdots S_n$$

has entries

$$S = \begin{pmatrix} c_{12}(k) & -c_{11}(k) \\ c_{22}(k) & c_{21}(-k) \end{pmatrix}$$

Example A reflectionless potential

We want to solve the Schrödinger equation in the special case $q(x) = -2\mathrm{sech}^2 x$. The solution of this problem is accomplished by a trick due to Darboux (1882). We consider the two problems

$$\begin{aligned} y'' &+& (k^2 - u(x))\, y = 0 \\ z'' &+& (k^2 - v(x))\, z = 0 \end{aligned}$$

and try to relate the two functions y and z. We look for a relationship of the form $z = A(x,k)y + y'$. Upon substituting z into the equation $z'' + (k^2 - v(x))z = 0$, we set the coefficients of y and y' to zero and obtain

$$\begin{aligned} A'' &+& u' + A(u - v) = 0 \\ 2A' &+& (u - v) = 0. \end{aligned}$$

We eliminate $u - v$ to find $A'' + u' - 2AA' = 0$ which can be integrated once to

$$A' + u - A^2 = c^2 = \text{ constant.}$$

This last equation is a Riccati equation and can be converted to a Schrödinger equation by setting $A = -w'/w$. The resulting equation is

$$w'' + (c^2 - u)w = 0$$

and we have that $v = u - 2(\ln w)''$. The upshot of this is that if y is the general solution of $y'' + (k^2 - u(x))y = 0$, then $z = y' - yw'/w$ is the general solution of $z'' + (k^2 - u + 2(\ln w)'')z = 0$ where w is any particular form of $y(x; k)$.

As an example, if we take $u = 0$, we find that $y = Ae^{ikx} + Be^{-ikx}$. For a particular solution, we take $k = i$, $A = B = 1/2$ so that $w = \cosh x$. Then $(\ln w)'' = \text{sech}^2 x$ and

$$z = Ae^{ikx}(ik - \tanh x) - Be^{-ikx}(ik + \tanh x)$$

is the general solution of $z'' + (k^2 + 2\text{sech}^2 x)z = 0$.

An examination of this solution reveals that for the potential $q(x) = -2\text{sech}^2 x$, $c_{11} \equiv 0$, $c_{12} = (ik+1)/(ik-1)$, so that the reflection coefficient is identically zero, the transmission coefficient is

$$T_r = \frac{ik - 1}{ik + 1}.$$

There is one bound state at $k = i$ given by $z(x) = u_1(x, i) = \text{sech} x$. The potential $q(x) = -2\text{sech}^2 x$ is called a REFLECTIONLESS POTENTIAL.

Since $c_{11} \equiv 0$ for a reflectionless potential, the foregoing proof that the roots of $c_{12}(k) = 0$ are simple fails. Nonetheless, for a reflectionless potential the roots $k = k_0$ are indeed simple. Can you supply the correct proof?

Example   A crystal lattice

The ideas of scattering theory can also be applied to the problem of an infinite discrete lattice. Suppose we have a lattice structure (such as a crystal) which we model as an infinite number of objects connected by springs and satisfying the equations of motion

$$m_j \frac{d^2 u_j}{dt^2} = k_j(u_{j+1} - u_j) + k_{j-1}(u_{j-1} - u_j) \quad , \quad -\infty < j < \infty.$$

The variables u_j represents the displacement of the jth object from its equilibrium.

To study the vibrations of this lattice we make the substitution $u_j = v_j e^{-i\omega t}$ to obtain the infinite set of algebraic difference equations

$$k_j(v_{j+1} - v_j) + k_{j-1}(v_{j-1} - v_j) = -\omega^2 m_j v_j.$$

Case 1   The Ideal Crystal

To study the solutions of this equation we assume that we know something about the constants k_j, m_j. Suppose, to begin with, that $k_j = k$, $m_j = m$, independent of j, that is, the lattice is homogeneous. We try a solution of the form $v_j = e^{ji\theta}$ and obtain the characteristic equation

$$\sin^2 \theta/2 = \frac{\omega^2 m}{4k}.$$

If $\frac{\omega^2 m}{4k} < 1$, there is a unique value $\theta = \theta_0(\omega)$, $0 < \theta_0 < \pi$ so that $\theta = \pm\theta_0$ satisfies this equation. In view of our notation $u_j = e^{-i\omega t}v_j$, we see that the solution $v_j = e^{-ji\theta_0}$ corresponds to a left moving wave and $v_j = e^{ji\theta_0}$ corresponds to a right moving wave, with "speed" dependent on ω.

Other solutions of the characteristic equation are found by setting $z = e^{i\theta}$ from which we obtain

$$z^2 + \left(\frac{\omega^2 m}{k} - 2\right) z + 1 = 0.$$

The solutions for $\frac{\omega^2 m}{4k} < 1$ we already know lie on the unit circle $z = e^{\pm i\theta_0}$. If $\omega^2 m/4k > 1$ there are two real roots, one with $|z| < 1$ and one with $|z| > 1$. As a result the only bounded solutions of this problem occur when $\omega^2 m/4k < 1$ and correspond to left and right moving waves (i.e., continuous spectrum of the difference operator).

Case 2 The Slightly Imperfect Crystal

Suppose $m_j = m$ for all j and that $k_j = k^*$ for $j \neq 0$. That is, the only inhomogeneity in the lattice occurs at $j = 0$.

To understand the result of a scattering experiment, we take $v_j = e^{-ji\theta} + Re^{ji\theta}$ for $j > 0$ to represent an incoming wave of unit amplitude from $j = \infty$, and a reflected outgoing wave of amplitude R. For $j \leq 0$, we take $v_j = Te^{-ij\theta}$. We substitute the assumed solutions into the difference equation and find two equations for the unknowns R, T,

$$k_0\left(Rz + \frac{1}{z} - T\right) + k^*(T - T/z) = 0 \qquad (j = 0)$$

$$k^*\left(-1 - R + Rz + \frac{1}{z}\right) + k_0(T - Rz - 1/z) = 0 \qquad (j = 1)$$

where $z = e^{i\theta}$, $-m\omega^2 = k^*\left(z - 2 + \frac{1}{z}\right)$. We solve these for R and T to find

$$R(z) = \frac{(k_0 - k^*)(z - 1)}{z\left(2k_0 z + k^*(1 - z)\right)}$$

$$T(z) = \frac{k_0(z + 1)}{2k_0 z + k^*(1 - z)}.$$

A bound state occurs for $z = \frac{k^*}{k^* - 2k_0}$ provided $k_0 > k^*$, with corresponding resonant frequency

$$\omega^2 = \frac{4}{m}\frac{k_0^2}{2k_0 - k^*}.$$

Observe that when the lattice has an inhomogeneity, one can detect it by looking at the reflection coefficient. If $k_0 > k^*$, there is a frequency at which a bound state exists, and one can experimentally determine this frequency by trying to resonate the crystal. The resonant frequency is a signature of the imperfection.

If the imperfection has a more complicated structure, there may be many bound states. However, the reflection coefficient provides a signature for the imperfection which allows the nature of the imperfection to be determined. The details of this reconstruction are discussed in Chapter 9.

As was true with the previous differential operators, we can find the spectral representation of the delta function for the Schrödinger operator by an examination of its Green's function. For the Schrödinger operator, the Green's function satisfies $G'' + (\lambda - q(x))G = -\delta(x - \xi)$ and we find

$$
\begin{aligned}
G(x,\xi;\lambda) &= -\big[u_1(\xi,k)u_2(x,k)H(\xi - x) \\
&\quad + u_1(x,k)u_2(\xi,k)H(x - \xi)\big]/2ikc_{12}(k), \\
&\equiv -\phi(x,\xi;k)/2ikc_{12}(k),
\end{aligned}
$$

where $k^2 = \lambda$. This function is outgoing at infinity, or equivalently, exponentially decaying as $x \to \infty$ when $\operatorname{Im} k > 0$.

We now integrate $G(x,\xi;\lambda)$ around a large circle in the complex λ-plane in the counterclockwise direction. We find that

$$
\delta(x - \xi) = \frac{-1}{2\pi i}\int_C G(x,\xi;\lambda)d\lambda = -\frac{1}{\pi i}\int_{C'} G(x,\xi;k^2)k\,dk
$$

where C' is a large semicircle in the upper half k plane, traversed in the counterclockwise direction. Since ϕ is analytic in the upper half k-plane, it follows that

$$
\delta(x - \xi) = \frac{1}{2\pi}\int_{-\infty}^{\infty} \frac{\phi(x,\xi;k)}{c_{12}(k)}dk + \sum_j \frac{\phi(x,\xi;k_j)}{i\dot{c}_{12}(k_j)}
$$

where $c_{12}(k_j) = 0$, $\operatorname{Im} k_j > 0$, $j = 1, 2, \ldots, n$. This representation consists of two parts, an integral along the real k axis, and a discrete sum over the roots of $c_{12}(k) = 0$.

The discrete sum can be simplified by noting that if $c_{12}(k_j) = 0$ then $u_2(x,k_j) = c_{11}(k_j)u_1(x,k_j)$ and $\dot{c}_{12}(k_j) = -ic_{11}(k_j)\int_{-\infty}^{\infty} u_1^2(x,k_j)dx$. It follows that

$$
\sum_j \frac{\phi(x,\xi;k_j)}{\dot{c}_{12}(k_j)} = \sum_j \frac{u_1(x,k_j)u_1(\xi,k_j)}{\alpha_j}
$$

where $\alpha_j = \int_{-\infty}^{\infty} u_1^2(x,k_j)dx$. From this we see that the discrete sum is the usual sum over all normalized eigenfunctions, or bound states, of the problem. We define $\psi(x,k)$ to be the improper eigenfunction with asymptotic behavior

$$
\psi(x,k) \sim e^{ikx} + S_{\pm}e^{i|kx|}
$$

for large x. That is, $\psi(x, k)$ represents the result of a scattering experiment. In fact, for $k > 0$,

$$\psi(x, k) = \frac{u_1(x, k)}{c_{21}(k)}$$

while for $k < 0$,

$$\psi(x, k) = \frac{u_2(x, -k)}{c_{12}(-k)}.$$

Using the defining relationship between $u_1(x, k)$, $u_2(x, k)$ and the coefficients c_{ij}, it is a straightforward matter to verify that

$$\int_{-\infty}^{\infty} \psi(x, k)\overline{\psi(\xi, k)}dk \;=\; \int_0^{\infty} \frac{u_2(x, k)\overline{u_2(\xi, k)}}{c_{12}(k)\overline{c_{12}(k)}}dk$$

$$+ \; \int_0^{\infty} \frac{u_1(x, k)\overline{u_1(\xi, k)}}{c_{21}(k)\overline{c_{21}(k)}}dk$$

$$= \; \int_{-\infty}^{\infty} \frac{\phi(x, \xi; k)}{c_{12}(k)}dk$$

and we conclude that

$$\delta(x - \xi) = \frac{1}{2\pi}\int_{-\infty}^{\infty} \psi(x, k)\overline{\psi(\xi, k)}dk + \sum_j \frac{u_1(x, k_j)u_1(\xi, k_j)}{\alpha_j}$$

where $\psi(x, k) \sim e^{ikx} + S_\pm e^{i|kx|}$ for large x, and $\alpha_j = \int_{-\infty}^{\infty} u_1^2(x, k_j)dx$. In other words, the delta function is represented as a sum of discrete, normalized eigenfunctions (bound states) and a continuous integral of improper eigenfunctions, normalized to resemble the result of a scattering experiment.

Some examples may help to clarify this formula. If we take $q(x) = 0$, the two fundamental solutions can be written down explicitly as

$$u_1(x, k) = e^{ikx} \quad , \quad u_2(x, k) = e^{-ikx}$$

and $c_{12}(k) \equiv 1$. Clearly, there are no roots of $c_{12}(k) = 0$ and the representation of the delta function that results is one we have seen before

$$\delta(x - \xi) = \frac{1}{2\pi}\int_{-\infty}^{\infty} e^{ik(x-\xi)}dk$$

from which the Fourier Transform results, as expected.

A slightly less trivial example is $q(x) = A\delta(x)$. We know from before that

$$u_2(x, k) = \begin{cases} e^{-ikx} & x < 0 \\ \frac{A}{2ik}e^{ikx} + \frac{k+iA/2}{k}e^{-ikx} & x > 0 \end{cases}$$

and in a similar way

$$u_1(x,k) = \begin{cases} \frac{k+iA/2}{k}e^{ikx} + \frac{A}{2ik}e^{-ikx} & x < 0 \\ e^{ikx} & x > 0 \end{cases}$$

The bound state is given by $u = e^{\frac{A|x|}{2}}$ provided $A < 0$, and the eigenvalue is at $\lambda = k_0^2 = -A^2/4$. The representation of the delta function that results is

$$\delta(x - \xi) = \frac{H(-A)}{2|A|} \exp(-|Ax| - |A\xi|) + \frac{1}{2\pi} \int_{-\infty}^{\infty} \psi(x,k)\bar{\psi}(\xi,k)dk$$

where $\psi(x,k) = \frac{k}{k+iA/2}u_1(x,k)$ for $k > 0$ and $\psi(x,k) = \frac{k}{k-iA/2}u_2(x,-k)$ for $k < 0$.

Further Reading

A classical book on eigenfunction expansions is

1. E.C. Titchmarsh, *Eigenfunction Expansions Associated with Second-Order Differential Equations*, Oxford University Press, Oxford, 1946.

2. B. Friedman, *Principles and Techniques of Applied Mathematics*, Wiley and Sons, New York, 1956

contains a nice discussion of the main ideas of this chapter. The treatment of spectral theory given in

3. I. Stakgold, *Green's Functions and Boundary Value Problems*, Wiley and Sons, New York, 1979

is from a more modern perspective.
A proof of the Wiener-Paley theory can be found in

4. A.F. Timin, *Theory of Approximation of Functions of a Real Variable*, Macmillan, New York, 1963.

The recent book by

5. G.L. Lamb, Jr., *Elements of Soliton Theory*, *Wiley-Interscience*, New York, 1980

gives a readable introduction to scattering theory for the Schrödinger equation and is also relevant to the topics of Chapter 9.

Problems for Chapter 7

Section 7.1

1. Show that if λ is in the continuous spectrum of L, then $L - \lambda$ does not have closed range.

2. Classify the spectrum of the following operators acting on the space of complex valued vectors $x = (x_1, x_2, \ldots, x_n, \ldots)$, $\|x\|^2 = \sum_{i=1}^{\infty} |x_i|^2 < \infty$.

   a. $L_1 x = (0, x_1, x_2, \ldots)$

   b. $L_2 x = (x_2, x_3, x_4, \ldots)$

   c. $L_3 x = (x_1, x_2/2, x_3/3 \ldots)$

   Hint: Notice that L_2 is the adjoint of L_1.

3. For a vector x as in Problem 2, define the ith component of $L_4 x$ by

$$(L_4 x)_i = x_{i-1} + x_{i+1} \quad \text{(take } x_0 = 0\text{)}.$$

   Show that L_4 has no discrete spectrum and that every real λ with $-2 < \lambda < 2$ belongs to the continuous spectrum of L_4.
   Hint: Use improper eigenfunctions to show $(L_4 - \lambda)^{-1}$ is unbounded.

4. Show that every positive real number λ is in the point spectrum of $Lu = -\frac{d^2 u}{dx^2}$, $-a \le x \le a$, with $u(a) = -u(-a)$, $u'(a) = u'(-a)$.

5. Suppose $f(x)$ is a smooth function, $\lim_{x \to -\infty} f(x) = \alpha$, $\lim_{x \to \infty} f(x) = \beta < \alpha$. Let $Lu = u'' - cu' + f(x)u$. Show that for every real λ in the interval (β, α), $Lu = \lambda u$ has a bounded solution, that is, the continuous spectrum includes the interval (β, α).

Section 7.2

Find the spectral representation of the delta function for the following operators

1. $Lu = -\frac{d^2 u}{dx^2}$, $x \in [0, 1]$, $u(0) = u'(1) = 0$

2. $Lu = -\frac{d^2 u}{dx^2}$, $x \in [0, \infty)$, $u'(0) = 0$

3. $Lu = -\frac{d^2 u}{dx^2}$, $x \in [0, \infty)$, $u'(0) = \alpha u(0)$, α real

4. Prove that the operator in Problem 2 has continuous spectrum for λ real, $\lambda \in [0, \infty)$.

5. Find the Fourier transform of $f(t) = t^{1/2} e^{(i+c)t} H(t)$. What is the correct path of integration for the inverse transform?

6. Establish the relationship

$$f(x) = \sum_{n=1}^{\infty} \alpha_n J_\nu(\lambda_n x)$$

$$\alpha_n = \frac{2}{J_{\nu+1}^2(\lambda_n)} \int_0^1 tf(t) J_\nu(\lambda_n t) dt$$

where $J_\nu(\lambda_n) = 0$ for $n = 1, 2, \ldots$.

7. Suppose the Fourier transform of $f(x)$ is $F(\mu)$. Evaluate the Fourier transform of

a. $\frac{df}{dx}$ (assume that $\frac{df}{dx}$ is in $L^1(-\infty, \infty)$.)

b. $\int_a^x f(x) dx$

c. $g(x) = f(x - k)$

d. $e^{ikx} f(x)$

8. Evaluate $\frac{1}{2\pi} \int_{-\infty}^{\infty} \frac{1}{1+i\mu} (i\mu F(\mu)) e^{-i\mu x} d\mu$ where $F(\mu)$ is the Fourier transform of $f(x)$.

9. Solve the integral equation

$$4 \int_{-\infty}^{\infty} e^{-|x-y|} u(y) dy + u = f(x), \quad -\infty < x < \infty.$$

10. If $F(\mu)$ is the Fourier transform of $f(t)$, what function has transform $|F(\mu)|^2$?

11. Find the Fourier transform of $J_0(at)$. Show that $J_0(at)$ is band limited, that is, its Fourier transform is zero for $|\mu| > a$.
Answer: $F(\mu) = 2(a^2 - \mu^2)^{-1/2}$ for $|\mu| < a$.

Section 7.3

1. Denote the Laplace transform of f by $L(f)$. Show that

a. $L(f') = sL(f) - f(0)$

b. $L\left(\int_0^t f(\eta) d\eta\right) = \frac{1}{s} L(f)$

c. $L(e^{at} f(t)) = F(s - a)$

d. $L(H(t - a) f(t - a)) = e^{-as} F(s)$

e. $L\left(\int_0^t f(\tau) g(t - \tau) d\tau\right) = L(f) L(g)$

2. Evaluate the Laplace transforms of

a. t^α b. e^{at}

3. a. Use Laplace transforms to solve the Volterra integral equation

$$\int_0^x k(x-y)u(y)dy + u(x) = f(x).$$

b. Use Laplace transforms to solve the generalized Abel's integral equation

$$T(x) = \int_0^x \frac{f(y)dy}{(x-y)^\alpha}.$$

4. Denote the Mellin transform of $f(x)$ by $M[f(x)] = F(s)$. Show that

a. $M[f(ax)] = a^{-s}F(s)$

b. $M[x^\alpha f(x)] = F(s+\alpha)$

c. $M[\ln x f(x)] = \frac{d}{ds}F(s)$

d. $M\left[x\frac{df}{dx}\right] = -sF(s)$

e. $M\left[\int_0^x f(t)dt\right] = -\frac{1}{s}F(s+1)$

Evaluate the Mellin Transform for the following functions:

5. $f(x) = \begin{cases} 1 & 0 < x < 1 \\ 0 & x > 1 \end{cases}$

6. $f(x) = (1+x)^{-1}$

7. $f(x) = e^{-x}$

8. $f(x) = \begin{cases} \ln x & 0 < x < 1 \\ 0 & x > 1 \end{cases}$

9. $f(x) = e^{ix}$

10. $f(x) = \cos x$

11. $f(x) = x^{-\nu}J_\nu(x)$

Hint: Use the integral representation

$$J_\nu(x) = \frac{2(x/2)^\nu}{\sqrt{\pi}\Gamma(\nu+1/2)} \int_0^1 \cos xt(1-t^2)^{\nu-1/2}dt$$

with Re $\nu > -1/2$.

12. Find the transform appropriate for the three-dimensional spherically symmetric Laplacian operator

$$\nabla^2 u = u_{rr} + \frac{2}{r}u_r, 0 < r < \infty.$$

13. Find the Hankel transform $G(\rho)$ for the following functions

a. $g(r) = 1/r$

b. $g(r) = (a^2 - r^2)^{-1/2}$ for $r < a, g(r) = 0$ for $r > a$,

c. $g(r) = (1 - e^{-r})/r^2$ (answer: $G(\rho) = \sinh^{-1}(1/\rho)$)

Section 7.4

1. Find the transform appropriate to the difference operator

$$(\Delta u)_j = u_{i-1} + u_{i+1} - 2u_i$$

defined on the set of sequences $\{u_i\}_{i=1}^{\infty}$ with $u_0 = 0$ and $\sum_{i=1}^{\infty} |u_i|^2 < \infty$.

2. Find the solution of the initial value problem

$$(\Delta v)_n = f_n \quad , \quad (\Delta v)_n = v_{n+1} - \lambda v_n + v_{n-1}$$

where $v_n = f_n = 0$ for $n < 0$.

Hint: Solve $(\Delta v)_n = \delta_{nj}$ with $v_n = 0$ for $n < j$.

Section 7.5

1. Find two linearly independent solutions on $-\infty < x < \infty$ of $u'' + q(x)u = 0$ where

$$q(x) = \begin{cases} 1 & x > 0, \\ -1 & x < 0 \end{cases}$$

2. Discuss scattering theory on the half line. That is, discuss the behavior of solutions of the Schrödinger equation $u'' + (\lambda - q(x))\, u = 0$ for $0 \le x < \infty$ with the boundary condition $u(0) = 0$. Identify the reflection coefficient. (There is no transmission coefficient.) When are there bound states?

Find reflection coefficients, the bound states, and, if appropriate, transmission coefficients for the operator $Lu = -\frac{d^2 u}{dx^2} + q(x)u$ for the choices of $q(x)$ and domains in Problems 3-6.

3. $x \in [0, \infty)$, $u(0) = 0$, $q(x) = A\delta(x - 1)$

4. $x \in [0, \infty)$, $u(0) = 0$, $q(x) = -A$, $x < a$, $q(x) = 0$, $x > a$.

5. $x \in (-\infty, \infty)$, $q(x) = \begin{cases} A & 0 < x < a \\ -A & -a < x < 0 \\ 0 & \text{elsewhere} \end{cases}$

6. $x \in (-\infty, \infty)$, $q(x) = \alpha\delta(x) + \beta\delta(x - a)$.

7. Obtain the following estimates for the fundamental solutions of the Schrödinger equation

$$\left| u_1(x, k)e^{-ikx} - 1 + \frac{1}{2ik} \int_x^{\infty} q(y)\,dy \right| \le K_1/|k|^2$$

$$\left| u_2(x, k)e^{ikx} - 1 + \frac{1}{2ik} \int_{-\infty}^x q(y)\,dy \right| \le K_2/|k|^2$$

$$\left| c_{12}(k) - 1 + \frac{1}{2ik} \int_{-\infty}^{\infty} q(y)\,dy \right| \le K_3/|k|^2$$

for $\operatorname{Im} k > 0$, $|k|$ large.

8. a. Find the relationship between $R_r(k)$ and $R_L(k)$ and between $T_r(k)$ and $T_L(k)$ if $q(-x) = q(x + \phi)$.

    b. What is the relationship between the reflection and transmission coefficients of the two Schrödinger equations with potentials $q_1(x)$ and $q_2(x)$ where $q_1(x) = q_2(x + \phi)$?

9. a. Find reflection and transmission coefficients for the vibrations of an inhomogeneous string governed by $u_{tt} = c^2(x)u_{xx}$ where $c(x) = c_2$ for $0 < x < a$ and $c(x) = c_1$ elsewhere. Compare the cases $c_1 < c_2$ and $c_1 > c_2$. Discuss the existence of standing waves.

    b. Discuss the vibrations of a semi-infinite string with $u(0) = 0$ and $c(x) = c_1$ for $0 < x < a$ and $c(x) = c_2$ for $x > a$. Are there standing waves?

10. Show that $q(x) = -6\operatorname{sech}^2 x$ is a reflectionless potential. Find the transmission coefficient and all bound states for this potential.

11. Use the method of Darboux to show that $q(x) = -n(n + 1)\operatorname{sech}^2 x$ is a reflectionless potential of $y'' + (k^2 - q(x)) y = 0$ with n bound states. Show that the general solution is $y_n = 0_n 0_{n-1} \cdots 0_1 y_0$ where the operator $0_j = \frac{d}{dx} - j \tanh x$, $y_0 = Ae^{ikx} + Be^{-ikx}$.

12. What is the spectral representation of the delta function for the Schrödinger equation with the reflectionless potential $q(x) = -2\operatorname{sech}^2 x$?

13. Find all bound states of the infinite lattice with two imperfections $m_0 \neq m$ and $m_k \neq m$ for $k \neq 0$.

14. Use the Schwarz inequality to prove that

$$\int_{-\infty}^{\infty} x^2 |\psi|^2 dx \cdot \int_{-\infty}^{\infty} \left|\frac{d\psi}{dx}\right|^2 dx \geq \frac{1}{2}$$

provided $\psi(x)$ vanishes rapidly as $x \to \pm\infty$.
Remark: If ψ is a Schrödinger bound state, this is a statement of the Heisenberg Uncertainty Principle, where $\int_{-\infty}^{\infty} x^2 |\psi|^2 dx$ measures the spread of the density and $\int_{-\infty}^{\infty} |\frac{d\psi}{dx}|^2 dx$ measures the spread of the momentum of a particle.

15. Find the reflection coefficient for the Schrödinger equation on the domain $a \leq x \leq \infty$ with potential $q(x) = -2\operatorname{sech}^2 x$ and boundary condition $u(a) = 0$. When are there bound states?

8

PARTIAL
DIFFERENTIAL
EQUATIONS

The study of linear partial differential equations is in some respects no different than that of matrices or ordinary differential equations. Symbolically, if L is a partial differential operator, to solve $Lu = f$ we must find the operator L^{-1} which inverts the problem to $u = L^{-1}f$. The operator L^{-1} is the Green's operator. The main difficulty, however, is that the construction of this Green's operator is now an order of magnitude harder than before. If we have learned anything from all that precedes this section, it is that we want some way to reduce our problems to simpler equations which we already know how to solve. That is, after all, the whole point of transform theory.

We begin with a brief summary of relevant background information. The LAPLACIAN OPERATOR $\nabla^2 u$ is the second order operator

$$\nabla^2 u = \nabla \cdot (\nabla u)$$

which in Cartesian coordinates has the representation $\nabla^2 u = \sum_{i=1}^{n} \frac{\partial^2 u}{\partial x_i^2}$ where $(x_1, x_2, \ldots x_n)$ are n independent (Cartesian) variables. The simplest equations involving the Laplacian are

$$\nabla^2 u = 0 \qquad \text{LAPLACE'S EQUATION}$$

and

$$\nabla^2 u = f \qquad \text{POISSON'S EQUATION}$$

with some appropriate boundary conditions. Other equations involving the Laplacian are

$$\frac{\partial u}{\partial t} - \nabla^2 u \;=\; f \qquad \text{(heat equation)}$$

$$\frac{\partial^2 u}{\partial t^2} - \nabla^2 u \;=\; f \qquad \text{(wave equation)}$$

again, with boundary conditions specified. Laplace's and Poisson's equations typically represent the time independent (steady state) situation of which the heat and wave equations are the time dependent versions.

Other important uses of Laplace's and Poisson's equations are to two dimensional fluid flow (as we saw in Chapter 6) and to electrostatics and gravitation. In electrostatics, the equation $\nabla^2 u = \rho$ relates the electrical potential u to the density of charge distribution ρ, and the electrical field is the gradient of u. To study gravity, one identifies the function u with the gravitational potential, ∇u with the gravitational force field and ρ with the distribution of mass in space.

The Laplacian has many features similar to a Sturm-Liouville operator in one spatial dimension. For example, GREEN'S FORMULA

$$\int_R \left(u\nabla^2 v - v\nabla^2 u \right) dV = \int_{\partial R} \left(u\frac{\partial v}{\partial n} - v\frac{\partial u}{\partial n} \right) ds$$

where R is an n-dimensional region and ∂R its boundary, and $\frac{\partial}{\partial n}$ is the outward normal derivative. This is the higher dimensional analogue of the integration by parts formula

$$\int_a^b (uv'' - vu'')dx = (uv' - vu') \Big|_a^b$$

which we use so often for Sturm-Liouville theory. Similarly,

$$\int_R u\nabla^2 u \, dV = \int_{\partial R} u\frac{\partial u}{\partial n} ds - \int_R |\nabla u|^2 \, dV$$

is the analogue of the integration by parts formula

$$\int_a^b uu'' dx = uu' \Big|_a^b - \int_a^b u'^2 dx.$$

These formulae show that $-\nabla^2 u$ is a formally self adjoint, positive definite operator, with inner product $\langle u, v \rangle = \int_R uv \, dV$, provided the boundary conditions are chosen correctly. The boundary conditions are called DIRICHLET

CONDITIONS if u is specified on the boundary, or NEUMANN CONDITIONS if $\frac{\partial u}{\partial n}$ is specified on the boundary. If $u = 0$ on the boundary $-\nabla^2 u$ is positive definite. If the mixed condition $\frac{\partial u}{\partial n} + \alpha u = 0$, $\alpha > 0$, holds on the boundary, $-\nabla^2 u$ is also a positive definite operator, while if $\partial u/\partial n = 0$ on all of the boundary, $-\nabla^2 u$ is non-negative definite.

8.1   Poisson's Equation

8.1.1   Fundamental solutions

We begin by trying to find the Green's function for Laplace's operator for a variety of unbounded domains R. The Green's function is defined by the differential equation $-\nabla^2 G = \delta$. (The minus sign is a matter of convention since $-\nabla^2$ is a positive definite operator.) However, just what we mean by a delta function in higher dimensions is not yet clear.

To define higher dimensional distributions, we first define the set of test functions to be infinitely differentiable and to have compact support. Then distributions are linear functionals on the space of test functions denoted $\langle f, \phi \rangle = \int_R f\phi \, dV$, where integration is in the sense of Lebesgue, whenever that makes sense.

We can define the product of distributions as follows. If $f_1(x_1)$ and $f_2(x_2)$ are each locally integrable functions of one variable, then the product $f_1(x_1) f_2(x_2)$ is a locally integrable function of two variables and defines a two-dimensional distribution whose action can be defined iteratively by $\psi(x_1) = \langle f_2(x_2), \phi(x_1, x_2) \rangle$, $\langle f_1 f_2, \phi(x_1, x_2) \rangle = \langle f_1(x_1), \psi(x_1) \rangle$, or equivalently by $\chi(x_2) = \langle f_1(x_1), \phi(x_1, x_2) \rangle$, $\langle f_1 f_2, \phi(x_1, x_2) \rangle = \langle f_2(x_2), \chi(x_2) \rangle$. Certainly $\psi(x_1)$ and $\chi(x_2)$ are test functions if $\phi(x_1, x_2)$ is a two-dimensional test function.

If f_1 and f_2 are arbitrary one-dimensional distributions, we use these statements to define $f_1 f_2$ as a two-dimensional distribution, as long as the arguments of $f_1 f_2$ are different. In this way, one can readily show that the two-dimensional delta function $\delta(x_1, x_2)$, defined to have action $\langle \delta(x_1, x_2), \phi(x_1, x_2) \rangle = \phi(0, 0)$, can be expressed as $\delta(x_1, x_2) = \delta(x_1)\delta(x_2)$ where $\delta(x)$ is the usual one-dimensional delta distribution. Higher dimensional distributions are defined in the same way.

Other properties of higher dimensional distributions are also defined in such a way as to be consistent with usual integration rules. Thus, for example, one defines partial differentiation in x of a distribution $g(x, y)$ by $\langle g_x(x, y), \phi(x, y) \rangle = -\langle g(x, y), \phi_x(x, y) \rangle$. Similarly, a change of variables must leave the action of a distribution unchanged. Thus, under the change of variables $u = u(x)$ (x here is an n-dimensional vector of independent variables), if $f(x) = F(u)$ is a distribution, and $\phi(x) = \Phi(u)$ is a test function, it must be that $\int f(x)\phi(x)dx = \int F(u)\Phi(u)J(u)du$, where $J(u)$ is the Jacobian of $u(x)$, $J = |\det(\partial x/\partial u)|$. It follows, for example, that in polar coordinates $\delta(x, y) = \frac{\delta(r)}{2\pi r}$.

To find the Green's function for the Laplacian in three dimensions we seek to solve

$$-\nabla^2 G = \delta(x - \xi)\delta(y - \eta)\delta(z - \zeta).$$

It is convenient to introduce the change of variables $x' = x - \xi$, $y' = y - \eta$, $z' = z - \zeta$ so that the point (ξ, η, ζ) is translated to the origin, and then we must solve

$$\frac{\partial^2 G}{\partial x'^2} + \frac{\partial^2 G}{\partial y'^2} + \frac{\partial^2 G}{\partial z'^2} = -\delta(x')\delta(y')\delta(z').$$

What this actually means is that we seek a solution of

$$\iiint G\nabla^2 \Psi \, dV = -\Psi(0,0,0)$$

for all test functions $\Psi(x, y, z)$. (This is the weak formulation of the equation $\nabla^2 G = -\delta$.) We expect the Green's function to be spherically symmetric so we introduce spherical coordinates

$$z' = r \cos\theta, \quad x' = r\sin\theta\cos\phi, \quad y' = r\sin\theta\sin\phi,$$

in terms of which the Laplacian becomes

$$\nabla^2 G = \frac{1}{r^2}\frac{\partial}{\partial r}\left(r^2 \frac{\partial G}{\partial r}\right) + \frac{1}{r^2 \sin\theta}\frac{\partial}{\partial \theta}\left(\sin\theta \frac{\partial G}{\partial \theta}\right) + \frac{1}{r^2 \sin^2\theta}\frac{\partial^2 G}{\partial \phi^2}.$$

Furthermore, the weak formulation has the representation

$$\int_0^\infty \int_0^{2\pi} \int_0^\pi G(r)\,(\nabla^2\Psi)r^2 \, \sin\theta \, d\theta \, d\phi \, dr = -\Psi \big|_{r=0}$$

for all smooth Ψ with compact support.

Since Ψ has compact support, we can integrate by parts and find

$$-\Psi \big|_{r=0} = \int_0^{2\pi} \int_0^\pi \left[\lim_{r\to 0}(r^2 G_r \Psi - r^2 G\Psi_r) + \int_0^\infty (r^2 G_r)_r \Psi \, dr\right] \sin\theta \, d\theta \, d\phi.$$

Now we realize that G must satisfy $\frac{d}{dr}\left(r^2 \frac{dG}{dr}\right) = 0$ for $r > 0$ and

$$\lim_{r\to 0}\left(r^2 \frac{dG}{dr}\right) = -\frac{1}{4\pi}, \qquad \lim_{r\to 0}(r^2 G) = 0.$$

We find that $G = \frac{1}{4\pi r} + C$ where C is arbitrary. We pick $C = 0$, so that $\lim_{r\to\infty} G = 0$, with the result that $G = \frac{1}{4\pi r}$, or in the original coordinate system

$$G(x, y, z, \xi, \eta, \zeta) = \frac{1}{4\pi} \left[(x - \xi)^2 + (y - \eta)^2 + (z - \zeta)^2 \right]^{-1/2}.$$

As a consequence, the solution of Poisson's equation $\nabla^2 u = -f$ in $\mathbb{R}^3$ is given by

$$u(x, y, z) = \frac{1}{4\pi} \int_{-\infty}^{\infty} \int_{-\infty}^{\infty} \int_{-\infty}^{\infty} \frac{f(\xi, \eta, \zeta) d\xi \, d\eta \, d\zeta}{[(x - \xi)^2 + (y - \eta)^2 + (z - \zeta)^2]^{1/2}}.$$

Notice that this representation of the solution agrees with what we know from physics about gravity, namely, that the gravitational potential is found by summing over all sources, weighted by the inverse of the distance from the source. As a result, the force field obeys the inverse square law, being the gradient of the potential.

In two dimensions, a similar story unfolds. The Laplacian in two dimensional Cartesian coordinates is

$$\nabla^2 u = \frac{\partial^2 u}{\partial x^2} + \frac{\partial^2 u}{\partial y^2},$$

or in polar coordinates $x = r\cos\theta$, $y = r\sin\theta$

$$\nabla^2 u = \frac{1}{r} \frac{\partial}{\partial r} \left(r \frac{\partial u}{\partial r} \right) + \frac{1}{r^2} \frac{\partial^2 u}{\partial \theta^2}.$$

The two dimensional Green's function must satisfy

$$\int_0^\infty \int_0^{2\pi} G(r) \nabla^2 \Psi r \, dr \, d\theta = -\Psi \Big|_{r=0}$$

for all smooth test functions Ψ. We integrate this weak formulation by parts to obtain

$$\int_0^{2\pi} \left(\lim_{r\to 0} (r G_r \Psi - r G \Psi_r) + \int_0^\infty (r G_r)_r \Psi \, dr \right) d\theta = -\Psi \Big|_{r=0}.$$

It follows that the Green's function must satisfy

$$\frac{d}{dr} \left(r \frac{dG}{dr} \right) = 0, \qquad r > 0,$$

and

$$\lim_{r\to 0} \left(r \frac{dG}{dr} \right) = \frac{-1}{2\pi}, \qquad \lim_{r\to 0} (rG) = 0.$$

We find that $G = \frac{-1}{2\pi} \ln r$, or, in original (untranslated) coordinates

$$G(x, y, \xi, \eta) = \frac{-1}{4\pi} \ln \left[(x - \xi)^2 + (y - \eta)^2 \right].$$

Accordingly, the solution of Poison's equation $\nabla^2 u = f$ in two dimensions is

$$u(x, y) = \frac{1}{4\pi} \int_{-\infty}^{\infty} \int_{-\infty}^{\infty} f(\xi, \eta) \ln \left[(x - \xi)^2 + (y - \eta)^2 \right] d\xi \, d\eta.$$

8.1.2 The Method of Images

The Green's functions just constructed are useful only for infinite domain problems. For problems with finite boundaries, the Green's function must satisfy homogeneous boundary conditions as well. For certain simple problems, this can be done directly using a technique called the METHOD OF IMAGES. For more complicated problems, the method of images is theoretically possible, but too cumbersome to be practical.

Consider Poisson's equation $\nabla^2 u = f$ on the half space $y > 0$, $-\infty < x < \infty$, subject to the Dirichlet condition $u = 0$ at $y = 0$. We seek a Green's function for this problem that satisfies the homogeneous boundary condition on the x-axis.

The idea of the method of images is to seek a function $H(x, y)$ which satisfies $\nabla^2 H = 0$ in the upper half plane and for which

$$G = H(x, y) + G_0(x - \xi, y - \eta)$$

satisfies the boundary condition $G(x, y) \mid_{y=0} = 0$, where G_0 is the Green's function for the full plane problem. We know that $G_0(x, y)$ satisfies $\nabla^2 G_0 = 0$ at all points except $x = y = 0$, so a candidate for H is the function $G_0(x - \xi, y - \eta)$ with ξ and η in the lower half plane, $\eta < 0$. In fact, if we take

$$G(x, y, \xi, \eta) = G_0(x - \xi, y - \eta) - G_0(x - \xi, y + \eta)$$

we see that $G(x, 0, \xi, \eta) = G_0(x-\xi, -\eta) - G_0(x-\xi, \eta) = 0$, since G_0 is an even function of its arguments. What we have done is added together two Green's functions, one with influence at $x = \xi$, $y = \eta$ and the other with influence at the mirror image about the x-axis at $x = \xi$, $y = -\eta$ (hence, the name the method of images). The result is that the sum is zero on the x-axis.

The method of images works quite well for half space and quarter space problems. For example, the Green's function for Poisson's equation with the Neumann condition $\frac{\partial u}{\partial y} = 0$ specified on the x-axis $y = 0$, $-\infty < x < \infty$, is

$$G(x, y, \xi, \eta) = G_0(x - \xi, y - \eta) + G_0(x - \xi, y + \eta).$$

The Green's function for the quarter space problem $\nabla^2 u = f$, $0 \leq x < \infty$, $0 \leq y < \infty$ with boundary condition $u = 0$ on $y = 0$, $0 \leq x < \infty$ and $\frac{\partial u}{\partial x} = 0$ on $x = 0$, $0 \leq y < \infty$ is also easy to write down. It is

$$
\begin{aligned}
G(x,y,\xi,\eta) \;=\;& G_0(x - \xi, y - \eta) - G_0(x - \xi, y + \eta) \\
+\;& G_0(x + \xi, y - \eta) - G_0(x + \xi, y + \eta).
\end{aligned}
$$

The reason this method works so nicely is because of the symmetry of the domain with respect to reflection. There are, however, many problems where this symmetry is lost. Consider for example Poisson's equation on an infinite strip $\nabla^2 u = f$, $-\infty < x < \infty$, $0 \leq y \leq a$, with Dirichlet conditions $u = 0$ at $y = 0$, and $y = a$, $-\infty < x < \infty$, and suppose we wish to construct the Green's function with influence at $x = \xi$, $y = \eta$, $0 < \eta < a$. The form of our Green's function is

$$
G = G_0(x,y,\xi,\eta) + H(x,y)
$$

where H must satisfy $\nabla^2 H = 0$ in the strip $0 < y < a$, $-\infty < x < \infty$, and G must satisfy the homogeneous boundary conditions. Motivated by the method of images, we might reflect the point ξ, η about the x-axis and choose H proportional to $G_0(x - \xi, y + \eta)$ so that $G = 0$ on $y = 0$. To satisfy $G = 0$ at $y = a$, we reflect the points (ξ, η) and $(\xi, 2a + \eta)$. We must add to G terms proportional to G_0 with influence at these points. In adding these terms, however, the boundary condition at $y = 0$ is no longer satisfied and more reflected points must be added. The upshot of this is that there are an infinite number of images that must included, and the form of the Green's function is

$$
G(x,y,\xi,\eta) = \sum_{i=0}^{\infty} \alpha_i \ln \left[(x - \xi)^2 + (y - y_i)^2 \right]
$$

where y_i are all points of the form $y = 2ia \pm \eta$, and α_i are chosen so that the boundary conditions are satisfied with $\alpha_0 = -\frac{1}{4\pi}$.

8.1.3   Transform Methods

We quickly exhaust the usefulness of the method of images and are in need of other ways to find Green's functions. The Green's function for Poisson's equation is always defined to satisfy $\nabla^2 G = -\delta$ with any homogeneous boundary conditions that may apply. If we need to solve $\nabla^2 u = f$ with inhomogeneous conditions, we can still use this Green's function to obtain the solution. This is possible because of Green's formula, which for Green's functions reduces to

$$
-u \;=\; \int_R Gf\, dV + \int_{\partial R} \left(u \frac{\partial G}{\partial n} - G \frac{\partial u}{\partial n} \right) ds.
$$

This reduces further for specific boundary conditions. For example, with Dirichlet conditions

$$-u = \int_R Gf\,dV + \int_{\partial R} u\frac{\partial G}{\partial n}\,ds$$

while with Neumann conditions we have

$$-u = \int_R Gf\,dV - \int_{\partial R} G\frac{\partial u}{\partial n}\,ds.$$

With this in mind we are always able to focus our attention solely on finding the Green's function with homogeneous boundary conditions. Keep in mind, however, that in specific situations it may be easier to directly solve a problem with inhomogeneous conditions, than to find the general inverse and then to apply it to solve a problem with inhomogeneous boundary data. (This is analogous to the fact that with a matrix A, it is often preferable from a computational viewpoint to directly solve $Ax = b$ via Gaussian elimination than to first find A^{-1} and then calculate $x = A^{-1}b$.)

There are a number of equivalent ways to do the computations necessary to find a Green's function, and, of course they all give the same answer. To get an overview of these (only slightly) different methods we consider first the problem $\nabla^2 u = 0$ on a rectangle $0 \le x \le a, 0 \le y \le b$ with Dirichlet boundary data $u(x, 0) = f(x)$, $u(x, b) = g(x)$, $u(0, y) = u(a, y) = 0$. This is a problem that is usually solved via SEPARATION of VARIABLES in elementary texts. The idea of separation of variables is to try as a first guess a solution of the form $u = X(x)Y(y)$ and to observe that this implies that

$$\frac{X''(x)}{X(x)} = -\frac{Y''(y)}{Y(y)}.$$

The usual argument is that a function of x can equal a function of y only if both functions are equal to the same constant. Thus we take

$$X'' + kX = 0$$

$$Y'' - kY = 0.$$

Here we must make a choice, to choose k either positive or negative. The two possibilities lead to two different representations of the solution. If we choose $k = \lambda^2 > 0$, we find

$$X = A\sin\lambda x + B\cos\lambda x$$

and

$$Y = C\sinh\lambda y + D\cosh\lambda y.$$

If $U = XY$ is to satisfy the homogeneous boundary conditions $u(0, y) = u(a, y) = 0$, we must choose $B = 0$ and $\lambda = \frac{n\pi}{a}$.

Since n is not determined, we take u to be a superposition of all possible solutions

$$u(x,y) = \sum_{n=1}^{\infty} \sin \frac{n\pi x}{a} \left(C_n \sinh \frac{n\pi y}{a} + D_n \cosh \frac{n\pi y}{a} \right).$$

Now we choose C_n and D_n so that $u(x,0) = f(x)$, $u(x,b) = g(x)$. This is accomplished by requiring

$$\sum_{n=1}^{\infty} D_n \sin \frac{n\pi x}{a} = f(x)$$

and

$$\sum_{n=1}^{\infty} \left(C_n \sinh \frac{n\pi b}{a} + D_n \cosh \frac{n\pi b}{a} \right) \sin \frac{n\pi x}{a} = g(x).$$

Since these are Fourier Sine series, these equalities hold (in L^2, of course) if we use the Fourier coefficients

$$D_n = \frac{2}{a} \int_0^a f(x) \sin \frac{n\pi x}{a} dx$$

and

$$C_n \sinh \frac{n\pi b}{a} + D_n \cosh \frac{n\pi b}{a} = \frac{2}{a} \int_0^a g(x) \sin \frac{n\pi x}{a} dx.$$

from which the solution representation follows.

This representation motivates another way to find the same solution. We could have realized from the start that the solution we seek is of the form

$$u(x,y) = \sum_{n=1}^{\infty} U_n(y) \sin \frac{n\pi x}{a},$$

that is, it is represented in terms of certain eigenfunctions and we merely need to determine the coefficient functions $U_n(y)$. These coefficient functions are given by

$$U_n(y) = \frac{2}{a} \int_0^a u(x,y) \sin \frac{n\pi x}{a} dx,$$

and represent a transform of coordinates, exchanging the x dependence for n dependence. The transform can be applied directly to the governing partial differential equation if we multiply the equation by $\sin \frac{n\pi x}{a}$ and integrate by parts with respect to x from 0 to a, to obtain

$$U_n'' - \frac{n^2\pi^2}{a^2} U_n = \frac{n\pi}{a} \left((-1)^n u(a,y) - u(0,y) \right)$$

with boundary conditions

$$U_n(0) \; = \; \frac{2}{a} \int_0^a f(x) \sin \frac{n\pi x}{a} \, dx$$

$$U_n(b) \; = \; \frac{2}{a} \int_0^a g(x) \sin \frac{n\pi x}{a} \, dx.$$

This system of ordinary differential equations, with boundary conditions, is the transformed version of the original partial differential equation, and although it is an infinite system, the equations are separated in that there is no interconnection between the jth and kth equation for $j \neq k$. The transformed equations are, in fact, the projection of the original equation onto the orthogonal basis functions. For the problem as posed, $u(a, y) = u(0, y) = 0$, although, in general, the boundary conditions may be inhomogeneous since the basis functions are complete even with inhomogeneous conditions. With homogeneous boundary conditions $u(a, y) = u(0, y) = 0$, the solution of this system is

$$U_n(y) = \frac{1}{\sinh \frac{n\pi b}{a}} \left\{ U_n(b) \sinh \frac{n\pi y}{a} + U_n(0) \sinh \frac{n\pi}{a}(b - y) \right\}$$

which is another way of writing exactly what we found before.

What we have learned from this second approach is that the method of separation of variables is really a transform method, where, once we decide what transform to apply, the partial differential equation is reduced to an infinite system of (separated) ordinary differential equations. For this problem the transform was given by

$$U_n(y) \; = \; \frac{2}{a} \int_0^a u(x, y) \sin \frac{n\pi x}{a} \, dx$$

and its inverse

$$u(x, y) \; = \; \sum_{n=1}^{\infty} U_n(y) \sin \frac{n\pi x}{a}.$$

We learn another important fact from this approach, namely that separation of variables, or transform techniques, have the equivalent effect of replacing a differential operator by a constant operator, and the only remnant of the operator

$$\frac{\partial^2}{\partial x^2}$$

in the transformed equation is the appearance of the constant

$$\frac{-n^2\pi^2}{a^2}.$$

We could equally well apply the second natural transform for this problem

$$u(x,y) \;=\; \sum_{n=1}^{\infty} V_n(x) \sin \frac{n\pi y}{b}$$

with

$$V_n(x) \;=\; \frac{2}{b} \int_0^b u(x,y) \sin \frac{n\pi y}{b} \, dy.$$

To apply this transform we multiply the equation $\nabla^2 u = 0$ by $\sin \frac{n\pi y}{b}$ and integrate from $y = 0$ to $y = b$, to find the equation

$$V_n'' - \frac{n^2\pi^2}{b^2} V_n = \frac{2n\pi}{b^2} \left((-1)^n g(x) - f(x) \right)$$

which must be solved subject to boundary conditions $V_n(0) = V_n(a) = 0$. We find $V_n(x)$ by using the Green's function

$$G_n(x,\xi) = \begin{cases} \dfrac{\sinh kx \sinh k(\xi - a)}{k \sinh ka} & 0 \le x < \xi \\[2ex] \dfrac{\sinh k\xi \sinh k(x - a)}{k \sinh ka} & \xi < x \le a \end{cases}$$

where $k = \frac{n\pi}{b}$. It follows that

$$u(x,y) = \sum_{n=1}^{\infty} \left(\frac{2n\pi}{b^2} \int_0^a G_n(x,\xi) \left((-1)^n g(\xi) - f(\xi) \right) d\xi \right) \sin \frac{n\pi y}{b}.$$

We now turn our attention to Green's functions. Suppose we want to find the Green's function for the Laplacian on the rectangle $0 \le x \le a, 0 \le y \le b$ with Dirichlet conditions on $x = 0$ and $x = a$ and on $y = 0$ and $y = b$. We first look for a good candidate for a transform. We know that the operator

$$\frac{\partial^2 u}{\partial x^2}$$

with Dirichlet conditions has the transform generated by its eigenfunctions

$$\sin \frac{n\pi x}{a}$$

and eigenvalues

$$\lambda_n = \frac{-n^2\pi^2}{a^2},$$

so we apply the transform

$$U_n(y) \;=\; \frac{2}{a} \int_0^a u(x,y) \sin \frac{n\pi x}{a} \, dx$$

$$u(x,y) \;=\; \sum_{n=1}^{\infty} U_n(y) \sin \frac{n\pi x}{a}$$

to the equation

$$\frac{\partial^2 u}{\partial x^2} + \frac{\partial^2 u}{\partial y^2} = -\delta(x - \xi)\delta(y - \eta).$$

After transformation, the equation is

$$U_n'' - \frac{n^2\pi^2}{a}U_n = -\frac{2}{a}\sin\frac{n\pi\xi}{a}\delta(y - \eta)$$

$$U_n(0) = U_n(b) = 0.$$

The solution of the transformed equation is straightforward, being

$$U_n(y) = \begin{cases} \dfrac{-2\sinh\frac{n\pi y}{a}\sinh\frac{n\pi}{a}(\eta - b)\sin\frac{n\pi\xi}{a}}{n\pi\sinh\frac{n\pi b}{a}} & 0 \le y < \eta \le a \\[3mm] \dfrac{-2\sinh\frac{n\pi\eta}{a}\sinh\frac{n\pi}{a}(y - b)\sin\frac{n\pi\xi}{a}}{n\pi\sinh\frac{n\pi b}{a}} & 0 \le \eta < y \le a, \end{cases}$$

and, of course,

$$G(x, y, \xi, \eta) = \sum_{n=1}^{\infty} U_n(y)\sin\frac{n\pi x}{a}.$$

This can be summarized in operational notation as follows. We have two differential operators L_1 and L_2, in this case

$$L_1 u = \frac{\partial^2 u}{\partial x^2}$$

and

$$L_2 u = \frac{\partial^2 u}{\partial y^2}$$

with boundary conditions $u = 0$ at $x = 0$ and a for L_1, and $u = 0$ at $y = 0$ and b for L_2. We seek solutions of $L_1 G + L_2 G = -\delta(x - \xi)\delta(y - \eta)$ with G in the intersection of the domains of L_1 and L_2. The key step is to realize that we can replace the operator L_1 by a constant operator by projecting it onto the appropriate eigenfunction. If L_1 is a self-adjoint operator with a complete set of normalized eigenfunctions $\{\phi_k\}$ with corresponding eigenvalues λ_k, $L_1\phi_k = \lambda_k\phi_k$, and if $L_2\langle\phi_k, u\rangle = \langle\phi_k, L_2 u\rangle$ (i.e. projection commutes with L_2), then we transform the equation $L_1 G + L_2 G = -\delta(x - \xi)\delta(y - \eta)$ by taking the inner product of the equation with ϕ_k. That is,

$$\langle\phi_k, L_1 G + L_2 G\rangle = \langle L_1\phi_k, G\rangle + \langle\phi_k, L_2 G\rangle = -\phi_k(\xi)\delta(y - \eta)$$

or

$$(\lambda_k + L_2)\langle\phi_k, G\rangle = -\phi_k(\xi)\delta(y - \eta).$$

Notice that if the problem we are transforming has inhomogeneous boundary data, these will appear in the transformed equation because of the integration by parts formula $\langle \phi_k, Lu \rangle = \langle L\phi_k, u \rangle - J(\phi_k, u)$. We let $g(y, \eta; \lambda_k)$ be the Green's function for the operator $L_2 + \lambda_k$, $(L_2 + \lambda_k)g = -\delta(y - \eta)$, and then,

$$\langle \phi_k, G \rangle = g(y, \eta; \lambda_k)\phi_k(\xi),$$

Since $\langle \phi_k, G \rangle$ are Fourier coefficients of G relative to ϕ_k, we write

$$G = \sum_{k=1}^{\infty} g(y, \eta; \lambda_k)\phi_k(x)\phi_k(\xi),$$

which is a valid representation provided $g(y, \eta; \lambda_k)$ is well behaved at each of the eigenvalues λ_k of the operator L_1.

There is no operational reason to choose the transform associated with L_1 over that one for L_2. For example, if L_2 is self adjoint and has a complete set of normalized eigenfunctions $\{\Psi_k(y)\}$ and corresponding eigenvalues $\{\mu_k\}$, we let $h(x, \xi; \mu_k)$ be the Green's function for $L_1 + \mu_k$

$$L_1 h + \mu_k h = -\delta(x - \xi).$$

We transform the equation using the eigenfunctions of L_2 and obtain

$$(L_1 + \mu_k) \langle \Psi_k, G \rangle = -\Psi_k(\eta)\delta(x - \xi)$$

or

$$\langle \Psi_k, G \rangle = h(x, \xi; \mu_k)\Psi_k(\eta).$$

It follows that

$$G = \sum_{k=1}^{\infty} h(x, \xi; \mu_k)\Psi_k(y)\Psi_k(\eta)$$

provided $h(x, \xi; \mu_k)$ is well behaved at $\mu = \mu_k$. In other words, there are two equivalent representations of the solution of $L_1 u + L_2 u = f(x, y)$, depending on which eigenfunction expansion one chooses as the transform.

It would be incorrect to view this transform method as simply substituting the supposed solution $u = \sum \langle \phi_k, u \rangle \phi_k$ into the governing equation. In general, the function $\sum \langle \phi_k, u \rangle \phi_k$ is in L^2 but not necessarily differentiable, and cannot be substituted into the partial differential equation. The solution is only an L^2 representation of a C^2 function and may differ from the C^2 solution on a set of measure zero, in particular, at the boundary, if inhomogeneous boundary conditions are applied. Since the transform is a projection method, we are actually finding a weak solution of the equation, or if we take only a finite rather than infinite sum, a Galerkin approximation.

This method to solve partial differential equations should come as no surprise. After all, a major point of this text is that if a linear self-adjoint operator L has an orthonormal basis of eigenfunctions $L\phi_k = \lambda_k \phi_k$, then the equation $Lu = f$ can be diagonalized and solved componentwise. If we take the inner product of the equation with ϕ_k we learn that $\langle f, \phi_k \rangle = \langle Lu, \phi_k \rangle = \lambda_k \langle u, \phi_k \rangle$ and, of course, $u = \sum \langle u, \phi_k \rangle \phi_k$. This formalism is correct whether the operator is a matrix, integral, differential or difference operator. For partial differential operators, the right-hand side f may itself be another operation, say $f = L_2 u$ in which case we require L_2 to commute with ϕ_k.

As another example, consider Poisson's equation on an infinite strip $-\infty < x < \infty, 0 \le y \le a$ with Dirichlet boundary conditions at $y = 0, a$. The operator

$$L_1 = \frac{\partial^2}{\partial x^2}$$

has the domain $L^2(-\infty, \infty)$ whereas the operator

$$L_2 = \frac{\partial^2}{\partial y^2}$$

has as its domain the functions which are zero at $y = 0, a$. The transform for L_2 is generated by the trigonometric functions $\sin \frac{n\pi y}{a}$ with eigenvalues

$$\lambda_k = \frac{-n^2 \pi^2}{a^2}.$$

If we apply the Fourier Transform in x to the equation

$$G_{xx} + G_{yy} = -\delta(x - \xi)\delta(y - \eta)$$

we find that (multiply the equation by e^{ikx} and integrate from $x = -\infty$ to $x = \infty$)

$$U_{yy}(y, k) - k^2 U(y, k) = -e^{ik\xi}\delta(y - \eta)$$

where

$$U(y, k) = \int_{-\infty}^{\infty} e^{ikx} G(x, y) dx$$

is the transform of $G(x, y)$.

We can solve the transformed ordinary differential equation for $U(y, k)$ using standard techniques (it has the same form as for the previous problem)

$$U(y, k) = \begin{cases} \dfrac{-\sinh ky \sinh k(\eta - a)}{k \sinh ka} e^{ik\xi}, & 0 \le y < \eta \le a, \\[3mm] \dfrac{-\sinh k\eta \sinh k(y - a)}{k \sinh ka} e^{ik\xi}, & 0 \le \eta < y \le a. \end{cases}$$

The representation of $G(x, y)$ follows from the inverse Fourier Transform

$$G(x, y) = \frac{1}{2\pi} \int_{-\infty}^{\infty} U(y, k) e^{-ikx} dk.$$

The second representation of the Green's function can be found in two ways. The first way is to use the eigenfunctions of the operator

$$L_2 = \frac{\partial^2}{\partial y^2}$$

as the basis of a transform. These eigenfunctions are

$$\left\{\sin\frac{n\pi y}{a}\right\}_{n=1}^{\infty}$$

with corresponding eigenvalues

$$\lambda_n = \frac{-n^2\pi^2}{a^2}.$$

We transform the equation $G_{xx} + G_{yy} = \delta(x-\xi)\delta(y-\eta)$ by setting

$$g_k(x) = \frac{2}{a}\int_0^a G(x,y)\sin\frac{k\pi y}{a}dy$$

and the transformed equation is (multiply the equation by $\sin\frac{k\pi y}{a}$ and integrate from $y = 0$ to $y = a$)

$$g_k'' - \frac{k^2\pi^2}{a^2}g_k = -\frac{2}{a}\sin\frac{k\pi\eta}{a}\delta(x-\xi).$$

The L^2 solutions of this transformed equation are

$$g_k(x) = \frac{1}{k\pi}\sin\frac{k\pi\eta}{a}e^{\frac{-k\pi}{a}|x-\xi|}$$

and the Green's function is

$$G(x,y) = \sum_{k=1}^{\infty}\frac{1}{k\pi}e^{\frac{-k\pi}{a}|x-\xi|}\sin\frac{k\pi\eta}{a}\sin\frac{k\pi y}{a}.$$

A second way to obtain this representation is to change the first representation via contour integration. For example, if $y < \eta$, $x < \xi$ we have

$$G(x,y) = \frac{-1}{2\pi}\int_{-\infty}^{\infty}\frac{\sinh ky\sinh k(\eta-a)}{k\sinh ka}e^{ik(\xi-x)}dk$$

Since $x < \xi$, we can close the path of integration in the complex plane with a large semicircle in the upper half plane, and evaluate it using the residue theorem. The only poles of the integrand are simple poles at $k = \frac{n\pi i}{a}$ for $n = 1, 2, \ldots$ and the residue theorem yields

$$G(x,y) = -\sum_{n=1}^{\infty}\frac{\sinh\frac{n\pi i}{a}y\sinh\frac{n\pi i}{a}(\eta-a)e^{\frac{-n\pi}{a}(\xi-x)}}{\pi n\cosh n\pi i}$$

which reduces to

$$G(x,y) = \sum_{n=1}^{\infty} \frac{1}{n\pi} \sin \frac{n\pi y}{a} \sin \frac{n\pi \eta}{a} e^{\frac{-n\pi}{a}(\xi - x)},$$

in agreement with the result of the first calculation.

The principal difference between the various transforms is that when the domain has a finite extent, the transform is a discrete sum whereas when the domain is infinite, the transform requires an integral representation.

To see further how integral transforms work, consider the equation

$$G_{xx} + G_{yy} - \alpha^2 G = -\delta(x)\delta(y), \quad -\infty < x < \infty, \quad -\infty < y < \infty.$$

We can take the delta influence to be at the origin by making a shift of coordinate system.

If we use Fourier transforms

$$\widehat{G}(k_1, y) = \int_{-\infty}^{\infty} e^{ik_1 x} G(x,y) dx$$

we obtain the transformed equation (multiply the equation by $e^{ik_1 x}$ and integrate over x, $-\infty < x < \infty$)

$$\widehat{G}_{yy} - \left(k_1^2 + \alpha^2 \right) \widehat{G} = -\delta(y).$$

We now take a second Fourier Transform, this time in y

$$\widehat{\widehat{G}}(k_1, k_2) = \int_{-\infty}^{\infty} e^{ik_2 y} \widehat{G}(k_1, y) dy$$

to obtain the algebraic equation

$$\left(k_1^2 + k_2^2 + \alpha^2 \right) \widehat{\widehat{G}} = 1$$

from which it follows that

$$G(x,y) = \frac{1}{4\pi^2} \int_{-\infty}^{\infty} \int_{-\infty}^{\infty} \frac{e^{-i(k_1 x + k_2 y)}}{k_1^2 + k_2^2 + \alpha^2} dk_1 dk_2.$$

In the polar coordinates $x = \rho \cos\theta$, $y = \rho \sin\theta$, $k_1 = r\cos\phi$, $k_2 = r\sin\phi$, this integral has the representation

$$G(x,y) = \frac{1}{4\pi^2} \int_0^{\infty} \int_0^{2\pi} \frac{e^{-ir\rho\cos(\phi - \theta)}}{r^2 + \alpha^2} r \, d\phi \, dr.$$

From Chapter 6, we know that the Bessel function $J_0(z)$ has the integral representation

$$J_0(z) = \frac{1}{2\pi} \int_0^{2\pi} e^{-iz\sin\theta} d\theta$$

so that

$$G(x, y) = \frac{1}{2\pi} \int_0^\infty \frac{rJ_0(r\rho)}{r^2 + \alpha^2} dr, \qquad \rho^2 = x^2 + y^2.$$

The appearance of the Bessel function J_0 suggests that there is a more direct way to find this representation. In polar coordinates, $x = r\cos\theta$, $y = r\sin\theta$, the original equation takes the form

$$\frac{1}{r}\frac{\partial}{\partial r}\left(r\frac{\partial G}{\partial r}\right) + \frac{1}{r^2}\frac{\partial^2 G}{\partial \theta^2} - \alpha^2 G = -\frac{\delta(r)}{2\pi r}.$$

The solution we seek will be independent of θ, $G_{\theta\theta} = 0$. For the operator $LG = \frac{1}{r}(rG_r)_r$, the natural transform is the Hankel transform pair

$$\hat{G}(\mu) = \int_0^\infty rJ_0(\mu r)G(r)dr$$

$$G(r) = \int_0^\infty \mu J_0(\mu r)\hat{G}(\mu)d\mu.$$

Applying this transform to the equation we obtain the algebraic equation

$$-(\mu^2 + \alpha^2)\hat{G} = \frac{-1}{2\pi}$$

so that

$$\hat{G}(\mu) = \frac{1}{2\pi(\mu^2 + \alpha^2)}$$

or

$$G(r, \theta) = \frac{1}{2\pi} \int_0^\infty \frac{\mu J_0(\mu r)}{\mu^2 + \alpha^2} d\mu$$

which, of course, agrees with our first answer.

To recover the Green's function for Poisson's equation from this representation we cannot simply set $\alpha = 0$, since then the integral does not exist. Instead we differentiate with respect to r, to obtain

$$\frac{\partial G}{\partial r}(r, \theta) = \frac{1}{2\pi} \int_0^\infty \frac{\mu^2 J_0'(\mu r)d\mu}{\mu^2 + \alpha^2}$$

which for $\alpha = 0$ reduces to

$$G_r(r, \theta) = \frac{1}{2\pi} \int_0^\infty J_0'(\mu r)d\mu$$

$$= -\frac{1}{2\pi r}.$$

It follows that $G(r, \theta) = \frac{-1}{2\pi} \ln r$, which agrees with our previous calculation.

Consider now Poisson's equation on the half space $-\infty < x < \infty$, $y \geq 0$ with Dirichlet data $u = 0$ at $y = 0$. This is the same problem for which we were previously able to use the method of images. The two relevant transforms for this problem are the Fourier transform in x and the Fourier-Sine transform in y.

Applying the Fourier transform in x to the equation $G_{xx} + G_{yy} = -\delta(x - \xi)\delta(y - \eta)$ we obtain

$$g''(y, k) - k^2 g(y, k) = -e^{ik\xi}\delta(y - \eta)$$

for which we seek $L^2[0, \infty)$ solutions with $g(0, k) = 0$ for all real values of k. We find that

$$g(y, k) = \begin{cases} \dfrac{\sinh(|k| y)\, e^{-|k|\eta} e^{ik\xi}}{|k|}, & 0 \leq y < \eta < \infty \\[2ex] \dfrac{\sinh(|k| \eta)\, e^{-|k|y} e^{ik\xi}}{|k|}, & 0 \leq \eta < y < \infty \end{cases}$$

so that

$$G(x, y) = \frac{1}{2\pi} \int_{-\infty}^{\infty} g(y, k) e^{-ikx}\, dk.$$

Similarly, if we apply the Fourier-Sine Transform for the y variable, we obtain

$$h''(x, k) - k^2 h(x, k) = -\sin(k\eta)\delta(x - \xi)$$

where $h(x, k)$ is the sine transform of $G(x, y)$

$$h(\dot{x}, k) = \int_0^{\infty} G(x, y) \sin ky\, dy$$

with inverse transform

$$G(x, y) = \frac{2}{\pi} \int_0^{\infty} h(x, k) \sin ky\, dk.$$

The solution $h(x, k)$ must be an $L^2(-\infty, \infty)$ function so that

$$h(x, k) \;=\; \frac{1}{2k} e^{-k|x - \xi|} \sin k\eta$$

and

$$G(x, y) \;=\; \int_0^{\infty} \frac{1}{k\pi} e^{-k|x - \xi|} \sin ky \sin k\eta\, dk.$$

The equivalence of these two representations for $G(x,y)$ can be checked by contour integration. For $0 \le y < \eta$, the first representation of $G(x,y)$ is

$$G(x,y) = \frac{1}{2\pi} \int_{-\infty}^{\infty} \sinh(|k|y) e^{-|k|\eta} e^{ik(\xi-x)} \frac{dk}{|k|}$$

$$= \frac{1}{2\pi} \int_{0}^{\infty} \sinh ky\, e^{-k\eta} e^{-ik(\xi-x)} \frac{dk}{k}$$

$$+ \frac{1}{2\pi} \int_{0}^{\infty} \sinh ky\, e^{-k\eta} e^{ik(\xi-x)} \frac{dk}{k}$$

For $\xi > x$, the first of these two integrals can be closed in the lower half plane and the second can be closed in the upper half plane. Doing so enables us to change variables and integrate along the imaginary axis, ($k \to -ik$ in the first and $k \to ik$ in the second) so that

$$G(x,y) = \frac{1}{\pi} \int_{0}^{\infty} e^{-k(\xi-x)} \sin ky \sin k\eta \frac{dk}{k}$$

which checks with the second representation found before.

Recall that this Green's function was also found by the method of images to be

$$G(x,y) = \frac{1}{4\pi} \ln \left[\frac{(x-\xi)^2 + (y-\eta)^2}{(x-\xi)^2 + (y+\eta)^2} \right].$$

We leave it to the reader to verify that these two functions are indeed the same. (Problem 8.1.12)

Examples need not be restricted to domains of rectangular geometry. Consider, for example, Laplace's equation on the interior of the circular domain $0 \le r \le R$, $0 \le \theta < 2\pi$ with Dirichlet data $u = f(\theta)$ on the boundary $r = R$. The equation in polar coordinates is

$$r \frac{\partial}{\partial r}\left(r \frac{\partial u}{\partial r} \right) + \frac{\partial^2 u}{\partial \theta^2} = 0,$$

$$u(R, \theta) = f(\theta).$$

The easiest solution to construct is the one associated with the eigenfunctions of the operator

$$\frac{\partial^2}{\partial \theta^2}$$

with 2π periodic conditions. To find this representation, we seek a solution of the form

$$u(r, \theta) = u_0(r) + \sum_{j=1}^{\infty} (u_j(r) \sin n\theta + v_j(r) \cos n\theta)$$

where

$$u_0(r) = \frac{1}{2\pi} \int_0^{2\pi} u(r, \theta) d\theta,$$

$$u_j(r) = \frac{1}{\pi} \int_0^{2\pi} u(r, \theta) \sin j\theta \, d\theta,$$

$$v_j(r) = \frac{1}{\pi} \int_0^{2\pi} v(r, \theta) \cos j\theta \, d\theta.$$

Since this defines a transform for the equation, we determine (multiply the equation by the eigenfunction $\cos n\theta$ or $\sin n\theta$ and integrate from $\theta = 0$ to $\theta = 2\pi$) that

$$r\frac{d}{dr}\left(r\frac{du_0}{dr}\right) = 0, \qquad u_0(R) = \frac{1}{2\pi} \int_0^{2\pi} f(\theta) d\theta,$$

$$r\frac{d}{dr}\left(r\frac{du_j}{dr}\right) - j^2 u_j = 0, \qquad u_j(R) = \frac{1}{\pi} \int_0^{2\pi} f(\theta) \sin j\theta \, d\theta,$$

$$r\frac{d}{dr}\left(r\frac{dv_j}{dr}\right) - j^2 v_j = 0, \qquad v_j(R) = \frac{1}{\pi} \int_0^{2\pi} f(\theta) \cos j\theta \, d\theta.$$

The solution must be bounded at the origin, so it follows that

$$u_0(r) = u_0(R) = \frac{1}{2\pi} \int_0^{2\pi} f(\theta) d\theta$$

and that

$$u_j(r) = u_j(R)\left(\frac{r}{R}\right)^j,$$

$$v_j(r) = v_j(R)\left(\frac{r}{R}\right)^j$$

As a result,

$$u(r, \phi) = \frac{1}{2\pi} \int_0^{2\pi} f(\theta) d\theta + \frac{1}{\pi} \sum_{j=1}^{\infty} \left(\frac{r}{R}\right)^j \int_0^{2\pi} f(\theta) \cos j(\phi - \theta) d\theta.$$

In Chapter 6, we found another representation of this solution using analytic function theory. The reader should check that these two are indeed the same (Problem 8.1.13).

The Green's function for Poisson's equation in cylindrical coordinates with homogeneous Dirichlet data at $r = R$ has two representations, one based on Fourier series and the second using a variant of Mellin transforms. To find the Fourier series representation, we multiply the defining equation (see Problem 8.1.1)

$$\nabla^2 G = \frac{1}{r}(rG_r)_r + \frac{1}{r^2}\frac{\partial^2 G}{\partial\theta^2} = -\frac{1}{r}\delta(r - r_0)\delta(\theta - \theta_0)$$

by $\cos n\theta$ or $\sin n\theta$ and integrate with respect to θ, $0 \leq \theta \leq 2\pi$, to obtain

$$\frac{1}{r}(ru'_n)' - \frac{n^2}{r^2}u_n = -\frac{1}{r}\delta(r - r_0)\cos n\theta_0$$

or

$$\frac{1}{r}(rv'_n)' - \frac{n^2}{r^2}v_n = -\frac{1}{r}\delta(r - r_0)\sin n\theta_0$$

with Dirichlet boundary data $u_n(R) = v_n(R) = 0$. We find that

$$u_n(r) = \begin{cases} \dfrac{1}{2}\left(\left(\dfrac{r}{r_0}\right)^n - \left(\dfrac{rr_0}{R^2}\right)^n\right)\cos n\theta_0 & r < r_0 \\[3mm] \dfrac{1}{2}\left(\left(\dfrac{r_0}{r}\right)^n - \left(\dfrac{rr_0}{R^2}\right)^n\right)\cos n\theta_0 & r > r_0 \end{cases}$$

$$u_0(r) = \begin{cases} \ln(R/r_0) & r < r_0 \\ \ln(R/r) & r > r_0 \end{cases}$$

and similarly for $v_n(r)$. It follows that

$$G(r,\theta,r_0,\theta_0) = \frac{1}{2\pi}u_0(r) + \frac{1}{\pi}\sum_{n=1}^{\infty} u_n(r)\cos n(\theta - \theta_0).$$

The second representation for this Green's function is found most readily by first making the change of variables

$$r = Re^{-t}$$

for which the governing equation becomes (see Problem 8.1.21)

$$G_{tt} + G_{\theta\theta} = -\delta(t - t_0)\delta(\theta - \theta_0)$$

with $G(t,\theta)$ 2π periodic in θ, bounded at $t = \infty$, and satisfying the Dirichlet condition $G(0,\theta) = 0$. Clearly we want to apply the Fourier-Sine transform

$$\widehat{G}(k,\theta) = \int_0^{\infty} \sin kt\, G(t,\theta)dt$$

$$G(t,\theta) = \frac{1}{\pi}\int_0^{\infty} \sin kt\, \widehat{G}(k,\theta)dk$$

from which we learn that $\widehat{G} = g \sin kt_0$ where $g_{\theta\theta} - k^2 g = -\delta(\theta - \theta_0)$. The solution of this transformed equation is

$$
g(k, \theta, \theta_0) = \begin{cases} \dfrac{-\sinh k(\theta - \theta_0) + \sinh k(\theta - \theta_0 + 2\pi)}{4k \sinh^2 k\pi} & 0 \le \theta < \theta_0 \le 2\pi \\[3mm] \dfrac{\sinh k(\theta - \theta_0) - \sinh k(\theta - \theta_0 - 2\pi)}{4k \sinh^2 k\pi} & 0 \le \theta_0 < \theta \le 2\pi \end{cases}
$$

and the Green's function can be written down using the inverse Fourier-Sine transform

$$
G(t, \theta) = \frac{1}{\pi} \int_0^\infty \sin kt \sin kt_0 g(k, \theta, \theta_0) dk.
$$

Reversing the change of variables $r = Re^{-t}$ we find

$$
G(t, \theta) = \frac{1}{\pi} \int_0^\infty \sin \left(k \ln \frac{R}{r} \right) \sin \left(k \ln \frac{R}{r_0} \right) g(k, \theta, \theta_0) dk
$$

as the representation of the Green's function.

8.1.4 Eigenfunctions

The eigenfunctions of Laplace's operator are defined as nontrivial solutions of $-\nabla^2 u = \lambda u$ satisfying appropriate homogeneous boundary conditions. (This equation is often called the Helmholtz equation.) We have already examined one dimensional eigenfunctions in some detail. The two most important one dimensional cases are

1. **Scalar Domain,** $a \le x \le b$ for which $\nabla^2 u = u_{xx}$ with self-adjoint boundary conditions. The eigenfunctions are trigonometric functions.

2. **Cylindrical geometry** with rotational symmetry,

$$
\nabla^2 u = \frac{1}{r}(r u_r)_r + \frac{1}{r^2} u_{\theta\theta}
$$

where $u_{\theta\theta}$ is assumed equal to zero. The equation $\nabla^2 u + \lambda u = 0$ reduces to

$$
\frac{1}{r}(r u_r)_r + \lambda u = 0
$$

which is Bessel's equation of order zero. Solutions are $u = J_0 \left(\sqrt{\lambda} x \right)$ and the eigenvalues λ are chosen to satisfy a homogeneous boundary condition on a circular boundary $r = R$.

We know that these eigenfunctions form a complete basis on the domain of the operator.

In higher dimensions, we use transform methods and our knowledge of one dimensional problems to obtain eigenfunctions. The easiest example is the Laplacian on a rectangular domain with homogeneous Neumann or Dirichlet boundary data, the calculation of which we leave to the reader (Problem 8.1.16). A slightly more complicated example has a circular domain.

We seek nontrivial solutions of

$$-\nabla^2 u = \lambda u$$

on the circular domain $0 \le r \le R$, $0 \le \theta \le 2\pi$ with u bounded and $u(R, \theta) = 0$. In this geometry the equation has the form

$$\nabla^2 u + \lambda u = \frac{1}{r} \frac{\partial}{\partial r} \left(r \frac{\partial u}{\partial r} \right) + \frac{1}{r^2} \frac{\partial^2 u}{\partial \theta^2} + \lambda u = 0.$$

We observe that the operator $\frac{\partial^2}{\partial \theta^2}$ with periodic boundary conditions on $0 \le \theta \le 2\pi$ has the trigonometric functions as eigenfunctions. Thus we propose the transform

$$u(r, \theta) = U_0(r) + \sum_{n=1}^{\infty} U_n(r) \cos n\theta + V_n(r) \sin n\theta$$

where

$$U_0(r) = \frac{1}{2\pi} \int_0^{2\pi} u(r, \theta) d\theta,$$

$$U_n(r) = \frac{1}{\pi} \int_0^{2\pi} u(r, \theta) \cos n\theta \, d\theta,$$

$$V_n(r) = \frac{1}{\pi} \int_0^{2\pi} u(r, \theta) \sin n\theta \, d\theta.$$

After transformation, the eigenvalue problem becomes a decoupled set of ordinary differential equations

$$\frac{1}{r}(rU_n')' + \left(\lambda - \frac{n^2}{r^2} \right) U_n = 0 \qquad n \ge 0$$

$$\frac{1}{r}(rV_n')' + \left(\lambda - \frac{n^2}{r^2} \right) V_n = 0 \qquad n \ge 1.$$

These are Bessel's equations of order n with solutions

$$U_n(r) = A_n J_n \left(\sqrt{\lambda} r \right)$$

$$V_n(r) = B_n J_n \left(\sqrt{\lambda} r \right).$$

The boundary condition $u(R, \theta) = 0$ is satisfied if and only if $J\left(\sqrt{\lambda}R\right) = 0$ so we must pick $\lambda = (\lambda_{nm}/R)^2$ where $J_n(\lambda_{nm}) = 0$, λ_{nm} is the mth zero of the nth Bessel function, $\lambda_{nm} < \lambda_{n,m+1}$. The numbers λ_{nm} are tabulated in standard reference works. We can now write the eigenfunctions as

$$u_{nm}(r, \theta) = J_n\left(\frac{\lambda_{nm}r}{R}\right) \cos n\theta \qquad n \geq 0$$

$$v_{nm}(r, \theta) = J_n\left(\frac{\lambda_{nm}r}{R}\right) \sin n\theta \qquad n \geq 1$$

and the eigenvalues are $\lambda = (\lambda_{nm}/R)^2$. We know from the theory of self adjoint compact operators (Chapter 3) that this doubly infinite set of eigenfunctions forms a complete basis for the domain of this operator.

As we shall see later, the structure of eigenfunctions contains important physical information. An examination of these functions shows that all except $u_{01}(r, \theta) = J_0(\lambda_{01}r/R)$ have nodal curves, that is, curves in the interior of the domain on which u is zero. For example, the nodal curves of $u_{0m}(r, \theta)$ are the circles $r = R\lambda_{0j}/\lambda_{0m}$, $j < m$, and the nodal curves of $u_{nm}(r, \theta)$ or $v_{nm}(r, \theta)$ are the circles $r = R\lambda_{nj}/\lambda_{nm}$, $j < m$ and radial lines separated by an angle π/n.

The eigenvalues λ_{nm} do not have a particularly nice order, nor are they simple ratios of each other (compare this with the eigenvalues of $-u''$ on $0 \leq x \leq 1$ with Dirichlet data). It can be readily shown (Problem 5.4.2) that $\lambda_{nm} < \lambda_{n+1,m}$. In Figure 8.1 is shown the nodal structure of the twelve lowest eigenvalues. The relative frequency of these modes is the ratio of the eigenvalue λ_{nm} to the smallest eigenvalue $\lambda_{01} = 2.40482$. So, for example, $\lambda_{32}/\lambda_{01} = .406$ since $\lambda_{32} = 9.76120$.

As a second example, we look for the eigenfunctions of the Laplacian in spherical geometry, where

$$\nabla^2 u = \frac{1}{r^2}\frac{\partial}{\partial r}\left(r^2\frac{\partial u}{\partial r}\right) + \frac{1}{r^2 \sin\theta}\frac{\partial}{\partial \theta}\left(\sin\theta\frac{\partial u}{\partial \theta}\right) + \frac{1}{r^2 \sin^2\theta}\frac{\partial^2 u}{\partial \phi^2},$$

with $0 \leq r \leq R$, $0 \leq \theta \leq \pi$, $0 \leq \phi \leq 2\pi$. The first obvious thing to do is to replace the operator

$$\frac{\partial^2}{\partial \phi^2}$$

by $-n^2$, equivalent to a trigonometric transformation. We then have the transformed equation

$$\frac{\partial}{\partial r}\left(r^2\frac{\partial U}{\partial r}\right) + \lambda r^2 U + \frac{1}{\sin\theta}\frac{\partial}{\partial \theta}\left(\sin\theta\frac{\partial U}{\partial \theta}\right) - \frac{n^2}{\sin^2\theta}U = 0.$$

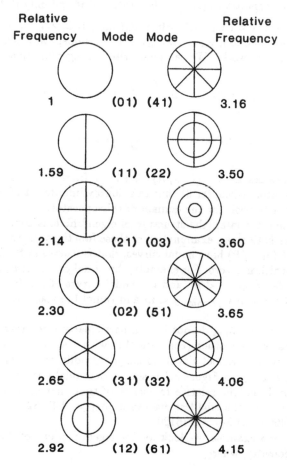

FIGURE 8.1   Nodal lines for the first twelve eigenfunctions of the Laplacian in polar coordinates.

Notice that this equation is the sum of two operators

$$L_1 u + L_2 u = 0$$

where

$$L_1 u \;=\; \frac{\partial}{\partial r}\left(r^2 \frac{\partial u}{\partial r}\right) + \lambda r^2 u$$

and

$$L_2 u = \frac{1}{\sin\theta}\frac{\partial}{\partial\theta}\left(\sin\theta\frac{\partial u}{\partial\theta}\right) - \frac{n^2 u}{\sin^2\theta}$$

which each depend on a single variable. This suggests that an eigenfunction representation is possible. In fact, if we make the change of variables $x = \cos\theta$, the operator L_2 becomes

$$L_2 u = \frac{\partial}{\partial x}\left((1-x^2)\frac{\partial u}{\partial x}\right) - \frac{n^2 u}{(1-x^2)}.$$

Its only bounded eigenfunctions are the associated Legendre functions (Chapter 4)

$$P_m^n(x) = (1-x^2)^{n/2}\frac{d^n P_m(x)}{dx^n}$$

which are nontrivial provided $m > n$, and satisfy $-L_2 P_m^n = \lambda P_m^n$, $\lambda = m(m+1)$. For each fixed n there are an infinite number of associated Legendre functions.

We transform the equation again by setting $U = P_m^n(\cos\theta)V(r)$ from which we learn that

$$(r^2 V')' + \lambda r^2 V - m(m+1)V = 0$$

subject to the requirement that $V(R) = 0$. This final equation looks very much, but not exactly, like a Bessel's equation. We try a solution of the form $V = r^\alpha w$. Upon substitution we find

$$r^2 w'' + (2\alpha+2)rw' + (r^2\lambda + \alpha(\alpha+1) - m(m+1))\,w = 0.$$

The number α can be chosen to make this into a Bessel's equation. In fact, for $\alpha = -1/2$ we have

$$\frac{1}{r}(rw')' + \left(\lambda - \frac{(m+1/2)^2}{r^2}\right)w = 0$$

which is Bessel's equation of order $m + 1/2$. Its bounded solutions are

$$w = J_{m+1/2}\left(\sqrt{\lambda}r\right)$$

and we realize that λ must be picked to satisfy $J_{m+1/2}\left(\sqrt{\lambda}R\right) = 0$. The zeros of these fractional order Bessel functions are tabulated in standard reference works. We denote the kth root by $\mu_{m+1/2,k}$, where $J_{m+1/2}(\mu_{m+1/2,k}) = 0$.

We are finally in a position to write down the eigenfunctions. They are

$$u(r,\theta,\phi) = r^{-1/2}J_{m+1/2}\left(\frac{\mu_{m+1/2,k}r}{R}\right)P_m^n(\cos\theta)\cos n(\phi-\phi_0)$$

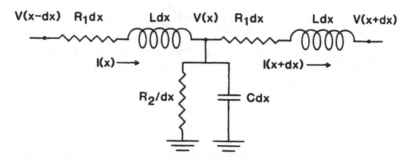

FIGURE 8.2   Circuit diagram for a segment of coaxial cable.

and the eigenvalues are $\lambda_{m,k} = (\mu_{m+1/2,k}/R)^2$, provided $m > n$. The functions

$$j_n(z) = \left(\frac{\pi}{2z}\right)^{1/2} J_{n+1/2}(z)$$

are referred to as Spherical Bessel functions of the first kind. Recall from Chapter 6 that the fractional order Bessel function $J_{n+1/2}(z)$ has a representation in terms of trigonometric functions.

8.2   The Wave Equation

8.2.1   Derivations

The wave equation is most often derived as an equation governing small vibrations of a string (Chapter 5) or acoustic vibrations of a gas. Another interesting derivation is for propagation of electrical signals in a coaxial cable transmission line. We model a coaxial cable as a long array of coupled inductors and capacitors each of which are infinitesimally short. In addition, we assume that there are resistive losses R_1 in the axial direction and R_2 in the transverse direction. A segment of the cable circuit is shown in Figure 8.2.

We use Kirkhoff's laws to write down the balance of currents and voltages. It follows that

$$\frac{V(x+dx) - V(x)}{dx} = -\left(L\frac{dI}{dt} + R_1 I\right)$$

and

$$\frac{I(x+dx) - I(x)}{dx} = -\left(C\frac{dV}{dt} + \frac{V}{R_2}\right)$$

Taking the limit $dx \to 0$ we find that

$$L\frac{\partial I}{\partial t} + R_1 I = -\frac{\partial V}{\partial x}$$

$$C\frac{\partial V}{\partial t} + \frac{V}{R_2} = -\frac{\partial I}{\partial x}$$

where V is the voltage and I the axial current in the cable. We can combine these two equations into one to obtain

$$CL\frac{\partial^2 I}{\partial t^2} + \left(CR_1 + \frac{L}{R_2}\right)\frac{\partial I}{\partial t} + \frac{R_1}{R_2}I = \frac{\partial^2 I}{\partial x^2},$$

called the TELEGRAPHER'S EQUATION. From it we obtain the wave equation if we take $R_1 = 0$ (no conduction resistance in the axial direction) and $R_2 = \infty$ (no transverse current). We obtain the heat equation if we take $C = 0$, and $R_1 = 0$, or $L = 0$ and $R_2 = \infty$.

One should never try to solve an equation such as this with all its parameters in sight as it involves more work than is necessary, and increases the chances of making an error. To limit the number of free parameters one should always introduce dimensionless space and time scales. For this problem we can set $t = \alpha\tau$, $x = \beta\xi$, where $\alpha = (CLR_2/R_1)^{1/2}$, $\beta = (R_2/R_1)^{1/2}$ and the equation becomes

$$\frac{\partial^2 I}{\partial \tau^2} + \gamma\frac{\partial I}{\partial \tau} + I = \frac{\partial^2 I}{\partial \xi^2}$$

which has only the one free parameter $\gamma = (CR_1 + L/R_2)(R_2/R_1CL)^{1/2}$. If either α or β are zero, this change of space and time scale does not work and a different scaling must be devised. In any case, we can always reduce the original number of free parameters. For example, if $R_1 = 0$ and $R_2 = \infty$, any combination of α, β with $-\alpha/\beta = (CL)^{1/2}$ reduces the original equation to the wave equation with no free parameters.

Another derivation of the telegrapher's equation comes from considering the random motion of bugs. You can think of lady bugs wandering along a one-dimensional hedge in search of aphids, although this description is not quite correct for lady bugs since lady bugs can also fly. We suppose the bugs move with constant velocity c and that a certain number of them $p^+(x,t)$ are moving to the right while $p^-(x,t)$ of them are moving to the left. The bugs can change direction so that at any time some bugs are moving from the group $p^+(x,t)$ to the group $p^-(x,t)$ or vice versa. It follows that

$$\frac{\partial p^+}{\partial t} = \lambda p^- - \lambda p^+ - c\frac{\partial p^+}{\partial x},$$

$$\frac{\partial p^-}{\partial t} = \lambda p^+ - \lambda p^- + c\frac{\partial p^-}{\partial x},$$

where λ is the rate at which direction reversals are made, and the terms $c\partial p^+/\partial x$ and $c\partial p^-/\partial x$ represent the flux of bugs along the hedge. We define $p(x,t) = p^+(x,t) + p^-(x,t)$ to be the total number of bugs at position x at time t. Then

$$\frac{\partial p}{\partial t} + c\frac{\partial}{\partial x}\left(p^+ - p^-\right) = 0,$$

$$\frac{\partial}{\partial t}\left(p^+ - p^-\right) + 2\lambda\left(p^+ - p^-\right) = -c\frac{\partial p}{\partial x}.$$

We eliminate $p^+ - p^-$ from these equations by cross-differentiation, to find the telegrapher's equation

$$\frac{\partial^2 p}{\partial t^2} + 2\lambda\frac{\partial p}{\partial t} = c^2\frac{\partial^2 p}{\partial x^2}.$$

The wave equation also has application to the study of blood flow. It has been known for a long time that blood flow is pulsatile. The beating heart produces a pressure wave that travels through the circulatory system and can be felt at various points such as the wrist, and neck. This pressure wave is not the same as the sound wave (lub-dub) heard in a stethoscope. The heart sounds result from fluid turbulence created when the bicuspid and tricuspid valves close. The pressure wave is a consequence of the elasticity of the arteries whose expansion and contraction is coupled to the blood flow. For simplicity, we suppose we have an elastic tube through which an incompressible, inviscid fluid is flowing. We denote by $A(x)$ the cross-sectional area of the circular tube, and let $u(x,t)$, $p(x,t)$ be the velocity and pressure of the fluid, assumed to be parallel to the tube axis x.

The conservation law for this fluid is easy to derive. The amount of fluid contained in a small section of tube can change only if there is an imbalance between the influx and efflux. Hence, for a section of tube length dx

$$\frac{d}{dt}(A\,dx) = u(x)A(x) - u(x + dx)A(x + dx)$$

or, taking $dx \to 0$

$$\frac{\partial A}{\partial t} + \frac{\partial(Au)}{\partial x} = 0.$$

The equation of motion for the fluid is found by balancing forces. Since the force exerted by a fluid is assumed to be pressure force only (no shear, non-normal forces) the force in the x direction on a volume of fluid at position x is

$$p(x)A(x) - p(x + dx)A(x + dx) + p_0\frac{\partial A}{\partial x}dx$$

where p_0 is the pressure exerted by the wall, so $p_0 \frac{\partial A}{\partial x} dx$ is its component in the axial direction. Thus, Newton's law can be expressed as

$$\rho A \left(\frac{\partial u}{\partial t} + u \frac{\partial u}{\partial x} \right) = \frac{\partial}{\partial x} \left((p_0 - p) A \right)$$

where ρ is the density of the fluid. Notice that the acceleration of this slug of fluid is the total derivative

$$\frac{Du}{Dt} = \frac{\partial u}{\partial t} + \frac{\partial u}{\partial x} \frac{\partial x}{\partial t} = \frac{\partial u}{\partial t} + u \frac{\partial u}{\partial x},$$

called the *substantial*, or MATERIAL DERIVATIVE of u. We need one more equation to complete this model. The correct thing to do is to include the inertial effects of the wall of the tube (done by Korteweg in 1878) but a simpler idea is to assume that the adjustment between pressure and cross-sectional area occurs instantaneously with no delays due to inertia of the wall. For example, it is reasonable to take

$$p - p_0 = \frac{Eh}{r_0} \left(1 - \frac{r_0}{r} \right) = \frac{Eh}{r_0} \left(1 - \left(\frac{A_0}{A} \right)^{1/2} \right)$$

where r is the radius of the tube, r_0 the radius of the tube in equilibrium, E is Young's modulus and h is the tube thickness. This relation expresses Hooke's law since $(p - p_0)r$, which is the same as the tension in the membrane wall, is linearly related to $r - r_0$.

Now we assume that u, $p - p_0$, and $A - A_0$ are small and linearize the three equations. The result is

$$\frac{\partial A}{\partial t} + A_0 \frac{\partial u}{\partial x} = 0$$

$$\rho \frac{\partial u}{\partial t} = -\frac{\partial p}{\partial x}$$

$$p - p_0 = \frac{Eh}{2 r_0 A_0} (A - A_0).$$

We eliminate u from these equations by differentiating the first with respect to t and the second with respect to x and obtain

$$\frac{\partial^2 p}{\partial t^2} = \frac{Eh}{2 \rho r_0} \frac{\partial^2 p}{\partial x^2}$$

and an identical equation for A. The constant

$$c = \left(\frac{Eh}{2 \rho r_0} \right)^{1/2}$$

is the speed of pressure waves in the circulatory system (not to be confused with the actual speed of the blood flow of about .3m/sec). Typical values of the parameters in dogs are

$$E \sim 4.3 \times 10^6 \text{dyne/cm}^2, \quad h/r \sim 0.105, \quad \rho \sim 1.06\text{g/cm}^3$$

so that $c \sim 4.6$m/sec. You should be able to estimate this speed for yourself by noting the difference in arrival times of the pressure wave at your neck (the carotid artery) and your foot (the posterior tibial artery).

8.2.2 Fundamental Solutions

The wave equation $u_{tt} = u_{xx}$ has several important simple solutions. If we introduce a change of coordinate system $\xi = x - t$, $\eta = x + t$ we find that the equation $u_{tt} - u_{xx} = 0$ transforms to the equation $u_{\xi\eta} = 0$. The general solution of this equation is $u = f(\xi) + g(\eta)$ where f and g are arbitrary, twice continuously differentiable functions. Using the theory of distributions, one can weaken the smoothness requirement on f and g and allow f and g to be distributions (Problem 8.2.2). In any case, the solution we obtain is

$$u(x,t) = f(x - t) + g(x + t).$$

The physical significance of this is that u is always the superposition of the two forms f and g, which, as time progresses are translated to the right and left respectively, without change of shape. The only way that u can change its shape is when the underlying components f and g are translated. The translating shape $f(x - t)$ is referred to as a right moving wave while $g(x + t)$ is a left moving wave.

We can solve some specific initial value problems rather easily. For example, the wave equation on the entire real line $-\infty < x < \infty$ with prescribed initial data $u(x, 0) = F(x)$, $u_t(x,0) = 0$ is solved if

$$f(x) + g(x) = F(x)$$
$$f'(x) - g'(x) = 0.$$

This uniquely determines f and g (up to additive constants) as $f = g = \frac{F}{2}$ and $u(x, t)$ is uniquely determined to be

$$u(x,t) = \frac{1}{2}F(x - t) + \frac{1}{2}F(x + t).$$

In other words, on an infinitely long string, an initial shape propagates both to the left and right with half its initial amplitude propagating in each direction.

We can also solve the wave equation with initial data $u(x, 0) = 0$, $u_t(x, 0) = G(x)$. We require

$$f(x) + g(x) = 0$$
$$-f'(x) + g'(x) = G(x).$$

Again f and g are determined uniquely up to additive constants, and the solution $u(x,t)$ is specified uniquely as

$$u(x,t) = \frac{1}{2} \int_{x-t}^{x+t} G(\eta)d\eta.$$

Using linear superposition we learn that

$$u(x,t) = \frac{1}{2} \int_{x-t}^{x+t} G(\eta)d\eta + \frac{1}{2}F(x-t) + \frac{1}{2}F(x+t)$$

satisfies initial data $u(x,0) = F(x)$, $u_t(x,0) = G(x)$. This solution is important because it demonstrates CAUSALITY, or specifically, the fact that information travels with a finite speed, and the solution at point x_0 at the time t is influenced solely by the initial data in the domain $x_0 - t \le x \le x_0 + t$, called the REGION OF INFLUENCE for the point x_0.

These solutions apply to an infinite string. For a finite or semi-infinite string, boundary conditions must be satisfied. Consider, as an example, the domain $0 \le x < \infty$ with Dirichlet condition $u = 0$ at $x = 0$ for all time. The form of the solution remains unchanged as $u(x,t) = f(x-t) + g(x+t)$. Suppose f_0 and g_0 are determined so that $u(x,t) = f_0(x-t) + g_0(x+t)$ satisfies the required initial specifications but not the boundary condition, and that $f_0(x)$ and $g_0(x)$ are zero for negative arguments. It is easy to see that $u(x,t) = f_0(x-t) - f_0(-t-x) + g_0(x+t) - g_0(t-x)$ satisfies both the initial conditions and the boundary condition. Note that to satisfy the boundary condition we have added a solution corresponding to the negative of the unperturbed solution, reflected in the line $x = 0$. Because of this symmetry in the line $x = 0$, any shape that hits the line $x = 0$ from the right is reflected with negative amplitude as a right moving shape. By a similar argument

$$u(x,t) = f_0(x-t) + f_0(-t-x) + g_0(x+t) + g_0(t-x)$$

satisfies initial requirements and the Neumann condition $\frac{\partial u}{\partial x} = 0$ at $x = 0$. This solution shows that, with Neumann conditions, left-moving shapes hitting the boundary at $x = 0$ are reflected as right moving waves with amplitude preserved.

These arguments are the analogue of the method of images discussed earlier. As before, we could proceed to find solutions on bounded domains by adding the appropriate reflections of shapes, but there will be an infinite number of reflected terms to add, so that insight is quickly lost.

There are two ways to find the Green's function for the wave equation

$$G_{tt} - G_{xx} = \delta(x)\delta(t)$$

with zero initial data $G_t(x,0) = G(x,0) = 0$, on the infinite line $-\infty < x < \infty$. The first is to use Fourier Transforms in x

$$g(t,k) = \int_{-\infty}^{\infty} e^{ikx} G(x,t)dx.$$

Taking the Fourier transform of the equation we find

$$g''(t, k) + k^2 g(t, k) = \delta(t)$$

which we solve with zero initial data. The solution is

$$g(t, k) = \frac{1}{k} \sin kt \, H(t)$$

where $H(t)$ is the Heaviside function. The inverse Fourier transform yields

$$
\begin{aligned}
G(x, t) &= \frac{H(t)}{2\pi} \int_{-\infty}^{\infty} \sin kt \, e^{-ikx} \frac{dk}{k} \\
&= \frac{H(t)}{2\pi} \int_{-\infty}^{\infty} \left(e^{ik(t-x)} - e^{-ik(t+x)} \right) \frac{dk}{2ik}
\end{aligned}
$$

Now notice (or recall) that

$$H(x) = \frac{1}{2\pi} \int_{-\infty}^{\infty} e^{-ikx} \frac{dk}{ik}$$

when the path of integration is chosen to pass above the origin $k = 0$. Thus we have that

$$
\begin{aligned}
G(x, t) &= \frac{1}{2} H(t) \left[H(t + x) - H(x - t) \right] \\
&= \frac{1}{2} H\left(t - |x| \right).
\end{aligned}
$$

For influence at the points $x = \xi$, $t = \tau$, we have that

$$G(x, t; \xi, \tau) = \frac{1}{2} H\left(t - \tau - |x - \xi| \right).$$

Notice that this Green's function expresses the physical fact that a disturbance at point $x = \xi$, time $t = \tau$ can influence the medium only in the forward cone $t - \tau \geq |x - \xi|$.

The second way to obtain this Green's function is with Laplace Transforms. To solve the equation $G_{tt} - G_{xx} = \delta(x - \xi)\delta(t - \tau)$ we introduce the Laplace transform of G

$$h(x, s) = \int_0^{\infty} G(x, t) e^{-st} dt$$

with inverse transform

$$G(x, t) = \frac{i}{2\pi} \int_{-i\infty}^{i\infty} e^{st} h(x, s) ds$$

where the vertical path of integration lies to the right of all singularities of $h(x, s)$. When we transform the equation, we obtain

$$-h''(x, s) + s^2 h(x, s) = e^{-s\tau} \delta(x - \xi)$$

for which we seek an $L^2(-\infty, \infty)$ solution for Re $s > 0$. The solution is

$$h(x, s) = \frac{e^{-s|x - \xi|} e^{-s\tau}}{2s}$$

and the Green's function is

$$G(x, t; \xi, \tau) = \frac{1}{2\pi} \int_{-i\infty}^{i\infty} e^{-s(|\xi - x| + \tau - t)} \frac{ds}{2s}$$

where the vertical contour passes to the right of $s = 0$. This integral we evaluate as

$$G(x, t; \xi, \tau) = \frac{1}{2} H(t - \tau - |x - \xi|)$$

which is the same answer as before.

To find the Green's function for the wave equation on the semi-infinite domain $0 \leq x < \infty$ with Dirichlet data $u(0, t) = 0$ and zero initial data $u(x, 0) = u_t(x, 0) = 0$ we can use either the Fourier-Sine or Laplace Transform. However, it is much easier to use the method of images. We leave it to the reader to verify that transform methods yield the correct result. Using the method of images, we look for a point in the nonphysical domain $-\infty < x \leq 0$ at which to place a source that will exactly balance the unperturbed Green's function at $x = 0$. Clearly, this point is the reflection of the influence point ξ about the axis $x = 0$, and the Green's function is

$$G(x, t; \xi, \tau) = \frac{1}{2} H(t - \tau - |x - \xi|) - \frac{1}{2} H(t - \tau - |x + \xi|)$$

For the wave equation on a bounded domain $0 \leq x \leq a$ with, say, Dirichlet (or Neumann) data, there are no surprises. The choices to find the Green's function are the method of images, Laplace transforms or Fourier series.

With Fourier series we represent the Green's function as

$$G(x, t; \xi, \eta) = \sum_{n=1}^{\infty} g_n(t) \sin \frac{n\pi x}{a}$$

and

$$g_n(t) = \frac{2}{a} \int_0^a G(x, t) \sin \frac{n\pi x}{a} dx.$$

Here we have taken Dirichlet data $u = 0$ at $x = 0$, a. (With Neumann data, we use a Fourier cosine series.) We transform the equation $G_{tt} - G_{xx} = \delta(x - \xi)\delta(t - \tau)$ to obtain

$$\ddot{g}_n + \frac{n^2\pi^2}{a^2}g_n = \frac{2}{a}\sin\frac{n\pi\xi}{a}\delta(t - \tau)$$

which we must solve with zero initial data. We have seen this equation before; its solution is

$$g_n = \frac{2}{n\pi}\sin\frac{n\pi}{a}(t - \tau)\sin\frac{n\pi\xi}{a}H(t - \tau)$$

so that the Green's function is

$$G(x, t; \xi, \tau) = \frac{2}{\pi}H(t - \tau)\sum_{n=1}^{\infty}\frac{1}{n}\sin\frac{n\pi(t - \tau)}{a}\sin\frac{n\pi\xi}{a}\sin\frac{n\pi x}{a}.$$

It is interesting to re-express this in terms of Heaviside functions. The relevant trigonometric identity is

$$\sin x \sin y \sin z = \frac{1}{4}\big(\sin(x + y - z) + \sin(x + z - y)$$
$$+ \sin(y + z - x) - \sin(x + y + z)\big)$$

from which we observe that

$$G(x, t; \xi, \tau) = -\frac{1}{2\pi}H(t - \tau)\sum_{n=1}^{\infty}\frac{1}{n}\Big[\sin\frac{n\pi}{a}(t - \tau + x + \xi)$$
$$- \sin\frac{n\pi}{a}(t - \tau + x - \xi) - \sin\frac{n\pi}{a}(t - \tau + \xi - x)$$
$$+ \sin\frac{n\pi}{a}(t - \tau - x - \xi)\Big].$$

We can sum this infinite series by using that

$$\frac{2}{\pi}\sum_{n=1}^{\infty}\frac{1}{n}\cos\frac{n\pi\xi}{a}\sin\frac{n\pi x}{a} = H(x - \xi) - \frac{x}{a}$$

on the interval $0 \le x \le a$ and is an odd periodic function on the infinite interval $-\infty < x < \infty$. If we define $H_p(x) = H(x) - H(-x)$ on $-a \le x \le a$ and periodically extended elsewhere, then $G(x, t; \xi, \tau)$ can be expressed as

$$G(x, t; \xi, \tau) = \frac{1}{4}H(t - \tau)\Big[H_p(t - \tau + x - \xi) + H_p(t - \tau + \xi - x)$$
$$- H_p(t - \tau - x - \xi) - H_p(t - \tau + x + \xi)\Big].$$

From this expression we see that the Green's function consists of the super-position of four identical odd periodic functions, two of which are left moving and two of which are right moving. Furthermore, their functions are "cen-tered" at $x = \xi$ or its reflected image $x = -\xi$, in order to satisfy boundary conditions. This expression is the same as found directly by the method of images.

As a final example in this section consider the telegrapher's equation $u_{tt} + au_t + bu = u_{xx}$ with positive constants a, b. Suppose signals are sent from position $x = 0$ so that $u(0,t) = f(t)$ and $u_x(0,t) = g(t)$ are specified. We take a Fourier transform in time

$$U(x,k) = \int_{-\infty}^{\infty} u(x,t)e^{ikt} dt$$

with inverse

$$u(x,t) = \frac{1}{2\pi} \int_{-\infty}^{\infty} U(x,k)e^{-ikt} dk$$

and find after integrating by parts that the transformed equation is

$$U_{xx} = (b - ika - k^2)U$$

with specified data $U(0,k) = F(k)$, $U_x(0,k) = G(k)$ where $F(k)$ and $G(k)$ are the Fourier Transforms of $f(t)$, $g(t)$, respectively. If we set $\mu^2 = (k^2 - ika - b)$, we find

$$U(x,k) = F(k)\cos\mu(k)x + \frac{G(k)}{\mu(k)}\sin\mu(k)x$$

and, of course,

$$u(x,t) = \frac{1}{2\pi} \int_{-\infty}^{\infty} \left(F(k)\cos\mu(k)x + \frac{G(k)}{\mu(k)}\sin\mu(k)x \right) e^{-ikt} dk.$$

We will examine this solution further in Chapter 10.

8.2.3 Vibrations

The vibrations of a string or membrane are governed (Chapter 5) by the equation

$$\rho\frac{\partial^2 u}{\partial t^2} = E\nabla^2 u$$

where ρ is the density of the vibrating membrane, E is Hooke's constant and u is the transverse displacement of the membrane from equilibrium. To eliminate free parameters we introduce the change of length scale $x = L\xi$ for a string of length L or $r = R\xi$ for a circular membrane. We also make the change of time scale $t = \alpha\tau$ with $\alpha = L\sqrt{\rho/E}$ for a string and $\alpha = R\sqrt{\rho/E}$ for a circular membrane. Once these changes of independent variables are made, the wave equation $u_{\tau\tau} = \nabla^2 u$ results.

To study the free (i.e. unforced) vibrations of the wave equation on a bounded domain we use what we already know about the eigenfunctions of the Laplacian. With homogeneous Neumann or Dirichlet boundary data, the Laplacian is a non-positive definite, self-adjoint operator. It is a straightforward matter to show that the eigenfunctions form a complete set for the relevant domain. If $\{\phi_n\}_{n=1}^{\infty}$ are the normalized eigenfunctions corresponding to eigenvalue λ_n, $-\nabla^2\phi_n = \lambda_n\phi_n$, any function in the domain of the operator can be represented as

$$u = \sum_{n=1}^{\infty} a_n\phi_n,$$

$$a_n = \int_R u\phi_n dV = \langle u, \phi_n \rangle$$

where R is the relevant bounded spatial domain.

By allowing the components α_n to depend on t, the representation of u in terms of the eigenfunctions of the Laplacian becomes a transform for the wave equation. We obtain the transformed equation by taking the inner product of the original wave equation $u_{tt} = \nabla^2 u$ with eigenfunction ϕ_k to obtain

$$\langle \phi_k, u \rangle_{tt} = \langle \phi_k, \nabla^2 u \rangle = \int_R \phi_k \nabla^2 u \, dV$$

$$= \int_R u\nabla^2\phi_k dV + \int_{\partial R} \left(\phi_k \frac{\partial u}{\partial n} - u\frac{\partial \phi_k}{\partial n} \right) ds$$

or

$$\langle \phi_k, u \rangle_{tt} = -\lambda_k \langle \phi_k, u \rangle + \int_{\partial R} \left(\phi_k \frac{\partial u}{\partial n} - u\frac{\partial \phi_k}{\partial n} \right) ds,$$

using Green's theorem. If u satisfies homogeneous boundary data, there is no boundary contribution to the transformed equation. When we take $a_n(t) = \langle \phi_k, u \rangle$, the transformed equation becomes

$$\ddot{a}_n = -\lambda_n a_n,$$

and the solutions are

$$u(t, x) = \sum_{n=1}^{\infty} A_n\phi_n(x) \cos\sqrt{\lambda_n}(t - t_n) \qquad x \in R.$$

The numbers $\sqrt{\lambda_n}$ correspond to the frequencies of vibration of the nth eigenmode ϕ_n, and the constants A_n, t_n can be determined from initial data. In terms of original (unscaled) variables, the frequencies of vibration are

$$\omega_n = \frac{1}{L}\left(\lambda_n \frac{E}{\rho}\right)^{1/2}$$

where L is the appropriate length scale for the problem. It is immediately apparent that frequencies of vibration are increased by decreasing the density of material, by decreasing the size of the object or by choosing a stronger material (larger E).

If vibrations are forced,

$$u_{tt} = \nabla^2 u + f(x,t),$$

we represent $f(x,t)$ in terms of the eigenfunctions ϕ_n as

$$f(x,t) = \sum_{n=1}^{\infty} f_n(t)\phi_n$$

and the transformed equations become

$$\ddot{a}_n = -\lambda_n a_n + f(t).$$

The solution of this ordinary differential equation depends, of course, on the nature of $f_n(t)$. However, we have successfully reduced the original partial differential equation to a system of uncoupled ordinary differential equations which can be solved.

As a specific example, we consider the vibrations of a circular membrane $0 \leq r \leq 1$, $0 \leq \theta \leq 2\pi$ with Dirichlet boundary condition $u(1,\theta) = 0$, $0 \leq \theta \leq 2\pi$. (The physical radius R of the membrane was absorbed into the change of time and space scale.) We know that the eigenfunctions are

$$\phi_{nm}(r,\theta) = J_n\left(\lambda_{nm} r\right)\cos m(\theta - \theta_0)$$

with eigenvalues $\lambda = \lambda_{nm}^2$ so the free vibrations are superpositions of the eigenmodes. We have already examined the nodal structure of these eigenfunctions. An important observation to make is that the ratios of λ_{nj} to λ_{mk} are not integers, so that overtones are not multiples of the fundamental frequency. For this reason, the sound of a vibrating drum is not as "clean" as the sound of a vibrating string.

8.2.4 Diffraction Patterns

Suppose acoustic (or light) waves are incident from $z = -\infty$ onto a screen at $z = 0$ and that the screen is solid except for a small hole at the origin. We want to know how waves behave after they pass through the hole.

To formulate this problem we seek solutions of the wave equation $u_{tt} = c^2 \nabla^2 u$ in the three dimensional half-space $z > 0$. Here the variable u represents the condensation, that is, the relative change in density from the steady state density, and ∇u is the velocity field.

At the wall we expect $\frac{\partial u}{\partial z} = 0$ since nothing can pass through the solid barrier. On the other hand, we suppose that for $z < 0$ waves are propagating in from $z = -\infty$ and so have the representation

$$u = \frac{u_0}{ik} e^{i(kz - \omega t)}, \qquad k = \omega/c.$$

As a result, at the opening we take

$$\frac{\partial u}{\partial z} = u_0 e^{-i\omega t}.$$

To find u for $z > 0$ we look for solutions of the form $u = \psi e^{-i\omega t}$, ψ depending only on space, so that ψ must satisfy the Helmholtz equation

$$\nabla^2 \psi + k^2 \psi = 0$$

with boundary conditions ψ outgoing,

$$\frac{\partial \psi}{\partial z} = 0 \quad \text{on the wall,}$$

$$\frac{\partial \psi}{\partial z} = u_0 \quad \text{at the opening.}$$

To solve this problem we need the fundamental solution for the Helmholtz equation. This should be similar to the Green's function for Poisson's equation in three dimensions. We solve

$$\nabla^2 G + k^2 G = \frac{\delta(r)}{4\pi r^2}$$

by seeking a radially symmetric solution, that is G must satisfy $(r^2 G')' + k^2 r^2 G = 0$ for $r > 0$ and $\lim_{r \to \infty} (r^2 G) = 0$.

The easiest way to solve this equation is to make the change of variables $G = f/r$ (recall that this worked nicely with spherical Bessel functions) so that $f'' + k^2 f = 0$, or

$$G = \frac{-e^{\pm ikr}}{4\pi r}$$

(Notice that for $k = 0$, this reduces exactly to the known fundamental solution for Poisson's equation.) We will use the function

$$G = \frac{-e^{ikr}}{4\pi r}$$

because it represents outgoing waves, since, on physical grounds, the solution should be radiating away from the source.

The function $G(r)$ is the fundamental solution for all of three space. If we put a source at a point ξ then

$$G(\mathbf{x}, \xi) = \frac{-e^{ik|\mathbf{x}-\xi|}}{4\pi |\mathbf{x} - \xi|}$$

(Here $\mathbf{x}$ and ξ are position vectors in $I\!\!R^3$.) Now we can use the method of images to construct Green's functions for the half space problems

$$H^{(1)}(\mathbf{x}, \xi) = G(\mathbf{x}, \xi) - G(\mathbf{x}, \xi^*)$$

and

$$H^{(2)}(\mathbf{x}, \xi) = G(\mathbf{x}, \xi) + G(\mathbf{x}, \xi^*).$$

If we take ξ^* to be the reflection of ξ across the $z = 0$ plane, that is, $\xi^* = (x_1, y_1, -z_1)$ when $\xi = (x_1, y_1, z_1)$, then the function $H^{(1)}(\mathbf{x}, \xi)$ satisfies the homogeneous Dirichlet boundary condition $H^{(1)}|_{z=0} = 0$ and $H^{(2)}(\mathbf{x}, \xi)$ satisfies the homogeneous Neumann condition

$$\frac{\partial H^{(2)}}{\partial z}\bigg|_{z=0} = 0.$$

For the problem at hand, the only sources are on the boundary. We can use the Green's function $H^{(2)}(\mathbf{x}, \xi)$ and Green's formula to write

$$\psi = -\int_{\partial R} H^{(2)}(\mathbf{x}, \xi)\frac{\partial \psi}{\partial n}d\xi$$

where $\frac{\partial \psi}{\partial n}$ is the normal outward derivative of ψ on the boundary of the domain. Since ξ is restricted to the boundary $z = 0$, we have that

$$\psi = \frac{-u_0}{2\pi}\int_D \frac{e^{ik|\mathbf{x}-\xi|}}{|\mathbf{x} - \xi|}d\xi$$

where D is the domain of the hole at $z = 0$. Recall that $\frac{\partial \psi}{\partial n} = 0$ everywhere except at the hole.

We would be done with this problem except for the fact that we do not yet know what it all means. To get an interpretation of this solution we will approximate the integral in the limit that $\mathbf{x}$ is a long way from the hole, that is, when $|\mathbf{x}|$ is much larger than the radius of the hole. By the law of cosines

$$|\mathbf{x} - \xi| = \left(|\mathbf{x}|^2 + |\xi|^2 - 2|\mathbf{x}||\xi|\cos\eta\right)^{1/2}$$

$$\cong |x|\left(1 - \frac{|\xi|}{|\mathbf{x}|}\cos\eta + \cdots\right)$$

where η is the angle between the two vectors $\mathbf{x}$ and $\boldsymbol{\xi}$. We have ignored terms of order $|\boldsymbol{\xi}|^2 / |\mathbf{x}|$ which we assume to be quite small. We calculate the angle η by noting that if $\boldsymbol{\xi}$ lies along the unit vector $(\cos\theta_1, \sin\theta_1, 0)$ and $\mathbf{x}$ has its direction along the unit vector $(\sin\phi\cos\theta, \sin\phi\sin\theta, \cos\phi)$ (θ, ϕ are usual angular variables in spherical coordinates), then

$$\cos\eta = \sin\phi\cos(\theta - \theta_1).$$

Thus, for $|\mathbf{x}|$ large enough we approximate the integral by

$$\psi(\mathbf{x}) \cong \frac{-u_0}{2\pi} \frac{e^{ik|\mathbf{x}|}}{|\mathbf{x}|} \int_D e^{-ik|\boldsymbol{\xi}|\cos\eta} d\boldsymbol{\xi}.$$

We now specify that the hole is circular with radius a, so that

$$\psi(\mathbf{x}) \cong \frac{-u_0}{2\pi} \frac{e^{ik|\mathbf{x}|}}{|\mathbf{x}|} \int_0^{2\pi} \int_0^a e^{-ikr\sin\phi\cos(\theta_1-\theta)} r\, dr\, d\theta_1.$$

Recall that the zeroth order Bessel function can be expressed as

$$J_0(x) = \frac{1}{2\pi} \int_0^{2\pi} e^{ix\cos\theta} d\theta$$

so that

$$\psi(\mathbf{x}) \cong -u_0 \frac{e^{ik|\mathbf{x}|}}{|\mathbf{x}|} \int_0^a J_0(kr\sin\phi) r\, dr.$$

Finally, we use that

$$\frac{d}{dx}(xJ_1(x)) = xJ_0(x)$$

and find that

$$\psi(\mathbf{x}) \cong \frac{-u_0 a^2 e^{ik|\mathbf{x}|}}{|\mathbf{x}|} \frac{J_1(ka\sin\phi)}{ka\sin\phi}.$$

In other words, the solution $u = \psi e^{-i\omega t}$ is well approximated by radially radiating waves with amplitude

$$|u| = \frac{u_0 a^2}{|\mathbf{x}|} \frac{J_1(ka\sin\phi)}{ka\sin\phi}.$$

How does this amplitude vary with angle ϕ? First of all, notice that

$$\lim_{x\to 0} \frac{J_1(x)}{x} = \frac{1}{2}$$

so that $|u|$ does not vanish at $\phi = 0$. Indeed, $|u|$ is a maximum at $\phi = 0$, which agrees with our physical intuition. However, this amplitude may vanish at other values of ϕ. In particular, since J_1 has an infinite number of zeros $J_1(\lambda_{1j}) = 0$, the amplitude vanishes if

$$ka \sin\phi = \lambda_{1j}$$

or whenever $\sin\phi = \frac{\lambda_{1j}}{ka}$. The first few zeros of $J_1(x)$ are $\lambda_{11} = 3.832$, $\lambda_{12} = 7.016$, $\lambda_{13} = 10.173$, etc. Thus, if $ka > 3.832$ there is at least one ring at which there is no wave, that is, a silent ring. As ka increases there are more silent rings, however for ka too small there are none. This means that if the wavelength $2\pi/k$ is sufficiently small relative to the size of the hole there will be a diffraction pattern of silent rings. If k is very large, however, there are so many rings that they will be too close together to be experimentally resolved, unless the wave source is monochromatic (as with a finely tuned laser).

8.3   The Heat Equation

8.3.1   Derivations

The heat equation can be derived from two physical principles. Suppose we have an object whose temperature we can measure. If there are no internal sources of heat, the change in heat content per unit time must equal the flux of heat across its boundaries. Thus, for any region R

$$\frac{d}{dt}\int_R \rho c u\, dV = -\int_{\partial R} q \cdot n\, ds$$

where u is the temperature, c is the heat capacity of the object, ρ is the density, q is the flux of heat (a vector) and n is outward normal vector to the boundary. The divergence theorem states that for any differentiable vector field q

$$\int_R \operatorname{div} q\, dV = \int_{\partial R} q \cdot n\, ds.$$

Thus our conservation law is expressed as

$$\int_R \left[\frac{\partial}{\partial t}(\rho c u) + \nabla \cdot q\right] dV = 0.$$

As this must be true for every domain, we conclude that

$$\frac{\partial}{\partial t}(\rho c u) + \nabla \cdot q = 0$$

everywhere inside the object. This relationship between temperature and heat is called a CONSERVATION LAW.

The second piece of information we need is how q relates to u. This relationship, called a CONSTITUTIVE RELATIONSHIP, is assumed to be $q = -k\nabla u$ where k is the coefficient of thermal conductivity. The assumed relationship is called FOURIER'S LAW of COOLING and has been experimentally verified in many types of material such as gases and liquids. There are, however, other materials, such as porous media, for which this law does not work well. If Fourier's law can be assumed, then, in combination with the conservation law, we find the heat equation

$$\rho c \frac{\partial u}{\partial t} = k\nabla^2 u.$$

The ratio $D = k/\rho c$ is called the DIFFUSION COEFFICIENT. It is not hard to determine that if there are additional internal sources of heat, this equation must be modified to

$$\rho c \frac{\partial u}{\partial t} = k\nabla^2 u + f,$$

and if the material is a reacting material, then f may depend on the temperature u and the equation becomes nonlinear. We shall see examples of nonlinear heat equations in Chapter 12.

The heat equation is used in chemistry where u represents the concentration of some chemical in a mixture. In this context, the constitutive relationship $q = -k\nabla u$ is called Fick's law but it still gives statement to the physical reality that heat and chemicals move in the direction of decreasing gradients.

An alternate derivation is to take $L = 0$ in the circuit diagram of Figure 8.2, and find the equations governing voltage. Variants of this model are used in physiology to study propagation of electrical action potentials along nerve axons (see Chapter 12).

The heat equation

$$\frac{\partial u}{\partial t} = D \frac{\partial^2 u}{\partial x^2}$$

can (and should) always be scaled so that $D = 1$. If we have a finite domain of length L, we first scale space by setting $x = L\xi$, and then ξ varies from 0 to 1. Next we scale time by setting $t = L^2/D\tau$ so that τ is a dimensionless variable and the resulting partial differential equation is

$$\frac{\partial u}{\partial \tau} = \frac{\partial^2 u}{\partial \xi^2}.$$

If we are on an infinite domain we can either leave x unscaled or pick L to be some "typical" length scale of the problem.

8.3.2   Fundamental Solutions and Green's Functions

The fundamental solution (i.e. Green's function) for the heat equation, satisfying $G_t - G_{xx} = \delta(x - \xi)\delta(t - \tau)$ on the infinite domain, can be found using either Fourier or Laplace Transforms. If we use the Fourier Transform we take

$$g(t,k) = \int_{-\infty}^{\infty} G(x,t)e^{ikx}\,dx$$

and find that the governing transformed equation is

$$\dot{g} + k^2 g = e^{ik\xi}\delta(t - \tau).$$

The function $g(t, k)$ is a Green's function for the time dependent problem with zero initial data $g(0, k) = 0$, so that

$$g(t, k) = e^{-k^2(t-\tau)}e^{ik\xi}H(t - \tau)$$

and then

$$G(x,t;\xi,\tau) = \frac{1}{2\pi}H(t - \tau)\int_{-\infty}^{\infty} e^{ik(\xi-x)}e^{-k^2(t-\tau)}\,dk.$$

We can evaluate this integral explicitly by completing the square of the exponential argument

$$\int_{-\infty}^{\infty} \exp\left[-k^2 t - ikx\right]\,dk \;=\; \int_{-\infty}^{\infty} \exp\left[-t\left(k + \frac{ix}{2t}\right)^2\right]\exp\left(\frac{-x^2}{4t}\right)\,dk$$

$$=\; \left(\frac{\pi}{t}\right)^{1/2}\exp\left(\frac{-x^2}{4t}\right).$$

As a result,

$$G(x,t;\xi,\tau) = \frac{1}{2\sqrt{\pi(t - \tau)}}H(t - \tau)\exp\left[-\frac{(x - \xi)^2}{4(t - \tau)}\right]$$

is the fundamental solution of the heat equation.

As we would expect, the fundamental solution can be used to solve the heat equation with sources $u_t = u_{xx} + f(x,t)$ whereby

$$u(x,t) = \int_0^t \int_{-\infty}^{\infty} \frac{f(\xi,\tau)}{\sqrt{4\pi(t - \tau)}}\exp\left[\frac{-(x - \xi)^2}{4(t - \tau)}\right]\,d\xi\,d\tau.$$

Equally important, the fundamental solution is used to solve the initial value problem $u_t = u_{xx}$ with initial data $u(x,0) = g(x)$. The Fourier transform of the heat equation (with no sources) is $U_t = -k^2 U$ with initial data

$$U(0,k) = G(k) = \int_{-\infty}^{\infty} g(x)e^{ikx}\,dx,$$

and its solution is

$$U(t,k) = G(k)e^{-k^2 t}$$

To reconstruct the solution we use the convolution theorem. We know the inverse Fourier transform of $G(k)$ is $g(x)$ and from our previous calculation the inverse Fourier transform of $e^{-k^2 t}$ is $\left(\frac{1}{4\pi t}\right)^{1/2} \exp\left(\frac{-x^2}{4t}\right)$. The convolution theorem tells us that

$$u(x,t) = \frac{1}{\sqrt{4\pi t}} \int_{-\infty}^{\infty} g(\xi) \exp\left(\frac{-(x-\xi)^2}{4t}\right) d\xi$$

solves the initial value problem.

As a special example, suppose the initial data is given by the bell-shaped curve

$$g(x) = \left(\frac{a}{\pi}\right)^{1/2} e^{-ax^2}$$

The integral expression for $u(x,t)$ can be evaluated explicitly as

$$u(x,t) = \left(\frac{a}{(4ta+1)\pi}\right)^{1/2} \exp\left[\frac{-ax^2}{4ta+1}\right]$$

We can see from this expression that the effect of diffusion is to spread and lower the bell-shaped curve while conserving total area (the total heat) under the curve. Notice also that this solution is proportional to the fundamental solution with input at time $t = -1/4a$.

The fundamental solution can also be used to solve a heat equation with proportional loss

$$v_t = v_{xx} - \alpha v.$$

If we set $v = e^{-\alpha t}u$, then u must satisfy the original heat equation $u_t = u_{xx}$ which we already know how to solve.

For bounded domains, transforms are based on the eigenfunction expansion of the spatial operator. For example, a bar with heat losses along its length and perfectly insulated ends has a temperature governed by

$$u_t = u_{xx} - \alpha u$$

where $u_x(0,t) = u_x(1,t) = 0$. (Here we have scaled the length to be one and have scaled time so that $D = 1$.) We introduce a change of representation based on the eigenfunctions of the differential operator $Lu = u_{xx}$ with Neumann boundary conditions

$$u(x,t) = \sum_{n=1}^{\infty} a_n(t) \cos n\pi x$$

from which we obtain

$$\dot{a}_n = -(n^2 + \alpha)a_n.$$

It follows that

$$u(x,t) = \sum_{n=1}^{\infty} a_n e^{-(n^2+\alpha)t} \cos n\pi x$$

where the coefficients a_n are determined from initial data,

$$a_0 = \int_0^1 u(x,0)dx,$$

$$a_n = 2\int_0^1 u(x,0)\cos n\pi x \, dx.$$

In higher dimensional bounded media, we use the normalized eigenfunctions ϕ_n of the Laplacian, with appropriate boundary conditions, to write

$$u = \sum_{n=1}^{\infty} a_n(t)\phi_n.$$

To solve the heat equation with proportional loss

$$u_t = \nabla^2 u - \alpha u$$

we project the equation onto the eigenfunction ϕ_n and obtain (assuming boundary conditions are homogeneous)

$$\dot{a}_n = -(\lambda_n + \alpha)a_n$$

where $-\nabla^2 \phi_n = \lambda_n \phi_n$. The solution of this equation is easily found with the result that

$$u(x,t) = \sum_{n=1}^{\infty} A_n e^{-(\lambda_n+\alpha)t}\phi_n(x).$$

The coefficients A_n are determined from initial data as

$$A_n = \langle u(x,0), \phi_n \rangle.$$

Example

Suppose we wish to know how long to cook a spherical potato in a large pot of water of constant temperature. We want to solve the equation

$$u_t = D\nabla^2 u$$

on a spherical domain $0 \leq r \leq R$, subject to boundary data $u(R,t) = T_1$, and initial data $u(r, t = 0) = T_0$. The solution will not depend on the angular coordinates ϕ, θ, so we need to know the eigenfunctions of

$$\frac{1}{r^2}(r^2 u_r)_r + \lambda u = 0,$$

with u bounded and $u(R) = 0$. We know that the solutions of this equation are the spherical Bessel functions, found by making the change of variables

$$u = v/r$$

from which it follows that

$$v'' + \lambda v = 0.$$

In other words, $u_n = \frac{1}{r}\sin\sqrt{\lambda_n}r$ and we require $\lambda_n = \frac{n^2\pi^2}{R^2}$. This set of eigenfunctions is complete so we write

$$u(r,t) = \sum_{n=1}^{\infty} \alpha_n(t)\frac{1}{r}\sin\frac{n\pi r}{R} + T_1$$

and the governing heat equation transforms to

$$\dot{\alpha}_n = \frac{-Dn^2\pi^2}{R^2}\alpha_n.$$

It follows that

$$u(r,t) = \sum_{n=1}^{\infty} A_n\frac{1}{r}\exp\left[-\frac{Dn^2\pi^2 t}{R^2}\right]\sin\frac{n\pi r}{R} + T_1$$

where

$$A_n = \frac{2(T_0 - T_1)}{R}\int_0^R r\sin\frac{n\pi r}{R}dr$$

$$= 2(-1)^{n+1}\frac{(T_0 - T_1)R}{n\pi}$$

The temperature at the center of the potato is found by setting $r = 0$,

$$u(0,t) = \sum_{n=1}^{\infty} A_n\exp\left(\frac{-Dn^2\pi^2 t}{R^2}\right)\frac{n\pi}{R} + T_1.$$

Since the amplitudes of the eigenmodes are exponentially decaying, a good approximation for the temperature at the center is found by keeping only the first mode,

$$u(0,t) \cong T_1 + \frac{A_1 \pi}{R} \exp\left[-\frac{D\pi^2}{R^2}t\right]$$

$$\cong T_1 + 2(T_0 - T_1)\exp\left(-\frac{D\pi^2}{R^2}t\right).$$

If we want to achieve a temperature $u(0,t) = T^*$, we must keep the potato in the pot for time

$$t = \frac{R^2}{D\pi^2}\ln\frac{2(T_0 - T_1)}{T^* - T_1}.$$

To see how the numbers come out, suppose the pot is boiling so $T_1 = 212°$ and the potato was initially at room temperature $T_0 = 70°$. It is reasonable to use $D = .17 \times 10^{-6}m^2/s$ for the diffusion coefficient of a potato, and a typical size of a potato is about $R = 5$cm. Using these numbers, we learn that it takes about 35 minutes to raise the center temperature to $140°$. To understand why it takes much longer to bake a potato than to boil it, see Problems 8.3.8, 9, 10.

Example

We can illustrate the usefulness of Fourier time series with the "root cellar" problem (or wine cellar, if you prefer). Suppose $u(x,t)$ is the deviation of the temperature of soil from its annual mean, and suppose the surface temperature varies as a periodic function $u(0,t) = f(t)$ over the course of one year. We want to find the subsurface temperature of the soil.

We model this problem as simply as possible with the one dimensional heat equation

$$u_t = Du_{xx}$$

with depth x, $0 \le x < \infty$ and require $u(0,t) = f(t)$, $\lim_{x\to\infty} u(x,t) = 0$. Since $f(t)$ is periodic it can be represented as a Fourier series

$$f(t) = \sum_{n=-\infty}^{\infty} a_n e^{in\omega t}$$

and the solution will be a superposition of the relevant terms. We will examine only one such term, $f(t) = Ae^{i\omega t}$. The solution is of the form

$$u(x,t) = AU(x)e^{i\omega t}$$

where $U(x)$ satisfies

$$DU'' = i\omega U, \quad U(0) = 1, \quad \lim_{x\to\infty} U(x) = 0$$

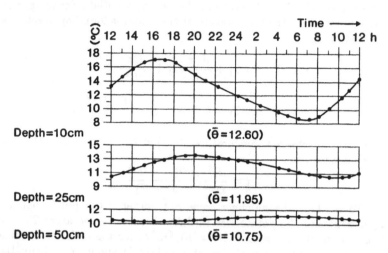

FIGURE 8.3 Typical diurnal variation of soil temperature at depths of 10, 25, and 50 cm below the surface.

so that

$$u(x,t) = Ae^{-\left(\frac{\omega}{2D}\right)^{1/2}x} e^{i\left(\omega t - \left(\frac{\omega}{(2D)}\right)^{1/2}x\right)}.$$

Of course, the full solution for general $f(t)$ is a superposition of such solutions. The main observation is that $u(x,t)$ is oscillatory but out of phase from the surface temperature, and it attenuates with depth. Higher frequency modes attenuate more rapidly so that daily temperature fluctuations have a penetration depth which is $(365)^{1/2} \cong 19$ times smaller than the penetration depth of annual temperature fluctuations.

The fluctuations are out of phase from the surface temperature by

$$\phi = \left(\frac{\omega}{2D}\right)^{1/2} x.$$

To build a root cellar whose temperature is π out of phase from the surface temperature (i.e. warmer in winter, cooler in summer) we should make the depth

$$x = \pi \left(\frac{2D}{\omega}\right)^{1/2}.$$

For typical soils, $D \cong 10^{-6} m^2/s$ and $\omega = 2.0 \times 10^{-7} s^{-1}$ so that $x = 9.9$ meters. However, at this depth

$$x \left(\frac{\omega}{2D}\right)^{1/2} = \pi$$

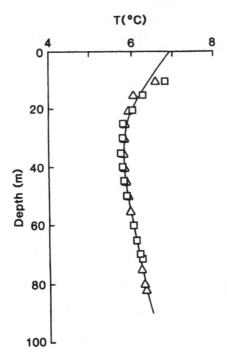

FIGURE 8.4 Subsurface temperature 3.5 years after clear cutting of the surface foliage in the Uinta Mountains.

so that the relative amplitude of the oscillation is only $e^{-\pi} = .04$, that is, the amplitude of oscillation is only 4% of the oscillation at the surface. At twice that depth, the relative variation in temperature is about .2%, which is negligible.

Daily oscillations have much less of an effect. For daily oscillations, $\omega = 7 \times 10^{-5} s^{-1}$ so that the temperature is π out of phase with the surface temperature at $x = .5$ meters where the oscillation has an amplitude of only 4% of the surface oscillation and at $x = 1.0$ meters, the oscillation is negligible.

In Figure 8.3 is shown the typical diurnal temperature variation in sand with no vegetation $(D \sim 1.1 \times 10^{-6} m^2/s)$. The agreement with the calculated solution is quite good.

Example

Another "hot" problem that is of major importance in geophysics is to determine the effect of climactic environmental temperature changes on the subsurface temperature. For example, if a forest or area of thick vegetation is cleared, then the average yearly surface temperature will suddenly increase, with the result that the subsurface temperature will increase as well.

To model this problem we suppose the temperature is $T = T_0$ at the surface for time $t < 0$ and that $\frac{dT}{dx} = $ constant < 0 as the depth $x \to \infty$. This corresponds to the fact that the heat flux from the core of the earth is a constant. The temperature profile for $t < 0$ is assumed to be at steady state. At time $t = 0$ an abrupt change is made in the surface temperature to $T = T_1$. The goal is to find the temperature profile for $t > 0$. This problem is left as an exercise (Problem 8.3.6).

In Figure 8.4 we have reproduced temperature measurements taken from wells in the Uinta Mountains of Utah, 3.5 years after the surface area was cleared for construction of the Stillwater Dam. The triangles and boxes are data points and the solid line is the solution of the heat equation using a thermal diffusivity of $2.0 \times 10^{-6} m^2/s$ which is appropriate for Uinta Mountain quartzite. The original surface temperature was taken to be $5.2^\circ C$, the increase in temperature was $1.8^\circ C$, and the thermal gradient was assumed to be $14.7^\circ C/km$. As is evident, the agreement between data and theory is good except for the points near the surface where annual variations are still in evidence. Below 20 meters, however, annual variations have no perceptible effect on the measurements.

8.4   Differential-Difference Equations

Differential-difference equations occur naturally in at least two contexts. In the first, one may want to model the dynamics of large electrical systems, with many component subsystems such as in Figure 8.2. In Section 8.2 we took a continuum limit to obtain the telegrapher's equation, but it may be that the continuum limit is inappropriate and it may be preferable to solve the discrete system. An example of a system of this form is that for the coupled network in Figure 8.2

$$LC\frac{d^2 v_n}{dt^2} + \left(\frac{L}{R_2} + R_1 C\right) \frac{dv_n}{dt} = v_{n+1} - 2v_n + v_{n-1} - \frac{R_1}{R_2} v_n.$$

If the system is finite, it can be written in matrix form

$$LC\frac{d^2 V}{dt^2} + \left(\frac{L}{R_2} + R_1 C\right) \frac{dV}{dt} = AV$$

where V is the vector of values v_n and the matrix A is the appropriate tridiagonal matrix.

The second way that differential-difference equations occur is via discretization of an already continuous model. For example, if one wants to solve the heat equation on a finite domain $0 \le x \le 1$, a common idea is to set $u_n = u(nh)$, $h = 1/N$ and replace the spatial differential operator

$$\frac{\partial^2}{\partial x^2}$$

by the difference operator $(\Delta_h u)_n = \frac{1}{h^2}(u_{n+1} - 2u_n + u_{n-1})$, $1 \le n \le N - 1$. In addition, one must specify end conditions such as $u_0 = u_N = 0$.

One might solve this system of ordinary differential equations on a computer or analyze them by other means. The process of replacing the original partial differential equation by ordinary differential equations which estimates the solution along straight lines in the $x - t$ plane is called the METHOD OF LINES.

Differential-difference equations are generally written in the form

$$\frac{du}{dt} = Au$$

or

$$\frac{d^2 u}{dt^2} = Au$$

where, if u is an $(N - 1)$-vector, A is an $(N - 1) \times (N - 1)$ matrix, or if u is in ℓ^2, then A is a difference operator on ℓ^2.

In the case that $u \in {I\!\!R}^{N-1}$, the solution process is straightforward. We suppose there is a basis that diagonalizes A, say $(u_1, u_2, \ldots u_{N-1})$, where $Au_k = \lambda_k u_k$. We write

$$u = \sum_{k=1}^{N-1} \alpha_k u_k$$

so that

$$\frac{d\alpha_k}{dt} = \lambda_k \alpha_k.$$

The solution is, naturally,

$$u = \sum_{k=1}^{N-1} \alpha_k(0) e^{\lambda_k t} u_k.$$

For example, for the difference operator

$$(\Delta u)_n = u_{n+1} - 2u_n + u_{n-1}$$

$$u_0 = u_N = 0$$

we already know that the appropriate diagonalizing basis vectors are the vectors $\phi^{(k)}$ with entries

$$\phi_j^{(k)} = \sin \frac{\pi j k}{N}, \qquad j = 1, 2, \ldots N - 1.$$

To numerically evaluate the solution of the discretized heat equation we could use the fast Sine transform on the initial data $u_n(t = 0)$ to find the coefficients $\alpha_k(0)$ and then for a few values of time apply the fast inverse Sine transform to the vector $\alpha_k(0)e^{\lambda_k t}$ where, as we know, λ_k is the eigenvalue corresponding to the eigenvector $\phi^{(k)}$, $\lambda_k = -4 \sin^2 \frac{\pi k}{2N}$.

Another example of a discretized equation is the two dimensional Poisson equation discretized in both spatial dimensions, subject to homogeneous Dirichlet conditions, for example. If we take $u_{jk} \sim u(jh, kh)$ where $h = 1/N$ to be the value of u at the mesh points, the discretized Poisson equation becomes

$$(u_{j+1,k} - 2u_{j,k} + u_{j-1,k}) + (u_{j,k+1} - 2u_{j,k} + u_{j,k-1}) = h^2 f_{jk},$$

$$1 \leq j, \quad k \leq N,$$

with boundary conditions $u_{0k} = u_{Nk} = u_{j0} = u_{jN} = 0$.

One way to numerically solve this system of algebraic equations is to view the unknowns u_{jk} as elements of a vector with $(n-1)^2$ entries and to solve the system using some matrix inversion routine. The matrix has a banded structure so a regular Gaussian elimination technique is rather inefficient, but routines that exploit the banded structure are available, as are indirect, iteration techniques.

One very efficient way to solve this set of equations is via fast Sine transforms. We recognize that the difference operators can be diagonalized using the eigenvectors of the difference operator $(\Delta u)_n = u_{n+1} - 2u_n + u_{n-1}$. Thus we write

$$f_{jk} \;=\; \frac{4}{N^2} \sum_{p=1}^{N-1} \sum_{q=1}^{N-1} \beta_{pq} \phi_j^{(p)} \phi_k^{(q)}$$

and

$$u_{jk} \;=\; \frac{4}{N^2} \sum_{p=1}^{N-1} \sum_{q=1}^{N-1} \alpha_{pq} \phi_j^{(p)} \phi_k^{(q)}.$$

Upon substituting into the discrete Poisson equation, we learn that

$$\lambda_{pq} \alpha_{pq} = h^2 \beta_{pq}$$

where

$$\lambda_{pq} = -4 \left(\sin^2 \frac{\pi p}{2N} + \sin^2 \frac{\pi q}{2N} \right)$$

are associated eigenvalues.

Now we see how to solve the discrete Poisson equation using fast Sine transforms. For each fixed k we find the fast Sine transform of f_{jk}, denoted $\widehat{\beta}_{jk}$. Then for each fixed j we find the fast Sine transform of the vectors $\widehat{\beta}_{jk}$, denoted β_{jk}. Said another way, if f_{jk} is viewed as an element of a matrix F, we first find the fast Sine transform of the columns of F and then find the fast Sine transform of the rows of the once transformed matrix. This twice transformed matrix has entries β_{jk}.

The coefficients α_{jk} are given by

$$\alpha_{jk} = h^2 \beta_{jk}/\lambda_{jk}$$

and we can now find the matrix of solution values u_{jk} by taking the fast inverse Sine transform of all the columns and then all the rows, and multiply everything by the scale factor $4/N^2$. The resulting transformed matrix has entries u_{jk}.

A simple operation count shows why this procedure is to be preferred over direct solution techniques. If N is a power of 2, the fast Sine transform of a vector of length N requires on the order of $N \log_2 N$ operations. The double fast Sine transform of F then requires on the order of $2N^2 \log_2 N$ operations as does the inverse transform. Converting β_{jk} to α_{jk} and normalization by $4/N^2$ requires an additional N^2 operations so that, there are about $4N^2(\log_2 N)+ N^2$ operations required. For large N this is significantly fewer than the (order of) $N^6/3$ operations required by direct Gauss elimination of an $N^2 \times N^2$ matrix.

This method is not the fastest way to solve the Poisson equation numerically. It is faster to perform only one Fast Sine Transform to reduce the equations to a collection of tridiagonal problems which can be solved using a tridiagonal matrix routine. Then, the system can be transformed back to find the solution.

If the discrete system of differential equations is infinite dimensional, we use the transform appropriate to the difference operator. As an example, consider the finite difference approximation to the wave equation

$$\frac{d^2 u_n}{dt^2} = \frac{1}{h^2} (u_{n+1} - 2u_n + u_{n-1}), \qquad -\infty < n < \infty,$$

where $u_n(t) \cong u(nh, t)$. For this operator, the appropriate transform is the z-transform

$$U(z) = \sum_{n=-\infty}^{\infty} u_n z^n$$

with inverse

$$u_n = \frac{1}{2\pi i} \int_{|z|=1} U(z) z^{-n-1} dz.$$

We transform the equation by multiplying by z^n and summing over n, or equivalently, by taking the inner product of the equation with the improper eigenfunction $\phi_n(z) = z^n$, to obtain

$$\frac{d^2 U}{dt^2} = \frac{1}{h^2}\left(z - 2 + \frac{1}{z}\right) U(z).$$

The solution is

$$U = U_0(z)\exp\left(\left(\sqrt{z} - \frac{1}{\sqrt{z}}\right)\frac{t}{h}\right) + U_1(z)\exp\left(-\left(\sqrt{z} - \frac{1}{\sqrt{z}}\right)\frac{t}{h}\right).$$

To get some idea of the behavior of this solution, we evaluate the inverse transform in the special case $U_0(z) = 1$, $U_1(z) = 0$. We find

$$
\begin{aligned}
u_n(t) &= \frac{1}{2\pi i}\int_{|z|=1} \exp\left(\left(\sqrt{z} - \frac{1}{\sqrt{z}}\right)\frac{t}{h}\right) z^{-n}\frac{dz}{z} \\
&= \frac{1}{2\pi}\int_0^{2\pi} \exp\left(\frac{2it}{h}\sin\theta/2 - in\theta\right) d\theta \\
&= \frac{1}{\pi}\int_0^{\pi} \exp\left(\frac{2it}{h}\sin\theta - 2ni\theta\right) d\theta \\
&= J_{2n}\left(\frac{2t}{h}\right)
\end{aligned}
$$

where $J_n(x)$ is the nth order Bessel function.

We can use this information to solve the initial value problem with initial data

$$u_n(t = 0) = \delta_{n0}, \qquad u_n'(t = 0) = 0.$$

This corresponds to having an initial disturbance at the origin with no initial velocity. We know that the solution of the wave equation is

$$u(x, t) = \frac{1}{2}f(x - t) + \frac{1}{2}f(x + t)$$

when $u(x, 0) = f(x)$, $u_t(x, 0) = 0$, that is, the initial data evolve into two invariant wave shapes, one traveling left, one traveling right with constant speed 1. For the discretized problem, the solution is somewhat different. We find that

$$u_n(t) = J_{2n}\left(\frac{2t}{h}\right)$$

is the solution of the initial value problem.

Notice that there is definitely *not* an invariant wave form. What happens at each location n is different from what happens at its neighbors. The solution is, however, symmetric about $n = 0$ so that information is propagating both forward and backward in n.

At each position n, the solution increases from zero gradually until its first maximum is reach followed by a damped oscillation. The arrival of the first peak corresponds, in some sense, to the first arrival of the wave. One can show that the location of the first maximum of the nth order Bessel function for large n occurs at

$$j'_n = n + .80861n^{1/3} + \text{ terms of order } n^{-1/3}$$

and at the first maximal value

$$J_n(j'_n) = .67488n^{-1/3} + \text{ terms of order } n^{-1}.$$

As a result, the first maximum reaches position n at time

$$t_1 \cong nh$$

for large n, corresponding to waves with speed one, and the value of the solution when the maximum arrives is

$$u_n(t_1) \cong .67488(2n)^{-1/3}$$

so that the height of the disturbance is decreasing as n increases. The fact that the wave solution appears to be breaking up is due to DISPERSION, namely that this solution is a superposition of different wave shapes which move with different characteristic speeds. It is actually quite unusual to have all wave shapes move with exactly the same speed as in the wave equation, since wave propagation in most physical systems is dispersive. Here we see that the discretization of the Laplacian introduces dispersion to the solution.

Further Reading

There are numerous books on partial differential equations. Three which are recommended are (in increasing order of difficulty)

1. H. F. Weinberger, *A First Course in Partial Differential Equations*, Blaisdell, New York, 1965.

2. E. Zauderer, *Partial Differential Equations of Applied Mathematics*, Wiley and Sons, New York, 1983.

3. P. R. Garabedian, *Partial Differential Equations*, Wiley and Sons, New York, 1964.

In this text we have not included a discussion of the solution of first order partial differential equations and the method of characteristics. This technique is extremely important and can be found in the above mentioned books or in

4. G. B. Whitham, *Linear and Nonlinear Waves*, Wiley and Sons, New York, 1974,

which also contains a nice discussion of dispersive waves. A nice treatment of separation of variables is given by

5. B. Friedman, *Principles and Techniques of Applied Mathematics*, Wiley and Sons, New York, (1956).

There are many ways to solve partial differential equations numerically. A discussion of the use of spectral methods for numerical solutions can be found in

6. M. Pickering, *An Introduction to Fast Fourier Transform Methods for Partial Differential Equations, with Applications*, Wiley and Sons, New York, 1986.

7. D. Gottlieb and S. A. Orszag, *Numerical Analysis of Spectral Methods; Theory and Applications*, SIAM, Philadelphia, 1977.

A lovely discussion of vibrations of a kettledrum showing pictures of the eigenmodes is

8. T. D. Rossing, The Physics of Kettledrums, *Scientific American*, Nov. 1982.

References on the geophysics problems discussed in this chapter are

9. O. Kappelmeyer and R. Häenel, *Geothermics with Special Reference to Application*, Gebrüder Borntraeger, Berlin, 1974.

10. M. S. Bauer and D. S. Chapman, Thermal regime at the Upper Stillwater damsite, Uinta Mountains Utah, Tectonophysics, **128**, 1-20, 1986.

Finally, the "standard reference works" which have tables of the special functions we use here are

11. M. Abramowitz and I. A. Stegun, *Handbook of Mathematical Functions*, Dover, New York, 1965,

12. E. Jahnke and F. Emde, *Tables of Functions*, Dover, New York, 1945.

One can readily calculate the special functions numerically using standard routines (available on most mainframe computers) or using the algorithms found in

13. W. H. Press, B. P. Flannery, S. A. Teukolsky, and W. T. Vetterling, *Numerical Recipes: The Art of Computing*, Cambridge University Press, Cambridge, 1986.

Problems for Chapter 8

Section 8.1

1. a. Show that $\delta(x, y) = \frac{\delta(r)}{2\pi r}$.

   b. Show that $\delta(x - x', y - y') = \frac{\delta(r-r')\delta(\theta-\theta')}{r}$, $r' > 0$.

2. a. Show that $\delta(x, y, z) = \frac{\delta(r)}{4\pi r^2}$.

   b. Show that $\delta(x - x', y - y', z - z') = \frac{\delta(r-r')\delta(\theta-\theta')\delta(\phi-\phi')}{r^2 \sin \theta}$.

3. a. Find the weak formulation for the partial differential equation $\frac{\partial u}{\partial t} + c\frac{\partial u}{\partial x} = 0$.

   b. Show that $u(x, t) = f(x - ct)$ is a generalized solution of $u_t + cu_x = 0$ for any one-dimensional distribution f.

4. Find the Green's functions for the Laplacian on the unit disc with inhomogeneous Dirichlet data via the method of images.
   Hint: Use a conformal mapping to map the half plane problem to the problem on the disc.

5. Find two representations for the solution of

$$\nabla^2 u + k^2 u = f(x, y), \quad 0 < x < \infty, \quad 0 < y < a$$

$$u_y(x, 0) = u_y(x, a) = 0$$

$$u_x(0, y) = 0, u \text{ outgoing as } x \to \infty$$

   Show the equivalence of these representations using contour integration.

6. Find the Green's function for Poisson's equation on a rectangle $0 \le x \le a$, $0 \le y \le b$, with boundary data

$$u(x = 0, y) = u(x = a, y) = u(x, y = 0) = 0$$

$$u_y(x, y = b) = 0.$$

7. Find two representations of the Green's function for Poisson's equation on the infinite strip, $-\infty < x < \infty$, $0 \le y \le a$, with boundary data $u(x, 0) = 0$, $u_y(x, a) = 0$.

8. Find the solution of Laplace's equation on an annulus $0 < a \le r \le b < \infty$, $0 \le \theta < 2\pi$, with Dirichlet data $u(a, \theta) = f(\theta)$, $u(b, \theta) = g(\theta)$.

9. Find the bounded solution of Laplace's equation on the exterior domain $r \ge a$ with Dirichlet data $u(a, \theta) = f(\theta)$.

10. Solve $\nabla^2 u = 0$ on the wedge shaped region $0 < \theta < \alpha$, $0 < r < a$ (in polar coordinates) subject to boundary conditions $u(a, \theta) = 0$, $0 < \theta \le \alpha$ and $u(r, 0) = f(r)$, $u(r, \alpha) = 0$ for $0 < r \le a$.

11. Find the Green's function for Poisson's equation on the wedge shaped region $0 \leq r \leq R$, $0 \leq \theta \leq \theta_0 < 2\pi$, with Dirichlet boundary data.

12. Show that $\int_0^\infty e^{-kx} \sin ky \sin k\eta \frac{dk}{k} = \frac{1}{4} \ln \left[\frac{x^2 + (y - \eta)^2}{x^2 + (y + \eta)^2} \right]$

    Hint: Evaluate $f(x) = \int_0^\infty e^{-kx} \frac{dk}{k}$ by showing that $f'(x) = -\frac{1}{x}$.

13. Verify that

$$\frac{1}{2\pi} \int_0^{2\pi} f(\theta) d\theta + \frac{1}{\pi} \int_0^{2\pi} \sum_{j=1}^\infty r^j f(\theta) \cos j(\phi - \theta) d\theta =$$

$$\frac{1}{2\pi} \int_0^{2\pi} f(\theta) \frac{1 - r^2}{1 - 2r \cos(\phi - \theta) + r^2} d\theta \quad \text{for} \quad 0 \leq r < 1.$$

    Hint: Evaluate $\sum_{j=1}^\infty r^j \cos j\phi$ using Euler's formula $\cos j\phi = \frac{1}{2}(e^{ij\phi} + e^{-ij\phi})$ and geometric series.

14. Solve Laplace's equation on the half space $-\infty < x < \infty$, $0 < y < \infty$ with mixed boundary data, $u(x, 0) = f(x)$ for $0 < x < \infty$ and $u_y(x, 0) = g(x)$, $-\infty < x < 0$.
    Hint: View this domain as a wedge shaped region $0 \leq r < \infty$, $0 \leq \theta \leq \pi$, and assume that f and g are the restrictions of analytic functions whose only singularities are at 0 and ∞.

15. Find the solution of $\nabla^2 u - \alpha^2 u = 0$ on a circular domain $0 \leq r \leq R$ with inhomogeneous Dirichlet boundary data $u(R, \theta) = f(\theta)$.

16. Find the eigenvalues and eigenfunctions for the Laplacian on a rectangular domain $0 \leq x \leq a$, $0 \leq y \leq b$ with homogeneous Dirichlet data. Describe the nodal structure of the eigenfunctions and the relative size of corresponding eigenvalues.

17. Find the eigenvalues and eigenfunctions for the Laplacian on the semicircular domain $0 \leq r \leq R$, $0 \leq \theta \leq \pi$ with Dirichlet boundary conditions. Compare the eigenvalues of this problem with those of the full circle $0 \leq \theta \leq 2\pi$.

18. Find the eigenvalues and eigenfunctions for the Laplacian on the cylindrical domain $0 \leq r \leq R$, $0 \leq \theta \leq 2\pi$, $0 \leq z \leq a$ with homogeneous Dirichlet boundary data.

19. Find the eigenfunctions and eigenvalues for the Laplacian operator restricted to a spherical surface of radius R,

$$\nabla^2 u = \frac{1}{R^2 \sin \theta} \frac{\partial}{\partial \theta} \left(\sin \theta \frac{\partial u}{\partial \theta} \right) + \frac{1}{R^2 \sin^2 \theta} \frac{\partial^2 u}{\partial \phi^2}.$$

    Discuss the nodal structure of the eigenfunctions.

20. Use Hankel transforms to show that the solution of $u_{rr} + \frac{1}{r}u_r + u_{xx} = 0$, $0 \le r < \infty$, $x > 0$ (with u vanishing rapidly as $r^2 + x^2 \to \infty$) satisfies

$$u(x,r) = \frac{1}{\pi} \int_0^\pi u(x + ir\cos\theta, 0)d\theta.$$

21. Show that $\delta(x - x_0)\delta(y - y_0) = \delta(t - t_0)\delta(\theta - \theta_0)$ where $x = \mathrm{Re}^{-t}\cos\theta$, $y = \mathrm{Re}^{-t}\sin\theta$.

22. Describe the nodal structure and relative frequencies corresponding to the 5 smallest eigenvalues for the Laplacian in the interior of a sphere with homogeneous Dirichlet boundary data.

23. Use the method of images to find the Green's function for the Laplacian on a wedge shaped region with interior angle $2\pi/3$, and homogeneous Dirichlet boundary conditions.

Section 8.2

1. Suppose $f(x)$ is given for all x. Find the solution of the wave equation of the form
$$u(x,t) = f(x+t) + h(t-x)$$
satisfying the boundary condition $u_x + \alpha u = 0$ at $x = 0$. Interpret the solution for the special case $f(x) = \sin kx$.

2. Show that $u(x,t) = f(x+t) + g(x-t)$ is a distributional solution of the wave equation $u_{tt} = u_{xx}$ when f and g are one-dimensional distributions.

3. Use the method of images to find the Green's function for the wave equation on $0 \le x < \infty$ with zero initial data and Neumann boundary data $u_x(0,t) = 0$.

4. Use transform methods to find the Green's function for the wave equation on $0 \le x < \infty$ with zero initial data and Dirichlet data $u(0,t) = 0$.

5. Find Green's function for the wave equation on $0 \le x < \infty$ with zero initial data and the mixed boundary condition $u_x + \alpha u = 0$, α real, at $x = 0$. Interpret the solution physically.

6. Which has a purer sound (i.e. smaller overtone amplitudes)? A plucked string or a hammered string?
   (Hint: A plucked string has $u_t(x,0) = 0$ and a hammered string has $u(x,0) = 0$.)

7. Show that among all vibrating rectangular membranes of area A with homogeneous boundary data, the square has the smallest fundamental frequency.

8. Compare the fundamental frequencies of a vibrating square and circular membrane with the same area and homogeneous Dirichlet boundary data.

9. Find the frequencies of vibration of a rectangular membrane $0 \le x \le a$, $0 \le y \le b$ with two sides fixed $u(0, y) = u(a, y) = 0$ and two sides free $u_y(x, 0) = u_y(x, b) = 0$. How do these compare with the frequencies of a string of the same length a.

10. Find the fundamental solution for the Helmholtz equation in two dimensions. Show that it is related to the Hankel function $H_0^{(1)}(z)$.

11. Find the diffraction pattern for acoustic waves through a long thin slit. Hint: Assume the pattern is independent of the length coordinate of the slit and reduce this to a two dimensional problem. Use asymptotic properties of Hankel functions (Chapter 6) to estimate silent bands.

12. Find the diffraction pattern for the full eclipse of a flat circular disc.

Section 8.3

1. Derive the equation governing the voltage in the continuous line of circuit elements depicted in Figure 8.2 with $L = 0$.

2. What is the effect of changing the diffusion coefficient D on the solution of the heat equation $u_t = D u_{xx}$?

3. Find the Green's function for the initial value problem

$$u_t - u_{xx} = \delta(t)\delta(x), \qquad 0 \le x \le 1$$

with zero initial data and Neumann boundary data, $u_x(0, t) = u_x(1, t) = 0$.

4. A volume V of fluid is to be heated in a cylindrical container with constant temperature at the boundary. What should the dimensions of the cylinder be to heat the center as quickly as possible? Is this faster or slower than heating the fluid in a spherical container of the same volume?

5. The sun is hottest at noon, but the air temperature outside is hottest at 2–3 pm. Why the difference? (Make a model to explain your ideas.)

6. The temperature profile of a section of earth is at steady state with the surface at $T = T_0$ and $\frac{dT}{dx} = $ constant as the depth $x \to \infty$. At the time $t = 0$ the surface temperature is suddenly increased to $T = T_1$. Find the temperature profile for $t > 0$.

7. The thermal diffusivity of water is $0.14 \times 10^{-6} m^2/s$. At what depth in a deep lake is the annual temperature variation 50% of that of the surface temperature? At what time of the year is the maximum attained at this depth?

8. It takes much longer to bake a potato than to boil it. Explain why this might be true and how to change the model to reflect the difference between baking and boiling.

9. a. An object with diffusion coefficient D_1 and thermal conductivity k_1 is immersed in a larger domain with diffusion coefficient D_2, and thermal conductivity k_2. The temperature at the boundary of the exterior medium is assumed constant. What are the equations governing the temperature of the two media? In particular, what is the correct condition at the interface between the two domains? (Hint: What quantity is conserved?)

   b. Use these equations to discuss the baking of a potato in a large (spherical) oven. Does this new model resolve the dilemma raised in Problem 8? How do the eigenvalues for this problem change as the size of the oven increases or as the diffusion coefficient in the oven changes?

   c. The coefficient of thermal conductivity for a potato is about 10 times larger than that for air at $600° K$. The diffusion coefficients are $D_1 = .17 \times 10^{-6} m^2/s$ for a potato and $D_2 = 75 \times 10^{-6} m^2/s$ for air at $600° K$. Use these numbers to estimate typical times to bake a potato in a spherical oven.

10. Discuss the efficacy of heating the inside of a hemispherical tent with a warm hemispherical rock. How would you model this problem?

Section 8.4

1. Solve the discretized heat equation

$$u_{n_t} = \frac{1}{h^2}(u_{n+1} - 2u_n + u_{n-1}), \qquad -\infty < n < \infty,$$

with initial data $u_n(t = 0) = \sin \frac{2\pi n}{k}$, k integer. Interpret your solution. How does the solution change as k is varied? How does this differ from the continuous problem $u_t = u_{xx}$?
Hint: Express your solution in terms of $I_n(z)$, the modified Bessel function of first kind.

2. Solve the equation

$$u_{n_t} = \frac{1}{2h}(u_{n+1} - u_{n-1}), \qquad -\infty < n < \infty$$

with initial data $u_n(t = 0) = \delta_{n0}$. Describe the behavior of the solution. Compare your solution with the solution of the continuous equation $u_t = u_x$.

3. Show that application of a Fast Sine Transform to the discretized Poisson equation with homogeneous Dirichlet condition reduces the equations to tridiagonal matrix equations. Describe how to fully solve these discretized equations.

4. Describe how to use Fast Transforms to solve the discretized Poisson equation on a rectangle with homogeneous Neumann conditions.

5. Solve the discretized wave equation $u_{n_{tt}} = \frac{1}{h^2}(u_{n+1} - 2u_n + u_{n-1})$, $-\infty < n < \infty$, with $u_n = \delta_{no}$, $u_{n_t} = 0$, at time $t = 0$ using Fourier transforms in time.

6. Find special solutions of $u_{n_{tt}} = \frac{1}{h^2}(u_{n+1} - 2u_n + u_{n-1})$ of the form $u(x,t) = e^{i(wt - k(w)n)}$. How fast do such solutions travel?

8.5 Appendix: Thermal Diffusivity for Some Common Materials

Thermal Diffusivity for Some Common Materials

Material	Thermal diffusivity $[\text{cm}^2/\text{s}]$
Air at $300°K$ (atmospheric pressure)	0.159
Air at $600°K$ (atmospheric pressure)	0.527
Apple	0.0017
Banana	0.0015
Clay	0.01
Concrete	0.0069
Cotton	0.0058
Granite	0.014–0.021
Loamy soil	0.008
Sand, dry	0.012
Sand, wet	0.009
Snow	0.005
Water	0.0014

8.6 Appendix: Coordinate Transformation with REDUCE

The representations of the Laplacian in different coordinate systems are easy to find if you have the right handbook available or if the coordinate system is not too unusual (i.e., not toroidal coordinates). If perchance you wish to find the representation of the Laplacian for some coordinate system by hand, the procedure is straightforward to describe (just use the chain rule) but often tedious to implement. The following is how to do the calculation using REDUCE. Other symbolic languages can be used as well.

```
matrix a(3,3), ai(3,3);
operator u, du;
for all k, m, n, r, s, p let
df(u(k,m,n,r,s,p),r)=u(k+1,m,n,r,s,p);
for all k, m, n, r, s, p let
df(u(k,m,n,r,s,p),s)=u(k,m+1,n,r,s,p);
for all k, m, n, r, s, p let
df(u(k,m,n,r,s,p),p)=u(k,m,n+1,r,s,p);
for all u let
du(u,x)=ai(1,1)*df(u,r)+ai(1,2)*df(u,s)+ai(1,3)*df(u,p);
for all u let
du(u,y)=ai(2,1)*df(u,r)+ai(2,2)*df(u,s)+ai(2,3)*df(u,p);
for all u let
du(u,z)=ai(3,1)*df(u,r)+ai(3,2)*df(u,s)+ai(3,3)*df(u,p);

procedure LAPLAC(f,g,h,r,s,p);
begin
a(1,1):=df(f,r);
a(2,1):=df(f,s);
a(3,1):=df(f,p);
a(1,2):=df(g,r);
a(2,2):=df(g,s);
a(3,2):=df(g,p);
a(1,3):=df(h,r);
a(2,3):=df(h,s);
a(3,3):=df(h,p);
ai:=1/a;
u:=u(0,0,0,r,s,p);
return du(du(u,x),x)+du(du(u,y),y)+du(du(u,z),z);
end;
```

To use this procedure, it is necessary to specify the new coordinate system
f, g, h as functions of r, s, p. For example, for cylindrical coordinates, take

```
f:=r*cos(s);
g:=r*sin(s);
h:=p;
```

The response to the statement

```
LAPLAC(f,g,h,r,s,p);
```

will be the Laplacian in cylindrical coordinates.

9

INVERSE SCATTERING TRANSFORM

9.1 Inverse Scattering

In the mid-1960s, an important class of transform techniques was discovered. This transform, called the INVERSE SCATTERING TRANSFORM, is capable of producing exact solutions for a small group of nonlinear evolution equations.

The idea of this transform is to view the solution $q(x,t)$ of the equation to be solved as the potential of the Schrödinger equation

$$-\frac{d^2u}{dx^2} + q(x,t)u = \lambda u.$$

For each fixed time t, we study the scattering problem and find the reflection and transmission coefficients and the bound states. Then, as time evolves, the potential $q(x,t)$ (and hence its scattering data) evolves as well. Now, the amazing thing we will discover is that there are certain equations for which the evolution of the scattering data is diagonalized, that is, the scattering data for a given frequency evolves independent of all other frequencies. Thus, instead of studying directly the evolution of $q(x,t)$, we study the evolution of the scattering data, which is described by separated, linear differential equations whose solutions are easily obtained, and from this information we can construct $q(x,t)$.

To make any transform work, we must know how to transform functions and also how to invert the transform. For the Schrödinger equation, we know how, at least in principle, to transform potentials $q(x)$ by finding the scattering data. A natural question to ask is if one can reconstruct a potential $q(x)$ from knowledge of its scattering data. This is an important problem, not just in the context of the Schrödinger equation, but in many other applications such as x-ray tomography, seismic exploration, radar imaging, etc., where the shape of an object must be determined from some indirect information, such as the way waves are reflected or transmitted by the object. In this section we will discuss only the problem of reconstructing the potential of the Schrödinger equation, with the realization that this is but a small part of the important field of inverse problems in which there is still much active research and development.

Consider the Schrödinger equation

$$-u'' + q(x)u = \lambda u$$

with absolutely integrable potential $q(x)$, $\int_{-\infty}^{\infty} |q(x)|\, dx < \infty$. Recall from Chapter 7 that there are solutions of this equation $u_1(x,k)$, $u_2(x,k)$ where $u_1 \sim e^{ikx}$ for large $x > 0$ and $u_2 \sim e^{-ikx}$ for large $x < 0$, with $k^2 = \lambda$. We define the function $A(x,y)$ by

$$u_1(x,k) = e^{ikx} + \int_x^{\infty} A(x,y)e^{iky}\, dy.$$

Although we know that $u_1(x,k)$ is perfectly well defined, it is not immediately clear that $A(x,y)$ exists. To check its existence we substitute $u_1(x,k)$ into the Schrödinger equation and find after integration by parts, that

$$u_1'' - q(x)u_1 + k^2 u_1 = \int_x^{\infty} \left(A_{xx}(x,y) - A_{yy}(x,y) - q(x)A(x,y) \right) e^{iky}\, dy$$

$$+ \lim_{y \to \infty} \left(A_y(x,y) - ikA(x,y) \right) e^{iky}$$

$$- \left(2\frac{d}{dx}A(x,x) + q(x) \right) e^{ikx}.$$

We require the kernel $A(x,y)$ to satisfy

$$A_{xx} - A_{yy} - q(x)A = 0$$

$$\frac{d}{dx}A(x,x) = -\frac{1}{2}q(x)$$

$$\lim_{y\to\infty}A(x,y) = \lim_{y\to\infty}A_y(x,y) = 0.$$

It was proved by Gelfand and Levitan that the function $A(x,y)$ exists and is unique.

Recall from Chapter 7 that $u_1(x,k)$ and $u_2(x,k)$ are related through

$$u_2(x,k) = c_{11}(k)u_1(x,k) + c_{12}(k)u_1(x,-k),$$

and that both $u_1(x,k)e^{-ikx}$ and $u_2(x,k)e^{ikx}$ are analytic for k in the upper half complex k plane, $\operatorname{Im} k \geq 0$. It follows that

$$\frac{1}{2\pi}\int_C T(k)u_2(x,k)e^{iky}\,dk = \frac{1}{2\pi}\int_C R(k)u_1(x,k)e^{iky}\,dk$$

$$+ \frac{1}{2\pi}\int_C u_1(x,-k)e^{iky}\,dk,$$

where C is any horizontal contour in the upper half k plane starting at $k = -\infty+i0^+$ and ending at $k = +\infty+i0^+$ while passing above all zeros of $c_{12}(k)$. Here

$$T(k) = \frac{1}{c_{12}(k)}$$

and

$$R(k) = \frac{c_{11}(k)}{c_{12}(k)}$$

are the transmission and reflection coefficients, respectively.

Since $u_1(x,k)$, $u_2(x,k)$ can be obtained via Neumann iterates, it is not difficult (Problem 7.5.7) to show that

$$\frac{u_2(x,k)}{c_{12}(k)}e^{ikx} = 1 + O\left(\frac{1}{k}\right), \qquad \text{as} \qquad |k| \to \infty.$$

(The notation $O(1/k)$ is defined in Chapter 10 to mean a term that is bounded by $K\,|1/k|$ for some constant K and all k sufficiently large.) Therefore, since $u_2(x,k)e^{ikx}$ is analytic in the upper half plane,

$$\frac{1}{2\pi}\int_C T(k)u_2(x,k)e^{iky}\,dk = \frac{1}{2\pi}\int_C \frac{u_2(x,k)e^{ikx}}{c_{12}(k)}e^{ik(y-x)}\,dk$$

$$= \delta(y-x) + \int_C O\left(\frac{1}{k}\right)e^{ik(y-x)}\,dk.$$

Using Jordan's lemma for $y > x$ we see that

$$\frac{1}{2\pi} \int_C T(k) u_2(x,k) e^{iky} dk = 0 \qquad \text{provided} \qquad y > x.$$

From the representation of $u_1(x,k)$ in terms of $A(x,y)$, we find that

$$0 = \frac{1}{2\pi} \int_C R(k) e^{ik(x+y)} dk \ + \ \frac{1}{2\pi} \int_x^\infty \int_C R(k) A(x,s) e^{ik(s+y)} dk \, ds$$

$$+ \ \frac{1}{2\pi} \int_x^\infty \int_C A(x,s) e^{-ik(s-y)} dk \, ds$$

again, provided $y > x$. We define $r(x)$ by

$$r(x) \ = \ \frac{1}{2\pi} \int_C R(k) e^{ikx} dk$$

and obtain the equation

$$0 = r(x+y) + \int_x^\infty r(s+y) A(x,s) ds + A(x,y) \qquad \text{provided} \qquad y > x.$$

This equation, called the Gelfand-Levitan-Marchenko (GLM) equation is important since, if we can solve it for $A(x,y)$ given $r(x)$, then $q(x) = -2\frac{d}{dx}A(x,x)$. In other words, given the reflection coefficient $R(k)$, defined in the upper half k plane, we can reconstruct the potential $q(x)$ whenever we can solve the GLM integral equation for $A(x,y)$.

We do not actually need to know $R(k)$ in the entire complex plane. Since $R(k) = c_{11}(k)/c_{12}(k)$ we use the residue theorem to reduce $r(x)$ to

$$r(x) \ = \ \frac{1}{2\pi} \int_{-\infty}^\infty R(k) e^{ikx} dk - i \sum_{k_j} \frac{c_{11}(k_j)}{c_{12}(k_j)} e^{ik_j x},$$

where $k_j = i\mu_j$ are the locations of the bound states $c_{12}(k_j) = 0$. Equivalently,

$$r(x) \ = \ \frac{1}{2\pi} \int_{-\infty}^\infty R(k) e^{ikx} dk + \sum_j \frac{e^{-\mu_j x}}{\beta_j},$$

where

$$\beta_j \ = \ \int_{-\infty}^\infty u_1^2(x,k_j) \, dx$$

is the normalization constant for the bound state. Thus, if we know the reflection coefficient on the real line and the location and normalization of the bound states, we can reconstruct the potential $q(x)$ by solving the Gelfand-Levitan-Marchenko integral equation.

There is an alternate, more physically motivated, derivation of the GLM equation. We note that the Schrödinger equation is the Fourier transform of the wave equation

$$u_{tt} = u_{xx} - q(x)u.$$

Physically, the term $q(x)u$ represents "bracing" or an additional local restoring force for a vibrating string. We suppose that there is a left moving delta function $u_0 = \delta(x + t)$ incident from $x \to \infty$ and after the delta function has passed along the string the solution has a right moving reflected component so that

$$u \sim u_\infty \equiv \delta(x + t) + r(x - t), \qquad x \to +\infty.$$

We propose to find a complete solution of the wave equation in the form

$$u(x,t) = u_\infty(x,t) + \int_x^\infty A(x,y)u_\infty(y,t)dy.$$

This is equivalent to the crucial step in the first derivation, and it will work provided

$$A_{yy} - A_{xx} + q(x)A = 0$$

$$q(x) = -2\frac{d}{dx}A(x,x)$$

$$\lim_{y\to\infty} A(x,y) = \lim_{y\to\infty} A_y(x,y) = 0,$$

as before. (To check this, substitute the assumed form of the solution into the vibrating string equation.)

Because of causality, we know that $u(x,t)$ must be zero before the delta function influence arrives so that

$$u_\infty(x,t) + \int_x^\infty A(x,y)u_\infty(y,t)dy = 0, \qquad x + t < 0.$$

Substituting the definition of $u_\infty(x,t)$ we obtain that

$$r(x - t) + A(x,t) + \int_x^\infty A(x,s)r(s - t)ds = 0, \qquad x + t < 0.$$

With $y = -t$ this is precisely the GLM equation. Furthermore we now see that the reflection coefficient $R(k)$ and the backscattered wave $r(x - t)$ are related through the Fourier transform.

Example

Consider the reflection coefficient $R(k) = \frac{-ia}{k+ia}$ which we know from Chapter 7 to be the reflection coefficient for a delta function potential. We first calculate

$$r(x) = \frac{1}{2\pi}\int_C \frac{-ia}{k + ia}e^{ikx}dk.$$

The contour C always lies above the pole at $k = -ia$ so when $x > 0$ we can close the contour above C and obtain $r(x) = 0$ whereas when $x < 0$ we close the contour below C and obtain $r(x) = -ae^{ax}$. Thus we write $r(x)$ more compactly as $r(x) = -H(-x)ae^{ax}$. Substituting $r(x)$ into the GLM equation we obtain

$$0 = aH(-x - y)e^{a(x+y)} + ae^{ay}\int_x^\infty H(-s - y)e^{as}A(x,s)ds - A(x,y) = 0$$

provided $y > x$. If $x + y > 0$, then $H(-s - y) = 0$ for all $s > x$ and the equation reduces to $A(x, y) = 0$. If $x + y < 0$, the equation reduces to

$$0 = ae^{a(x+y)} + ae^{ay}\int_x^{-y} e^{as}A(x,s)ds - A(x,y) = 0,$$

and it is easy to verify that $A(x,y) = a$. We conclude that $A(x,y) = aH(-x - y)$ so that $q(x) = -2a\frac{d}{dx}H(2x) = -2a\delta(x)$ as expected.

Example

Consider the case $r(x) = me^{-\mu x}$. This corresponds to a reflectionless potential since $r(x)$ consists only of a bound state contribution and no integral along the k axis. To solve the GLM equation

$$0 = r(x + y) + \int_x^\infty r(y + s)A(x,s)ds + A(x,y), \qquad x < y$$

we try a solution of the form $A(x,y) = a(x)e^{-\mu y}$ and find

$$a(x)\left[1 + m\int_x^\infty e^{-2\mu s}ds\right] + me^{-\mu x} = 0$$

or

$$a(x) = \frac{-me^{-\mu x}}{1 + \frac{m}{2\mu}e^{-2\mu x}}.$$

From this we learn that

$$A(x, x) = \frac{-m}{m/2\mu + e^{2\mu x}}$$

and that

$$q(x) = -2\mu^2 \operatorname{sech}^2(\mu x - \phi), \qquad \phi = \frac{1}{2}\ln\left(\frac{m}{2\mu}\right).$$

The important observation to make here is that the potential $q(x)$ is a reflectionless potential (which we already knew from Chapter 7) and that the parameters of the potential are determined by the location of the bound state and the normalization constant. It is noteworthy that the shape of the potential is determined solely by the bound state location; the normalization constant m influences only the phase shift.

Example

We can generalize the previous example by considering

$$r(x) = \sum_{j=1}^{n} m_j e^{\mu_j x},$$

which also corresponds to a reflectionless potential. To solve the GLM equation we define two vector functions

$$\phi(t) = \left(e^{-\mu_1 t}, e^{-\mu_2 t}, \ldots, e^{-\mu_n t}\right)$$

$$\psi(t) = \left(m_1 e^{-\mu_1 t}, m_2 e^{-\mu_2 t}, \ldots, m_n e^{-\mu_n t}\right)$$

and note that $r(x + y) = \psi^T(x) \phi(y)$. We are led to try a solution of the form $A(x, y) = \chi^T(x) \phi(y)$ and, upon substitution into the GLM equation we find that

$$0 = \psi^T(x) \phi(y) + \chi^T(x)\left[I + \int_x^\infty \phi(s)\psi^T(s)ds\right] \phi(y).$$

We define the matrix $V = I + \int_x^\infty \phi(s)\psi^T(s)ds$ and then

$$\chi^T(x) = -\psi^T(x)V^{-1}(x)$$

so that

$$A(x, x) = -\psi^T(x)V^{-1}(x)\phi(x).$$

Now notice that, if V^{-1} has entries v_{ij}^{-1}

$$A(x, x) = -\sum_{i,j=1}^{n} \psi_i v_{ij}^{-1} \phi_j = -\sum_{i,j=1}^{n} \psi_i \phi_j v_{ij}^{-1}$$

$$= \sum_{i,j=1}^{n} \frac{d(v_{ji})}{dx} v_{ij}^{-1} = \text{Trace}\left(\frac{dV}{dx} \cdot V^{-1}\right).$$

Furthermore, for a system of differential equations $dV/dx = BV$, we know from Abel's formula (Problem 4.4.10) that

$$\frac{d(\det V)}{dx} = \text{Trace } B(\det V).$$

Since V satisfies the differential identity

$$\frac{dV}{dx} = \left(\frac{dV}{dx} \cdot V^{-1}\right) V$$

we conclude that

$$A(x, x) = \text{Trace}\left(\frac{dV}{dx} \cdot V^{-1}\right) = \frac{\frac{d}{dx}(\det V)}{\det V}.$$

That is, $q(x) = -2\frac{d^2}{dx^2}\ln(\det V)$ where $V(x) = I + \int_x^\infty \phi(s)\psi^T(s)ds$.

It takes a fair amount of work to make sense of this representation of $q(x)$. In the case that $n = 2$, one can show (after considerable algebra) that

$$q(x) = 2\left(\mu_2^2 - \mu_1^2\right)\frac{\mu_1^2\,\text{csch}^2\,\xi + \mu_2^2\,\text{sech}^2\,\eta}{(\mu_1\coth\xi - \mu_2\tanh\eta)^2},$$

where $\xi = \mu_1 x + \frac{\phi-\psi}{2}$, $\eta = \mu_2 x + \frac{\phi+\psi}{2}$

$$\phi = \frac{1}{2}\ln\left[\frac{4\mu_1\mu_2}{m_1m_2}\left(\frac{\mu_1+\mu_2}{\mu_1-\mu_2}\right)^2\right], \qquad \psi = \frac{1}{2}\ln\left(\frac{m_1\mu_2}{m_2\mu_1}\right).$$

If ξ is order one and $|\eta| \gg 1$, $q(x)$ is approximately

$$q(x) \sim -2\mu_1^2\,\text{sech}^2(\xi \pm \Delta), \qquad \eta \to \pm\infty$$

while if η is order one, and $\xi \to \pm\infty$, then

$$q(x) \sim -2\mu_2^2\,\text{sech}^2(\eta \pm \Delta),$$

where

$$\Delta = \frac{1}{2}\ln\left(\frac{\mu_1 - \mu_2}{\mu_1 + \mu_2}\right).$$

Thus, in certain parameter ranges $q(x)$ appears to be a nonlinear superposition of simple single bound state solutions.

9.2   Isospectral Flows

With scattering theory and the GLM equation, we have the possibility of a new transform method. For any function $q(x,t)$ we define its scattering transform to be the reflection data

$$R(k) = \frac{c_{11}(k)}{c_{12}(k)}$$

for any fixed time t, and using the GLM equation we can invert the transform and reconstruct the potential from knowledge of the reflection data. Now the important question is whether or not this transform is of any practical use. To be useful the evolution of the transformed data should be simpler to describe (and solve) than the original untransformed evolution equation.

The amazing fact is that there are equations for which this transform works remarkably well. It is no surprise that the list of such equations is small since we have already seen in great detail that when one changes the differential operator one must also change the appropriate transform.

The general approach to determine when this transform works was described by Lax. We suppose that we have some eigenvalue problem $Lu = \lambda u$ where the operator L may be changing in time. For example, in the case we have discussed so far $Lu = -u'' + q(x,t)u$ is the Schrödinger operator. As time changes we expect the solutions u (i.e., eigenfunctions of L) to change according to some evolution equation, say $u_t = Mu$ where the linear operator M again depends in general on t. We differentiate the eigenvalue problem with respect to time and find that

$$(L_t + LM - ML)u = \lambda_t u.$$

(The expression $LM - ML$ is called the COMMUTATOR of M and L.) Now the crucial step is to require $\lambda_t = 0$, that is, we require that as time changes, although the eigenfunctions u may change, the eigenvalues λ do not. The resulting requirement is that we must have

$$L_t + LM - ML = 0$$

which, since L and M depend on the function $q(x,t)$, is an evolution equation for $q(x,t)$, not $u(x,t)$. In other words, if $q(x,t)$ is chosen to satisfy the equation $L_t + LM - ML = 0$, the eigenvalues of the spectral problem $Lu = \lambda u$ will not change as time evolves and we have an ideal candidate on which to try our scattering transform. Since $\lambda_t = 0$, we call this an ISOSPECTRAL FLOW.

What kinds of equations are of this form? The first equation to which the method was applied (and for which the method was originally discovered) was the KORTEWEG-DEVRIES (KdV) EQUATION

$$q_t - 6qq_x + q_{xxx} = 0.$$

We take the Schrödinger operator $Lv = -v'' + q(x,t)v$ as the spectral operator and allow its eigenfunctions to evolve according to

$$v_t = -q_x v + (4\lambda + 2q)v_x \equiv Mv.$$

A short calculation of $L_t + LM - ML = 0$ shows that the potential q satisfies the KdV equation. We shall study this equation in more detail in the next section.

Other equations for which the inverse-spectral transform method work require examination of different spectral operators. Suppose, for example, we consider the more general scattering problem

$$u_x = -iku + qv$$

$$v_x = ikv - ru.$$

The special case of this problem $r = -1$ reduces exactly to the Schrödinger equation $v_{xx} - q(x)v = -k^2v$. This new formulation is referred to as the TWO COMPONENT SCATTERING PROBLEM. The scattering theory for this problem follows a similar program to that of the Schrödinger equation, that is, we find four different solutions with asymptotic behavior

$$\phi_1 \sim \begin{pmatrix} e^{-ikx} \\ 0 \end{pmatrix}, \quad \phi_2 \sim \begin{pmatrix} 0 \\ e^{ikx} \end{pmatrix} \quad \text{as} \quad x \to -\infty$$

and

$$\psi_1 \sim \begin{pmatrix} 0 \\ e^{ikx} \end{pmatrix}, \quad \psi_2 \sim \begin{pmatrix} e^{-ikx} \\ 0 \end{pmatrix} \quad \text{as} \quad x \to +\infty,$$

and scattering coefficients c_{ij} are defined by the relations

$$\phi_1 = c_{11}\psi_1 + c_{12}\psi_2$$

$$\psi_1 = c_{21}\phi_2 + c_{22}\phi_1.$$

From these defining relationships we then find reflection and transmission coefficients and bound states. Finally, we can derive a GLM theory from which to reconstruct the potential functions from knowledge of the scattering coefficients c_{ij}.

We can use two component scattering theory to find more equations that are isospectral flows. If we consider the two component spectral problem

$$v_{1_x} + ikv_1 = -\frac{q_x}{2}v_2$$

$$v_{2_x} - ikv_2 = \frac{q_x}{2}v_1$$

and suppose that the evolution of eigenfunctions is governed by

$$v_{1_t} = \left(\frac{i}{4k}\cos q\right)v_1 + \left(\frac{i}{4k}\sin q\right)v_2,$$

$$v_{2_t} = \left(\frac{i}{4k}\sin q\right)v_1 - \left(\frac{i}{4k}\cos q\right)v_2,$$

then the Lax calculation of the isospectral flow yields

$$q_{xt} = \sin q.$$

This equation, called the SINE-GORDON EQUATION, has important uses in differential geometry, crystal dislocation theory, the theory of coherent lasers, to name only a few.

If we take the eigenvalue problem

$$v_{1_x} + ikv_1 \;=\; qv_2$$

$$v_{2_x} - ikv_2 \;=\; -q^*v_1$$

and suppose that the evolution of eigenfunctions is governed by

$$v_{1_t} \;=\; \left(2ik^2 - i|q|^2\right)v_1 - \left(iq_x + 2kq\right)v_2$$

$$v_{2_t} \;=\; -\left(iq_x^* + 2kq^*\right)v_1 - \left(2ik^2 - i|q|^2\right)v_2$$

then the Lax calculation gives an isospectral flow governed by

$$iq_t \;=\; q_{xx} + 2|q|^2 q.$$

This equation, called the NONLINEAR SCHRÖDINGER EQUATION is important in problems of nonlinear optics. This equation is of the form of the time dependent Schrödinger equation

$$iq_t \;=\; q_{xx} + v(x)q,$$

where the potential function $v(x) = 2|q|^2$ depends in a nonlinear way on the solution q.

Our last example of an isospectral flow comes from considering the infinite lattice of discrete elements connected by nonlinear springs

$$\ddot{u}_n = e^{-(u_n - u_{n-1})} - e^{-(u_{n+1} - u_n)}, \quad -\infty < n < \infty$$

called the TODA LATTICE. Here u_n is the displacement of the nth mass from its equilibrium position and the nonlinear springs have exponential, rather than linear, restoring forces.

To show that this is an isospectral flow, we must first formulate a discrete scattering problem and decide how it should evolve in time so that the Toda Lattice equation describes the evolution of the potential of the scattering problem.

We define the associated scattering problem by the difference equation

$$a_n v_{n+1} + a_{n-1} v_{n-1} + b_n v_n = \lambda v_n$$

where

$$a_n \;=\; \frac{1}{2} e^{-(u_n - u_{n-1})/2}$$

$$b_n \;=\; -\frac{1}{2}\dot{u}_{n-1}$$

The evolution of eigenfunctions is taken to satisfy

$$\dot{v}_n = 2a_n v_{n+1} + B_n v_n,$$

where $B_n = \sum_{k=-\infty}^{n-1} \frac{\dot{a}_k}{a_k}$. It is a straightforward calculation to show that the isospectral flow is governed by the system of equations

$$\dot{a}_n = a_n (b_{n+1} - b_n)$$

$$\dot{b}_n = 2 (a_n^2 - a_{n-1}^2),$$

where $a_n = \frac{1}{2} e^{-(u_n - u_{n-1})/2}$, $b_n = -\frac{1}{2}\dot{u}_{n-1}$, and, as promised, this reduces to the Toda lattice. To completely solve this problem, it remains to develop a scattering theory and inverse scattering reconstruction for the discrete scattering problem, which we shall do in Sec. 9.4.

9.3 Korteweg-deVries Equation

The Korteweg-deVries equation was derived in 1895 to model one-dimensional water waves in shallow water with long wave length (i.e., waves in a long narrow channel). In its original form the equation is

$$\frac{\partial \eta}{\partial \tau} = \frac{3}{2} \left(\frac{g}{l}\right)^{1/2} \frac{\partial}{\partial \xi} \left\{ \frac{1}{2}\eta^2 + \frac{2}{3}\alpha\eta + \frac{\sigma}{3} \frac{\partial^2 \eta}{\partial \xi^2} \right\},$$

where ξ is the spatial variable along the one-dimensional channel, τ is time, $\eta(\xi, \tau)$ is the height of the surface above its equilibrium level l, g is the gravitation constant, α is a constant related to uniform motion of the fluid and $\sigma = \frac{1}{3}l^3 - Tl/\rho g$, where T is the surface tension and ρ is the density of the fluid. It is helpful to introduce rescaled variables

$$t = \frac{1}{2}(g/l\sigma)^{1/2}\tau, \qquad x = -\xi/\sqrt{\sigma}$$

$$q = -\frac{1}{2}\eta - \frac{1}{3}\alpha$$

in terms of which we have

$$\frac{\partial q}{\partial t} - 6q\frac{\partial q}{\partial x} + \frac{\partial^3 q}{\partial x^3} = 0.$$

There is no particular significance to the factor 6 since that factor can be chosen anyway one likes, but 6 is the generally accepted value. Notice also that if $\alpha = 0$ then waves above equilibrium level have q negative.

It is somewhat customary in introductions to this equation to include the now-famous quotation of J. Scott-Russell's "Report on Waves" (1844) describing his chase on horseback of a wave in a channel:

I was observing the motion of a boat which was rapidly drawn along a narrow channel by a pair of horses, when the boat suddenly stopped—not so the mass of water in the channel which had put it in motion; it accumulated round the prow of the vessel in a state of violent agitation, then suddenly leaving it behind, rolled forward with great velocity, assuming the form of a large solitary elevation, a rounded, smooth and well defined heap of water, which continued its course along the channel apparently without change of form or diminution of speed. I followed it on horseback, and overtook it still rolling on at a rate of some eight or nine miles an hour, preserving its original figure some thirty feet long and a foot to a foot and a half in height. Its height gradually diminished, and after a chase of one or two miles I lost it in the windings of the channel. Such, in the month of August 1834, was my first chance interview with that singular and beautiful phenomenon...

What Scott-Russell saw was what is nowadays called a SOLITARY WAVE, that is, a wave that propagates without change of shape for long distances or ideally, forever. It is an easy matter to observe solitary waves for yourself. In a pinch, you can use a wallpaper hanger's tray but your best bet is to find a fountain with a long narrow channel. Late at night after the fountain is turned off and the water is calm, go make some waves.

In their original paper, Korteweg and deVries derived their equation and showed that there are solitary wave solutions. But the amazing fact is that the equation can be solved completely and as we will see in a bit, the solution is quite remarkable. In fact, while you are making waves in the park try to see how two or more waves interact. If you create a small wave followed by a larger one, you will see that the two propagate with very little change in shape, the larger one overtaking the smaller. The two waves interact and then separate with the large one moving ahead of the smaller with the original shape of two waves completely restored. The only remnant of the interaction (this will be hard to observe without careful measurements) is that the larger wave is somewhat ahead and the smaller one somewhat behind where they would have been if they were propagating separately without interaction.

The simplest solutions of the Korteweg-deVries equation are solitary waves. We look for solutions of the form

$$q(x,t) = U(x - ct)$$

and substitute into the KdV equation to find

$$cU' + 6UU' - U''' \;=\; 0.$$

We can integrate this once to get

$$-cU - 3U^2 + U'' = K_1 \;=\; \text{constant}.$$

Multiplying by U' and integrating once more we find

$$U'^2 - 2U^3 - cU^2 - 2K_1 U + K_2 = 0,$$

where K_2 is also a constant. It is an easy matter (draw the solution trajectories in the $U - U'$ plane) to conclude that there are periodic solutions. These can be described in terms of the Jacobi elliptic functions cn and are called cnoidal waves.

Solitary waves have $U' = 0$ at $\pm\infty$. Without loss of generality we can take $K_1 = K_2 = 0$ so that

$$U'^2 = U^2(2U + c).$$

This equation we integrate directly to yield

$$q(x,t) = U(x - ct) = -\frac{1}{2}c\,\text{sech}^2\left[\frac{1}{2}\sqrt{c}\,(x - ct + x_0)\right].$$

It is interesting to note from this solution that the amplitude is related to the speed and width of the wave. Larger amplitude waves are narrower and move faster than smaller waves. This observation certainly agrees with our experience playing in puddles.

More general solutions must be found using scattering theory ideas. Before we plunge ahead (other idioms might be jump in, get our feet wet, get in over our heads) we review the idea of the scattering transform. We wish to solve some evolution equation

$$L_t + LM - ML = 0,$$

for $q(x,t)$ starting from some initial data $q(x,0) = q_0(x)$. Rather than solving it directly, which could presumably be done, we view the solution $q(x,t)$ as the potential of the spectral problem

$$Lv = \lambda v$$

where the operator L depends on $q(x,t)$. As $q(x,t)$ evolves in time, so also will its reflection data. The crucial step of the transform technique is to determine from the evolution equation how the scattering data evolve. Then, starting from $q_0(x)$ we find the initial scattering data, allow them to evolve until time t and then invert the scattering data to find the function $q(x,t)$ at time t. This process we summarize with the (by now familiar) canonical diagram of

transform theory

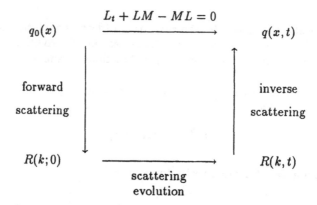

For the KdV equation we have already determined that the spectral problem is the Schrödinger equation $-u'' + q(x,t)u = k^2 u$ and that the eigenfunctions evolve according to

$$
\begin{aligned}
u_t &= -q_x u + (4\lambda + 2q)u_x, \qquad \lambda = k^2 \\
&= -4u_{xxx} + 3q_x u + 6qu_x.
\end{aligned}
$$

We know that the fundamental solutions of the Schrödinger equation satisfy

$$
u_2(x, k) = c_{11}(k)u_1(x, k) + c_{12}(k)u_1(x, -k)
$$

but now the functions u_1, u_2 c_{11}, c_{12} are changing in time.

Since $u_2(x, k) \sim e^{-ikx}$ as $x \to -\infty$, independent of time, we take $u = h(t)u_2(x, k)$ and note that if q and q_x vanish as x becomes large, then for large x, the eigenfunction u must satisfy

$$
u_t = -4u_{xxx}.
$$

Taking $x \to -\infty$ we find that

$$
h_t = -4ik^3 h,
$$

that is,

$$
h(t) = h_0 e^{-4ik^3 t}.
$$

On the other hand, for $x \to \infty$,

$$
\begin{aligned}
&(h_t c_{11} + h c_{11_t})\, e^{ikx} + (h_t c_{12} + h c_{12_t})\, e^{-ikx} \\
&= -4h \left(c_{11}(ik)^3 e^{ikx} + c_{12}(-ik)^3 e^{-ikx} \right).
\end{aligned}
$$

It follows that

$$c_{11_t}(k) \;\;=\;\; 8ik^3 c_{11}(k)$$

and

$$c_{12_t}(k) \;\;=\;\; 0,$$

are the evolution equations in the transformed domain. The fact that $c_{12_t}(k) = 0$ was precisely what we were hoping for when we sought out isospectral flows. It guarantees that bound state locations do not change as time evolves. Equally noteworthy is the fact that the evolution of $c_{11}(k)$ is diagonalized. This also is a major goal of a good transform theory, that the transformed equations should be diagonalized.

We now solve the transformed evolution equations and find

$$c_{11} = c_{11}^0 e^{8ik^3 t}, \qquad c_{12} = c_{12}^0,$$

where c_{11}^0, c_{12}^0 are the initial values of c_{11} and c_{12}, respectively. We also find that

$$r(x,t) \;\;=\;\; \frac{1}{2\pi} \int_C R(k) e^{ikx} dk,$$

where $R(k) = c_{11}/c_{12}$, is given by

$$r(x,t) \;\;=\;\; \frac{1}{2\pi} \int_{-\infty}^{\infty} R_0(k) e^{i(8k^3 t + kx)}$$

$$+ \;\; \sum_j m_j(0) e^{8\mu_j^3 t - \mu_j x},$$

where $c_{12}(k_j) = 0$ at $k_j = i\mu_j$, $R_0(k) = \frac{c_{11}^0(k)}{c_{12}^0(k)}$. Finally, we must use the function $r(x,t)$ in the GLM equation to find the potential $q(x,t)$ as a function of t, and the solution is completely prescribed, starting from arbitrary initial data.

What does the solution look like? To get some insight into the behavior of the solution we consider the simplest case of reflectionless initial data. With $r(x,t)$ in the form

$$r(x,t) \;\;=\;\; \sum_{j=1}^{n} m_j(t) e^{-\mu_j x}$$

$$m_j(t) \;\;=\;\; m_j(0) e^{8\mu_j^3 t}$$

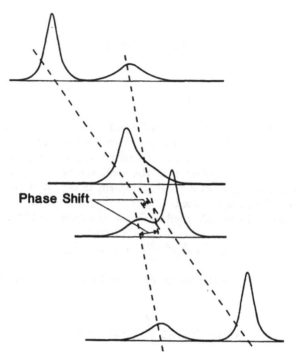

FIGURE 9.1   Two soliton solution of the Korteweg-deVries equation at three values of time.

we already know how to reconstruct the potential $q(x, t)$. If there is precisely one bound state, we find

$$q(x, t) \quad = \quad -2\mu^2 \operatorname{sech}^2(\mu x - \phi)$$

$$\phi \quad = \quad \frac{1}{2} \ln \left(\frac{m(t)}{2\mu} \right)$$

or

$$q(x, t) \quad = \quad -\frac{1}{2} c \operatorname{sech}^2 \left(\frac{1}{2} \sqrt{c}(x - ct) + \phi_0 \right),$$

where $c = 4\mu^2 > 0$. This is exactly the solitary pulse solution found by direct integration of the KdV equation.

The solution with two bound states is given exactly by

$$q(x, t) = 2 \left(\mu_2^2 - \mu_1^2 \right) \frac{\mu_1^2 \operatorname{csch}^2 \xi + \mu_2^2 \operatorname{sech}^2 \eta}{\left(\mu_1 \coth \xi - \mu_2 \tanh \eta \right)^2},$$

where $\xi = \mu_1 x - 4\mu_1^3 t + \xi_0$, $\eta = \mu_2 x - 4\mu_2^3 t + \eta_0$. Thus, $q(x,t)$ represents the nonlinear superposition of two forms, one traveling with speed $4\mu_1^2$, the other traveling with speed $4\mu_2^2$. If we suppose $\mu_1 > \mu_2$, then for t much less than zero for those values of x where ξ is about one and η is very negative, $q(x) \sim -2\mu_1^2 \operatorname{sech}^2(\xi + \Delta)$ while in regions of x where η is about one and ξ is very positive $q(x) \sim -2\mu_2^2 \operatorname{sech}^2(\eta - \Delta)$. That is, for very large negative t the solution looks like two solitary pulses, the large one to the left of the small one. After a long time, when t is large positive, if η is order one and ξ is very large negative, then $q(x) \sim -2\mu_2^2 \operatorname{sech}^2(\eta + \Delta)$ while if ξ is order one, η very large positive, $q(x) \sim -2\mu_1^2 \operatorname{sech}^2(\xi - \Delta)$. That is, after a long time the large solitary pulse is to the right of the small solitary pulse. They have coalesced and reemerged with their shape unscathed. The only remnant of the interaction is the phase shift $\Delta = \frac{1}{2}\ln(\frac{\mu_1 + \mu_2}{\mu_1 - \mu_2})$. That is, the large pulse is moved forward by an amount $2\Delta/\mu_1$, relative to where it would have been in the absence of an interaction, and the small pulse is retarded by an amount $2\Delta/\mu_2$ relative to where it would have been in an unperturbed situation. In Figure 9.1 we show a sketch of this solution at three successive times.

In general, the reflectionless solution with n bound states has a similar behavior. With n bound states the solution resembles the superposition of n solitary pulses whose speeds and amplitudes are determined by the values μ_j. The solitary pulses emerge unscathed from interaction except for a phase shift, given by the sum of phase shifts from all possible pairwise interactions.

For general initial data one cannot expect reflectionless potentials, so we ask what happens with general initial data. Notice that the function $r(x,t)$ consists of two parts, the discrete sum from the bound states and the integral term

$$\frac{1}{2\pi} \int_{-\infty}^{\infty} R_0(k) e^{i(8k^3 t + kx)} dk.$$

Using the method of steepest descents (which we will learn about in Chapter 10), it is not hard to show that for large t, x bounded, the integral term is of order $t^{-1/3}$. That is, the contribution of the continuous spectrum to the function $r(x,t)$ decays like $t^{-1/3}$. As a result, after a long time the influence of continuous data disappears and the solution that remains resembles that of the n bound state solution.

We now know that the number of solitary pulses (often called solitons) that emerge from general initial data is determined precisely by the number of initial bound states. Thus, the initial profile $q(x,t) = V_0 \delta(x)$ will evolve into one solitary pulse if V_0 is negative. A square well initial profile can give rise to many solitons, depending, of course, on its width and depth.

9.4   The Toda Lattice

To illustrate further the ideas of inverse scattering theory, we will discuss the solution of the Toda lattice equations

$$\ddot{u}_n = e^{-(u_n - u_{n-1})} - e^{-(u_{n+1} - u_n)}, \qquad -\infty < n < \infty.$$

The ideas for this problem are exactly the same as for the Korteweg-deVries equation, although, naturally, the details are different, because now we are working with difference rather than differential operators.

We know from Sec. 9.2 that we want to examine scattering theory for the difference equation

$$a_n v_{n+1} + a_{n-1} v_{n-1} + b_n v_n = \lambda v_n.$$

Another situation in which this difference equation occurs is the simple linear lattice

$$\ddot{u}_n = k_n \left(u_{n+1} - u_n \right) + k_{n-1} \left(u_{n-1} - u_n \right).$$

If we look for solutions of the form $u_n = v_n e^{-i\omega t}$ we find the difference equation

$$-\omega^2 v_n = k_n v_{n+1} + k_{n-1} v_{n-1} - \left(k_n + k_{n-1} \right) v_n.$$

By assuming $\lim_{|n| \to \infty} k_n = k^*$, we can write this in the form

$$\left(2k^* - \omega^2 \right) v_n = k_n v_{n+1} + k_{n-1} v_{n-1} + \left(2k^* - k_n - k_{n-1} \right) v_n$$

which is, of course, in the form of the original difference equation when we make the identification

$$\lambda = 2k^* - \omega^2, \quad a_n = k_n, \quad b_n = 2k^* - k_n - k_{n-1}.$$

To study scattering for this problem, we suppose that the numbers $a_n - 1/2$ and b_n decay rapidly to zero as $|n| \to \infty$. For large $|n|$ the difference equation reduces to

$$\frac{1}{2} \left(v_{n+1} + v_{n-1} \right) = \lambda v_n$$

and solutions (improper eigenfunctions) are

$$v_n = z^{\pm n},$$

where $\lambda = \frac{1}{2}(z + 1/z)$. For $|z| = 1$ we set $z = e^{i\theta}$ and observe that $v_n = z^{-n}$ corresponds to a left moving wave. For $|z| = 1$, λ lies on the real axis between

-1 and 1, and the numbers v_n are bounded for all n. We define two fundamental solutions of the scattering problem $\phi_n(z)$, $\psi_n(z)$ by the asymptotic
requirements

$$\phi_n(z) \quad \sim \quad z^n \qquad \text{as} \qquad n \to +\infty$$

$$\psi_n(z) \quad \sim \quad z^{-n} \qquad \text{as} \qquad n \to -\infty.$$

One must show that $\phi_n(z)$ and $\psi_n(z)$ indeed exist. In fact, they not only exist,
but $\phi_n z^{-n}$ and $\psi_n z^n$ are analytic for $|z| < 1$, so that they have power series
representations valid for $|z| < 1$ (see Problems 9.4.1,3). Using a calculation
akin to the method of Frobenius (assume a power series solution and match
coefficients term by term), one can show that if

$$\phi_n(z) \quad = \quad A_n z^n + O\left(z^{n+1}\right)$$

for $|z|$ small, then $A_n = \frac{A_{n+1}}{2a_n}$, and since $\lim_{n \to \infty} A_n = 1$, we determine that

$$\phi_n(z) \quad = \quad \frac{z^n}{\prod_{j=n}^{\infty}(2a_j)} + O\left(z^{n+1}\right).$$

In a similar way,

$$\psi_n(z) \quad = \quad \frac{z^{-n}}{\prod_{j=-\infty}^{n-1}(2a_j)} + O\left(z^{-n+1}\right).$$

The functions $\phi_n(z)$ and $\phi_n\left(z^{-1}\right)$ are linearly independent unless $z = \pm 1$.
When they are linearly independent, it must be that

$$\psi_n(z) = \beta(z)\phi_n(z) + \alpha(z)\phi_n\left(z^{-1}\right),$$

for $\alpha(z)$, $\beta(z)$ appropriately chosen, since a second order difference equa-
tion can have at most two linearly independent solutions. The ratio $R(z) = \beta(z)/\alpha(z)$ is the reflection coefficient. Roots of $\alpha(z) = 0$ with $|z| < 1$, if they
exist, are the eigenvalues for this problem. If $\alpha(z_k) = 0$ and $|z_k| < 1$ then
$\psi_n(z_k) = \beta(z_k)\phi_n(z_k)$ is the eigenfunction, since $\psi_n(z_k) \sim z_k^{-n}$ as $n \to -\infty$
and $\psi_n(z_k) = \beta(z_k)\phi_n(z_k) \sim \beta(z_k) z_k^n$ as $n \to +\infty$. (Here we assumed that
$\beta(z)$ is not identically zero, as would be the case for a reflectionless problem.)
    One can show (Problem 9.4.2) that, if they exist, roots of $\alpha(z) = 0$ are real
and discrete. Thus, the spectrum of this difference operator is the real λ axis
$-1 \leq \lambda \leq 1$, corresponding to continuous spectrum, and there is discrete
spectrum at the points

$$\lambda_k = \frac{1}{2}\left(z_k + \frac{1}{z_k}\right)$$

where $\alpha(z_k) = 0$, $|z_k| < 1$.

Example

Suppose $a_n = 1/2$ for $n \neq 0$. Determine the scattering coefficients and bound states for the difference equation $a_n v_{n+1} + a_{n-1} v_{n-1} = \lambda v_n$, $\lambda = \frac{1}{2}\left(z + \frac{1}{z}\right)$, in terms of a_0.

We know that $\psi_n = z^{-n}$ for $n \leq 0$ and assume that $\psi_n = \beta z^n + \alpha z^{-n}$ for $n \geq 1$. We substitute these representations into the difference equation to obtain two equations for the two unknowns α, β

$$a_0(\beta z + \alpha/z) = \frac{1}{2z} \qquad (n = 0)$$

$$a_0 = \frac{1}{2}(\alpha + \beta) \qquad (n = 1)$$

which we solve to find

$$\alpha = \frac{1}{1 - z^2}\left(\frac{1}{2a_0} - 2a_0 z^2\right)$$

$$\beta = \frac{1}{1 - z^2}\left(2a_0 - \frac{1}{2a_0}\right)$$

so that the reflection coefficient is

$$R(z) = \frac{\beta(z)}{\alpha(z)} = \frac{4a_0^2 - 1}{1 - 4a_0^2 z^2}.$$

Bound states occur at $z = \mp\frac{1}{2a_0}$. Thus, if $a_0 > 1/2$ there are two bound states given by

$$\psi_n^1 = \left(\frac{1}{2a_0}\right)^{|n|}$$

$$\psi_n^2 = \left(\frac{-1}{2a_0}\right)^{|n|}.$$

For $a_0 < 1/2$, there are no bound states. Notice that the bound state ψ_n^1 is positive for all n, while the bound state ψ_n^2 is oscillatory, i.e., alternating in sign.

The inverse scattering theory follows the same program as with the Schrödinger equation. We propose to write $\phi_n(z)$ as

$$\phi_n(z) = \sum_{j=n}^{\infty} K(n, j) z^j, \qquad K(n, n) = 1$$

and note that if $\phi_n(z)$ is to satisfy the difference equation

$$a_n v_{n+1} + a_{n-1} v_{n-1} + b_n v_n = \lambda v_n, \qquad \lambda = \frac{1}{2}\left(z + \frac{1}{z}\right)$$

then $K(n, j)$ must satisfy (independent of z)

$$
\begin{aligned}
0 \;=\; & a_n K(n+1, j) + a_{n-1} K(n-1, j) \\
+ \;& b_n K(n, j) - \frac{1}{2} K(n, j-1) - \frac{1}{2} K(n, j+1)
\end{aligned}
$$

for $j > n$, and

$$a_n \;=\; \frac{1}{2} \frac{K(n+1, n+1)}{K(n, n)}$$

$$b_n \;=\; \frac{1}{2} \frac{K(n, n+1) + 2a_{n-1} K(n-1, n)}{K(n, n)}.$$

We multiply the equation

$$\frac{\psi_n(z)}{\alpha(z)} = R(z)\phi_n(z) + \phi_n\left(z^{-1}\right)$$

by $\frac{1}{2\pi i} z^{m-1} dz$ and integrate counterclockwise around a very small circle containing the origin, to obtain

$$
\begin{aligned}
I \;\equiv\; & \frac{1}{2\pi i} \int_C \frac{\psi_n(z)}{\alpha(z)} z^{m-1} dz \\
=\; & \frac{1}{2\pi i} \int_C R(z)\phi_n(z) z^{m-1} dz + \frac{1}{2\pi i} \int_C \phi_n\left(z^{-1}\right) z^{m-1} dz \\
=\; & \sum_{j=n}^{\infty} K(n, j) \frac{1}{2\pi i} \int_C R(z) z^{j+m-1} dz + K(n, m), \quad m \geq n
\end{aligned}
$$

since $\frac{1}{2\pi i} \int_C z^{n-m-1} dz = \delta_{nm}$. To evaluate the first integral we note that $\psi_n(z)z^n$ and $\phi_n(z)z^{-n}$ are analytic functions for $|z| < 1$, and near the origin

$$\psi_n(z)z^n \;=\; \frac{1}{\prod_{j=-\infty}^{n-1} (2a_j)} + O(z)$$

$$\phi_n(z)z^{-n} \;=\; \frac{1}{\prod_{j=n}^{\infty} (2a_j)} + O(z)$$

so that the Wronskian $W_n(\phi_n, \psi_n) \equiv \phi_{n+1}\psi_n - \psi_{n+1}\phi_n$ is, to leading order in z,

$$W_n(\phi_n, \psi_n) = \frac{-1}{2a_n z P}, \qquad P = \prod_{j=-\infty}^{\infty} (2a_j).$$

On the other hand (see Problem 9.4.5), $a_n W_n(\phi_n, \psi_n) = \frac{\alpha(z)}{2}\left(z - \frac{1}{z}\right)$. It follows that $\alpha(z) \sim \frac{1}{P}$,

$$\frac{\psi_n(z)z^n}{\alpha(z)} = \prod_{j=n}^{\infty} (2a_j) + O(z),$$

and that

$$I = \prod_{j=n}^{\infty} (2a_j)\,\delta_{nm}.$$

Since

$$2a_n = \frac{K(n+1, n+1)}{K(n, n)}$$

it follows that,

$$\prod_{j=n}^{\infty} 2a_n = \frac{K(\infty, \infty)}{K(n, n)}.$$

We define

$$\eta(n) = \frac{1}{2\pi i} \int_C R(z) z^{n-1} dz$$

and learn that

$$\sum_{j=n}^{\infty} K(n, j)\eta(j + m) + K(n, m) = \frac{K(\infty, \infty)}{K(n, n)}\delta_{nm}, \quad m \geq n$$

which is the analog of the Gelfand-Levitan-Marchenko equation for this discrete problem. If we are given $\eta(n)$, then we use this equation to find a_n and b_n, which solves the inverse problem.

The contour C, which is a small circle about the origin, can be deformed into the unit circle $|z| = 1$ if we take into account that $\alpha(z)$ may have simple zeros in the interior of the unit circle. In general, $\eta(n)$ can be written in the form

$$\eta(n) = \frac{1}{2\pi i} \int_{|z|=1} R(z) z^{n-1} dz + \sum_{j=1}^{k} c_j z_j^n,$$

where $\alpha(z_j) = 0$, $j = 1, 2, \ldots, k$. Notice that the integral term is related to the inverse z-transform of $R(z)$.

To give a full solution for the Toda lattice, we must now determine how the scattering coefficients $\alpha(z)$, $\beta(z)$ evolve in time. We know that the Toda lattice is an isospectral flow with evolution of the eigenfunctions governed by

$$\frac{dv_n}{dt} = 2a_n v_{n+1} + B_n v_n$$

$$B_n = \sum_{j=-\infty}^{n-1} \frac{d}{dt} \ln(a_j).$$

Since $a_n = \frac{1}{2} e^{-(u_n - u_{n-1})/2}$ it follows that

$$B_n = \frac{-1}{2} \frac{du_{n-1}}{dt}.$$

Therefore, if the lattice is motionless as $n \to \pm\infty$, the evolution equation becomes

$$\frac{dv_n}{dt} = v_{n+1}$$

for large $|n|$. We apply this equation to $v_n(t) = h(t)\psi_n(z)$ in the limit $n \to -\infty$ from which we learn that $\dot{h} = h/z$. Since $\psi_n = \beta\phi_n(z) + \alpha\phi_n(z^{-1})$, using the differential equation for v_n in the limit $n \to +\infty$, we see that

$$\frac{d\beta}{dt} = (z - 1/z)\beta$$

$$\frac{d\alpha}{dt} = 0.$$

These are the evolution equations in the transform domain, which are separated, linear equations. It follows that

$$R(z,t) = R(z,0)\exp\left((z - 1/z)t\right)$$

$$c_j(t) = c_j(0)\exp\left((z - 1/z)t\right).$$

It is typical of many problems that although one can describe the solution technique with relative ease, to actually implement the technique in general is often very difficult. Such is the case here as well. For this problem, we carry out the details of the solution only in the special case where $R(z) \equiv 0$, i.e., the reflectionless case.

Example

We suppose that at time $t = 0$, $\eta(n) = cz^n$, z fixed, corresponding to reflectionless initial data. We seek a solution of the inverse problem,

$$c\sum_{j=n}^{\infty} K(n,j)z^{j+m} + K(n,m) = \frac{K(\infty,\infty)}{K(n,n)}\delta_{nm}, \quad m \geq n.$$

We try a solution of the form $K(n,m) = A_n z^m$ for $m > n$ and from the inverse equation for $m > n$, we learn that

$$\frac{K(n,m)}{K(n,n)} = \frac{-cz^{n+m}}{1 + C^2 z^{2n+2}}, \qquad C^2 = \frac{c}{1 - z^2}.$$

From the equation for $n = m$, we learn that

$$K^2(n,n) = K(\infty,\infty)\left(\frac{cz^{2n+2} + 1 - z^2}{cz^{2n} + 1 - z^2}\right)$$

from which it follows that

$$e^{-(u_n - u_{n-1})} = (2a_n)^2 = \frac{K^2(n+1,n+1)}{K^2(n,n)} = \frac{(z - 1/z)^2}{\left(Cz^{n+1} + (Cz^{n+1})^{-1}\right)^2} + 1.$$

We now set $z = e^{-\omega}$ if $z > 0$ (or $z = -e^{-\omega}$ if $z < 0$) and find that

$$e^{-(u_n - u_{n-1})} = \sinh^2 \omega \, \text{sech}^2\left(n\omega - t \sinh \omega + k_0\right),$$

where k_0 is a constant determined by the initial value of c. This is the single soliton solution for the Toda lattice. Notice that the solution is shape invariant, that is every particle experiences the same motion, and the wave form propagates with constant velocity $n/t = \frac{\sinh \omega}{\omega}$.

A multiple soliton solution can be described following a similar method. The details of this solution can be found in the paper by Flaschka (1974) and we leave this pursuit to the interested reader.

Further Reading

A very nice paper on the general area of inverse problems is

1. J. B. Keller, Inverse Problems, Am. Math. Monthly **83**, 107–118 (1976).

In recent years, a number of books on the inverse scattering transform have been published. Three examples of these are

2. G. L. Lamb, Jr., *Elements of Soliton Theory*, John Wiley and Sons, New York, 1980,

3. M. J. Ablowitz and H. Segur, *Solitons and the Inverse Scattering Transform*, SIAM, Philadelphia, 1981,

4. W. Eckhaus and A. Van Harten, *The Inverse Scattering Transformation and the Theory of Solitons*, North Holland, Amsterdam, 1981,

all of which include extensive bibliographies.

Two of the fundamental papers in this field are

5. I. M. Gelfand and B. M. Levitan, On the determination of a differential equation from its spectral function, Am. Math. Soc. Transl., Series 2, pp. 259–309 (1955),

which develops the inverse scattering theory, and

6. P. Lax, Integrals of nonlinear equations of evolution and solitary waves, Comm. Pure Appl. Math. **21**, 467–490 (1968),

in which the idea of an isospectral flow is developed. The original derivation of the Kortweg-deVries equation is in their paper

7. D. J. Kortweg and G. deVries, On the change of form of long waves advancing in a rectangular canal, and on a new type of long stationary waves, Phil. Mag. **39**, 422–443, (1895).

Finally the multisoliton solution for the Toda Lattice is given by

8. H. Flaschka, On the Toda Lattice, II—Inverse Scattering Solution, Prog. Theor. Phys. **51**, 703–716 (1974).

Problems for Chapter 9

Section 9.1

1. Derive the Gelfand-Levitan-Marchenko equation using scattering data $R(k) = \frac{c_{22}(k)}{c_{21}(k)}$,

$$0 = r(x + y) + A(x, y) + \int_{-\infty}^{x} r(y + s)A(x, s)ds \qquad y < x,$$

where

$$r(x) = \frac{1}{2\pi} \int_{C} R(k)e^{-ikx}\, dk$$

$$q(x) = 2\frac{d}{dx}A(x, x).$$

2. Show that the potential $q(x) = -6\,\mathrm{sech}^2(x)$ is a reflectionless potential with two bound states at $k_1 = i$, $k_2 = 2i$.

3. Find the potential $q(x)$ if

$$c_{11} = \frac{-1}{2k(k + i)}, \qquad c_{12} = \frac{2k^2 + 1}{2k(k + i)}.$$

Section 9.2

1. Find the equation describing the isospectral flow for the Schrödinger equation

$$-v'' + q(x,t)v = \lambda v$$

and the evolution equation

$$v_t = av.$$

Explain in geometrical terms why this is an isospectral flow.

2. Find the equation describing the isospectral flow for the spectral problem

$$a_n v_{n+1} + a_{n-1} v_{n-1} = \lambda v_n$$

and the evolution equation

$$v_{n_t} = A_n v_{n+1} + B_n v_n,$$

where

$$A_n = a_n \lambda + a_n b_n, \qquad B_n = \left(1 - a_n^2\right) + \sum_{k=-\infty}^{n-1} \frac{a_{k_t}}{a_k}.$$

Section 9.3

1. Solve the equation $q_t = a q_x$ via the scattering transform.
Hint: Consider the spectral operator $Lv = -v'' + qv$ and the evolution equation $v_t = av$. Find the evolution of the scattering data and then reconstruct the potential from the evolution data. Show that if $q(x,0) = q_0(t)$, then $q(x,t) = q_0(x + at)$.

2. Suppose $q(x,0) = A\delta(x)$. Find the amplitude and speed of the soliton that will emerge from this initial data for the KdV equation.

3. Solve the KdV equation with initial data $q(x,0) = -6 \operatorname{sech}^2 x$.

Section 9.4

1. Prove that $\psi_n z^n$ and $\phi_n z^{-n}$ exist and are analytic for $|z| < 1$, provided $a_n - 1/2$ and b_n vanish fast enough as $n \to \pm\infty$.

2. Define $\alpha(z)$, $\beta(z)$ by

$$\psi_n(z) = \beta(z)\phi_n(z) + \alpha(z)\phi_n(z^{-1}).$$

Show that all roots of $\alpha(z) = 0$ lie on the real axis and are simple.

3. Find asymptotic representations of $\phi_n(z)$, $\psi_n(z)$ valid for $|z|$ small.

4. Find the scattering coefficients for the difference equation

$$a_n v_{n+1} + a_{n-1} v_{n-1} + b_n v_n - \lambda v_n = 0$$

when

a. $a_n = 1/2$, $b_n = B \delta_{nj}$.

b. $b_n = 0$, $a_n = 1/2 + A$, $0 \leq n \leq i$, $a_n = 1/2$ otherwise.

c. $a_n = 1/2$, $b_n = B$, $0 \leq n \leq j$, $b_n = 0$ otherwise.

When are there bound states?

5. Define the Wronskian $W_n = W_n (\phi_n, \psi_n)$ as

$$W_n = \det \begin{pmatrix} \phi_{n+1} & \phi_n \\ \psi_{n+1} & \psi_n \end{pmatrix}.$$

If ϕ_n and ψ_n are solutions of $a_n v_{n+1} + a_{n-1} v_{n-1} + b_n v_n - \lambda v_n = 0$, show that $a_n W_n = \text{constant}$.

6. A Toda Lattice initially at rest has one of its masses displaced to the right, and then it is released. How many solitons emerge and in what direction and with what speed do they travel?

7. A Toda Lattice is initially at rest when one of its masses is suddenly given an initial rightward velocity. How many solitons emerge and in what direction and with what speed do they travel?

10

ASYMPTOTIC EXPANSIONS

10.1 Definitions and Properties

Exact solutions to differential equations are marvelous if they can be found, but even when they can be found, it is rarely the end of the study. A complete understanding of a problem always requires interpretation, and most often this requires approximation or numerical evaluation of complicated expressions.

In this chapter we show how to get meaningful approximations of complicated integral expressions. The typical approximation we find will be a power series, say

$$f(z) \sim f_n(z) = \sum_{k=1}^{n} \frac{a_k}{z^k}.$$

We usually think of a power series as representing another function if it converges, that is, if

$$\lim_{n \to \infty} |f(z) - f_n(z)| = 0$$

for all z with $|z| > R$, say. The difficulty with this concept is that to get a "good" approximation one may need to include many terms. The concept of an asymptotic series is quite different. We say that $f_n(z)$ is asymptotic to $f(z)$

if

$$\lim_{z \to \infty} |z^n(f_n(z) - f(z))| = 0 \text{ for } n \text{ fixed.}$$

This concept has nothing at all to do with convergence, since finding a good approximation does not require taking more terms. Although most of our mathematical training probably involved convergent series, and not asymptotic series, it is quite likely that series are actually used as though they are asymptotic. For example, we know that $\cos x \sim 1 - x^2/2$ for small x. Of course, we can better approximate $\cos x$ by taking more terms of its Taylor series, but that really does little to enhance our geometrical understanding of the function. In fact, taking many terms fails to reveal that $\cos x$ is a 2π periodic function, and even though it is convergent for all x, the Taylor series converges slowly for large x. Similarly, although it is true that $e^{-x} \sim 1 - x + x^2/2$ for small x, from this representation it is not obvious that e^{-x} is a monotone, decreasing function. The usefulness of this approximation comes from taking just a few terms and keeping x small.

A series can be asymptotic, even though it may be divergent. In fact asymptotic series are often divergent, so that taking more terms is not simply more work, it is actually damaging!

Definition

Suppose the functions $f(z)$ and $\phi(z)$ are defined in some region R in the complex plane.

1. We say that $f(z)$ is of the order of $\phi(z)$ as z goes to z_0 (or f is "big oh" of ϕ), denoted $f(z) = O(\phi(z))$, if there is a positive constant $A > 0$ and a neighborhood U of z_0 so that $|f(z)| \le A|\phi(z)|$ for all z in the intersection of U and R.

2. We say that $f(z) = o(\phi(z))$, (pronounced "little oh" of ϕ) if for any $\epsilon > 0$ there is a neighborhood U of z_0 so that $|f(z)| \le \epsilon|\phi(z)|$ for all z in the intersection of U and R.

In other words, if $f(z) = O(\phi(z))$, $|f(z)/\phi(z)|$ is bounded in a neighborhood of z_0 whereas if $f(z) = o(\phi(z))$, then $|f(z)/\phi(z)|$ goes to zero in the limit $z \to z_0$ with z in R. The function $\phi(z)$ is called a GAUGE FUNCTION, being a function against which the behavior of $f(z)$ is gauged.

Definition

1. A sequence of gauge functions $\{\phi_j(z)\}_{j=0}^{\infty}$ is an ASYMPTOTIC SEQUENCE as $z \to z_0$, if

$$\phi_{j+1} = o(\phi_j), \text{ as } z \to z_0, j = 0, 1, 2, \ldots .$$

For **example,** the sequence $\ln x$, $x \ln x$, x, $x^2 \ln^2 x$, $x^2 \ln x$, $x^2, \ldots$ is an asymptotic sequence as $x \to 0$.

2. The function $f(z)$ has an asymptotic representation

$$f(z) \sim \sum_{k=0}^{n} a_k \phi_k(z) = f_n(z), \quad z \to z_0$$

if for each fixed n,

$$\lim_{z \to z_0} \left| \frac{f(z) - f_n(z)}{\phi_n(z)} \right| = 0.$$

Since $f_n(z) = f_{n-1}(z) + a_n \phi_n(z)$, it follows that

$$a_n = \lim_{z \to z_0} \left(\frac{f(z) - f_{n-1}(z)}{\phi_n(z)} \right), \quad f_{-1}(z) = 0.$$

Example

Consider the asymptotic expansion of e^{-z} in terms of powers of z^{-1} for large z, $z \to \infty$. Note that

$$a_0 = \lim_{z \to \infty} e^{-z} = 0,$$

$$a_1 = \lim_{z \to \infty} z e^{-z} = 0,$$

and inductively $a_n = \lim_{z \to \infty} z^n e^{-z} = 0$, so that $e^{-z} \sim 0$. The function e^{-z} is said to be TRANSCENDENTALLY SMALL for large z.

For a given asymptotic sequence, $\phi_n(z)$, the following properties are easily verified:

1. The asymptotic representation of $f(z)$ is unique, although it depends strongly on the choice of asymptotic sequence. For example,

$$\frac{1}{z-1} \sim \sum_{n=1}^{\infty} z^{-n}$$

and

$$\frac{1}{z-1} = \frac{z+1}{z^2-1} \sim \sum_{n=1}^{\infty} \frac{z+1}{z^{2n}}, \quad \text{as } z \to \infty.$$

2. Two distinct functions $f(z)$, $g(z)$ may have the same asymptotic representation. For example, $f(z)$ and $f(z) + e^{-z}$ have the same representation in terms of the gauge functions $\phi_n(z) = z^{-n}$ as $z \to \infty$.

3. The sum of asymptotic representations of two functions is the asymptotic representation of the sum of the functions. If $f(z) \sim \sum_{n=1}^{\infty} a_n \phi_n(z)$ and $g(z) \sim \sum_{n=1}^{\infty} b_n \phi_n(z)$ then $f(z) + g(z) \sim \sum_{k=1}^{\infty} (a_n + b_n) \phi_n(z)$.

4. If $f(x)$ is continuous and $f(x) \sim \sum_{n=0}^{\infty} a_n x^{-n}$ as $x \to \infty$ then

$$F(x) = \int_x^{\infty} \left(f(t) - a_0 - \frac{a_1}{t} \right) dt \sim \sum_{n=1}^{\infty} \frac{a_{n+1}}{n x^n}.$$

To prove this, note that if

$$f(x) \sim \sum_{n=0}^{\infty} \frac{a_n}{x^n},$$

then

$$f(x) = \sum_{n=0}^{k} \frac{a_n}{x^n} + O\left(\frac{1}{x^{k+1}} \right)$$

so that

$$F(x) = \int_x^{\infty} \left(\sum_{n=2}^{k} \frac{a_n}{t^n} + O\left(\frac{1}{t^{k+1}} \right) dt \right) \quad \text{for } k \geq 2.$$

$$= \sum_{n=2}^{k} \frac{a_n}{(n-1)x^{n-1}} + \int_x^{\infty} O\left(\frac{1}{t^{k+1}} \right) dt$$

Now $O\left(\frac{1}{x^{k+1}} \right)$ is bounded by $\frac{A}{x^{k+1}}$ for all large x. Therefore,

$$\left| \int_x^{\infty} O\left(\frac{1}{t^{k+1}} \right) dt \right| \leq \int_x^{\infty} \left| O\left(\frac{1}{t^{k+1}} \right) \right| dt$$

$$\leq A \int_x^{\infty} \frac{1}{t^{k+1}} dt = \frac{A}{k} \frac{1}{x^k} = O\left(\frac{1}{x^k} \right),$$

so that

$$F(x) = \sum_{n=1}^{k-1} \frac{a_{n+1}}{n x^n} + O\left(\frac{1}{x^k} \right).$$

10.2   Integration by Parts

Probably the easiest way to find an asymptotic representation is using integration by parts. We illustrate the technique with some typical examples.

		Number of Terms				
x	$xe^x E_1(x)$	2	3	4	5	10
2	.72265	.5	1.0	.25	1.75	-581.37
3	.78265	.6667	.8889	.6667	.96296	13.138
5	.85221	.8	.88	.832	.8704	.73099
10	.91563	.9	.92	.914	.9164	.91545
20	.95437	.95	.955	.95425	.95441	.95437
100	.990194	.99	.9902	.990194	.990194	.990194

TABLE 10.1   Asymptotic Approximation of $xe^x E_1(x)$ compared with exact values.

To approximate the exponential integral

$$E_1(x) = \int_x^\infty \frac{e^{-t}}{t}\, dt,$$

we integrate by parts and learn that

$$E_1(x) = e^{-x} \sum_{n=1}^k (-1)^{n+1} \frac{(n-1)!}{x^n} + (-1)^k k! \int_x^\infty \frac{e^{-t}}{t^{k+1}}\, dt.$$

The remainder term

$$R_n(x) = (-1)^n n! \int_x^\infty \frac{e^{-t}}{t^{n+1}}\, dt$$

is bounded for large x by

$$|R_n(x)| \le \frac{n!}{x^{n+1}} \int_x^\infty e^{-t}\, dt = \frac{n!\, e^{-x}}{x^{n+1}}.$$

As a result,

$$E_1(x) \sim e^{-x} \sum_{n=1}^\infty (-1)^{n+1} \frac{(n-1)!}{x^n}.$$

Notice that although this is an asymptotic representation, it is divergent. Its usefulness, therefore, is not with many terms, but rather with a small number of terms and x large. Adding more terms eventually makes the approximation worse.

In Table 10.1 we display the function $f(x) = xe^x E_1(x)$ and its asymptotic approximation for various values of x. Notice that the best approximation occurs with a different number of terms for each value of x, and the approximation always deteriorates when many terms are added. However, for a fixed number of terms, the approximation always improves as x increases.

As a second example, consider the incomplete gamma function

$$\gamma(a, x) = \int_0^x e^{-t} t^{a-1} \, dt.$$

Recall that the gamma function

$$\Gamma(a) = \int_0^\infty e^{-t} t^{a-1} \, dt$$

satisfies $\Gamma(a) = (a-1)\Gamma(a-1)$ and $\Gamma(1) = 1$ so that $\Gamma(k) = (k-1)!$. For small x, we use that $e^{-x} = \sum_{n=1}^\infty (-1)^n \frac{x^n}{n!}$, so that

$$\gamma(a, x) = \int_0^x \sum_{n=0}^\infty \frac{(-1)^n t^{n+a-1}}{n!} \, dt = \sum_{n=0}^\infty \frac{(-1)^n x^{n+a}}{n!(n+a)}$$

which converges for all x. In fact the power series converges for all bounded complex numbers. However, its usefulness as an approximation is limited to small x, since it converges very slowly (it is an alternating series) for large positive x.

To find a representation that is useful for large x, we define

$$\Gamma(a, x) = \Gamma(a) - \gamma(a, x) = \int_x^\infty e^{-t} t^{a-1} \, dt$$

which we can integrate by parts to obtain

$$\Gamma(a, x) = \sum_{k=1}^n \frac{\Gamma(a)}{\Gamma(a-k+1)} e^{-x} x^{a-k} + \frac{\Gamma(a)}{\Gamma(a-n)} \Gamma(a-n, x).$$

Now,

$$|\Gamma(a-n, x)| = \left| \int_x^\infty e^{-t} t^{a-n-1} \, dt \right| \le x^{a-n-1} \int_x^\infty e^{-t} \, dt$$

$$= x^{a-n-1} e^{-x}$$

for large x, provided $n > a - 1$. Thus,

$$\Gamma(a, x) \sim \sum_{k=1}^\infty \frac{\Gamma(a)}{\Gamma(a-k+1)} e^{-x} x^{a-k}$$

provided a is not an integer.

As a specific case, consider the complementary error function,

$$\text{Erfc}(x) = \frac{2}{\sqrt{\pi}} \int_x^\infty e^{-t^2} \, dt.$$

This function occurs frequently in probability theory in connection with the Gaussian distribution. If X is a normally distributed random variable with zero mean and standard deviation one, then the probability that $X > x$ is $P\{X > x\} = \text{Erfc}(x)$. If we set $t^2 = s$, then

$$\text{Erfc}(x) = \frac{1}{\sqrt{\pi}} \int_{x^2}^{\infty} e^{-s} s^{-1/2}\, ds = \frac{1}{\sqrt{\pi}} \Gamma\left(\frac{1}{2}, x^2\right)$$

so that

$$\text{Erfc}(x) \sim e^{-x^2} \sum_{k=1}^{\infty} \frac{1}{\Gamma(3/2 - k)} \frac{1}{x^{2k-1}} \quad \text{as } x \to \infty.$$

Once again it needs to be emphasized that the computational value of this formula is with one or two terms, not with many terms, in spite of the notational convenience of an infinite sum.

10.3   Laplace's Method

The workhorse of asymptotic expansions of integrals is Laplace's method and is based on a very simple observation, namely, if x is large, then e^{-xt^2} is extremely small except near $t = 0$. If the integrand contains a factor of e^{-xt^2}, the main contribution to the integral occurs near the origin.

As an example, consider the integral

$$\psi(x) = \int_{-\infty}^{\infty} \exp(-x \cosh \theta)\, d\theta.$$

The integrand is a "bell-shaped" curve of height e^{-x} whose width decreases as x increases. Since the integrand is inconsequentially small for large θ and large x we replace $\cosh \theta$ by $\cosh \theta \sim 1 + \theta^2/2$ which for small θ is accurate, but for large θ, where it does not matter, is in error. With this substitution we calculate that

$$\psi(x) \sim \int_{-\infty}^{\infty} \exp(-x(1 + \theta^2/2))\, d\theta$$

$$= e^{-x} \int_{-\infty}^{\infty} e^{-x\theta^2/2}\, d\theta = e^{-x} \left(\frac{2\pi}{x}\right)^{1/2}$$

To see that this is an asymptotic approximation we need to calculate more terms in the expansion. If we make the exact change of variables

$$\cosh \theta = 1 + t^2$$

suggested by our original approximation, we find

$$\psi(x) = \sqrt{2}e^{-x} \int_{-\infty}^{\infty} \frac{e^{-xt^2}}{(1+t^2/2)^{1/2}} \, dt.$$

Once again, we realize that far from the origin the integrand is inconsequential, so we need only approximate the integrand near the origin. For small t, the Taylor series of $1/(1+t^2/2)^{1/2}$ is as accurate as we wish and for large t, even though it may diverge, it does no harm. Therefore, we replace $1/(1+t^2/2)^{1/2}$ by the first few terms of its Taylor series, to obtain

$$\psi(x) = \sqrt{2}e^{-x} \int_{-\infty}^{\infty} e^{-xt^2} \left(1 - t^2/4 + \frac{3}{32}t^4 - \frac{5}{128}t^6 + \cdots \right) dt,$$

from which we evaluate

$$\psi(x) \sim \left(\frac{2\pi}{x} \right)^{1/2} e^{-x} \left(1 - \frac{1}{8x} + \frac{9}{128x^2} - \frac{75}{1024x^3} + O\left(\frac{1}{x^4} \right) \right).$$

The complete asymptotic representation can be found by making use of the binomial theorem

$$(1+x)^{\alpha} = \sum_{n=0}^{\infty} \frac{\Gamma(\alpha+1)x^n}{\Gamma(\alpha+1-n)n!}$$

which is valid for all x if α is an integer or for $|x| < 1$ if α is not an integer. It follows that

$$\psi(x) \sim \left(\frac{2\pi}{x} \right)^{1/2} e^{-x} \sum_{n=0}^{\infty} \frac{\Gamma\left(n+\frac{1}{2}\right)(2x)^{-n}}{\Gamma\left(\frac{1}{2}-n\right)n!}.$$

Another example is to approximate the gamma function

$$\Gamma(x+1) = \int_0^{\infty} e^{-t}t^x \, dt$$

for large x. In its current form, it is not obvious how to apply Laplace's method. However, if we write

$$\Gamma(x+1) = \int_0^{\infty} e^{x \ln t - t} \, dt$$

and observe that the function $x \ln t - t$ has a maximum at $x = t$, then the change of variables $t = xs$ is suggested. With this change of variables

$$\Gamma(x+1) = x^{x+1} \int_0^{\infty} e^{-x(s - \ln s)} \, ds$$

is now in a form to which Laplace's method is applicable. The function $-s + \ln s$ has a maximum at $s = 1$ and is concave downward away from $s = 1$. Therefore, the integrand is "bell-shaped" (although not symmetric) and for large x is exceedingly narrow. Thus, we replace the function $s - \ln s$ by its local approximation

$$s - \ln s \sim 1 + \frac{(1-s)^2}{2}$$

and obtain

$$\Gamma(x+1) \sim x^{x+1} e^{-x} \int_0^\infty e^{-x(1-s)^2/2} \, ds.$$

We still cannot evaluate this integral, but for large x it can be approximated by

$$\Gamma(x+1) \sim x^{x+1} e^{-x} \int_{-\infty}^\infty e^{-x(1-s)^2/2} \, ds$$

$$= x^{x+1} e^{-x} \left(\frac{2\pi}{x}\right)^{1/2}.$$

This approximation of $\Gamma(x+1)$ is known as Stirling's formula.

To find higher order terms, we are motivated by the first approximation to make the exact change of variable

$$s - \ln s = 1 + \tau^2$$

so that $\tau^2 = \sum_{n=2}^\infty \frac{(1-s)^n}{n}$. We must invert this power series to find s as a power series in τ

$$s = 1 + \sqrt{2}\tau + \frac{2}{3}\tau^2 + \frac{1}{9\sqrt{2}}\tau^3 - \frac{2}{135}\tau^4 + \frac{1}{540\sqrt{2}}\tau^5 + \cdots$$

so that

$$\Gamma(x+1) = \sqrt{2} x^{x+1} e^{-x} \int_{-\infty}^\infty e^{-x\tau^2} \left(1 + \frac{\tau^2}{6} + \frac{\tau^4}{216} + O(\tau^6)\right) d\tau$$

$$= \left(\frac{2\pi}{x}\right)^{1/2} x^{x+1} e^{-x} \left(1 + \frac{1}{12x} + \frac{1}{288x^2} + O\left(\frac{1}{x^3}\right)\right).$$

More generally, if we know the power series representation for s as a function of τ,

$$s = \sum_{n=0}^\infty a_n \tau^n$$

then we have the full asymptotic representation as

$$\Gamma(x+1) \sim x^{x+1} e^{-x} \sum_{n=0}^\infty (2n+1) a_{2n+1} \Gamma\left(n + \frac{1}{2}\right) x^{-(n+\frac{1}{2})}.$$

The difficulty, of course, is that it is not an easy matter to calculate the coefficients a_n. In Section 10.6, we suggest some REDUCE procedures that can be used to find these coefficients easily.

The idea of Laplace's method should by now be clear. We seek an asymptotic expansion for large x for the integral

$$f(x) = \int_a^b e^{xh(t)} g(t)\, dt$$

where $h(t)$ has a local maximum at $t = t_0$ in the interval $[a, b]$. We can approximate $h(t)$ by

$$h(t) = h(t_0) + h^{(n)}(t_0) \frac{(t-t_0)^n}{n!} + \cdots$$

where $h^{(n)}(t_0)$, the nth derivative of h with respect to t at t_0, is negative, and n is even.

A first order approximation to $f(x)$ is found by replacing $h(t)$ by

$$h(t_0) + h^{(n)}(t_0) \frac{(t-t_0)^n}{n!},$$

$g(t)$ by $g(t_0)$, and expanding the limits of integration to the infinite interval $(-\infty, \infty)$. From this approximation we obtain

$$
\begin{aligned}
f(x) \quad &\sim \quad e^{xh(t_0)} \int_{-\infty}^{\infty} e^{\frac{xh^n(t_0)(t-t_0)^n}{n!}} g(t_0)\, dt \\
&= \quad 2g(t_0) \left(\frac{n!}{x|h^{(n)}(t_0)|} \right)^{1/n} \Gamma\left(\frac{1}{n}+1\right) e^{xh(t_0)}.
\end{aligned}
$$

Higher order approximations come from the exact change of variables $h(t) = h(t_0) - s^n$, where n is the order of the first nonvanishing derivative of $h(t)$ at t_0. Since $h(t)$ has a Taylor series in the neighborhood of t_0, $h(t) = h(t_0) + \sum_{k=n}^{\infty} a_k (t-t_0)^k$, we invert the power series

$$s^n = \sum_{k=n}^{\infty} a_k (t-t_0)^k$$

to obtain an expression of the form

$$t = t_0 + \sum_{j=1}^{\infty} b_j s^j.$$

Substituting this expression into the integral, we obtain

$$f(x) = e^{xh(t_0)} \int_{\tilde{a}}^{\tilde{b}} e^{-xs^n} g\left(t_0 + \sum_{j=1}^{\infty} b_j s^j \right) \left(\sum_{j=1}^{\infty} j b_j s^{j-1} \right) ds.$$

If g is also analytic in a neighborhood of t_0, this whole mess can be represented in terms of one power series, namely

$$f(x) = e^{xh(t_0)} \int_{\hat{a}}^{\hat{b}} e^{-xs^n} \left(\sum_{j=0}^{\infty} c_j s^j \right) ds.$$

If necessary we replace $\hat{a}$ and $\hat{b}$, by $-\infty$ and ∞, respectively, and evaluate the integrals term by term. The representation we obtain is

$$f(x) \sim \frac{2e^{xh(t_0)}}{n} \sum_{j=0}^{\infty} \frac{c_{2j} \Gamma\left(\frac{2j-1}{n}\right)}{x^{\frac{2j-1}{n}}}.$$

If the maximum of $h(t)$ is attained at one of the endpoints then n need not be even and we obtain

$$f(x) \sim \frac{e^{xh(t_0)}}{n} \sum_{j=1}^{\infty} c_{j-1} \frac{\Gamma(j/n)}{x^{j/n}}.$$

All that we have done so far with Laplace's method seems correct, but we can actually prove that the results are asymptotic without much more work. The statement that this technique is valid is known as WATSON'S LEMMA.

Theorem 10.1

Suppose $\phi(t)$ is analytic in the sector $0 < |t| < R$, $|\arg t| \le \delta < \pi$ (with a possible branch point at the origin) and suppose

$$\phi(t) = \sum_{k=1}^{\infty} a_k t^{\frac{k}{n}-1} \quad \text{for} \quad |t| < R,$$

and that

$$|\phi(t)| \le Ke^{bt} \text{ for } R \le t \le T,$$

then

$$\int_0^T e^{-zt} \phi(t)\, dt \sim \sum_{k=1}^{\infty} \frac{a_k \Gamma\left(\frac{k}{n}\right)}{z^{k/n}}$$

as $|z| \to \infty$ in the sector $|\arg z| \le \delta < \pi/2$.

The form of $\phi(t)$ is motivated by the above general discussion of Laplace's method, where the integer n is the order of the first nonvanishing derivative of the function $h(t)$. This lemma actually shows that Laplace's method can be extended into a sector in the complex plane, and is not restricted to real valued functions.

The proof is a direct application of the stated hypotheses. We divide the interval of integration into two parts

$$f(z) = \int_0^R e^{-zt} \phi(t)\, dt + \int_R^T e^{-zt} \phi(t)\, dt.$$

The second integral we bound by

$$\left| \int_R^T e^{-zt} \phi(t)\, dt \right| \leq \int_R^T e^{-xt} |\phi(t)|\, dt, \quad x = \mathrm{Re}\, z > 0$$

$$\leq K \int_R^T e^{-(x-b)t}\, dt \leq K_1 e^{-xR}$$

for x sufficiently large. Thus, the second integral is of the order of e^{-xR} which is transcendentally small.

For the first integral we note that since the power series of $\phi(t)$ is convergent on the real axis,

$$\phi(t) = \sum_{k=1}^{m-1} a_k t^{\frac{k}{n}-1} + R_m(t)$$

where the remainder term has the bound

$$|R_m(t)| \leq C t^{\frac{m}{n}-1}$$

on the interval $0 < t \leq R$. For the first integral we have

$$\int_0^R e^{-zt} \phi(t)\, dt \;=\; \int_0^\infty e^{-zt} \sum_{k=1}^{m-1} a_k t^{\frac{k}{n}-1}\, dt$$

$$-\int_R^\infty e^{-zt} \sum_{k=1}^{m-1} a_k t^{\frac{k}{n}-1}\, dt$$

$$+\int_0^R e^{-zt} R_m(t)\, dt.$$

Of these three integrals, we evaluate the first and estimate the second and third

$$\int_0^\infty e^{-zt} \sum_{k=1}^{m-1} a_k t^{\frac{k}{n}-1}\, dt \;=\; \sum_{k=1}^{m-1} \frac{a_k \Gamma\left(\frac{k}{n}\right)}{z^{k/n}}$$

$$\int_R^\infty e^{-zt} \sum_{k=1}^{m-1} a_k t^{\frac{k}{n}-1}\, dt \;=\; \sum_{k=1}^{m-1} \frac{a_k \Gamma\left(\frac{k}{n}, zR\right)}{z^{k/n}}$$

The function $\Gamma\left(\frac{k}{n}, zR\right)$ is the incomplete gamma function discussed earlier, and $\Gamma\left(\frac{k}{n}, zR\right) = O(e^{-xR})$ for $x = \mathrm{Re}\, z > 0$. Finally,

$$\left| \int_0^R e^{-zt} R_m\, dt \right| \;\leq\; C \int_0^R e^{-xt} t^{\frac{m}{n}-1}\, dt$$

$$= \frac{C}{x^{m/n}} \int_0^{xR} e^{-t} t^{\frac{m}{n}-1}\, dt$$

$$= O(x^{-m/n}).$$

Piecing all of these estimates together we have a verification of Watson's Lemma as stated.

It is often useful to have Watson's Lemma stated in a slightly different, but equivalent, form.

Theorem 10.2

Suppose $\phi(t)$ is analytic in a neighborhood of $t = 0$,

$$\phi(t) = \sum_{k=0}^\infty a_k t^k, \quad |t| \leq R$$

and suppose that

$$|\phi(t)| \leq k e^{bt^n} \text{ for } R \leq t \leq T,$$

then

$$\int_0^T e^{-zt^n} \phi(t)\, dt \sim \frac{1}{n} \sum_{m=1}^\infty \frac{a_{m-1} \Gamma\left(\frac{m}{n}\right)}{z^{\frac{m}{n}}}$$

as $|z| \to \infty$ in the sector $|\arg z| \leq \delta < \pi/2$.

Example

Consider the integral

$$F(z) = \int_0^\infty \frac{e^{-t}}{t+z}\, dt$$

for complex valued z. If we make the change of variables $t = |z|s$, we find

$$F(z) = \frac{|z|}{z} \int_0^\infty \frac{e^{-|z|s}\, ds}{1 + \frac{|z|}{z}s}$$

The function $\phi(s) = \left(1 + \frac{|z|}{z}s\right)^{-1}$ is analytic for $|s| < 1$, and bounded for large real s, so using its power series expansion we have

$$F(z) \sim \frac{|z|}{z} \int_0^\infty e^{-|z|s} \sum_{n=0}^\infty \left(-\frac{|z|}{z}s\right)^n ds$$

$$= \frac{1}{z} \int_0^\infty e^{-\tau} \sum_{n=0}^\infty \left(\frac{-\tau}{z}\right)^n d\tau$$

$$= \sum_{n=0}^\infty (-1)^n \frac{n!}{z^{n+1}}$$

which is an asymptotic representation of $F(z)$ valid for $|z| \to \infty$.

10.4 Method of Steepest Descents

The asymptotic evaluation of contour integrals centers around one question and that is how best to deform the contour of integration to take advantage of Watson's Lemma. We begin with a specific example.

Consider the integral

$$\phi(x, a) = \frac{1}{2\pi i} \int_C \exp\left[x\left(za - \sqrt{z}\right)\right] \frac{dz}{z}$$

where a is real, positive, and the contour C is any vertical line of the form $z = x_0 + iy$, $x_0 > 0$, fixed, $-\infty < y < \infty$. It is sufficient to evaluate this integral with $a = 1$ since for any other value of a, $\phi(x, a) = \phi(x/a, 1)$. We drop the reference to a and set $\phi(x) \equiv \phi(x, 1)$. To know how best to deform the path C, we must first understand the structure of the function

$$f(z) = z - \sqrt{z}.$$

Note that $f(z)$ has a branch point at $z = 0$ and we take the branch cut to be along the negative real z axis. At $z_0 = \frac{1}{4}$, the function $f(z)$ has zero derivative, $f'(z_0) = 0$. The point z_0 is a SADDLE POINT of $f(z)$ (analytic functions cannot have local maxima or minima).

The idea of the method of steepest descents is to deform the path of integration into one for which the real part of the function $z - \sqrt{z}$ decreases as rapidly as possible. The hope is that, for x large, the integral is well approximated by the integrand near the saddle point and the rest can be ignored with impunity. In other words, we want to choose a path along which Laplace's method can be applied.

It is easiest to look at the behavior of f near the saddle point and to choose a path on which $\mathrm{Re}(f)$ decreases as fast as possible initially. Near the saddle point $z_0' = \frac{1}{4}$,

$$f(z) = -\frac{1}{4} + \left(z - \frac{1}{4}\right)^2 + \dots .$$

Clearly, to make $\mathrm{Re}(f)$ decrease we choose $z = \frac{1}{4} + iy$ for $-\infty < y < \infty$.

If we make this change of variables, and approximate $f(z)$ by $f(z) \sim -\frac{1}{4} - y^2$, we find that

$$\phi(x) \sim \frac{1}{2\pi} e^{-x/4} \int_{-\infty}^{\infty} \frac{\exp[-xy^2]\, dy}{\frac{1}{4} + iy}$$

$$\sim \frac{2}{\pi} e^{-x/4} \int_{-\infty}^{\infty} \exp[-xy^2]\, dy$$

$$= 2 \left(\frac{1}{\pi x}\right)^{1/2} e^{-x/4}.$$

The approximation we have just made is the first term of an expansion using Watson's lemma. This method of approximation is generally called the SADDLE POINT METHOD, although it is no more than the first term of the method of steepest descents (which we are about to discuss).

If one requires more information, an estimate of the error of this approximation, or a rigorous statement of its validity, it is necessary to find more terms. To do so we need a more accurate representation of the steepest descent path.

How do we find the steepest descent path? From calculus, we know that for a real valued function, the steepest paths are along gradient directions. For a complex analytic function $f(z) = u + iv$, $\nabla u \cdot \nabla v = 0$ because of the Cauchy-Riemann conditions. In other words, curves with $\mathrm{Im}(f)$ constant are steepest paths for $\mathrm{Re}(f)$ and vice versa. For the function $f(z) = z - \sqrt{z}$ the curves $\mathrm{Re}(f) = -\frac{1}{4}$ and $\mathrm{Im}(f) = 0$ are shown in Figure 10.1. These are the only level curves which pass through the saddle point. We can represent the curve $\mathrm{Re}(f) = -\frac{1}{4}$ in polar coordinates $z = Re^{i\theta}$, and we find

$$R = \frac{1}{4}\left(\frac{1}{1 \pm \sin\theta}\right)$$

are the two branches. The curve $\mathrm{Im}(f) = 0$, is found by setting $z = Re^{i\theta}$ so that $\mathrm{Im}(z - \sqrt{z}) = R\sin\theta - R^{1/2}\sin\theta/2$. It follows that either $\sin\theta/2 = 0$ or $R(1 + \cos\theta) = \frac{1}{2}$. Setting $z = x + iy$, we find that the second curve is represented by $x = \frac{1}{4} - y^2$, a parabola.

This parabola is the exact path into which we wish to deform the path of integration C. Notice that deformation is possible since the arcs at infinity

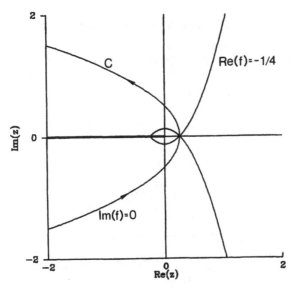

FIGURE 10.1   Paths of steepest descent and ascent for $f(z) = z - z^{1/2}$.

do not contribute to the integral ($\mathrm{Re}(f)$ is large and negative in the region of deformation – use Jordan's Lemma).

Motivated by these observations, we make the exact change of variables

$$z - \sqrt{z} = -\frac{1}{4} - \tau^2, \quad z = \left(\frac{1}{2} + i\tau\right)^2$$

where τ is real.

Every term in this change of variables is well motivated by geometrical reasoning. We want a curve parameterized by τ, passing through $z = 1/4$ when $\tau = 0$. Thus, $z - \sqrt{z} = -1/4$ when $\tau = 0$. We want a steepest descent path so we require τ to be real since then the imaginary part of $z - \sqrt{z}$ is constant. We pick $-\tau^2$ rather than $+\tau^2$ dependence to be sure the real part becomes negative (and hence we get a bell-shaped integrand along the path) and we take quadratic dependence on τ since the saddle point has locally quadratic behavior. In terms of the new variable τ we have

$$\phi(x) = \frac{2}{\pi} e^{-x/4} \int_{-\infty}^{\infty} \frac{\exp(-x\tau^2)\, d\tau}{1 + 2i\tau}.$$

Although this expression is exact, we cannot evaluate it exactly, and this is where Watson's lemma becomes useful. We expand the function $(1 + 2i\tau)^{-1}$ in a power series for small τ, keeping as many terms as we desire (but not too many!), and then evaluate the integrals term by term. This is guaranteed (rigorously!) by Watson's lemma to yield the asymptotic result

$$\phi(x) \sim 2 \left(\frac{1}{\pi x} \right)^{1/2} e^{-x/4} \left(1 - \frac{2}{x} + \frac{12}{x^2} - \frac{120}{x^3} + O\left(\frac{1}{x^4} \right) \right).$$

We can use the geometric series

$$(1 + 2i\tau)^{-1} = \sum_{n=0}^{\infty} (-2i\tau)^n$$

to write the full asymptotic representation as

$$\phi(x) \sim \frac{2}{\pi} e^{-x/4} \sum_{n=0}^{\infty} \frac{\Gamma(n + 1/2)(-4)^n}{x^{n+1/2}}.$$

A slight modification of the previous integral is

$$\psi(x) = \frac{1}{2\pi i} \int_C \frac{\exp x(za - \sqrt{z})}{z^2 + \omega^2} \, dz$$

where as before, the path of integration is the vertical line with $\mathrm{Re}\, z > 0$ running from $z = x_0 - i\infty$ to $z = x_0 + i\infty$, $x_0 > 0$. Once again we can set $a = 1$ since $\psi(x; a, \omega) = a^2 \psi(x/a; 1, a^2\omega)$. As before, we want to deform the path C into the steepest descent path $x = \frac{1}{4} - y^2$, a parabola. Now, however, there are two cases to consider depending on whether or not the deformation of C to the steepest descent path crosses the singularities at $z = \pm i\omega$. Notice that the steepest descent path crosses the imaginary z axis ($x = 0$) at $\mathrm{Im}\, z = y = \pm\frac{1}{2}$. Thus the two cases are $\omega < \frac{1}{2}$ or $\omega > \frac{1}{2}$.

Case 1

If $\omega < 1/2$, the deformation of C to the steepest descent path does not cross the singularity at $z = \pm i\omega$. The first term of the asymptotic expansion of the integral is

$$\psi(x) \sim \frac{8e^{-x/4}}{1 + 16\omega^2} \left(\frac{1}{x\pi} \right)^{1/2}.$$

Case 2

If $\omega > \frac{1}{2}$, the deformation of the path C to the steepest path crosses the singularity at $z = \pm i\omega$. As a result, there is a contribution due to the residue at $z = \pm i\omega$, and the asymptotic representation of the integral is

$$\psi(x) \sim \frac{8e^{-x/4}}{1 + 16\omega^2} \left(\frac{1}{x\pi} \right)^{1/2} + \frac{1}{\omega} e^{-x(\omega/2)^{1/2}} \sin x(\omega - (\omega/2)^{1/2}).$$

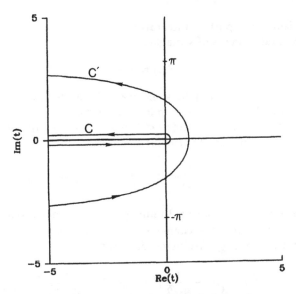

FIGURE 10.2 Path of steepest descent for Hankel's integral.

Higher order terms are found for both cases via Watson's lemma. Notice that the representations are continuous at $\omega = \frac{1}{2}$.

Now that the basic idea of the steepest descents method has been discussed, the remaining examples are for practice.

Example Hankel's integral

Consider

$$\frac{1}{\Gamma(z)} = \frac{1}{2\pi i} \int_C e^s s^{-z} \, ds$$

where the contour C traverses from $s = -i\epsilon - \infty$ to $s = i\epsilon - \infty$ around a branch cut on the $\mathrm{Re}\, s < 0$ axis, $\epsilon > 0$, (Figure 10.2).

To approximate this integral we first make a change of variables $s = zt$ so that

$$\frac{1}{\Gamma(z)} = \frac{1}{2\pi i} z^{1-z} \int_C e^{z(t-\ln t)} \, dt.$$

The function $f(t) = t - \ln t$ has a saddle point at $t = 1$, $f(1) = 1$, and can be represented as a power series near $t = 1$ by

$$f(t) = 1 + \sum_{n=2}^{\infty} \frac{(1-t)^n}{n}.$$

The steepest descent path through the saddle point $t = 1$ satisfies $\mathrm{Im}(t - \ln t) = 0$ which, with $t = x+iy$, can be represented as the curve $x = y \cot y$.

To deform the path of integration into the steepest descent path, we make the exact change of variables

$$f(t) = 1 - \eta^2.$$

Again, this change of variables is motivated by the behavior of the function $f(t) = t - \ln t$. We want a curve, parameterized by η, passing through $t = 1$ at $\eta = 0$, with $\text{Im}(f) = 0$, $\text{Re}(f) < 1$ and with locally quadratic behavior in η. With this change of variables the integral becomes

$$\frac{1}{\Gamma(z)} = \frac{1}{2\pi i} e^z z^{1-z} \int_{-\infty}^{\infty} e^{-z\eta^2} \left(\frac{dt}{d\eta} \right) d\eta$$

which is still exact. However, to find $\frac{dt}{d\eta}$ we must make some approximations. We first invert the change of variables to find $t = t(\eta)$ and then differentiate to find $\frac{dt}{d\eta}$. We determine that

$$t(\eta) = 1 + i\sqrt{2}\eta - \frac{2}{3}\eta^2 - \frac{i}{9\sqrt{2}}\eta^3 - \frac{2}{135}\eta^4 + \frac{i}{540\sqrt{2}}\eta^5 - \frac{4}{8505}\eta^6 + \cdots.$$

Now that $t(\eta)$ is known, we also know $\frac{dt}{d\eta}$ in power series form, and although it has only a finite radius of convergence, we invoke Watson's Lemma to integrate term by term. The result is

$$\frac{1}{\Gamma(z)} = \frac{1}{\pi\sqrt{2}} e^z z^{1-z} \int_{-\infty}^{\infty} e^{-z\eta^2} \left(1 - \frac{\eta^2}{6} + \frac{\eta^4}{216} + \cdots \right) d\eta$$

$$\sim e^z z^{-z} \left(\frac{z}{2\pi} \right)^{1/2} \left(1 - \frac{1}{12z} + \frac{1}{288z^2} + \cdots \right).$$

This answer is in agreement with Stirling's formula found earlier.

Example

Consider the integral

$$F(x) = \int_0^1 \exp[ix(t + t^2)] \frac{dt}{\sqrt{t}}$$

for large x. The path of integration is the straight line segment on the real t axis between zero and one which we intend to deform into steepest descent paths.

The function $f(t) = i(t + t^2)$ has a saddle point at $t = -1/2$ which is of no use to us; the "saddle point" method does not help here. Instead we look for steepest descent paths through $t = 0$, where $f(0) = 0$, so the steepest descent path has $\text{Im} f(t) = 0$. If we set $t = x + iy$ we find that

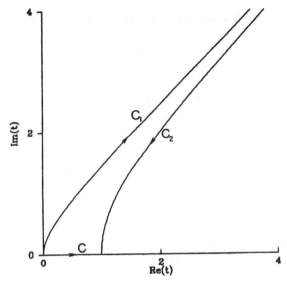

FIGURE 10.3 Paths of steepest descent for $f(t) = i(t + t^2)$.

$\mathrm{Re}f(t) = -y(2x+1)$ and $\mathrm{Im}f(t) = x+x^2-y^2$. The steepest descent path through $t = 0$ is the curve $y^2 = x + x^2$. Similarly the steepest descent path through $t = 1$ satisfies $\mathrm{Im}f(t) = 2$ and is given by $x + x^2 - y^2 = 2$. As a result of this computation, we deform the path of integration into two integrals, one on the path

$$C_1 : i(t + t^2) = -\tau, \ 0 \le \tau < \infty,$$

a steepest descent path through $t = 0$, and the other on the path

$$C_2 : i(t + t^2) = 2i - \eta, \ \ 0 \le \eta < \infty,$$

the steepest descent path through $t = 1$, shown in Figure 10.3. These exact changes of variables render our integral as

$$F(x) = \int_0^\infty \frac{ie^{-x\tau}\,d\tau}{\left(-\frac{1}{2} + \frac{1}{2}(1 + 4i\tau)^{1/2}\right)^{1/2}(1 + 4i\tau)^{1/2}}$$

$$-\frac{1}{3}\int_0^\infty \frac{e^{2ix}e^{-x\eta}i\,d\eta}{\left(-\frac{1}{2} + \frac{3}{2}(1 + 4i\eta/9)^{1/2}\right)^{1/2}(1 + 4i\eta/9)^{1/2}}$$

It is now time to apply Watson's Lemma. We expand the integrands, excluding the exponential, in powers of τ or η and then integrate term by term. After a tedious calculation (or better, use of REDUCE) we find that

$$F(x) = \sqrt{i} \int_0^\infty \frac{e^{-x\tau}}{\sqrt{\tau}} \left(1 - \frac{3i\tau}{2} - \frac{35}{8}\tau^2 + \frac{231}{16}\tau^3 + \cdots \right) d\tau$$

$$- \frac{i}{3}e^{2ix} \int_0^\infty e^{-x\eta} \left(1 - \frac{7i\eta}{18} - \frac{37\eta^2}{216} - \frac{911\eta^3}{11664} + \cdots \right) d\eta$$

with the result that

$$F(x) \sim \left(\frac{\pi}{x}\right)^{1/2} e^{i\pi/4} \left(1 - \frac{3i}{4x} - \frac{108}{32x^2} + \frac{3465i}{128x^3} \right)$$

$$- \frac{i}{3}e^{2ix} \left(\frac{1}{x} - \frac{7i}{18x^2} - \frac{37}{108x^3} \right) + O(x^{-4}).$$

Notice that the asymptotic sequence for this problem are the half powers of x^{-1}, i.e. $x^{-1/2}$, x^{-1}, $x^{-3/2}$ ....

Example

Consider the integral $F(z) = \int_0^1 e^{zt^3} dt$. There are two ways to approximate this integral, one for small z and one for large z. If z is small we expand the integrand in a Taylor series

$$e^{zt^3} = \sum_{n=0}^\infty \frac{(zt^3)^n}{n!}$$

and integrate term by term to obtain

$$F(z) = \int_0^1 \sum_{n=0}^\infty \frac{(zt^3)^n}{n!} dt = \sum_{n=0}^\infty \frac{z^n}{n!(3n+1)}.$$

Although this expansion is convergent for all z it provides a good approximation only for small z since for large z, it converges slowly and many terms are required for reasonable accuracy.

For large z we use the steepest descents method. We rewrite $z = |z|e^{i\theta}$, so that

$$F(z) = \int_0^1 e^{|z|e^{i\theta}t^3} dt$$

where $|z|$ is presumed to be large. The point $t = 0$ is a saddle point for the function $e^{i\theta}t^3$ and the steepest descent paths through the saddle point are the curves $t = re^{\frac{i n\pi - \theta}{3}}$, $r > 0$, $n = 1, 3, 5$. The curves with $n = 0, 2, 4$ are steepest ascent paths since there $\text{Re}(e^{i\theta}t^3) \geq 0$. The steepest descent path through the point $t = 1$ is the curve for which $\text{Im}(e^{i\theta}t^3) = \text{Im}(e^{i\theta}) = \sin\theta$.

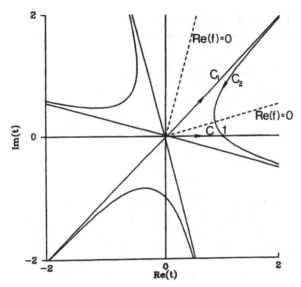

FIGURE 10.4 Paths of steepest descent for $f(t) = e^{i\theta}t^3$.

We deform the path of integration into the steepest descent paths by introducing the exact change of variables

$$C_1 : e^{i\theta}t^3 = -\tau, \ 0 \le \tau < \infty$$
$$C_2 : e^{i\theta}t^3 = e^{i\theta} - \tau, \ 0 \le \tau < \infty$$

and the integral becomes

$$F(z) = -\frac{1}{3}e^{-i\theta/3}e^{\pi 1/3}\int_0^\infty e^{-|z|\tau}\tau^{-2/3}d\tau + \frac{1}{3}e^{-i\theta}e^z\int_0^\infty \frac{e^{-|z|\tau}}{(1-e^{-i\theta}\tau)^{2/3}}d\tau$$

The paths C_1 and C_2 are shown in Figure 10.4 for $\theta = \pi/4$. The first integral we can evaluate exactly in terms of the gamma function; the second integral requires use of Watson's Lemma. We use the binomial theorem to express $(1-e^{-i\theta}\tau)^{-2/3}$ as a power series from which we obtain

$$F(z) = -\frac{\Gamma(4/3)}{z^{1/3}}e^{\pi 2/3} + \frac{e^z}{3}\int_0^\infty e^{-|z|\tau}\sum_{n=0}^\infty \frac{(-1)^n\Gamma(1/3)}{\Gamma(n+1)\Gamma(1/3-n)}e^{-(n+1)i\theta}\tau^n d\tau$$

We evaluate the integral term by term to find

$$F(z) \sim -\frac{\Gamma(4/3)e^{\pi i/3}}{z^{1/3}} + e^z\sum_{n=0}^\infty(-1)^n\frac{\Gamma(4/3)}{\Gamma(1/3-n)}z^{-(n+1)},$$

an expansion that is valid for all z with $|z|$ large, provided z is not real and positive ($\theta \ne 0$).

10.5   Method of Stationary Phase

The method of stationary phase takes note of the fact that if an integrand is rapidly oscillating, the integral is nearly zero except for contributions from those points where the oscillation ceases, or, as the name of the method suggest, where the phase of the integrand is stationary.

To illustrate the method, consider the integral

$$G(x) = \int_0^\infty \cos x(t^3/3 - t)\, dt.$$

For large x, the integrand is rapidly oscillating as long as the function $t^3/3 - t$ is changing in t. However, at the point $t = 1$, the function $t^3/3 - t$ has zero derivative, and thus, the phase of the integrand is stationary. The most significant contribution to the integral comes from the neighborhood of this point, and all other regions are inconsequential by comparison.

This argument leads us to make the approximation

$$t^3/3 - t = -\frac{2}{3} + (t-1)^2 \ldots$$

We truncate the approximation at the quadratic term and use this approximation in the integral

$$G(x) \sim \int_0^\infty \cos\left(-\frac{2x}{3} + x(t-1)^2\right) dt.$$

If our argument is correct, changing the lower limit from zero to $-\infty$ makes no substantial difference for large x, and we have

$$G(x) \sim \cos\frac{2x}{3} \int_{-\infty}^\infty \cos xt^2\, dt + \sin\frac{2x}{3} \int_{-\infty}^\infty \sin xt^2\, dt.$$

From Chapter 6 we know that

$$\int_{-\infty}^\infty \cos at^2\, dt = \int_{-\infty}^\infty \sin at^2 dt = \left(\frac{\pi}{2a}\right)^{1/2}$$

and so we have

$$G(x) \sim \left(\frac{\pi}{x}\right)^{1/2} \cos\left(\frac{2x}{3} - \pi/4\right)$$

plus less significant terms.

As a second example, consider the integral

$$J_0(x) = \frac{2}{\pi} \int_0^{\pi/2} \cos(x \cos\theta)\, d\theta$$

which, from Chapter 6, is an integral representation of the zeroth order Bessel function $J_0(x)$. For large x, the integrand is rapidly oscillating provided

$$\frac{d}{d\theta} \cos \theta \neq 0.$$

However, at $\theta = 0$, the phase of the integrand is stationary and we expect significant contributions to the integral. We approximate

$$\cos \theta \sim 1 - \theta^2/2$$

and obtain

$$J_0(x) \sim \frac{2}{\pi} \int_0^{\pi/2} \cos x(1 - \theta^2/2)\, d\theta.$$

Once again, there should be little change in this approximation if we move the upper limit from $\pi/2$ to ∞. The result is

$$J_0(x) \sim \frac{2}{\pi} \int_0^\infty \cos x(1 - \theta^2/2)\, d\theta = \left(\frac{2}{\pi x}\right)^{1/2} \cos(x - \pi/4)$$

for large x.

As we can see, the method of stationary phase gives a quick estimate of integrals of rapidly varying integrands. Unfortunately, the method provides no check on the validity of the answer, and no way to get a few more terms.

This shortcoming is remedied by the method of steepest descents and Watson's lemma. Observe that for any analytic function, a point of stationary phase is exactly a saddle point in the complex plane. Therefore it makes sense to deform the path of integration into a steepest descent path through the saddle point, invoking Watson's lemma along the way.

We return to our first example

$$G(x) = \int_0^\infty \cos x(t^3/3 - t)\, dt = \operatorname{Re} \int_0^\infty \exp[ix(t^3/3 - t)]\, dt.$$

We want to deform the path of integration (the real axis $0 \leq t < \infty$) into a steepest descent path. Observe that for the function $f(t) = i(t^3/3 - t)$ the steepest descent path through the origin is $t = iy$, $y < 0$ and through the saddle point it is $x(x^2 - 3 - 3y^2) = -2$ where $t = x + iy$. We deform the original path of integration into the two paths

$$C_1 : i(t^3/3 - t) = -\tau,$$

$$C_2 : i(t^3/3 - t) = -\frac{2i}{3} - \eta^2,$$

depicted in Figure 10.5. If our conjecture about the effect of the endpoint at zero is correct, the contribution from C_1 should be negligible. The choice of

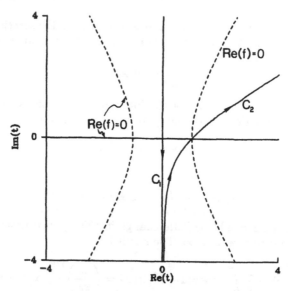

FIGURE 10.5   Paths of steepest descent for $f(t) = i(t^3/3 - t)$.

variables (linear for C_1, quadratic for C_2) is motivated by the fact that the behavior of $f(t)$ near the saddle point is quadratic while near $t = 0$ it is linear.

We use power series in t to invert these changes of variables

$$C_1 : t = i\left(-\tau + \tau^3/3 - \tau^5/3 + 4\tau^7/9 - \frac{55\tau^9}{81} + \cdots\right)$$

$$C_2 : t = 1 + \sqrt{i}\,\eta - \frac{i}{6}\eta^2 + \frac{5i\sqrt{i}}{72}\eta^3 - \frac{1}{27}\eta^4 + \frac{77\sqrt{i}}{3456}\eta^5 + \cdots .$$

Substituting this change of variables into the original integral and integrating term by term, we find

$$\text{Re}\left[\int_0^\infty \exp[ix(t^3/3 - t)]\,dt\right]$$

$$= \text{Re}\left[\int_0^\infty e^{-x\tau}\left(\frac{dt}{d\tau}\right)d\tau + e^{\frac{-2ix}{3}}\int_{-\infty}^\infty e^{-x\eta^2}\frac{dt}{d\eta}\,d\eta\right]$$

$$= \text{Re}\left[\left(\frac{\pi}{x}\right)^{1/2}e^{-i\left(\frac{2x}{3}-\pi/4\right)}\left(1 + \frac{5i}{48x} + \frac{385}{4608}\frac{1}{x^2} + \cdots\right)\right]$$

$$= \left(\frac{\pi}{x}\right)^{1/2}\left[\cos\left(\frac{2x}{3} - \frac{\pi}{4}\right)\left(1 - \frac{385}{4608x^2} + \cdots\right)\right.$$

$$\left. + \frac{5}{48x}\sin\left(\frac{2x}{3} - \frac{\pi}{4}\right) + \cdots\right].$$

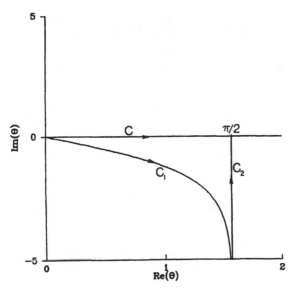

FIGURE 10.6 Paths of steepest descent for $f(\theta) = i \cos \theta$.

There is no contribution from the integral along C_1 since there the integrand is purely imaginary.

The second example

$$J_0(x) = \frac{2}{\pi} \mathrm{Re} \int_0^{\pi/2} \exp[ix \cos \theta] \, d\theta$$

can also be approximated using the method of steepest descents. The function $f(\theta) = i \cos \theta$ has a saddle point at $\theta = 0$. The steepest descent path through $\theta = 0$ is the curve $\cos x \cosh y = 1$, $\theta = x + iy$. The endpoint $\theta = \pi/2$ is not a saddle point; the steepest descent path through this endpoint is $\theta = \pi/2 - iy$, $y > 0$. We deform the original path of integration into the two paths

$$C_1 : i \cos \theta \;=\; i - \tau^2,$$

$$C_2 : i \cos \theta \;=\; -\eta,$$

shown in Figure 10.6. To invert these expressions for $\theta = \theta(\tau)$ and $\theta = \theta(\eta)$ we make use of the power series expansions

$$\cos \theta \;=\; \sum_{k=0}^{\infty} (-1)^k \frac{\theta^{2k}}{(2k)!}$$

and

$$\cos\theta \;=\; -\sum_{k=0}^{\infty}(-1)^k\frac{(\theta-\pi/2)^{2k+1}}{(2k+1)!}.$$

This enables us to find power series expansion for the inverse functions $\theta = \theta(\tau)$ and $\theta = \theta(\eta)$

$$C_1 : \theta \;=\; \frac{2}{\sqrt{2i}}\tau - \frac{i}{6\sqrt{2i}}\tau^3 - \frac{3}{8\sqrt{2i}}\tau^5 + \frac{5i}{448\sqrt{2i}}\tau^7 + \cdots$$

$$C_2 : \theta \;=\; \pi/2 + i\left(-\eta + \frac{1}{6}\eta^3 - \frac{3}{40}\eta^5 + \frac{5}{112}\eta^7 + \cdots\right).$$

As a consequence

$$J_0(x) \;=\; \frac{2}{\pi}\mathrm{Re}\left[e^{ix}\int_0^{\infty}e^{-x\tau^2}\frac{d\theta}{d\tau}\,d\tau - \int_0^{\infty}e^{-x\eta}\frac{d\theta}{d\eta}\,d\eta\right]$$

$$\sim \;\left(\frac{2}{\pi x}\right)^{1/2}\left[\cos(x-\pi/4)\left(1-\frac{45}{64x^2}+\cdots\right)\right.$$

$$\left. + \sin(x-\pi/4)\left(\frac{1}{8x}-\frac{525}{7168x^3}+\cdots\right)\right].$$

Using this approximation we can estimate the location of the large zeros of $J_0(x)$. In fact, it is not difficult to determine that $J_0(\lambda_k) = 0$ for

$$\lambda_k \sim \pi/4 + \frac{2k+1}{2}\pi + \frac{1}{(8k+6)\pi} + O\left(\frac{1}{k^2}\right).$$

As a final example, we consider the solution of the telegrapher's equation

$$u_{tt} = u_{xx} - u.$$

Using Fourier transforms in time, one can readily show that the general solution is

$$u(x,t) = \frac{1}{2\pi}\int_{-\infty}^{\infty}\left[A(\omega)e^{i(k(\omega)x-\omega t)} + B(\omega)e^{-i(k(\omega)x+\omega t)}\right]d\omega$$

where $k(\omega) = (\omega^2-1)^{1/2}$. To study the signaling problem we suppose signals are being sent from the fixed position $x = 0$ so that $u(0,t)$ and $u_x(0,t)$ are specified functions of time. Clearly, specifying $u(0,t)$ and $u_x(0,t)$ has the effect of specifying $A(\omega)$ and $B(\omega)$. Notice also that $A(\omega)$ is the amplitude of right moving waves, while $B(\omega)$ is the amplitude of left moving waves.

As a specific example we suppose $u(0,t) = \delta(t)$, $u_x(0,t) = 0$ so that $A(\omega) = B(\omega) = 1/2$, and we are interested in understanding the behavior of the right moving component

$$u_r(x,t) = \frac{1}{4\pi}\int_{-\infty}^{\infty}e^{i(k(\omega)x-\omega t)}\,d\omega.$$

The method of stationary phase gives an estimate of the behavior of $u_r(x,t)$. We suppose that t is large and that $x/t = \alpha$ is some fixed number, then

$$u_r(x,t) = \frac{1}{4\pi} \int_{-\infty}^{\infty} e^{it(\alpha k(\omega)-\omega)} \, d\omega$$

can be approximated by taking

$$f(\omega+ \equiv \alpha k(\omega) - \omega \approx f(\omega_0) + \frac{f''(\omega_0)}{2}(\omega - \omega_0)^2$$

where ω_0 is a point of stationary phase $f'(\omega_0) = 0$. It follows that

$$u_r(x,t) \sim \frac{1}{2}(2\pi |f''(\omega_0)|t)^{-1/2} e^{i(tf(\omega_0)+\pi/4)}$$

In other words, as one moves along the straight line $x/t = \alpha$ in the $x-t$ plane, one sees a decaying oscillatory function.

This representation of the solution, although correct, is somewhat misleading because ω_0 depends on x and t and is fixed only if we move along the line $x/t = \alpha$. It is difficult to visualize what the solution looks like for large t with x fixed, for example. To understand this we must re-express ω_0 in terms of x/t. For the specific example $k(\omega) = (\omega^2 - 1)^{1/2}$ we find

$$\begin{aligned} f(\omega) &= (\omega^2 - 1)^{1/2}\alpha - \omega \\ f'(\omega) &= 0 \text{ at } \omega_0 = (1 - \alpha^2)^{-1/2} \text{ provided } |\alpha| < 1 \end{aligned}$$

so that

$$\begin{aligned} f(\omega_0) &= -(1 - \alpha^2)^{1/2} \\ f''(\omega_0) &= \frac{-(1-\alpha^2)^{3/2}}{\alpha^2} \end{aligned}$$

provided $|\alpha| < 1$. As a result

$$u_r(x,t) \sim \frac{x}{2} \frac{e^{-i(t^2-x^2)^{1/2}+i\pi/4}}{(2\pi)^{1/2}(t^2 - x^2)^{3/4}} \text{ provided } t > x.$$

This solution is not valid for $x > t$. In this case the saddle point is on the imaginary axis at

$$\omega_0 = \frac{i}{(\alpha^2 - 1)^{1/2}}$$

and

$$\begin{aligned} f(\omega_0) &= i(\alpha^2 - 1)^{1/2}, \quad |\alpha| > 1 \\ f''(\omega_0) &= \frac{-i(\alpha^2 - 1)^{3/2}}{\alpha^2}. \end{aligned}$$

Now we use the saddle point method to obtain a similar answer

$$u_r(x,t) \sim \frac{xe^{i\pi/4}}{2(2\pi)^{1/2}} \frac{e^{-(x^2-t^2)^{1/2}}}{(x^2-t^2)^{3/4}}, \text{ provided } x > t.$$

Notice the important difference between these two representations. For $x < t$ the solution is oscillatory while for $x > t$ it is exponentially damped. The main signal is at $x = t$ where the solution is unbounded (since we started with a delta function). Ahead of the signal there is a small exponential precursor and behind the signal there is ringing, that is, damped oscillatory behavior. The modification of the wave equation into the telegrapher's equation was responsible for this smearing of the signal. Smearing of a signal is typical in wave propagation problems and occurs whenever the function $k(\omega)$ is not proportional to ω. As a result, waves of different frequencies travel with different speeds and a signal which is initially composed of a mixture of frequencies will break up, or disperse. The relationship $k = k(\omega)$ is called the DISPERSION RELATION, and determines how this dispersal occurs.

10.6   REDUCE Procedures

Many of the calculations shown in this chapter are quite tedious to do by hand, but can be done easily with REDUCE. In particular, calculation of a few terms of a power series can be done with a call to

```
PROCEDURE TAYLOR (F,X,X0,N);
   BEGIN
   SCALAR H, HH;
   H:=SUB(X=X0,F);
   HH:=F;
   FOR J:=1:N DO
      BEGIN
      HH:=DF(HH,X)/J;
      H:=H+SUB(X=X0,HH)*(X-X0)**J;
   RETURN H;
   END;
```

The procedure TAYLOR finds n terms of the Taylor series of $f(x)$ about x_0. For example, the statement

```
G:=TAYLOR(E**X,X,0,3);
```

will result in

$$G := (X^3 + 3X^2 + X + 6)/6.$$

To find inverse functions, use the procedure

```
PROCEDURE REVERS (F,X,X0,G,Y,N);
    BEGIN
    SCALAR XX,H,HH,J,K;
    OPERATOR A;
    ARRAY W(N);
    H:=TAYLOR(F,X,X0,N);
    XX:=FOR J:=0:N SUM A(J)*Y**J;
    A(0):=X0;
    HH:=SUB(X=XX,H)-G;
    COEFF(NUM(HH),Y,W);
    K:=0;
    FOR J:=1:N DO
        BEGIN
        IF W(J) NEQ 0 THEN
            BEGIN
            K:=K+1;
            SOLVE(W(J),A(K));
            A(K):=SOLN(1,1);
            END;
        END;
    XX:=FOR J:=0:K SUM A(J)*Y**J;
    RETURN XX;
    END;
```

This procedure will find n terms of the Taylor series of the solution $x = x(y)$ of $f(x) = g(y)$ where $f(x_0) = g(0)$ and $g(y)$ is a polynomial in y. So, for example, the statement

$$H:=REVERS (E**X,X,0,1+Y,Y,3);$$

will result in the first three terms of the Taylor series of $x = \ln(1 + y)$. The statement

$$H:=REVERS (X-LOG(X),X,1,1-Y**2,Y,3);$$

produces three terms of the power series solution of

$$x - \ln x = 1 - y^2$$

near $x = 1$ which we used to find an expansion of Hankel's integral representation of $1/\Gamma(z)$.

In general, the procedure REVERS will work in the neighborhood of a regular point or a quadratic or cubic saddle point, but will fail at higher order critical points.

Further Reading

Some books that discuss material related to this chapter are

1. N. Bleistein and R.A. Handelsman, *Asymptotic Expansions of Integrals*, Holt, Rinehart and Winston, New York, 1975.

2. E.T. Copson, *Asymptotic Expansions*, Cambridge University Press, Cambridge, 1967.

3. A. Erdelyi, *Asymptotic Expansions*, Dover, New York, 1956.

4. J.D. Murray, *Asymptotic Analysis*, Springer-Verlag, New York, 1984.

5. F.W. J. Olver, *Asymptotics and Special Functions*, Academic Press, New York, 1974.

Problems for Chapter 10

Section 10.1

1. Suppose $\sum_{k=0}^{\infty} a_k z^k$ is a convergent power series representation of $f(z)$ for $|z| \leq R$. Show that $\sum_{k=0}^{\infty} a_k z^k$ is also an asymptotic representation of $f(z)$ for $|z| \to 0$.

2. Show that the sum of asymptotic representations of two functions is the asymptotic representation of the sum of the functions.

Section 10.2

Use integration by parts to find asymptotic expansions for large x of the integrals in problems 1-3. In each case prove that the expansions are indeed asymptotic.

1. $E_n(x) = \int_1^{\infty} \frac{e^{-xt}}{t^n} dt$

2. $\int_0^1 (\cos t + t^2) e^{ixt} dt$

3. $\int_0^1 e^{ixt} t^{-1/2} dt$

4. Find asymptotic expansions valid as $x \to \infty$ for the Fresnel integrals

$$C(x) = \int_0^x \cos\left(\frac{\pi t^2}{x}\right) dt$$

and

$$S(x) = \int_0^x \sin\left(\frac{\pi t^2}{2}\right) dt.$$

5. How many terms of the asymptotic expansion should be used to best approximate $\int_8^\infty \frac{\sin t}{t}\, dt$?

Section 10.3

1. Use Watson's Lemma to obtain an asymptotic expansion of

$$E_1(x) = \int_x^\infty \frac{e^{-t}}{t}\, dt$$

Hint: Show that

$$E_1(x) = e^{-x} \int_0^\infty \frac{e^{-xt}}{1+t}\, dt$$

2. Find asymptotic representations for $f(z)$, $|z| \to \infty$ in the sector $0 < |\arg z| < \pi/2$ where

$$f(z) = \int_0^\infty \frac{e^{-zt}}{1+t^4}\, dt.$$

3. Show that $\int_0^\infty \frac{e^{-u^2}}{u+z}\, du \sim \sum_{k=0}^\infty (-1)^k \frac{\Gamma(\frac{k}{2}+\frac{1}{2})}{2z^{k+1}}$ as $|z| \to \infty$ in the sector $|\arg z| < 5\pi/4$.

Find at least three terms of the asymptotic expansions of the integrals in problems 4 through 8.

4. $\int_0^1 e^{-xt} \sin t\, dt$, $x \to \infty$

5. $\int_0^1 e^{-xt} t^{-1/2}\, dt$, $x \to \infty$

6. $\int_0^\infty e^{xt} t^{-t}\, dt$, $x \to \infty$

7. $\int_0^1 t^x \sin^2 \pi t\, dt$, $x \to \infty$

8. $\int_0^\pi \exp(-xt^2) t^{-1/3} \cos t\, dt$, $x \to \infty$

9. The modified Bessel function $I_n(x)$ has the integral representation

$$I_n(x) = \frac{1}{\pi} \int_0^\pi \exp(x \cos \theta) \cos n\theta\, d\theta.$$

Show that $I_n(x) \sim e^x/(2\pi x)^{1/2}$.

10. Estimate the Legendre function

$$P_n(x) = \frac{1}{\pi} \int_0^\pi (x + (x^2 - 1)^{1/2} \cos \theta)^n\, d\theta$$

for $n \to \infty$, $x > 1$.

11. **a.** Use that $k!n^{-k-1} = \int_0^\infty e^{-nx} x^k \, dx$ to show that

$$\sum_{k=0}^{n} \binom{n}{k} k! n^{-k} \sim \left(\frac{\pi n}{2}\right)^{1/2} \text{ as } n \to \infty.$$

b. Find the asymptotic behavior of

$$\sum_{k=0}^{n} \binom{n}{k} k! \lambda^k \text{ for } 0 < \lambda < 1, \ \lambda \text{ real.}$$

12. Find the first term of the asymptotic expansion of

$$F(k, a) = \int_{-\infty}^{a} e^{-kz^2} f(z) \, dz, \ k \to \infty,$$

for all a in $0 < \arg a < \pi/2$. The function $f(z)$ is assumed to be analytic and bounded in all regions of interest.

13. Find two terms of the asymptotic expansion of

$$I(x) = \int_0^\infty t^x e^{-t} \ln t \, dt, \text{ as } x \to \infty.$$

What is the asymptotic sequence for this integral?

14. Find the complete asymptotic expansion of

$$I(k) = \int_0^\infty e^{-kt - t^p/p} \, dt, \ p > 1, \ p \text{ real.}$$

Find the largest sector in the complex k plane for which this expansion is valid.

15. The Fresnel integrals $C(x)$ and $S(x)$ are defined by

$$C(k) + iS(x) = \int_0^x e^{it^2} \, dt, \ x \text{ real.}$$

Show that

$$C(x) = \frac{1}{2}\left(\frac{\pi}{2}\right)^{1/2} - P(x)\cos x^2 + Q(x)\sin x^2$$

$$S(x) = \frac{1}{2}\left(\frac{\pi}{2}\right)^{1/2} - P(x)\sin x^2 - Q(x)\cos x^2$$

where

$$P(x) = \frac{1}{2}\left(\frac{1}{2x^3} - \frac{15}{8x^7} + \cdots\right), \ Q(x) = \frac{1}{2}\left(\frac{1}{x} - \frac{3}{4x^5} + \frac{105}{16x^9} + \cdots\right).$$

Section 10.4

1. Find three terms of the asymptotic expansion of

$$I(k) = \int_C \frac{e^{k(z^2-1)}}{z - 1/2} \, dz$$

where C is the vertical path from $z = 1 - i\infty$ to $z = 1 + i\infty$.

2. The Hankel function $H_\nu^{(1)}(r)$ is

$$H_\nu^{(1)}(r) = \frac{1}{\pi} \int_C e^{ir \cos z} e^{i\nu(z - \pi/2)} \, dz$$

where the path C goes from $z = -\pi/2 + i\infty$ to $z = \pi/2 - i\infty$. Show that

$$H_\nu^{(1)}(r) \sim \left(\frac{2}{\pi r}\right)^{1/2} e^{i\left(r - \frac{\nu\pi}{2} - \pi/4\right)}, \quad \text{as } r \to \infty.$$

3. Find the asymptotic behavior of

$$I(x) = \int_{-\infty}^{\infty} \exp[ix(t + t^3/3)] \, dt, \quad x \to \infty.$$

4. Find the asymptotic behavior of the nth order Bessel function

$$J_n(z) = \frac{1}{\pi} \int_0^\pi \cos(z \sin \theta - n\theta) \, d\theta$$

for large z.

5. Find accurate approximations for the large zeros of the nth order Bessel function $J_n(x)$.

In problems 6 and 7, find asymptotic representations, valid as $x \to \infty$.

6. $\int_0^1 \cos(xt^p) \, dt$, p real, $p > 1$ (two terms).

7. $\int_0^{\pi/2} \left(1 - \frac{2\theta}{\pi}\right)^{-1/2} \cos(x \cos \theta) \, d\theta.$

8. Show that $A(x) = \int_0^\infty \cos(sx + s^3/3) \, ds$ satisfies the differential equation $A'' - xA = 0$. ($A(x)$ is called the Airy function). Approximate $A(x)$ for $|x| \to \infty$.

9. Estimate accurately the large zeros of the Airy function.

10. Estimate the first zero of $J_n(z)$ for large n.

11. Estimate the first zero of $J_n'(z)$ for large n.

Section 10.5

1. Suppose $g'(\alpha) = g''(\alpha) = 0$ and $g'''(\alpha) > 0$ for $a < \alpha < b$. Estimate

$$\int_a^b f(x)e^{ikg(x)}\,dx$$

for large k.

2. Use the method of stationary phase to find asymptotic representations of problems 10.4.6 and 10.4.7. Why does this procedure fail for problem 10.4.7?

3. Show that

$$\int_{-\infty}^{\infty} e^{its}(1+x^2)^{-t}\,ds \sim \left[\frac{\pi(1-\alpha)}{t}\right]^{1/2} e^{-\alpha t}(2\alpha)^{-t}$$

as $t \to \infty$, where $\alpha = -1 + \sqrt{2}$.

11

REGULAR PERTURBATION THEORY

11.1   The Implicit Function Theorem

It is probably the case that most of the scientific problems a mathematician might study are best described by nonlinear equations, while most of the existing mathematical theory applies to linear operators. This mismatch between mathematical information and scientific problems is gradually being overcome as we learn how to solve nonlinear problems. At present one of the best hopes of solving a nonlinear problem, or any problem for that matter, occurs if it is "close" to another problem we already know how to solve. We study the solution of the simpler problem and then try to express the solution of the more difficult problem in terms of the simpler one modified by a small correction.

As a simple example, suppose we wish to find the roots of the quadratic equation

$$x^2 - x + .01 = 0.$$

Of course, we can solve this equation exactly, but to do so would keep us from learning how to solve other problems for which exact solutions are not known. Instead we look for a technique that will generalize to harder problems.

The key step is to denote $\epsilon = .01$ and write our equation as $x^2 - x + \epsilon = 0$, and then to treat ϵ as a parameter. We notice that if $\epsilon = 0$ the solutions are $x = 0$ and $x = 1$ and we hope that while ϵ remains small, the solutions remain close to $x = 0$ or $x = 1$. In fact, the solutions depend continuously (in fact analytically) on ϵ, so we can write the solution $x = x(\epsilon)$ as a power series in ϵ

$$x(\epsilon) = \sum_{k=0}^{\infty} a_k \epsilon^k.$$

Upon substituting this guess into the polynomial equation, we obtain a power series in ϵ which must vanish identically. We know that a power series is identically zero if and only if each of its coefficients vanishes separately. Therefore, we write down the first few coefficients of this power series and require that they vanish term by term, that is,

$$a_0(a_0 - 1) = 0$$

$$(2a_0 - 1)a_1 + 1 = 0$$

$$(2a_0 - 1)a_2 + a_1^2 = 0$$

$$(2a_0 - 1)a_3 + 2a_1 a_2 = 0$$

and so on. Observe that we can solve these equations sequentially. If we know a_0 then

$$a_1 = \frac{-1}{2a_0 - 1}$$

$$a_2 = \frac{-a_1^2}{2a_0 - 1}$$

$$a_3 = \frac{-2a_1 a_2}{2a_0 - 1}$$

and so forth, provided $2a_0 - 1 \neq 0$. In the specific case that $a_0 = 0$ we find

$$x = \epsilon + \epsilon^2 + 2\epsilon^3 + \ldots$$

and if $a_0 = 1$, then
$$x = 1 - \epsilon - \epsilon^2 - 2\epsilon^3 + \ldots .$$

With $\epsilon = .01$, the two roots are

$$x = .010102, \quad x = .989898$$

to six significant figures. One can get as much accuracy as needed, but it is unlikely that one needs more accuracy than this.

There are a number of simple, but important, observations to make. First, the hierarchy of equations can be solved sequentially, that is, if $a_0, a_1, \ldots a_{k-1}$ are known, then a_k is determined by an equation of the form

$$(2a_0 - 1)a_k = f_k(a_0, a_1, \ldots a_{k-1}).$$

In fact, we can determine a_k if and only if the linear operator $L = (2a_0 - 1)$ is invertible. Notice that $L \equiv \frac{d}{dx}(x^2 - x)|_{x=a_0}$ is the derivative of the reduced problem evaluated at the known solution $x = a_0$. This observation, that the same linear operator is used at every step and that its invertibility is the key to solving the problem, is true in general. The general result is called the IMPLICIT FUNCTION THEOREM (which we state without proof).

Theorem 11.1

Suppose X, Y, Z are Banach Spaces and $f(x, \epsilon)$ is a C^1 mapping of an open set U of $X \times Y$ into Z. Suppose there is an $x_0 \epsilon X$ satisfying $f(x_0, 0) = 0$ and the partial derivative $f_x(x_0, 0)$ is an invertible linear map of X onto Z. Then there is an open set in Y, say W, with $0 \epsilon W$, an open set $V \subset U$ with $(x_0, 0) \epsilon V$ and a C^1 mapping g of W into Y, such that $x = g(\epsilon)$ for ϵ in W satisfies $f(x, \epsilon) = 0$ for x, ϵ in V. In other words, there is a solution of the implicit equation $f(x, \epsilon) = 0$ given by $x = g(\epsilon)$. This solution is unique in some open subset of W.

This theorem as stated is the best possible, since it only requires that $f(x, \epsilon)$ be a C^1 mapping. In point of fact, in most applications the mapping f has many derivatives, (often C^∞) and power series representations can be used.

The implicit function theorem is useful because it tells us that if we know how to solve a reduced problem, and the linearized operator of the reduced problem is invertible, then the more difficult problem has a solution locally. One does not actually use the implicit function theorem to compute the solution, since first, it does not suggest an algorithm, and second, the exact solution is probably not needed anyway. Sufficiently useful information is usually contained in the first few terms of the local power series representation, which, if it exists, is an asymptotic series for the solution (see problem 10.1.1).

Example

Solve the boundary value problem

$$u'' + \epsilon u^2 = 0, \quad u(0) = u(1) = 1$$

for small ϵ. We seek a power series solution $u(x, \epsilon) = \sum_{k=0}^{\infty} u_k(x)\epsilon^k$ where $u_k(0) = u_k(1) = 0$ for $k \geq 1$ and $u_0(0) = u_0(1) = 1$. We substitute the power series into the equation and determine that

$$u_0'' + \sum_{k=1}^{\infty} \left(u_k'' + \sum_{j=0}^{k-1} u_j u_{k-j-1} \right) \epsilon^k = 0.$$

This power series in ϵ must be identically zero so we require

$$u_0'' = 0, \quad u_0(0) = u_0(1) = 1,$$

$$u_k'' = - \sum_{j=0}^{k-1} u_j u_{k-j-1}, \quad u_k(0) = u_k(1) = 0, \quad k \geq 1.$$

The linear operator $Lu = u''$ with boundary conditions $u(0) = u(1) = 0$ is uniquely invertible on the Hilbert Space $L^2[0, 1]$ so we conclude that the implicit function theorem holds. We find the first few terms of the solution to be

$$u_0 \;=\; 1$$

$$u_1 \;=\; x(1-x)/2$$

$$u_2 \;=\; \frac{x}{12}(1-x)(x^2 - x - 1)$$

or $u(x, \epsilon) = 1 + \frac{1}{2}\epsilon x(1-x) + \frac{\epsilon^2}{12}x(1-x)(x^2 - x - 1) + O(\epsilon^3)$.

Example

Find a periodic solution of the forced logistic equation

$$u' = au(1 + \epsilon \cos t - u).$$

This equation can be viewed as determining the population size u of some living organism in an environment with a periodically varying resource. The expression $K(t) = 1 + \epsilon \cos t$ is the carrying capacity for the environment since when u is smaller than $K(t)$, u can grow, but if u is larger than $K(t)$, it must decrease. There are two steady solutions of this equation when $\epsilon = 0$, namely $u = 0$ and $u = 1$. The solution $u = 0$ means the population is extinct, and changing ϵ does not change this fact. In spite of the fact that this equation can be solved exactly (make the change of variables $u = 1/v$), it is advantageous to use an approximate solution technique. We look for a solution of the equation that is near one

$$u(t, \epsilon) = 1 + \epsilon u_1(t) + O(\epsilon^2)$$

and find that we must require

$$u_1' + au_1 = a\cos t.$$

The solution of this equation is

$$u_1 = \frac{a}{\sqrt{a^2+1}}\cos(t-\phi) + Ae^{-at}, \quad \tan\phi = \frac{1}{a}$$

where A is arbitrary. The periodic solution of the problem is

$$u = 1 + \frac{\epsilon a}{\sqrt{a^2+1}}\cos(t-\phi) + O(\epsilon^2).$$

Notice that the maximum population occurs after the maximum in carrying capacity has occurred and the minimum population occurs after the minimum in carrying capacity. Thus, for example, if time is scaled so that the period of forcing is one year and $a = 1$, the maximum population occurs six weeks after the maximal carrying capacity.

Example

The flow of an ideal incompressible irrotational fluid is parallel to the x axis at $x = \pm\infty$ with velocity U. Find the stream function for the flow around a nearly circular cylinder with radius $r(\theta, \epsilon) = a(1 - \epsilon\sin^2\theta)$.

We seek a solution of Laplace's equation $\nabla^2\psi = 0$ subject to the boundary condition $\psi = 0$ on the cylinder $r(\theta, \epsilon) = a(1 - \epsilon\sin^2\theta)$ and subject to the condition $\psi \to Uy$ as $x \to \pm\infty$.

From Chapter 6 we know how to solve this problem for $\epsilon = 0$. We take

$$\psi_0 \;=\; \mathrm{Im}(U(z + a^2/z)), \quad z = re^{i\theta}$$

or

$$\psi_0 \;=\; U(r - a^2/r)\sin\theta.$$

To solve this problem for $\epsilon \neq 0$ we try a solution of the form $\psi(r,\theta,\epsilon) = \sum_{j=0}^{\infty}\psi_j(r,\theta)\epsilon^j$. It is clear that ψ_1 must satisfy Laplace's equation $\nabla^2\psi_1 = 0$, but it is not so clear how to specify boundary conditions for ψ_1. Since we want

$$\psi(r(\theta,\epsilon),\theta,\epsilon) = \psi(a - a\epsilon\sin^2\theta,\theta,\epsilon) = 0,$$

we expand $\psi(r(\theta,\epsilon),\theta,\epsilon)$ in its Taylor series in ϵ and set each coefficient of ϵ to zero. The resulting hierarchy of equations is

$$\psi_0(a,\theta) \;=\; 0$$

$$\psi_1(a,\theta) \;=\; a\sin^2\theta\,\psi_{0_r}(a,\theta)$$

and so on. Since we already know ψ_0, we find that ψ_1 must satisfy Laplace's equation subject to the boundary condition

$$\psi_1(a,\theta) = 2aU \sin^3\theta,$$

and must vanish as $r \to \infty$.

If we use the trigonometric identity $\sin^3\theta = \frac{3}{4}\sin\theta - \frac{1}{4}\sin 3\theta$, separation of variables works quite well to solve this problem. Without writing out the details, we find that

$$\psi(r,\theta) = U(r - a^2/r)\sin\theta + \frac{\epsilon U}{2}\left(\frac{3a^2}{r}\sin\theta - \frac{a^4}{r^3}\sin 3\theta\right) + O(\epsilon^2).$$

Example

Determine the speed of swimming of a flagellate protozoan in a viscous fluid. We assume that relative to the coordinate system of the protozoan, the motion of the flagellum is given by

$$y = \epsilon \sin(kx - \omega t)$$

(that is, the fluid is moving past the protozoan). We want to find the fluid motion induced by the motion of the flagellum.

To get somewhere with this problem, we make the simplifying assumptions that the organism is infinitely long, and the motion is two dimensional, that is, independent of the third dimension. In reality, then, we have an infinite swimming corrogated sheet. The relevant equations are the equations for a Stokes' flow

$$\mu \nabla^2 u = \nabla p$$

$$\operatorname{div} u = 0,$$

where u is the velocity vector for the fluid, p is the pressure, and μ is the viscosity of the fluid. These equations are simplifications of the Navier-Stokes' equations for the motion of a viscous fluid in the limit that the Reynolds number is very small, or equivalently, the viscosity is very large. At the surface of the flagellum we apply a "no-slip" condition by requiring that a fluid particle at the surface have the same velocity as the surface.

It is convenient to introduce a stream function defined by $\psi_y = u_1$, $\psi_x = -u_2$, so that div $u = 0$. It follows immediately that p satisfies Laplace's equation and $\nabla^2(\nabla^2\psi) = 0$. This last equation is called the BIHARMONIC EQUATION.

To find the fluid motion we must solve the biharmonic equation subject to the no-slip condition on the surface of the flagellum. We seek a function $\psi(x, y)$ (also depending on t and ϵ) that satisfies the biharmonic equation and the boundary conditions

$$\psi_x(x, \epsilon \sin(kx - \omega t)) = \omega\epsilon \cos(kx - \omega t)$$

$$\psi_y(x, \epsilon \sin(kx - \omega t)) = 0.$$

Since ϵ is presumed to be small, we expand the surface conditions in powers of ϵ as

$$\psi_x(x, 0) + \epsilon\psi_{xy}(x, 0)\sin(kx - \omega t) + \frac{\epsilon^2}{2}\psi_{xyy}(x, 0)\sin^2(kx - \omega t) + \dots$$
$$= \epsilon\omega\cos(kx - \omega t)$$

$$\psi_y(x, 0) + \epsilon\psi_{yy}(x, 0)\sin(kx - \omega t) + \frac{\epsilon^2}{2}\psi_{yyy}(x, 0)\sin^2(kx - \omega t) + \dots$$
$$= 0.$$

Now we suppose that ψ is of order ϵ and can be expanded in a power series in ϵ as

$$\psi = \epsilon\psi_1(x, y) + \epsilon^2\psi_2(x, y) + \epsilon^3\psi_3(x, y) + \dots$$

It follows that each function $\psi_i(x, y)$ satisfies the biharmonic equation and the boundary conditions become the hierarchy of equations

$$\psi_{1x}(x, 0) = \omega \cos(kx - \omega t), \quad \psi_{1y}(x, 0) = 0,$$

$$\psi_{2x}(x, 0) + \psi_{1xy}(x, 0)\sin(kx - \omega t) = 0,$$

$$\psi_{2y}(x, 0) + \psi_{1yy}(x, 0)\sin(kx - \omega t) = 0,$$

and so on.

We have not encountered the biharmonic equation in this text before. Its solution for this problem is easily found using the Fourier transform in x. If we look for a solution of the biharmonic equation of the form $\psi(x, y) = e^{ikx}\phi(y)$, then $\phi(y)$ must satisfy the fourth order differential equation $\phi'''' - 2k^2\phi'' + k^4\phi = 0$. Taking $\phi(y) = e^{\lambda y}$, we find the characteristic equation $(k^2 - \lambda^2)^2 = 0$, so that $\lambda = \pm k$ are double roots of the characteristic equation. The corresponding solutions of the biharmonic equation are $\psi(x, y) = e^{ikx-ky}$, ye^{ikx-ky}, e^{ikx+ky}, and ye^{ikx+ky}. We are looking for solutions of the biharmonic equation whose derivatives (the fluid velocities) are bounded. Thus, in the upper half plane we take

$$\psi_1(x, y) = \omega(y + 1/k)\sin(kx - \omega t)e^{-ky},$$

which satisfies the required boundary conditions for ψ_1.

Now that ψ_1 is known, we find that ψ_2 must satisfy the boundary condition

$$\psi_{2x}(x,0) = 0, \quad \psi_{2y}(x,0) = \frac{\omega k}{2}(1 - \cos 2(kx - \omega t)).$$

We determine that in the upper half plane

$$\psi_2(x,y) = \omega k y(1 - e^{-2ky}\cos 2(kx - \omega t))/2.$$

We now know the solution to order ϵ^2.

How fast is this object swimming? Observe that for y very large, the fluid has a net horizontal velocity $u_1 \approx \epsilon^2 \omega k/2$. Since we originally assumed we were in the coordinate system moving with the flagellum, this fluid velocity is the speed of swimming. In comparison, the speed of the waves on the surface of the flagellum is $v = \omega/k$, so that the ratio of the two speeds is $s/v = \epsilon^2 k^2/2$.

This calculation is good to order ϵ^2. To get more terms requires some tedious bookkeeping. However, this bookkeeping is easily done with RE-DUCE, and one finds that $s/v = \epsilon^2 k^2/2 + \epsilon^4 k^4/4 + O(\epsilon^6)$. (See problems 11.1.11,12).

11.2   Perturbation of Eigenvalues

The failures of the implicit function theorem are in some sense more important, and certainly more interesting, than its successes. A simple example of its failure is with polynomial equations in the neighborhood of double roots. For example, we cannot use the implicit function theorem to solve $x^2 + \epsilon = 0$ since the reduced problem $x^2 = 0$ has a double root at $x = 0$ and the linearized operator $L = \frac{\partial}{\partial x}(x^2)|_{x=0}$ is identically zero and therefore not invertible.

In all cases, the failure of the implicit function theorem indicates there is some important geometrical or physical property of the system that needs to be considered further. As an example, suppose we wish to find the eigenvalues and eigenvectors of a matrix $A + \epsilon B$ where ϵ is presumed small.

Why might we want to do this? Well, for example, we might be interested in knowing the effect of small errors on the vibrations of a crystal lattice used as a timing device in some very precise clock. If we consider as a model the linear system of masses and springs

$$m_j \frac{d^2 u_j}{dt^2} = k_j(u_{j-1} - u_j) + k_{j+1}(u_{j+1} - u_j)$$

and we have that $m_j = m + \epsilon\eta_j$ and $k_j = k + \epsilon\kappa_j$, we are led to look for eigenvalues λ and eigenvectors u which satisfy the equation

$$[(A_0 + \epsilon A_1) + \lambda M]u = 0.$$

Here the matrices A_0 and A_1 are tridiagonal with elements determined by κ_j, the matrix M is diagonal with elements m_j, and $\lambda = \omega^2$ is the square of the frequency of vibration. This equation can be rewritten in the form of $(A + \epsilon B + \lambda)x = 0$ but it is just as easy to solve the problem without this change of form.

Returning to the problem of finding eigenvalues of $A + \epsilon B$, we could write out the polynomial equation $\det(A + \epsilon B - \lambda I) = 0$ and solve this equation using the implicit function theorem. However, this is not a practical way to solve the problem and no sane person finds pleasure in calculating determinants of large matrices.

A direct approach is preferable. We look for solutions of $(A + \epsilon B)x = \lambda x$ and suppose that they can be expressed as power series,

$$\lambda = \sum_{j=0}^{\infty} \lambda_j \epsilon^j, \quad x = \sum_{j=0}^{\infty} x_j \epsilon^j.$$

We substitute these expressions into the eigenvalue equation and collect like powers of ϵ, to obtain

$$(A - \lambda_0)x_0 + \sum_{k=1}^{\infty} \left((A - \lambda_0)x_k + Bx_{k-1} - \sum_{j=1}^{k} \lambda_j x_{k-j} \right) \epsilon^k = 0.$$

It follows that we want to solve

$$(A - \lambda_0)x_0 = 0$$

and

$$(A - \lambda_0)x_k = -Bx_{k-1} + \sum_{j=1}^{k} \lambda_j x_{k-j}, \quad k = 1, 2 \ldots .$$

The first of these equations we solve by taking x_0, λ_0 to be an eigenpair of A. The next equation to solve is

$$(A - \lambda_0)x_1 = \lambda_1 x_0 - Bx_0$$

and immediately we see the difficulty. The operator $L = A - \lambda_0 I$, which occurs in every equation, is not invertible, so we cannot solve this equation. The implicit function theorem appears to have failed.

With a little more reflection, however, we realize that all is not lost. Since $A - \lambda_0 I$ has a null space (which we assume to be one dimensional) the Fredholm Alternative implies that solutions can be found if and only if the right hand side is orthogonal to the null space of the adjoint operator. If y_0 spans the null space of $A^* - \bar{\lambda}_0 I$, we must require

$$\lambda_1 \langle x_0, y_0 \rangle - \langle Bx_0, y_0 \rangle = 0.$$

Since $\langle x_0, y_0 \rangle \neq 0$, this is an equation that uniquely determines λ_1 as

$$\lambda_1 = \frac{\langle Bx_0, y_0 \rangle}{\langle x_0, y_0 \rangle}$$

and for this value of λ_1, the solution x_1 exists and can be made unique by the further requirement $\langle x_1, x_0 \rangle = 0$. (It is known that $\langle x_0, y_0 \rangle \neq 0$. Why?)

All of the equations for x_k are of the same form and can be solved in the same way. We determine λ_k from

$$\lambda_k \langle x_0, y_0 \rangle + \sum_{j=1}^{k-1} \lambda_j \langle x_{k-j}, y_0 \rangle - \langle Bx_{k-1}, y_0 \rangle = 0$$

and then x_k exists and is unique if in addition we require $\langle x_k, x_0 \rangle = 0$.

Since this procedure works without a hitch, it must be that the implicit function theorem did not fail, but rather that we tried to use it in an inappropriate way. Suppose we set $x = x_0 + u$, $\lambda = \lambda_0 + \mu$ where $\langle u, x_0 \rangle = 0$. If we denote by Q the linear operator that projects a vector f onto the vector y_0,

$$Qf = \frac{\langle f, y_0 \rangle \, y_0}{\|y_0\|^2},$$

and by $P = I - Q$ the linear operator that projects a vector onto the range of $A - \lambda_0 I$, then the problem $(A + \epsilon B - \lambda I)x = 0$ can be rewritten as the pair of equations

$$(A - \lambda_0 I)u - \mu P(x_0 + u) + \epsilon PB(x_0 + u) = 0,$$

$$\mu \langle x_0 + u, y_0 \rangle - \epsilon \langle B(x_0 + u), \, y_0 \rangle = 0.$$

We view the pair (u, μ) as the unknown vector and note that $u = 0$, $\mu = \epsilon = 0$ is a solution of the equation. Moreover, we can identify an invertible linear operator L and rewrite the equations in terms of L as

$$L \begin{pmatrix} u \\ \mu \end{pmatrix} \equiv \begin{pmatrix} (A - \lambda_0 I) & -Px_0 \\ 0 & \langle x_0, y_0 \rangle \end{pmatrix} \begin{pmatrix} u \\ \mu \end{pmatrix}$$

$$= \begin{pmatrix} \mu Pu - \epsilon PB(x_0 + u) \\ \epsilon \langle B(x_0 + u), y_0 \rangle - \mu \langle u, y_0 \rangle \end{pmatrix}.$$

In this formulation the operator L is uniquely and boundedly invertible, so that the implicit function theorem does indeed apply to this problem as it is reformulated. There is therefore a solution $u = u(\epsilon)$, $\mu = \mu(\epsilon)$ with $u(0) = 0$, $\mu(0) = 0$, and this is precisely the solution whose asymptotic expansion we were able to compute explicitly.

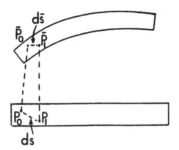

FIGURE 11.1   Deformation of a beam.

11.3   Nonlinear Eigenvalue Problems and Bifurcation Theory

The problem of the previous section was to find eigenvalues of one operator that is perturbed by a small linear operator. In this section we consider some of the other ways that eigenvalue problems may arise.

As a specific example we will consider the deformations of a thin elastic rod under end loading. To do so we need a short introduction into the theory of elastic deformations.

Suppose that a rod of length L, depth b, and height $2h$ in its undeformed state has a material point at position x, y, z, $0 \leq x \leq L$, $0 \leq y \leq b$, $-h \leq z \leq h$. If the rod is deformed, we suppose the material initially at point $P = (x, y, z)$ is moved to the new position $\bar{P} = (\bar{x}, \bar{y}, \bar{z})$ where $\bar{x} = x+u$, $\bar{y} = y$ and $\bar{z} = z + w$ (Figure 11.1). We calculate the change in length of a small arc due to this deformation by noting that

$$
\begin{pmatrix} d\bar{x} \\ d\bar{z} \end{pmatrix} = \begin{pmatrix} 1 + u_x & u_z \\ w_x & 1 + w_z \end{pmatrix} \begin{pmatrix} dx \\ dz \end{pmatrix} = (I + Q) \begin{pmatrix} dx \\ dz \end{pmatrix}
$$

so that

$$
d\bar{s}^2 = (dx \ dz)(I + Q^T)(I + Q) \begin{pmatrix} dx \\ dz \end{pmatrix}.
$$

To understand the physical meaning of $d\bar{s}^2$, imagine that a small rectangular box with sides dx and dz is deformed into a parallelepiped. The deformation consists of four types of action, namely, translation, rotation, stretching, and shearing. Translation and rotation have no effect on the shape of the object. A stretching changes the length of the sides of the rectangular box but keeps the corners square, and a shearing changes the angles at the corners but not the length of the sides. Any deformation of a small box is a product of rotations,

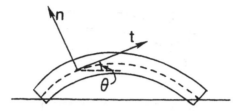

FIGURE 11.2 Tangent and normal to the midline of a thin rod after deformation.

stretching and shearing. To extract the effects of stretching and shearing we rewrite the matrix $(I + Q^T)(I + Q)$ as

$$(I + Q)^T(I + Q) = (I + N)^2$$

where N is called the **STRAIN MATRIX**. The elements of N, denoted ϵ_{ij}, are called strains. The values ϵ_{ii} (diagonal elements of N) are the normal strains, that is, the changes of length due to stretching in the coordinate directions and ϵ_{ij}, $i \neq j$ are shear strains, that is, deformations due to shearing.

To make these deformations more specific, we make some simplifying assumptions about the way the rod deforms. We suppose that the position of the midline of the bar ($z = 0$) after deformation can be represented by the position vector

$$R = \left(\begin{array}{c} X(x) \\ Z(x) \end{array} \right).$$

The position vector $R(x)$ has a unit tangent vector t and unit normal vector n, where $R' = |R'|t$, $t' = -|R'|\kappa n$, with κ being the curvature of the midline (Figure 11.2). Now we assume that the point x, z is moved by the deformation to the point $R(x) + zn$. In other words, we assume that points on a plane normal to the midline remain on a plane normal to the deformed midline after deformation, and hence, the point originally at x, z moves under deformation to a point $\bar{x}, \bar{z}$ which is z units along the normal from the deformed midline curve at coordinate position x.

Using this simplifying assumption, we can now directly calculate the strains for this rod. We find that

$$I + Q = (R' + zn', n)$$

and

$$(I + Q^T)(I + Q) = \left(\begin{array}{cc} (1 + \epsilon_{11})^2 & 0 \\ \\ 0 & 1 \end{array} \right)$$

The only nonzero strain is the normal strain $\epsilon_{11} = \epsilon - (1 + \epsilon)\kappa z$ where

$$\epsilon = (X'^2 + Z'^2)^{1/2} - 1$$

is the MIDLINE STRAIN and

$$\kappa = \frac{Z'X'' - X'Z''}{(X'^2 + Z'^2)^{3/2}}$$

is the curvature of the midline.

Once we know the strain produced by the deformation we calculate the energy required for that strain. Using Hooke's law, the total strain energy is

$$V = \frac{Eb}{2} \int_{-h}^{h} \int_0^L \epsilon_{11}^2 dx\, dz$$

$$= \frac{E}{2} \int_0^L (A\epsilon^2 + (1+\epsilon)^2 \kappa^2 I) dx$$

where $A = 2bh$ is the cross sectional area of the rod, $I = b \int_{-h}^{h} z^2 dz$ is its moment of inertia, and E is the Hooke's law constant (also called Young's modulus). Now, we set $X(x) = x + u(x)$ and $Z(x) = w(x)$ and make the approximation that the midline strain is not too large, that is,

$$\epsilon = \left((1+u_x)^2 + w_x^2\right)^{1/2} - 1 \sim u_x + \frac{w_x^2}{2}$$

and that the term $(1+\epsilon)\kappa$ is well represented by w_{xx} so that

$$V = \frac{E}{2} \int_0^L \left(A\left(u_x + \frac{w_x^2}{2}\right)^2 + I w_{xx}^2 \right) dx$$

is the total strain energy. Finally, we invoke Hamilton's principle, which in this case implies that the shape of the rod renders V a minimum. The Euler-Lagrange equations for this functional are

$$I w_{xxxx} - \frac{\partial}{\partial x} \left[A(u_x + w_x^2/2) w_x \right] = 0$$

$$\frac{\partial}{\partial x}(u_x + w_x^2/2) = 0,$$

subject to boundary conditions $w = w_{xx} = 0$ at $x = 0, L$ and $u(0) = \delta/2$, $u(L) = -\delta/2$. We can simplify this substantially by realizing that $\epsilon = u_x + w_x^2/2$ is a constant, so

$$L\epsilon = \int_0^L \epsilon\, dx = u\Big|_0^L + \int_0^L \frac{w_x^2}{2} dx$$

or

$$\epsilon = \frac{-\delta}{L} + \frac{1}{2L} \int_0^L w_x^2 dx$$

and we have the one equation for w

$$w_{xx} + \frac{A}{IL}\left(\delta - \frac{1}{2}\int_0^L w_x^2\,dx\right)w = 0.$$

It is quite helpful to introduce nondimensional variables. We set $x = Ls$, $y = \left(\frac{A}{I}\right)^{1/2}w$ and $\lambda = \frac{LA\delta}{I}$ and obtain the simplified equation

$$y'' + \left(\lambda - \frac{1}{2}\int_0^1 y'^2\,ds\right)y = 0$$

$$y(0) = y(1) = 0.$$

This equation is called the ELASTICA EQUATION or the EULER COLUMN EQUATION with endshortening proportional to λ.

Before we try to solve this equation we will derive a slightly different equation for the deformations of this beam without making the simplifying assumptions of small strains and curvature. We can give an exact formulation of the problem by letting θ be the angle made by the tangent to the midline with a horizontal line (Figure 11.2). Then

$$\sin\theta(x) \;=\; \frac{w_x}{1+\epsilon}$$

$$\cos\theta(x) \;=\; \frac{1+u_x}{1+\epsilon}$$

and the curvature is $\kappa = \frac{d\theta}{ds} = \frac{\theta_x}{1+\epsilon}$. Our goal is to represent the strain energy in terms of the variables $\theta(x)$ and $\epsilon(x)$,

$$V \;=\; \frac{E}{2}\int_0^L (A\epsilon^2 + I(1+\epsilon)^2\kappa^2)\,dx + P((u(L) - u(0))$$

$$=\; \int_0^L \left[\frac{EA}{2}\epsilon^2 + \frac{EI}{2}\theta_x^2 + P((1+\epsilon)\cos\theta - 1)\right]\,dx.$$

Here, the term $P(u(L)-u(0))$ represents the work done by the force P applied at the ends, and we use that

$$u(L) - u(0) = \int_0^L ((1+\epsilon)\cos\theta - 1)\,dx.$$

The condition $w(0) = w(L) = 0$ is equivalent to the restriction

$$\int_0^L (1+\epsilon)\sin\theta\,dx = 0.$$

Requiring the first variation of V to vanish, we find that

$$EA\epsilon + P\cos\theta = 0$$

$$\frac{EI}{2}\theta_{xx} + P(1+\epsilon)\sin\theta = 0$$

with

$$\theta_x = 0 \text{ at } x = 0, L.$$

We can eliminate ϵ to obtain one equation for θ,

$$EI\theta_{xx} + P\left(1 - \frac{P}{EA}\cos\theta\right)\sin\theta = 0$$

$$\theta_x(0) = \theta_x(L) = 0.$$

In the limit that θ is small so that $\cos\theta \sim 1 - \theta^2/2$ and $\sin\theta \sim \theta - \theta^3/6$, this equation reduces to DUFFING'S EQUATION

$$EI\theta_{xx} + P\left(\left(1 - \frac{P}{EA}\right)\theta + \left(\frac{4P}{EA} - 1\right)\theta^3/6\right) = 0.$$

$$\theta_x(0) = \theta_x(L) = 0.$$

Both of these model equations can be viewed as nonlinear eigenvalue problems. In the first Elastica equation we view the end shortening δ as an eigenvalue parameter and seek those values of δ for which there are nonzero solutions w. Similarly, for the last of these model equations we seek those values of P for which nontrivial solutions $\theta(x)$ exist.

The first elastica model is quite nice because it can be solved exactly. To solve

$$y'' + \left(\lambda - \frac{1}{2}\int_0^1 y'^2\,ds\right)y = 0, \quad y(0) = y(1) = 0,$$

we observe that $\int_0^1 y'^2\,ds$ is unknown but constant. If we set $\mu = \lambda - \frac{1}{2}\int_0^1 y'^2\,ds$, the only nontrivial solutions are

$$y(x) = a\sin n\pi x$$

$$\mu = n^2\pi^2.$$

For this solution to work, it must satisfy the consistency condition

$$\mu = n^2\pi^2 = \lambda - \frac{1}{2}\int_0^1 y'^2\,ds$$

or

$$\frac{a^2}{4} = \frac{\lambda}{n^2\pi^2} - 1.$$

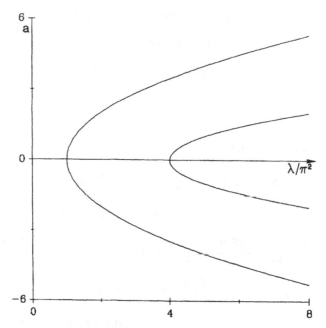

FIGURE 11.3 Bifurcation diagram for buckling of the elastica.

The consequences of this calculation are rather interesting. A plot of a against λ/π^2 is shown in Figure 11.3. We learn that for $\lambda < \pi^2$ the only solution is the trivial solution $y = 0$. At $\lambda = \pi^2$ the nontrivial solution $y = a \sin \pi x$ branches from the trivial solution and continues to exist for all $\lambda > \pi^2$. The point $\lambda = \pi^2$ is called a BIFURCATION POINT and the nontrivial solution is called a BIFURCATING SOLUTION BRANCH. There are also bifurcation points at $\lambda = n^2\pi^2$ from which the solution $y = a \sin n\pi x$ bifurcates.

Notice that this solution corresponds (crudely at least) to common experience. If you try to compress a flexible plastic ruler or rubber rod along its length, the rod compresses for small amounts of loading. As the loading increases there comes a point at which the rod "buckles" out of the straight configuration and larger loads increase the amplitude of the buckled solution. With a little work one can also observe one or two of the higher buckled states, but the lowest mode is the preferred mode, because it corresponds to the lowest energy state.

This model equation has the advantage that it can be solved exactly, so now we know the kinds of phenomena to expect, but we are at a serious disadvantage because we do not know (yet) how to uncover similar features in other problems without exact solutions.

To develop a more general approach, we reexamine the elastica model

$$y'' + \left(\lambda - \int_0^1 y'^2 \, ds\right) y = 0,$$

$$y(0) = y(1) = 0,$$

but we will not use the fact that we can solve it exactly. As a first guess, we might look for small solutions by ignoring nonlinear terms in y. We reduce the equation to the linear problem

$$y'' + \lambda y = 0,$$

$$y(0) = y(1) = 0.$$

The solution of this equation is $y(x) = A \sin n\pi x$ where A is arbitrary provided $\lambda = n^2 \pi^2$. This solution is physical nonsense. It tells us that except for very special values of end shortening λ, the only solution is zero. At precisely $\lambda = n^2 \pi^2$ there is a solution with shape $\sin n\pi x$ of arbitrary amplitude. In other words, according to this solution, if one loads a column, it will not buckle until $\lambda_1 = \pi^2$ at which point any amplitude is possible, but for λ slightly greater than λ_1, the column returns to its original unbuckled state, and this does not agree with our intuition.

The only way to get a result that is sensical is to use a nonlinear solution technique. We expect that for λ in the neighborhood of the bifurcation point, solutions exist but remain somewhat small. We are led to try a power series solution of the form

$$y \;=\; \epsilon y_1 + \epsilon^2 y_2 + \cdots$$

$$\lambda \;=\; \lambda_0 + \epsilon \lambda_1 + \epsilon^2 \lambda_2 + \cdots$$

where ϵ is some small parameter. We substitute this proposed solution into the elastic equation, collect like powers of ϵ and set each coefficient of ϵ to zero. The resulting hierarchy of equations begins like

$$y_1'' + \lambda_0 y_1 \;=\; 0$$

$$y_2'' + \lambda_0 y_2 + \lambda_1 y_1 \;=\; 0$$

$$y_3'' + \lambda_0 y_3 + \lambda_1 y_2 + \lambda_2 y_1 - \frac{1}{2} y_1 \int_0^1 y_1'^2 \, dx \;=\; 0.$$

The boundary conditions for all equations are $y_i(0) = y_i(1) = 0$.

The first equation of the hierarchy we already know how to solve. Its solution is $y_1 = A \sin n\pi x$ provided $\lambda_0 = n^2 \pi^2$. The second equation involves the linear operator $Ly = y'' + \lambda_0 y$ as do all the equations of the hierarchy.

This operator has a one dimensional null space and is invertible only on its range, which is orthogonal to its null space (the operator is self adjoint). We can find y_2 if and only if $\lambda_1 \langle y_1, y_1 \rangle = 0$. Since $y_1 \neq 0$, we take $\lambda_1 = 0$. We could take $y_2 \neq 0$ but to specify y_2 uniquely we require $\langle y_2, y_1 \rangle = 0$ for which $y_2 = 0$ is the unique solution.

The third equation of the hierarchy has a solution if and only if

$$\lambda_2 = \frac{1}{2} \int_0^1 y_1'^2 \, dx$$

$$= \frac{A^2 n^2 \pi^2}{4}$$

and then we find that $y_3 = 0$.

We summarize this information by

$$y = \epsilon A \sin n\pi x + \ldots$$

$$\lambda = n^2 \pi^2 (1 + \epsilon^2 A^2 / 4) + \ldots .$$

Interestingly, this solution is exact and needs no further correction, although we do not expect an exact solution in general. On the other hand, the technique works quite well in general and produces asymptotic expansions of the solution. Indeed, one can show that the power series solutions converge for $\epsilon \neq 0$ sufficiently small. In practice, however, a few terms of the series suffice to give as much information as is usually needed.

To get a picture of how bifurcation occurs in a more general setting, consider the nonlinear eigenvalue problem

$$Lu + \lambda f(u) = 0$$

subject to homogeneous separated boundary conditions. We suppose that L is a Sturm-Liouville operator, self adjoint with respect to weight function one, although these assumptions can be weakened considerably. We also suppose that $f(0) = 0$ and $f(u)$ has a polynomial representation

$$f(u) = a_1 u + a_2 u^2 + a_3 u^2 + \ldots ,$$

at least for small u, $a_1 \neq 0$.

We seek small nontrivial solutions of this problem, so we try a solution of the form

$$u = \epsilon u_1 + \epsilon^2 u_2 + \ldots$$

$$\lambda = \lambda_0 + \epsilon \lambda_1 + \epsilon^2 \lambda_2 + \ldots$$

which we substitute into the nonlinear equation. We collect like powers of ϵ, and equate each coefficient to zero. The resulting hierarchy of equations is

$$Lu_1 + \lambda_0 a_1 u_1 = 0$$

$$Lu_2 + \lambda_0 a_1 u_2 = -(a_1 \lambda_1 u_1 + \lambda_0 a_2 u_1^2)$$

$$Lu_3 + \lambda_0 a_1 u_3 = -(2a_2 \lambda_0 u_1 u_2 + a_1 \lambda_1 u_2 + a_3 \lambda_0 u_1^3 + a_2 \lambda_1 u_1^2 + a_1 \lambda_2 u_1)$$

and so on.

The first equation of this hierarchy we recognize as an eigenvalue problem. We suppose that the Sturm-Liouville operator L has simple eigenvalues μ_i and eigenfunctions ϕ_i satisfying $L\phi_i + \mu_i \phi_i = 0$. If we pick $u_1 = A\phi$, $\lambda_0 = \mu_i / a_1$ where $\phi = \phi_i$ for some particular i, we have solved the first equation.

The second equation is uniquely solvable if and only if $\langle u_2, \phi \rangle = 0$ and

$$\lambda_1 a_1 \langle \phi, A\phi \rangle + \lambda_0 a_2 \langle \phi, A^2 \phi^2 \rangle = 0.$$

In other words, we must pick

$$\lambda_1 = -\frac{\lambda_0 a_2 A \langle \phi, \phi^2 \rangle}{a_1 \langle \phi, \phi \rangle}$$

and then the solution u_2 can be obtained, and the process continues.

If, perchance $a_2 = 0$, we do not yet have any nontrivial information about the solution. We take $\lambda_1 = 0$, $u_2 = 0$ and try to solve the equation for u_3. This equation is uniquely solvable if we take $\langle u_3, \phi \rangle = 0$ and if we require

$$\lambda_2 a_1 \langle \phi, A\phi \rangle + a_3 \lambda_0 \langle \phi, A^3 \phi^3 \rangle = 0.$$

That is, we must take

$$\lambda_2 = \frac{-a_3 \lambda_0 A^2 \langle \phi, \phi^3 \rangle}{a_1 \langle \phi, \phi \rangle}$$

and then we can determine u_3.

We summarize by noting that when $a_2 \neq 0$ solutions are

$$u = \epsilon \phi + O(\epsilon^2)$$

$$\lambda = \lambda_0 + \epsilon \lambda_1 + O(\epsilon^2), \quad \lambda_1 = \frac{-\lambda_0 a_2 \langle \phi, \phi^2 \rangle}{a_1 \langle \phi, \phi \rangle}$$

(we can always set A=1) whereas if $a_2 = 0$ solutions are

$$u = \epsilon \phi + O(\epsilon^3)$$

$$\lambda = \lambda_0 + \epsilon^2 \lambda_2 + O(\epsilon^3)$$

$$\lambda_2 = \frac{-a_3 \lambda_0 \langle \phi, \phi^3 \rangle}{a_1 \langle \phi, \phi \rangle}.$$

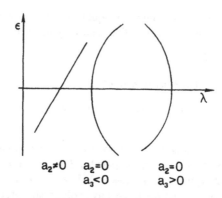

$$a_2 \neq 0 \qquad \begin{array}{c} a_2 = 0 \\ a_3 < 0 \end{array} \qquad \begin{array}{c} a_2 = 0 \\ a_3 > 0 \end{array}$$

FIGURE 11.4   Typical plots of bifurcating solutions.

The main geometrical difference (Figure 11.4) between these two solution branches is that if $a_2 \neq 0$, the bifurcating branch is locally linear, that is, there are nontrivial solutions for both $\lambda > \lambda_0$ and $\lambda < \lambda_0$. On the other hand, if $a_2 = 0$, the bifurcating branch is locally quadratic and the location of solutions depends on the sign of λ_2. If $\lambda_2 < 0$ there are two nontrivial solutions for $\lambda < \lambda_0$ and none for $\lambda > \lambda_0$ (locally, of course). If $\lambda_2 > 0$ the opposite occurs, namely there are two nontrivial solutions for $\lambda > \lambda_0$ and none for $\lambda < \lambda_0$. This situation ($a_2 = 0$) is called a PITCHFORK bifurcation. Higher order degeneracies can also occur and their representative solution structure can be determined (see Problem 11.3.5).

11.4   Oscillations and Periodic Solutions

Another problem related to the nonlinear eigenvalue problem is the calculation of periodic solutions of weakly nonlinear differential equations. In this section we will consider three specific examples.

11.4.1   Advance of the Perihelion of Mercury

In the first example we want to determine the relativistic effects on the orbit of planets. Using Newtonian physics and the inverse square law of gravitational attraction, one can show (Problem 11.4.1) that the motion of a satellite about a massive planet is governed by the equation

$$\frac{d^2 u}{d\theta^2} + u = a$$

where $a = GM/h^2$, M is the mass of the central planet, G is the gravitational constant $r = 1/u$ is the radius of the orbit, θ is the angular variable of the orbit and $r^2 \frac{d\theta}{dt} = h$ is the angular momentum per unit mass of the satellite. The solutions of this equation are given by

$$r = \frac{1}{a + A \sin \theta}$$

which is the equation in polar coordinates of an ellipse with eccentricity $e = A/a$. The period of rotation is

$$T = \frac{2\pi a}{h(a^2 - A^2)^{3/2}}.$$

Following Einstein's theory of general relativity, Schwarzschild determined that with the inclusion of relativistic effects the equation becomes

$$\frac{d^2 u}{d\theta^2} + u = a(1 + \epsilon u^2),$$

where $a\epsilon = \frac{3GM}{c^2}$, c^2 is the speed of light. (A derivation of this equation is given in Chapter 5.) We want to find periodic solutions of this equation for small ϵ. As a first attempt we try a solution of the form

$$u = a + A \sin \theta + \epsilon u_1 + \epsilon^2 u_2 + \cdots.$$

This guess incorporates knowledge of the solution for $\epsilon = 0$.

We substitute this proposed solution into the equation, collect like powers of ϵ, and find the hierarchy of equations

$$u_1'' + u_1 = a u_0^2$$

$$u_2'' + u_2 = a u_0 u_1$$

and so on, where $u_0 = a + A \sin \theta$. The first equation already shows the difficulty. The operator $Lu = u'' + u$ with periodic boundary conditions, has a two dimensional null space spanned by the two periodic functions $\sin \theta$ and $\cos \theta$. In order to find u_1, we must have $a u_0^2$ orthogonal to $\sin \theta$ and $\cos \theta$, which it is not. Therefore u_1 does not exist.

The difficulty is that, unlike the problems of the previous section, there is no free parameter to adjust in order that the orthogonality condition be satisfied. If there were a free parameter somewhere there might be a hope, so we need to look around for a free parameter.

After some reflection, the unknown parameter becomes apparent. Notice that we are looking for 2π-periodic solutions of this problem, but why, one might ask, should we expect the period to remain fixed at 2π as ϵ changes? Indeed, the logical choice for unknown parameter is the period of oscillation.

To incorporate the unknown period into the problem, we make the change of variables $x = \omega\theta$ and then seek fixed period 2π periodic solutions of the equation

$$\omega^2 u'' + u = a(1 + \epsilon u^2)$$

where ω is an unknown eigenvalue parameter.

Now the problem looks very much like a problem from the previous section so we try

$$u = u_0 + \epsilon u_1 + \epsilon^2 u_2 + \dots$$

$$\omega = 1 + \epsilon\omega_1 + \epsilon^2\omega_2 + \dots .$$

We substitute our guess into the governing equation, collect like powers of ϵ, and find the hierarchy of equations

$$u_0'' + u_0 = a$$

$$u_1'' + u_1 = au_0^2 - 2\omega_1 u_0''$$

$$u_2'' + u_2 = 2au_0 u_1 - 2\omega_2 u_0'' - \omega_1^2 u_0'' - 2\omega_1 u_1''$$

and so on.

We take $u_0 = a + A\sin x$, and then to solve for u_1, require that $au_0^2 - 2\omega_1 u_0''$ be orthogonal to both $\sin x$ and $\cos x$. We calculate that

$$au_0^2 - 2\omega_1 u_0'' = a(a^2 + 2aA\sin x + A^2\sin^2 x) + 2\omega_1 A\sin x$$

is orthogonal to $\sin x$ and $\cos x$ if and only if $\omega_1 = -a^2$. We can now write that

$$u_1 = a^3 + \frac{aA^2}{2} + \frac{aA^2}{6}\cos 2x,$$

which is uniquely specified by the requirement $\langle u_1, \sin x\rangle = \langle u_1, \cos x\rangle = 0$.

We could continue in this fashion generating more terms of the expansion. One can show that this procedure will work and in fact, converge to the periodic solution of the problem. For the physicist, however, one term already contains important information. We observe that the solution is

$$u = a + A\sin\left((a - a^2\epsilon + O(\epsilon)^2)\,\theta\right) + O(\epsilon).$$

The perihelion (the point most distant from the center) occurs when

$$(1 - a^2\epsilon)\theta = 2\pi n + \pi/2$$

so that two successive maxima are separated by

$$\Delta\theta = \frac{2\pi}{1 - a^2\epsilon}$$

which is greater than 2π. That is, on each orbit we expect the perihelion to advance by $\Delta\theta - 2\pi = \frac{2\pi a^2 \epsilon}{1-a^2\epsilon}$ radians. In terms of measurable physical quantities this is

$$\Delta\phi - 2\pi = 2\pi a^2 \epsilon = \frac{6\pi}{1-e^2}\frac{1}{c^2}\left(\frac{2\pi GM}{T}\right)^{2/3}$$

The perihelion of Mercury is known to advance 570 seconds of arc per century. Of this amount all but 43 seconds of arc can be accounted for as due to the effects of other planets. To see how well the above formula accounts for the remaining 43 seconds of arc, we must do a calculation. Using that the eccentricity of the orbit of Mercury is .205, the period of Mercury is 87.97 days, the mass of the sun is 1.990×10^{30} kg, the gravitational constant G is 6.67×10^{-11} Nm^2/kg and the speed of light c is 3.0×10^8 m/sec, we find that $\Delta\phi - 2\pi = 50.11 \times 10^{-8}$ radians/period. This converts to 42.9 seconds of arc per century, which is considered by physicists to be in excellent agreement with observation.

11.4.2 Van der Pol Oscillator

For a second example, we determine periodic orbits of the van der Pol equation. In the 1930's, the Dutch engineers van der Pol and van der Mark invented an electrical circuit exhibiting periodic oscillations that they proposed as a model for the pacemaker of the heart. Today we know that this model is far too crude to model the heartbeat in a reliably quantitative way, but in certain parameter regions it has some qualitative similarities with the cardiac pacemaker (the sinoatrial node).

The van der Pol oscillator is usually described as being the electrical circuit shown in Figure 11.5. It is a device with three parallel circuits, one a capacitor, the second a resistor, inductor and voltage source in series, and the last some nonlinear device, such as a tunnel diode. Kirkhoff's laws for this circuit give that

$$C\frac{dv}{dt} = i_1$$

$$Ri_2 + L\frac{di_2}{dt} = v + v_0$$

$$i_3 = f(v)$$

$$I_0 = i_1 + i_2 + i_3$$

where v is the voltage drop across the circuit, I_0 is the current input into the circuit, i_1, i_2, and i_3 are the currents in the three branches of the circuit, and $f(v)$ is the current-voltage response function of the nonlinear device.

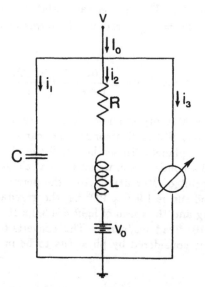

FIGURE 11.5   Circuit diagram for van der Pol equation.

We can reduce these equations to the two equations

$$C\frac{dv}{dt} = I_0 - i_2 - f(v)$$

$$L\frac{di_2}{dt} = v + v_0 - Ri_2$$

and, if $R = 0$ we have

$$C\frac{d^2v}{dt^2} + f'(v)\frac{dv}{dt} + \frac{v + v_0}{L} = \frac{dI_0}{dt}.$$

The important property of the tunnel diode is that it has a nonlinear current-voltage response function with a negative resistance region. A reasonable model is to take $f(v)$ to be the cubic polynomial

$$f(v) = Av\left(\frac{v^2}{3} - (v_1 + v_2)v/2 + v_1v_2\right).$$

We tune the voltage source so that $v_0 = -(v_1 + v_2)/2$, set

$$v = \left(\frac{v_2 - v_1}{2}\right)u + \frac{v_1 + v_2}{2}$$

and scale time so that $t = \sqrt{LC}\tau$. The resulting equation is van der Pol's equation

$$u'' + \epsilon u'(u^2 - 1) + u = 0 \quad \text{with} \quad \frac{dI_0}{dt} = 0.$$

where ϵ is the dimensionless parameter

$$\epsilon = A\left(\frac{v_2 - v_1}{2}\right)^2 \sqrt{\frac{L}{C}}.$$

The van der Pol circuit is not hard to actually build, but not with a tunnel diode. Tunnel diodes are obsolete and difficult to work with. A simpler device, using operational amplifiers (op-amps) is described in the problems at the end of the next chapter, and is easier to tinker with.

In this chapter we seek solutions of the van der Pol equation when the parameter $\epsilon > 0$ is small. The region ϵ very large is of more physiological interest, but requires significantly different techniques, which we defer until Chapter 12.

With our experience from the previous problem, we know to let the period of oscillation be a free parameter, and make the change of variables

$$x = \omega\tau$$

after which the equation becomes

$$\omega^2 \frac{d^2u}{dx^2} - \epsilon\omega\frac{du}{dx}(1 - u^2) + u = 0.$$

We expect solutions to depend on ϵ, and furthermore, we know how to solve the reduced problem $\epsilon = 0$. We try a power series solution of the form

$$u = u_0 + \epsilon u_1 + \epsilon^2 u_2 + \ldots$$

$$\omega = 1 + \epsilon\omega_1 + \epsilon^2\omega_2 + \ldots$$

and as usual, we find the hierarchy of equations

$$u_0'' + u_0 = 0$$

$$u_1'' + u_1 = -u_0'(u_0^2 - 1) - 2\omega_1 u_0''$$

$$u_2'' + u_2 = -\omega_1^2 u_0'' - \omega_1 u_0'(u_0^2 - 1) - 2\omega_1 u_1''$$

$$\qquad\qquad - (u_0^2 - 1)u_1' - 2u_0 u_0' u_1 - 2\omega_2 u_0''.$$

and so on.

The solution of the equation $u_0'' + u_0 = 0$ is

$$u_0 = A\cos(x + \phi).$$

Since the van der Pol equation is autonomous, the phase shift ϕ is arbitrary and it is convenient to take it to be zero. Actually, the calculations that we are about to do are much easier if we use complex exponentials rather than trigonometric functions. If we use trigonometric functions, we will be required to remember (or look up) a number of trigonometric identities, but with complex exponentials, the identities drop out on their own. Thus we take

$$u_0 = A_0(e^{ix} + e^{-ix}) = 2A_0 \cos x,$$

and find that u_1 must satisfy

$$u_1'' + u_1 = -A_0^3 i e^{3ix} - (A_0^3 i - A_0 i + 2A_0 \omega_1)e^{ix} + cc.$$

By cc we mean the complex conjugate of the preceding complex functions.

It is an easy matter to see what to do. The operator $Lu = u'' + u$ has a two dimensional null space spanned by $\sin x$ and $\cos x$ and in order to find u_1, the right hand side of the equation for u_1 must be orthogonal to these basis functions. One need not evaluate any integrals but simply observe that this occurs if and only if the coefficients of e^{ix} and e^{-ix} are identically zero. Thus, we require

$$A_0^3 - A_0 \;=\; 0$$

and

$$A_0 \omega_1 \;=\; 0.$$

We are surely not interested in the trivial solution $A_0 = 0$, so we take $A_0 = 1$ and $\omega_1 = 0$. It follows that

$$u_1 = \frac{i}{8}e^{3ix} - \frac{i}{8}e^{-3ix} + 2A_1 \cos x.$$

In general, it would be a mistake to take $A_1 = 0$. The operator $Lu = u'' + u$ has a two dimensional null space so at each step of the procedure we will need two free parameters to meet the solvability requirements. Indeed, to find u_1 we needed the two parameters A_0 and ω_1. The use of A_1 is equivalent to a restatement of our original assumption, namely that

$$u = A(\epsilon) \cos x + \epsilon u_1 + \epsilon^2 u_2 + \dots$$

where $A(\epsilon)$ has a power series representation. (Why were we able to get by with only one free parameter in the perihelion problem?)

Rather than writing down the full equation for u_2, we note that the right hand side has terms proportional to e^{ix} which must be set to zero. The coefficient of e^{ix} is $2A_1 i + 2\omega_2 + 1/8$ so that we take $A_1 = 0$, $\omega_2 = -\frac{1}{16}$.

We could continue calculating more terms but with little benefit. What we know now is that there is a periodic solution of the van der Pol equation of the form

$$u = 2\cos\omega\tau - \frac{\epsilon}{4}\sin 3\omega\tau + O(\epsilon^2)$$

where

$$\omega = 1 - \epsilon^2/16 + O(\epsilon^3).$$

In other words, as ϵ changes, the circular orbit becomes slightly noncircular with a slight decrease of frequency. The radius of the unique limit cycle is about 2 for ϵ small.

11.4.3 Knotted Vortex Filaments

The final example of this section is to determine the motion of a knotted vortex filament. A more prosaic way of describing this problem is to say we are going to find thin smoke rings which are trefoil knots. Exactly how to blow a knotted smoke ring is not known. To be sure, this example is a bit more complicated than the previous examples, but in spirit it is identical to the calculation of planetary orbits.

To begin this discussion we must know how a vortex filament moves in space. Vorticity is defined to be the curl of the velocity vector field of a fluid, and we suppose that the vorticity is confined to be nonzero only along a thin filament in space. Then the position vector R of the filament has a velocity given by $R_t = \kappa B$, where κ is the curvature, and B is the binormal vector of the filament. This equation is derived in the book by Batchelor listed in the references at the end of this chapter. The derivation actually shows that an infinitely thin vortex filament with nonzero curvature moves infinitely fast, but a change of time scale is made to make velocities finite and the equation $R_t = \kappa B$ results. How this equation describing the self induced motion of a vortex filament relates to real fluids is not completely resolved, but we shall use the equation of motion anyway.

We wish to find invariant solutions of the equation of motion $R_t = \kappa B$. By invariant solutions we mean solutions of the equation that move without change of shape. To do so we take the space-curve R to be represented by

$$R(s,t) = \rho(s,t)\mathbf{r}(\theta(s,t) + \Omega t) + (z(s,t) + Vt)\mathbf{k},$$

where $\mathbf{k}$ is a unit vector in the z direction, $\mathbf{r}$ is a unit vector in the $x-y$ plane, orthogonal to the z direction, and the argument of $\mathbf{r}$ denotes the angle in the $x-y$ plane. Thus, $\mathbf{r}'$ is a unit vector orthogonal to $\mathbf{r}$, and the vectors $\mathbf{r}, \mathbf{r}', \mathbf{k}$ form a right handed coordinate system. Here s is the arclength coordinate, so we must have

$$\rho_s^2 + \rho^2\theta_s^2 + z_s^2 = 1.$$

In this representation, the curve R is rotating at uniform rate Ω about the z axis, and is moving with uniform speed V in the z axis direction. Now we observe that $\kappa B = R_s \times R_{ss}$ (this follows from the Frenet-Serret equations), so that $R_t \cdot R_s = 0$, and $R_t \times R_s = R_{ss}$. In addition, we make the assumption that the functions ρ, θ and z are functions of the single variable $\zeta = s - ct$. This change of variables allows us to look for steadily progressing but invariant objects. It is an easy calculation to determine from $R_t \cdot R_s = 0$ that

$$-c + \rho^2 \theta' \Omega + V z' = 0$$

and from the k component of the equation $R_t \times R_s = R_{ss}$, we learn that

$$z'' = -\Omega \rho \rho'.$$

Finally, the arclength restriction becomes

$$\rho'^2 + \rho^2 \theta'^2 + z'^2 = 1.$$

It is this system of three equations that we wish to solve. Notice that the equation $z'' = -\Omega \rho \rho'$ can be integrated once to $z' = \Omega(A - \rho^2)/2$, where A is an unknown constant.

Some simple solutions of these equations have z', θ', and ρ constant. If $z' = 0$, we have a circle, and if z' is not zero we have a helix. It is an interesting exercise to find soliton solutions of these three equations. In this context a soliton solution corresponds to one wrap of a helix on an infinitely long filament. We will leave this as an exercise for the interested reader.

Our stated goal for this problem is to find solutions of these equations which are closed knotted curves. To do so we need to understand a few features about knotted curves. Probably the easiest way to construct a knot is to wrap a string around a donut. This may not be the way knots are tied by boyscouts, but for mathematicians it works well. A donut, or torus, can be described as the cross product of two circles of radius r_1 and r_2, with $r_1 > r_2$ and corresponding angular variables ϕ and ψ. A curve drawn on the surface of a torus has a winding number defined as follows:

Definition

The WINDING NUMBER for a curve on the surface of a torus, denoted as $\phi = \phi(t)$, $\psi = \psi(t)$, is the number $W = \lim_{t \to \infty} \psi(t)/\phi(t)$. The winding number exists and is unique if the curve is not self-intersecting, and is an invariant for the curve. That is, if the curve is deformed without self intersections into another curve, the winding number is unchanged.

The important fact about knots that we will use is

Theorem 11.2

A closed non-self-intersecting curve on the surface of a torus having winding number m/n, with m and n relatively prime, $m > n$, is a knot.

A trefoil knot is one with winding number $3/2$, that is three wraps around the small radius of the torus for two wraps around the large radius of the torus.

Now the bifurcation picture can be explored. How might we construct a trefoil knot? The answer is found by examining the opposite question of how a knot might be turned into some simpler structure. Notice that if the small radius of a torus is allowed to approach zero the torus becomes a circle and the trefoil knot becomes a double cover of the circle (twisted around itself). To undo this collapse we should look for a curve which oscillates with very small amplitude three times as a circle is traversed twice. A knot with winding number m/n will oscillate m times as the circle is traversed n times.

We examine the governing equations, and note that there is, of course, the trivial solution with constant radius and $z' = 0$, corresponding to a circle. For this solution we must also have $V = 1/\rho_0$, and $c = \rho_0 \Omega$, where the constant Ω is arbitrary. The three governing equations can be solved for z' and θ' in terms of ρ, and the one equation for ρ which results is of the form

$$\rho'^2 + f(\rho) = 0.$$

The function $f(\rho)$ is a big mess, but for ρ close to ρ_0, it has the form

$$f(\rho) = \left(\Omega^2 \rho_0^2 + \frac{1}{\rho_0^2} \right) (\rho - \rho_0)^2 + \cdots$$

which is quadratic to leading order in $\rho - \rho_0$. This equation is the first integral of the equation

$$\rho'' + f'(\rho) = 0,$$

which, since $f'(\rho)$ is locally linear with positive slope in a small neighborhood of ρ_0, has oscillatory solutions of the form $\rho = \rho_0 + a \sin \lambda \zeta$, where $\lambda^2 = \Omega^2 \rho_0^2 + 1/\rho_0^2$.

Now we see what to do. An n-cover of the circle with radius ρ_0 has total arclength $S = 2n\pi\rho_0$. If we pick $\Omega = \Omega_0$ with $\Omega_0^2 \rho_0^4 + 1 = m^2/n^2$, $m > n$, the solution of the linearized equation is

$$\rho = \rho_0 + a \sin \left(\frac{m}{n\rho_0} \zeta \right)$$

and both $\rho - \rho_0$ and z oscillate m times as the circle is traversed n times. If this curve is closed, it will be a knot.

We are finally in a position to find knotted solutions. We treat Ω as a parameter which is close to Ω_0, and look for small amplitude periodic solutions of the governing equations which correspond to closed knotted curves in space. To be assured that a space curve is closed we must require that ρ and z are periodic and that θ increases by $2n\pi$ per period. We seek power series solutions in the form

$$\rho(x) = \rho_0 + \epsilon\rho_1(x) + \epsilon^2\rho_2(x) + \cdots,$$

$$z(x) = \epsilon z_1(x) + \epsilon z_2(x) + \cdots,$$

$$\theta(x) = \frac{x}{\rho_0} + \epsilon\theta_1(x) + \epsilon\theta_2(x) + \cdots,$$

where $x = \omega\zeta$, $\omega = 1 + \epsilon\omega_1 + \epsilon\omega_2 + \cdots$. The constants Ω, V and A are also expanded in power series in ϵ, with $\Omega = \Omega_0 + \epsilon\Omega_1 + \epsilon^2\Omega_2 + \cdots$, where Ω_0 is chosen so that $\Omega_0^2\rho_0^4 + 1 = m^2/n^2$. Since ρ_0 is arbitrary, we also require $\rho - \rho_0$ to have zero average value.

The calculation that results is straightforward to describe, but it sure is helpful to use REDUCE (which is how this was done). We expand the governing equations in powers of epsilon, and collect the hierarchy of equations to be solved sequentially. We choose unknown parameters so as to guarantee that at each level, the solution is periodic. Notice that there are five unknown parameters which must be chosen to satisfy five constraints (what are the five constraints?). With endurance we can calculate as many terms as we want, with the result that

$$\rho = \rho_0 + \epsilon\cos\lambda x + \frac{\epsilon^2}{4}\frac{\Omega_0^2\rho_0^4 - 1}{\Omega_0^2\rho_0^4 + 1}\cos 2\lambda x + O(\epsilon^3),$$

$$\omega z' = -\epsilon\Omega_0\rho_0\cos\lambda x - \frac{\epsilon^2}{2}\frac{\Omega_0^3\rho_0^4}{\Omega_0^2\rho_0^4 + 1}\cos 2\lambda x + O(\epsilon^3),$$

$$\theta' = \frac{1}{\rho_0} - \frac{\epsilon}{\rho_0^2}\cos\lambda x + \frac{\epsilon^2}{2}\frac{\Omega_0^2\rho_0^4 + 2}{\rho_0^3(\Omega_0^2\rho_0^4 + 1)}\cos 2\lambda x + O(\epsilon^3),$$

$$\omega(\epsilon) = 1 + \frac{\epsilon^2}{4\rho_0^2}(1 - 2\Omega_0^2\rho_0^4) + O(\epsilon^4),$$

$$\Omega(\epsilon) = \Omega_0 + \frac{\epsilon^2\Omega_0}{4\rho_0^2}\frac{2 - 3\Omega_0^2\rho_0^4 - 2\Omega_0^4\rho_0^8}{\Omega_0^2\rho_0^4 + 1} + O(\epsilon^4),$$

$$V(\epsilon) = \frac{1}{\rho_0} + \frac{\epsilon^2}{4\rho_0^3}(1 - 4\Omega_0^2\rho_0^4) + O(\epsilon^4),$$

$$c(\epsilon) = \Omega_0\rho_0 + \frac{\epsilon^2\Omega_0}{4\rho_0}\frac{1 - 6\Omega_0^2\rho_0^4 - 4\Omega_0^4\rho_0^8}{\Omega_0^2\rho_0^4 + 1} + O(\epsilon^4),$$

where $\rho_0^2\lambda^2 = \Omega_0^2\rho_0^4 + 1 = m^2/n^2$. With this representation for the solution it is not difficult to verify that the curve R can be viewed as being wrapped on the surface of a torus with winding number m/n. However, the cross-section of the torus is not perfectly circular, since z and ρ oscillate with different amplitudes. If we could actually blow such a smoke ring, it would be observed to move along the z axis with speed V as it rotates about the z axis (its axis of symmetry) much like a propeller.

11.5   Hopf Bifurcations

Periodic orbits do not just appear out of nowhere. Much like the bifurcation of steady solutions of differential and algebraic equations, periodic orbits can be created or destroyed only by very special events. One of those events is the creation of a small periodic orbit out of a steady solution of a differential equation. This event, called a HOPF BIFURCATION, can be studied via perturbation methods.

Consider as an example the equation

$$\frac{d^2 u}{dt^2} + \frac{du}{dt}(u^2 - \lambda) + u = 0,$$

which is a variant of the van der Pol equation, and has as its only steady solution $u = 0$. The question we address is what solutions, if any, bifurcate from the trivial steady state as the parameter λ is varied.

We get some insight into the answer by looking at the equation linearized about $u = 0$

$$\frac{d^2 u}{dt^2} - \lambda u' + u = 0.$$

Solutions of this linear equation are easily found. They are $u = e^{\mu_i t}$, $i = 1, 2$ where μ_i satisfies the algebraic equation $\mu^2 - \lambda \mu + 1 = 0$. For $\lambda < 0$, both roots have negative real part and for $\lambda > 0$ they have positive real part. Thus, for $\lambda < 0$, the linear equation is stable while for $\lambda > 0$ it is unstable. At $\lambda = 0$ the roots are purely imaginary, $\mu = \pm i$ and the equation has periodic solutions $u = e^{\pm it}$.

The value $\lambda = 0$ is the only point at which the differential equation has periodic solutions. In light of our experience with bifurcations in Section 11.3, it is logical to ask if there are periodic solutions of the nonlinear equation which are created at $\lambda = 0$ and exist for other nearby values of λ. That is, we want to determine if $\lambda = 0$ is a bifurcation point for periodic solutions.

By now the idea should be more or less clear. We expect solutions near $\lambda = 0$, if they exist, to be small. Since we do not know the period the solutions will have, we make the change of variable $x = \omega t$ so that the equation becomes

$$\omega^2 u'' + \omega u'(u^2 - \lambda) + u = 0$$

and try a solution of the form

$$u = \epsilon u_1 + \epsilon^2 u_2 + \epsilon^3 u_3 + \ldots$$

$$\lambda = \epsilon \lambda_1 + \epsilon^2 \lambda_2 + \epsilon^3 \lambda_3 + \ldots$$

$$\omega = 1 + \epsilon \omega_1 + \epsilon^2 \omega_2 + \ldots$$

and seek solutions of fixed period 2π in x. We obtain the hierarchy of equations

$$u_1'' + u_1 = 0$$

$$u_2'' + u_2 = u_1' \lambda_1 - 2\omega_1 u_1''$$

$$u_3'' + u_3 = u_1' \lambda_2 + u_1' \lambda_1 \omega_1 + u_2' \lambda_1 - u_1' u_1^2$$

$$- 2u_2'' \omega_1 - u_1''(\omega_1^2 + 2\omega_2)$$

and so on.

We solve this system of equations sequentially. We take $u_1 = A(e^{ix} + e^{-ix})$ (again, it is easier to work with complex exponentials), but then to satisfy the equation for u_2 we must take both $\lambda_1 = 0$ and $\omega_1 = 0$, since the right hand side must be orthogonal to the two dimensional null space spanned by e^{ix} and e^{-ix}. With $\lambda_1 = \omega_1 = 0$, and $u_2 = 0$, the equation for u_3 becomes

$$u_3'' + u_3 = u_1' \lambda_2 - 2u_1'' \omega_2 - u_1' u_1^2$$

$$= -A\left(A^2 i e^{3ix} + (A^2 i - \lambda_2 i - 2\omega_2)e^{ix} + cc\right).$$

In other words, to find u_3 it is necessary and sufficient that $A^2 = \lambda_2$, $\omega_2 = 0$.

The upshot of this calculation is that there is a solution of the equation of the form

$$u = \epsilon \cos \omega t + O(\epsilon^3)$$

$$\lambda = \frac{\epsilon^2}{4} + O(\epsilon^3)$$

$$\omega = 1 + O(\epsilon^3).$$

There is little need or value in further calculations. We learn that as λ passes from $\lambda < 0$ to $\lambda > 0$ there is a periodic solution which bifurcates from the constant solution $u = 0$ and exists, locally at least, for $\lambda > 0$.

This calculation illustrates the HOPF BIFURCATION THEOREM, which we state without proof.

Theorem 11.3

Suppose the $n \times n$ matrix $A(\lambda)$ has eigenvalues $\mu_j = \mu_j(\lambda)$ and that for $\lambda = \lambda_0$, $\mu_1(\lambda_0) = i\beta$, $\mu_2(\lambda_0) = -i\beta$ and $\mathrm{Re}\mu_j(\lambda_0) \neq 0$ for $j > 2$. Suppose furthermore that

$$\mathrm{Re}\mu_1'(\lambda_0) \neq 0.$$

Then the system of differential equations

$$\frac{du}{dt} = A(\lambda)u + f(u)$$

with $f(0) = 0$, $f(u)$ a smooth function of u, has a branch (continuum) of periodic solutions emanating from $u = 0$, $\lambda = \lambda_0$.

The direction of bifurcation is not determined by the Hopf Bifurcation Theorem, but must be calculated by a local expansion technique.

11.6 REDUCE

Many of the calculations suggested in this chapter are quite routine but tedious, making them ideal candidates for REDUCE. The following sequence of commands will calculate the solution of the van der Pol equation given in Section 11.4. Follow along on your local terminal, or if you are using another language for symbolic manipulation, adapt this procedure into that language.

We use an operator U where $U(x, k)$ represents the kth term of the Taylor series expansion of $u(x, \epsilon)$ in powers expansion of ϵ.

```
REDUCE
N=3;
```

Of course you can take N larger if you want.

```
OPERATOR U,V,W;
FOR ALL X LET V(X):=FOR K:=0:N SUM U(X,K)*EP**K;
OM:=FOR K:=1:N SUM W(K)*EP**K;
OM:=OM+1;
RHS:=(OM**2-1)*DF(V(X),X,2)-EP*OM*DF(V(X),X)*(1-V(X)**2);
COEFF(RHS,EP,WW);
```

The terms WWJ represent the hierarchy of right hand sides of equations to be solved sequentially.

```
U(X,0):=A*E**(I*X)+A*E**(-I*X);
```

The first orthogonality condition is found by typing

```
COEFF(NUM(WW1),E**(I*X),WA);
WA4;
```

We must require $WA4 = 0$. The term $WA4$ is appropriate because RE-DUCE views e^{-ix} as $1/e^{ix}$ and expresses everything with the common denominator e^{3ix}. Thus, the term in the numerator of WW1 which is multiplied by e^{4ix} corresponds to the coefficient of e^{ix} which must be set to zero for solvability.

```
A:=1;
W(1):=0;
U(X,1):=-I*(E**(3*I*X)-E**(-3*I*X))/8+A1*U(X,0);
```

The expression $u_1'' + u_1 + WW1$ should now be identically zero.

```
COEFF(NUM(WW2),E**(I*X),WB);
WB6;
```

Set WB6 to zero to satisfy the second orthogonality condition. One can now go on with this calculation to get more terms, but at this point we have produced the same information as in Section 11.4, with the advantage that the computer did the dirty work.

Further Reading

An introduction to bifurcation theory can be found in the very nice survey article

1. I. Stakgold, "Branching of solutions of nonlinear equations", *SIAM Rev.* 13, 289 (1971).

Another introduction to the theory and a collection of problems is given in

2. J.B. Keller and S. Antman, *Bifurcation Theory and Nonlinear Eigenvalue Problems*, Benjamin, New York, 1969.

Two recent books on the topic, the second more abstract than the first, are

3. G. Iooss and D. Joseph, *Elementary Stability and Bifurcation Theory*, Springer-Verlag, New York, 1981,

4. S.N. Chow and J.K. Hale, *Methods of Bifurcation Theory*, Springer-Verlag, New York, 1982.

The implicit function theorem and its proof can be found in many analysis books. The version used here can be found in

5. Y. Choquet-Bruhat and C. DeWitt-Moretti, *Analysis, Manifolds and Physics*, North-Holland, Amsterdam, 1982.

A derivation of the elastica equation and a good introduction to elasticity theory is found in

6. J.J. Stoker, *Nonlinear Elasticity*, Gordon and Breach, New York, 1968.

An excellent text on electronic circuitry is

7. P. Horowitz and W. Hill, *The Art of Electronics*, Cambridge Press, Cambridge, MA, 1980,

and a van der Pol circuit that is easily built with inexpensive and readily available components is described in

8. J.P. Keener, Analog Circuitry for the van der Pol and FitzHugh-Nagumo Equations, IEEE Trans on Sys., Man, Cyber., *SMG-13*, 1010-1014, 1983.

The discussion of knotted vortex filaments takes as its starting point the equation of self-induced motion of a vortex filament, derived in the book

9. G.K. Batchelor, *An Introduction to Fluid Dynamics*, Cambridge University Press, Cambridge, 1967.

Exact solutions for this equation can be expressed in terms of Jacobi elliptic functions and are described in

10. S. Kida, "A vortex filament moving without change of form," J. Fluid Mech., 112, 397–409 (1981).

Solitons and the use of inverse scattering theory for these vortex filaments are discussed in the book

11. G.L. Lamb,*Elements of Soliton Theory*, Wiley, New York 1980.

A discussion of the Frenet-Serret equations can be found in books on differential geometry such as

12. J.J. Stoker, *Differential Geometry*, Wiley-Interscience, New York, 1969,

and discussion of knots and their relationship with tori can be found in

13. W.S. Massey, *Algebraic Topology: an Introduction*, Harcourt, Brace, and World, New York, 1967.

Finally, to see how the perturbation calculations of this chapter are done using MACSYMA (instead of REDUCE), refer to

14. R. H. Rand and D. Armbruster, *Perturbation Methods, Bifurcation Theory and Computer Algebra*, Applied Mathematical Sciences 65, Springer-Verlag, New York, 1987.

Problems for Chapter 11

Section 11.1

1. Approximate all roots of the equation

$$x^3 + x^2 + x + .01 = 0.$$

2. Approximate the root of $x^3 - x^2 + .01 = 0$ near $x = 1$. What difficulty is encountered for the roots near $x = 0$?

3. Approximate all roots of $\epsilon x^3 + x - 1 = 0$.
   Hint: To find large roots set $x = 1/y$ and find y as a function of ϵ.

4. Find the velocity potential and stream functions for a uniform plane flow (inviscid, incompressible) past a corrogated cylinder with radius $r(\epsilon) = a(1 + \epsilon \cos 20\theta)$.

5. Find the velocity potential and stream function for the flow past the infinite corrogated sheet at $y(x) = \epsilon \cos x$, which for large y is parallel to the x axis.

6. Find the leading order and order ϵ correction for the solution of $\nabla^2 u + \epsilon x u = 0$ on the square $0 \leq x,\ y \leq 1$ subject to boundary conditions $u_x(0, y) = u_x(1, y) = 0$, $u(x, 0) = a$, $u(1, a) = b$.

7. Suppose the eigenvalues μ of the matrix A are all such that $\operatorname{Re}\mu \neq 0$. Use the implicit function theorem to prove that the differential equation

$$\frac{du}{dt} = Au + \epsilon P(t) + \epsilon f(u)$$

   where $P(t + T) = P(t)$ $(i.e. P(t)$ is a period T periodic function) and $f(0) = 0$, has period T solutions.

8. Find the solutions of

$$\nabla^2 u + \epsilon u^2 \ = 0, \qquad 0 \leq r < 1,$$

$$u(1, \theta) \ = \cos \theta, \quad 0 \leq \theta \leq 2\pi.$$

9. Solve Laplace's equation on the interior of the region $r(\theta) = 1 + \epsilon \sin \theta$ subject to boundary condition $u(r(\theta), \theta) = \cos \theta$, $0 \leq \theta \leq 2\pi$.

10. Use the contraction mapping principle to prove the implicit function theorem for twice continuously differentiable functions $f(x, \epsilon)$.
    Remark: To solve $f(u) = 0$ with $f(0)$ small, rewrite $f(u) = 0$ as

$$u = f^{-1}(0)(f(u) - f'(0)u) \equiv g(u).$$

Apply the identity

$$f(u) - f(v) = (u - v) \int_0^1 f'(su + (1 - s)v) \, ds$$

to show that the map $u_{n+1} = g(u_n)$ is a contraction map, provided $f(0)$ is sufficiently small. Interpret the iteration scheme $u_{n+1} = g(u_n)$ in the case that $f(u)$ is a function $f : \mathbb{R} \to \mathbb{R}$.

11. Find the speed of a swimming flagellate (as formulated in the text) to order ϵ^4.

12. Find the speed of a swimming flagellate in bounded fluid. Assume that the fluid is contained between two walls and that the flagellate is exactly halfway between the walls.

Section 11.2

1. Estimate the eigenvalues and eigenvectors of the matrices

$$a. \quad \begin{pmatrix} 1 & 0 & 0 \\ 0 & 2 & 0 \\ 1-\epsilon & 0 & 3 \end{pmatrix} \quad b. \quad \begin{pmatrix} 1 & 1-\epsilon \\ \epsilon & 1 \end{pmatrix}$$

2. Approximate the eigenvalues and eigenfunctions of

$$y'' + \lambda(1 + \epsilon x)y = 0$$

$$y(0) = y(\pi) = 0$$

for all $\epsilon \neq 0$ small.

Section 11.3

1. Discuss the bifurcation of steady solutions for a pendulum rotating about its vertical axis.

   Remark: The equation of motion for a rotating pendulum is

$$\ddot{\theta} + (g/l - \Gamma^2 \cos \theta) \sin \theta = 0.$$

   (Derive this equation).

2. Find small nontrivial solutions of the nonlinear eigenvalue problem

$$u'' + \lambda u - u^2 = 0$$

$$u(0) = u(1) = 0$$

   Describe their behavior as a function of λ.

3. Find small nontrivial solutions of the elastica problem

$$\theta_{xx} + \frac{L^2 P}{EI}\left(1 - \frac{P}{EA}\cos\theta\right)\sin\theta = 0$$

$$\theta_x(0) = \theta_x(1) = 0.$$

Interpret your solution as a function of the parameter P.

4. Find nontrivial solutions of the equation

$$\nabla^2 u + \lambda u + \alpha u^3 = 0, \ 0 \le x \le 1, \ 0 \le y \le 2$$

subject to $u(0,y) = u(1,y) = u(x,0) = u(x,2) = 0$.
Discuss the two cases $\alpha = 1$ and $\alpha = -1$.

5. Describe the bifurcation of nontrivial solutions of

$$Lu + \lambda f(u) = 0$$

where L is a self-adjoint Sturm-Liouville operator with homogeneous boundary conditions and $f(u) = a_1 u + a_4 u^4 + a_5 u^5$. What happens if $a_4 = 0$?

6. The nonlinear integral equation

$$\phi(x) - \lambda \int_0^1 \phi^2(y)\,dy = 1$$

can be solved exactly. Describe its solution as a function of λ.

7. Nonlinear integral equations of the form

$$u(x) + \lambda \int_0^1 k(x,y)f\left(y, u(y)\right)\,dy = 0$$

are called Hammerstein integral equations. Discuss the behavior of solutions of this equation when $k(x,y) = \alpha(x)\alpha(y)$ and

a. $f(y,u) = \frac{1}{2}(1+u^2)$
b. $f(y,u) = u\sin u/\alpha(y)$.

8. Find small positive solutions of $u'' + \lambda u e^u = 0$ subject to boundary conditions $u(0) = u(1) = 0$.

9. Find even solutions of

$$u(x) = \exp\left[\lambda \int_{-\pi}^{\pi} u(y)\cos(y-x)\,dy\right].$$

Approximate the behavior of small amplitude and large amplitude solutions.

Remark: The equation can be reduced to a transcendental equation.

10. Reformulate the problem

$$u'' + \lambda u - u^2 = 0, \; u(0) = u(1) = 0$$

as a problem to which the implicit function theorem can be applied to find small nontrivial solutions.

Section 11.4

1. If $\vec{x}$ is the vectorial position of a satellite orbiting a massive planet of mass M, then

$$\frac{d^2 \vec{x}}{dt^2} = \frac{-MG \vec{x}}{\left|\vec{x}\right|^3},$$

where G is the gravitational constant. Show that

$$\frac{d^2 u}{d\theta^2} + u = a$$

where $r = \frac{1}{u}$, θ is the polar coordinate position of the satellite. What is the parameter a?

2. Find the period as a function of amplitude of a swinging, frictionless pendulum, for small amplitude.
Remark: The equation of motion is $\ddot{\theta} + g/l \sin\theta = 0$.

3. **a.** Find the a, b parameter curves along which Mathieu's equation

$$\frac{d^2 u}{dt^2} + (a + b \cos t)u = 0$$

has 2π-periodic solution for $|b|$ small. Sketch your results.
Remark: There are two such curves.

  b. Find parameter curves on which 4π-periodic solutions occur.

4. Find small amplitude period T periodic solutions of the equation

$$u_{tt} = u_{xx} + \lambda \sin u$$

subject to Dirichlet boundary data $u(0, t) = u(1, t) = 0$.

5. Find periodic solutions of period $2\pi/\omega$ of the forced Duffing's equation

$$u'' + au' + u + bu^3 = F \cos\omega t$$

when F is small, to leading order in F. Plot the amplitude vs. frequency response curve and interpret when $b > 0$ or $b < 0$. Do this first for $a = 0$, then for a small.

6. The periodic solution of $u'' + u = a(1 + \epsilon u^2)$ has a solution of the form $u = a + A \sin x + \epsilon u_1 + \cdots$, $x = \omega\theta$. Use knowledge of u_1 to find an improved estimate of the advance of the perihelion of Mercury. How much difference is there between this answer and the one presented in the text?

Section 11.5

1. Find Hopf bifurcation points and nearby periodic solutions for the FitzHugh-Nagumo equation

$$u' = u(u - 1/2)(1 - u) - v$$
$$v' = u + \lambda.$$

2. A continuous flow stirred tank reactor with exothermic chemical reaction $A \rightarrow B$ can be described by the differential equation

$$\frac{dT}{dt} = -\beta T + \alpha C e^T, \quad \alpha, \beta > 0$$

$$\frac{dC}{dt} = 1 - C - C e^T$$

where C is the concentration of chemical A, and T is the temperature of the mixture. Describe the steady solutions of this equation as a function of the parameter β/α. Are there any Hopf bifurcations?

3. The population of microoroganisms in a chemostat fed by a logistically growing nutrient can be described by the equations

$$\dot{s} = s(1 - s) - \frac{sx}{s + a}$$

$$\dot{x} = m \left(\frac{s - \lambda}{s + a} \right) x$$

where s is the concentration of nutrient, x is the population of microorganisms feeding on the nutrient. Describe the nontrivial steady state solutions of this problem and calculate any Hopf bifurcation branches.

4. The real matrix A has a one dimensional null space spanned by ϕ_0 and an eigenvector ϕ_1 with $A\phi_1 = i\lambda\phi_1$. Find the small periodic solutions of the matrix differential equation

$$\frac{dx}{dt} - Ax = \epsilon Bx + Q(x, x)$$

where $Q(x, x)$ is a bilinear form. Discuss the behavior of these solutions as functions of the entries of the matrix B for ϵ fixed.

5. a. Show that $u'' - u + u^2 = 0$ has a bounded, nontrivial solution $u_0(x)$ for which $u_0(x) \rightarrow 0$ as $x \rightarrow \pm\infty$.

    b. Under what conditions does

$$u'' - u + u^2 = \epsilon(au' + \cos x)$$

have bounded solutions with $u_0 - u$ small?
Remark: Discuss the behavior of solutions of

$$v'' - v + 2u_0(x)v = f(x).$$

When are such solutions bounded?

12

SINGULAR
PERTURBATION
THEORY

12.1   Initial Value Problems I

The solution of initial value problems is the first example of a singular perturbation problem that we will consider. To illustrate the difficulties and to understand why this is a singular, and not regular problem, consider the simple linear second order differential equation

$$\frac{d^2u}{dt^2} + 2\epsilon\frac{du}{dt} + u = 0, \qquad u(0) = a, \; u'(0) = 0$$

which describes the vibrations of a slightly damped mass-spring system. We suppose that ϵ is a small, positive number. The exact solution of this equation is given by

$$u(t) = Ae^{-\epsilon t}\cos\left(\sqrt{1 - \epsilon^2}\, t + \phi\right)$$

where $\tan\phi = \frac{-\epsilon}{\sqrt{1-\epsilon^2}}$, $A = \frac{a}{\sqrt{1-\epsilon^2}}$.

   If we try to find a power series representation for this solution in powers of ϵ, there are two ways to proceed. We could expand the known solution in its power series, but this method will not work for problems where the answer is not known beforehand. Alternately, we can use the ideas of the previous chapter, namely assume that $u(t, \epsilon)$ has a power series representation in powers of ϵ, and expand the governing equation. If we suppose that

$$u(t, \epsilon) = \sum_{j=0}^{\infty} \epsilon^j U_j(t),$$

substitute into the governing equation and then require that the coefficients of all powers of ϵ vanish, we find that

$$U_0'' + U_0 = 0, \qquad U_0(0) = a, \; U_0'(0) = 0,$$

$$U_j'' + U_j + 2U_{j-1}' = 0, \qquad U_j(0) = U_j'(0) = 0,$$

for $j = 1, 2, \ldots$. The solution $U_0(t)$ is clearly $U_0(t) = a \cos t$ and the equation for U_1 becomes

$$U_1'' + U_1 = 2a \sin t, \; U_1(0) = U_1'(0) = 0.$$

The solution $U_1(t)$ is

$$U_1(t) = -a\, t \cos t + a \sin t$$

and at this stage we have the approximate solution

$$u(t, \epsilon) = a \cos t - a\epsilon t \cos t + \epsilon a \sin t + O\left(\epsilon^2\right).$$

Now we see the problem. Although this approximate representation is valid for any fixed t and ϵ very small, it is not useful for large t. In fact, for large t with ϵ fixed, this representation is not indicative of what really happens. The terms of the form ϵt, which are growing in t, are not uniformly bounded. (The term ϵt is called a SECULAR TERM.)

There is nothing wrong with this solution representation for finite t and ϵ very small. In fact, it is precisely the Taylor series in ϵ of the exact solution. The problem, however, is that a useful representation of the solution should show the exponential decaying behavior of the solution and this information is lost when we have only a power series expansion of $e^{-\epsilon t}$. (It is not obvious from its power series representation that $e^{-t} = 1 - t + t^2/2 + \ldots$ is a decaying function). Another way to explain the dilemma is to notice that $e^{-\epsilon t}$ has an essential singularity at $\epsilon = 0$, $t = \infty$. That is, in the limit $\epsilon \to 0$, $t \to \infty$, the limiting value of $e^{-\epsilon t}$ is not uniquely defined. Thus, if we want to maintain the qualitative behavior of the exponential function $e^{-\epsilon t}$ for large t, we dare not expand this function in powers of ϵ!

A similar problem occurs with the function $\cos \sqrt{1 - \epsilon^2}\, t$, namely, its power series expansion in ϵ has secular terms that render the representation useless for qualitative interpretation for large t. Notice that

$$\cos \sqrt{1 - \epsilon^2}\, t \sim \cos t + \frac{\epsilon^2 t}{2} \sin t + O\left(\epsilon^4\right)$$

appears to be growing in t for large t.

The remedy to this dilemma is to make two observations. First, as we already noted, we do not want to expand $e^{-\epsilon t}$ in powers of ϵt. Second, we notice that the solution should behave like a periodic oscillator that is changing only very slowly. That is, the solution should be represented as

$$u(t, \epsilon) = A(\epsilon t) \cos(\omega(\epsilon) t + \phi)$$

where $A(\epsilon t)$ varies slowly for ϵ small, and the function $\omega(\epsilon)$ shows the ϵ dependence of the frequency of oscillation. We need to determine the two functions $A(\epsilon t)$ and $\omega(\epsilon)$.

To do this we could simply make the guess $u = A(\epsilon t) \cos(\omega(\epsilon) t + \phi)$ and substitute into the differential equation to find A and ω, although this is not a method that will always generalize to harder problems. Instead we make the guess that u is a function of two variables, a "fast" variable and a "slow" variable. We suppose $u = u(s, \tau, \epsilon)$ where $s = \omega(\epsilon) t$ is the fast variable, $\tau = \epsilon t$ is the slow variable, and the variables s and τ are treated as independent of each other.

With this as our guess, we seek a power series solution of the form

$$u(t) \;=\; u(s, \tau, \epsilon) = \sum_{j=0}^{\infty} U_j(s, \tau)\epsilon^j$$

$$\omega(\epsilon) \;=\; \sum_{j=0}^{\infty} \omega_j \epsilon^j.$$

First, however, we realize that if s and τ are new independent variables, then (from the chain rule)

$$\frac{du}{dt} \;=\; \frac{\partial u}{\partial s}\frac{ds}{dt} + \frac{\partial u}{\partial \tau}\frac{d\tau}{dt}$$

$$=\; \omega(\epsilon)\frac{\partial u}{\partial s} + \epsilon\frac{\partial u}{\partial \tau}$$

$$\frac{d^2 u}{dt^2} \;=\; \omega^2(\epsilon)\frac{\partial^2 u}{\partial s^2} + 2\epsilon\omega(\epsilon)\frac{\partial^2 u}{\partial s \partial \tau} + \epsilon^2\frac{\partial^2 u}{\partial \tau^2}.$$

Now, when we substitute the power series representation for u into the governing differential equation and require all coefficients of powers of ϵ to vanish, we obtain the hierarchy of equations

$$\omega_0^2\frac{\partial^2 U_0}{\partial s^2} + U_0 = 0, \qquad U_0(0,0) = a, \qquad \frac{\partial U_0}{\partial s}(0,0) = 0,$$

$$\omega_0^2\frac{\partial^2 U_1}{\partial s^2} + U_1 + 2\omega_0\frac{\partial^2 U_0}{\partial s \partial \tau} + 2\omega_0\omega_1\frac{\partial^2 U_0}{\partial s^2} + 2\omega_0\frac{\partial U_0}{\partial s} = 0,$$

$U_1(0,0) = 0,$

$$\omega_0 \frac{\partial U_1}{\partial s}(0,0) + \omega_1 \frac{\partial U_0}{\partial s}(0,0) + \frac{\partial U_0}{\partial \tau}(0,0) = 0,$$

and so on. The solution U_0 is given by

$$U_0 = A(\tau)\cos s$$

$$\omega_0 = 1$$

where now, since the equation for U_0 is a partial differential equation, A is a function of the independent variable τ with $A(0) = a$. Now we look for U_1, which is the solution of

$$\frac{\partial^2 U_1}{\partial s^2} + U_1 = 2(A_\tau + A)\sin s + 2\omega_1 A\cos s.$$

The function U_1 should be bounded for all time, specifically, periodic in s. If we solve this equation for U_1 without further restrictions we would find secular terms. In order to eliminate secular terms, we invoke the Fredholm Alternative theorem, namely if U_1 is to be periodic in s, then the right hand side must be orthogonal to the null space of the (self adjoint) differential operator $Lu = u_{ss} + u$. In other words, the right hand side must not contain terms proportional to $\sin s$ or $\cos s$. As a consequence we require

$$A_\tau + A = 0,$$

$$\omega_1 = 0,$$

from which it follows that $A = ae^{-\tau}$. At this stage, we have that

$$
\begin{aligned}
u(t) &= U_0(s,\tau) + O(\epsilon) \\
&= ae^{-\tau}\cos s + O(\epsilon) \\
&= ae^{-\epsilon t}\cos t + O(\epsilon).
\end{aligned}
$$

This solution representation looks much better and, in fact, has the correct qualitative behavior for large t. Furthermore, we can carry this calculation to higher order to find, $\omega(\epsilon) = 1 - \epsilon^2/2 + O(\epsilon^4)$. The method works quite well to give as good an approximate solution as we want, uniformly valid for all time.

12.1.1 Van der Pol Equation

Rather than going further on the damped linear oscillator, we study a problem for which the solution is not known ahead of time. We consider the van der Pol equation

$$u'' + \epsilon(u^2 - 1)u' + u = 0,$$

as an initial value problem. We know from the last chapter that this equation has periodic solutions for $\epsilon > 0$ small, and that these periodic solutions can be represented via Taylor series in ϵ. The initial value problem, however, is not regular, but singular, for large t.

We proceed as before, noting that for small ϵ, the solution should be a slowly varying oscillation. We guess that there are two independent timelike variables for this problem, $\tau = \epsilon t$ being the slow time and $s = \omega(\epsilon)t$ being the fast time. We again seek a solution of the form

$$u(t) \;=\; \sum_{j=0}^{\infty} U_j(s,\tau)\epsilon^j$$

$$\omega(\epsilon) \;=\; \sum_{j=0}^{\infty} \omega_j \epsilon^j, \qquad \omega_0 = 1,$$

which we substitute into the defining equation. Of course, the chain rule implies that

$$u' \;=\; \omega(\epsilon)\frac{\partial u}{\partial s} + \epsilon\frac{\partial u}{\partial \tau}$$

$$u'' \;=\; \omega^2(\epsilon)\frac{\partial^2 u}{\partial s^2} + 2\epsilon\omega(\epsilon)\frac{\partial^2 u}{\partial s\partial \tau} + \epsilon^2\frac{\partial^2 u}{\partial \tau^2}.$$

We substitute the power series representation of the solution into the defining equations and set to zero the coefficients of the powers of ϵ. We find the hierarchy of equations

$$\frac{\partial^2 U_0}{\partial s^2} + U_0 = 0$$

$$\frac{\partial^2 U_1}{\partial s^2} + U_1 + 2\omega_1\frac{\partial^2 U_0}{\partial s^2} + \frac{\partial U_0}{\partial s}\left(U_0^2 - 1\right) + 2\frac{\partial^2 U_0}{\partial s\partial \tau} = 0.$$

We find $U_0 = A\cos(s + \phi)$ where now, both A and ϕ are functions of τ. For computational reasons, it is preferable to express the solution U_0 in terms of complex exponentials

$$U_0 = Ae^{is} + \bar{A}e^{-is}$$

where A is a possibly complex function of τ. With U_0 known, the equation for U_1 becomes

$$\frac{\partial^2 U_1}{\partial s^2} + U_1 = \left\{\left[2\omega_1 A + i\left(A - A|A|^2 - 2A_\tau\right)\right]e^{is} - iA^3 e^{3is}\right\} + c.c.$$

where $c.c.$ denotes the complex conjugate of the previous term. Again, we need to invoke our friend the Fredholm Alternative Theorem since if U_1 is to be periodic in s, we cannot have terms proportional to e^{is} and e^{-is} on the right hand side of the equation for U_1. It follows that we must require

$$A_\tau = A \left(\frac{1}{2} - \frac{1}{2} |A|^2 - i\omega_1 \right).$$

Taking A in the form $A = \rho\, e^{i\phi}$, we require

$$\rho_\tau = \frac{1}{2}\rho \left(1 - \rho^2\right), \phi_\tau = -\omega_1.$$

The number ω_1 is arbitrary, but ϕ_τ exactly compensates for ω_1, so we might as well take $\omega_1 = \phi_\tau = 0$. The function $\rho(\tau)$ can be found explicitly, since the equation for ρ^2 is the logistic equation

$$\left(\rho^2\right)_\tau = \rho^2 \left(1 - \rho^2\right).$$

The full solution at this point is given by

$$u(t) = 2 \left(\frac{a}{a + e^{-\epsilon t}} \right)^{1/2} \cos\left(t + \phi_0\right) + O(\epsilon)$$

where a and ϕ_0 are determined from initial data.

The important observation to make is that here we have a representation of the solution that shows the slow transient approach to the known periodic solution. In fact, we have an indication that the periodic solution is stable.

In both of the previous problems there were two naturally occurring time scales in the problem. We exploited the near independence of these time scales to find asymptotic solution representations, valid for all time. This technique is called the MULTISCALE, or two-timing, TECHNIQUE.

A problem in which the two times scales are not immediately obvious is the variant of the van der Pol equation

$$v'' + v' \left(v^2 - \mu\right) + v = 0,$$

where μ is a parameter of the problem. Recall from the previous chapter that a periodic solution of this equation arises from a Hopf bifurcation as μ passes through zero. If possible, we would like to determine the stability of the bifurcating periodic solution.

We begin by identifying μ as a small number by setting $\mu = a\epsilon^2$ with $a = \pm 1$. We replace v by ϵu and the equation is changed into

$$u'' + \epsilon^2 u' \left(u^2 - a\right) + u = 0$$

which is precisely van der Pol's equation. (Try to understand why the choice $\mu = a\epsilon$ fails.) We already know how to find the approximate solution. Following the two-time scale procedure as before, we find that the solution is given by

$$v(t) = 2\epsilon\rho(\epsilon t) \cos(t + \phi) + O(\epsilon^2)$$

$$\mu = a\epsilon^2$$

where $\rho = \rho(\tau)$ satisfies the ordinary differential equation

$$\rho_\tau = \frac{1}{2}\rho\left(a - \rho^2\right).$$

We are now able to add to the information we obtained about this problem in the last chapter. We see from the differential equation for ρ that there is a stable steady solution at $\rho = 0$ if $a < 0$ whereas there are two steady solutions $\rho = 0$ and $\rho = \sqrt{a}$ if $a > 0$. Furthermore, when $a > 0$ the nontrivial solution $\rho = \sqrt{a}$ is stable. We conclude that as μ is changed from slightly less than zero to slightly greater than zero, the stable solution of the problem changes from the trivial solution to a nontrivial stable periodic solution. Furthermore, we now have an understanding of the slow transient behavior of the differential equation as it approaches the steady solution, since ρ^2 satisfies a logistic equation which we can solve.

12.1.2 Adiabatic Invariance

In many problems, it is apparent that two time scales are at work, but it is not immediately obvious what the precise choice of time scales should be. Consider the problem of a swinging pendulum with slowly varying length

$$u'' + f^2(\epsilon t)u = 0$$

where $f^2(\epsilon t)$ is proportional to the length of the pendulum. For example, one might imagine slowly shortening or lengthening a pendulum by drawing a weighted string through a hole in a table.

To use the two-time method we must select the appropriate time scales. The slow time scale is apparently $\tau = \epsilon t$ but a fast time scale is not obvious. Instead of guessing something wrong, we make a general guess and try to fill in the details later. The key step is knowing how to use the chain rule, so we suppose

$$\frac{du}{dt} = \frac{\partial u}{\partial s}g(\tau, \epsilon) + \epsilon\frac{\partial u}{\partial \tau},$$

that is, the fast time s satisfies $\frac{ds}{dt} = g(\tau, \epsilon)$. It follows that

$$\frac{d^2 u}{dt^2} = g^2(\tau, \epsilon)\frac{\partial^2 u}{\partial s^2} + 2\epsilon g(\tau, \epsilon)\frac{\partial^2 u}{\partial \tau \partial s} + \epsilon\frac{\partial u}{\partial s}\frac{\partial g}{\partial \tau}(\tau, \epsilon) + \epsilon^2\frac{\partial^2 u}{\partial \tau^2}.$$

Now we go through the usual step of expanding everything in sight in powers of ϵ

$$u = U_0(s, \tau) + \epsilon U_1(s, \tau) + ..$$

$$g(\tau, \epsilon) = g_0(\tau) + \epsilon g_1(\tau) + \ldots,$$

and substituting these power series into the governing differential equation and collecting coefficients of like powers of ϵ. From this process we find

$$g_0^2(\tau)\frac{\partial^2 U_0}{\partial s^2} + f^2(\tau)U_0 = 0$$

and

$$g_0^2(\tau)\frac{\partial^2 U_1}{\partial s^2} + f^2(\tau)U_1$$

$$+ \left[2g_0(\tau)g_1(\tau)\frac{\partial^2 U_0}{\partial s^2} + \left(\frac{\partial U_0}{\partial s}\right)^{-1}\frac{\partial}{\partial \tau}\left(g_0(\tau)\left(\frac{\partial U_0}{\partial s}\right)^2\right)\right] = 0$$

and so on. (In the spirit of asymptotic expansions, we hope not to need more terms.)

The solution for U_0 is apparent. It is natural to take

$$g_0(\tau) = f(\tau)$$

$$U_0(s, \tau) = A(\tau)\cos s.$$

The equation for U_1 now becomes

$$\frac{\partial^2 U_1}{\partial s^2} + U_1 = \frac{1}{f^2(\tau)}\left[2f(\tau)g_1(\tau)\cos s + \frac{\partial}{\partial \tau}\left(f(\tau)A^2\right)\sin s\right].$$

We observe that, as always, we have the same linear operator to invert at all orders. Here the linear operator is $Lu = u_{ss} + u$ and the equation $Lu = f$ has periodic solutions if and only if the right hand side f is orthogonal to both $\cos s$ and $\sin s$. In other words, for this equation, we must have

$$g_1(\tau) = 0$$

$$\frac{\partial}{\partial \tau}\left(fA^2\right) = 0.$$

The lowest order solution is completely determined. We find that

$$u = A_0\left(\frac{f(0)}{f(\epsilon\tau)}\right)^{1/2}\cos\left(\int_0^t f(\epsilon u)\,du + \phi_0\right) + O(\epsilon),$$

where A_0 and ϕ_0 can be chosen to satisfy initial data. The function $A^2(\tau)f(\tau) = A^2(0)f(0)$ is called an adiabatic invariant for this problem, and we now have a representation of the solution that shows both the slow amplitude variation and the frequency modulation of this oscillator.

12.1.3 Averaging

It would be a serious omission to end this section without mention of the averaging method (due to Krylov and Bogoliubov). The results of the averaging method are exactly the same as that of the multiscale method. There is a minor difference, however, in that it is easier to prove the validity of averaging, but easier to find higher order terms using the multiscale method.

The averaging method calculates the amplitude of a slowly varying oscillation in an average sense. The starting point of the averaging method is always polar coordinates. For example, to solve the van der Pol equation, we set $u = R\cos\theta$, $u' = R\sin\theta$ and find that

$$\frac{dR}{dt} = -\epsilon R\sin^2\theta\left(R^2\cos^2\theta - 1\right) = g(R,\theta)$$

$$\frac{d\theta}{dt} = -1 - \epsilon/2\sin^2\theta\left(R^2\cos^2\theta - 1\right).$$

For ϵ very small, $\frac{d\theta}{dt}$ is nearly -1, while R changes only slowly. Thus, we expect that $\frac{dR}{dt}$ should not differ significantly from its average value

$$\hat{g}(R) = \frac{1}{2\pi}\int_0^{2\pi} g(R,\theta)\,d\theta = \frac{\epsilon}{2}R(1 - R^2/4).$$

Furthermore, on average, the angular variable has $\frac{d\theta}{dt} = -1$. Thus, the averaged system of equations is

$$\frac{d\rho}{dt} = \frac{\epsilon\rho}{2}(1 - \rho^2/4), \quad \frac{d\theta}{dt} = -1.$$

The solution of the averaged system is the same, to leading order, as that given by the multiscale method. However, it is not hard to prove rigorously that averaging gives a valid result.

Theorem 12.1 (Averaging Theorem)

Suppose $f(x,t,\epsilon) : \mathbb{C}^n \times \mathbb{R} \times \mathbb{R}^+ \to \mathbb{C}^n$ is continuous, periodic with period T in t, and has continuous first partial derivatives with respect to x. There exists an $\epsilon_0 > 0$ and a function $u(x,t,\epsilon)$, continuous in x, t, ϵ on $\mathbb{C}^n \times \mathbb{R} \times (0,\epsilon_0)$, and periodic in t, such that the change of variables

$$x = y + \epsilon u(y,t,\epsilon)$$

applied to the differential equation

$$\dot{x} = f(x,t,\epsilon)$$

yields the equation

$$\dot{y} = \epsilon f_0(y) + \epsilon F(y, t, \epsilon)$$

where $f_0(y)$ is the average of $f(x, t, 0)$

$$f_0(x) = \frac{1}{T} \int_0^T f(x, t, 0)\, dt,$$

and $F(y, t, \epsilon)$ is continuous with continuous partial derivatives in x, $F(y, t, 0) = 0$.

The change of variables from x to y is called a near-identity transformation.

To use the averaging theorem, we first solve the reduced average system of equations $\dot{u} = \epsilon f_0(u)$. If (and this is an important if) the function $f_0(u)$ is stable to persistent disturbances, then the solution $y(t)$ of $\dot{y} = \epsilon f_0(y) + \epsilon F(y, t, \epsilon)$ is uniformly close to $u(t)$, $y(t) - u(t) = O(\epsilon)$, for all $t > 0$. ("Stable to persistent disturbances" means that the trajectories of $\dot{u} = \epsilon f_0(u)$ are not changed much by an arbitrarily small perturbation to the equation.) If $y(t)$ is bounded, it follows from the averaging theorem that $x(t)$, the solution of the original equation, is uniformly close to $y(t)$, and hence to $u(t)$, the solution of the averaged equation. Thus, we obtain a rigorous statement of the validity of the averaging method.

We can apply the averaging theorem to the van der Pol equation by setting $\theta = -t + \Psi$ from which we find

$$
\begin{aligned}
R' &= \epsilon R \sin^2(t - \Psi)\left(R^2 \cos^2(t - \Psi) - 1\right) \\
\Psi' &= \epsilon/2 \sin 2(t - \Psi)\left(R^2 \cos^2(t - \Psi) - 1\right).
\end{aligned}
$$

This system of equations can be averaged with the result that

$$
\begin{aligned}
\rho' &= \frac{\epsilon}{2}\left(\rho - \rho^3/4\right) \\
\phi' &= 0
\end{aligned}
$$

is the averaged system. The equation for ρ is stable to persistent disturbances and so the radius of the orbit of the van der Pol equation is approximated by $\rho(t)$ uniformly in t. Unfortunately, the equation for the angular variable is not stable to persistant disturbances and so Ψ is not approximated uniformly by the solution $\phi = t$ of the averaged equation. We conclude that the approximate solution is "orbitally valid" that is, the radius of the orbit is correct, but the averaged solution is out of phase from the exact solution.

12.2   Initial Value Problems II

The multiscale analysis of the previous section works when there is a slowly varying oscillation. It will not work, however, with the mass spring system

$$\epsilon u'' + u' + u = f(t)$$

with specified initial data, $u(0) = a$, $u'(0) = b/\epsilon$, where $\epsilon \ll 1$ indicates that the system is overdamped, having small mass and order one damping.

The naïve approach to this problem is to set $\epsilon = 0$ and see what happens. With $\epsilon = 0$ we have the reduced equation

$$u' + u = f(t)$$

whose solution we can find exactly as

$$u(t) = A\,e^{-t} + \int_0^t e^{s-t} f(s)\,ds.$$

As long as $u''(t)$ is of order one, $\epsilon u''$ is small and this is a good approximation. However, unless we are unreasonably lucky, we cannot satisfy two initial conditions with the one free parameter A, and we have no idea (yet) how best to pick A.

What went wrong? Obviously, in setting $\epsilon = 0$, we have reduced the order of the equation by one and cannot hope to satisfy both pieces of initial data. But something more subtle also happened. In setting $\epsilon = 0$, we replaced the characteristic equation $\epsilon \lambda^2 + \lambda + 1 = 0$ by the reduced characteristic equation $\lambda + 1 = 0$. That is, we replaced the exponential solutions

$$e^{\lambda_- t},\ e^{\lambda_+ t}, \qquad \text{where} \qquad \lambda_{\pm} = \frac{-1 \pm \sqrt{1 - 4\epsilon}}{2\epsilon}$$

by e^{-t} and $e^{-\infty} = 0$. We have discarded terms of the order of $e^{-t/\epsilon}$, which for small ϵ and finite t are vanishingly small, but at $t = 0$ contribute in a significant way to the solution.

The occurrence of the term $e^{-t/\epsilon}$ is the reason this problem is singular. There is no expansion of $e^{-t/\epsilon}$ which is valid in a neighborhood of $t = \epsilon = 0$. We refer to the term $e^{-t/\epsilon}$ as a boundary layer term because it undergoes a change of order one in a temporal region of order ϵ (very thin), and this takes place at the boundary of the temporal domain.

To find what goes on in the boundary layer we make a change of time scale that gives the boundary layer some significance. We take $\tau = t/\epsilon$ to be a "fast" time. In terms of τ, the governing equation becomes

$$\ddot{U} + \dot{U} + \epsilon U = \epsilon f(\epsilon \tau)$$

$$U(0) = a \qquad \dot{U}(0) = b$$

where $U(\tau) = u(\epsilon\tau)$. Setting $\epsilon = 0$, we find the "boundary layer" equation

$$\ddot{U} + \dot{U} = 0, \qquad U(0) = a, \dot{U}(0) = 6$$

whose solution is

$$U(\tau) = b\left(1 - e^{-\tau}\right) + a$$

or in terms of the original time variable $t = \epsilon\tau$,

$$u(t) = b\left(1 - e^{-t/\epsilon}\right) + a.$$

Higher order corrections to both approximations are found as regular perturbations.

Now it is possible to piece together what happens. According to the boundary layer solution near $t = 0$, the solution starts with correct initial data and rapidly decays to $a + b$. Thereafter, it follows (very nearly) the outer dynamics. Apparently, to make the two approximations match, we should take $A = a + b$. A solution that is valid, uniformly for all time, is the sum of the two approximations, with the common part (the contribution common to both approximations) subtracted out. That is,

$$u(t) = -be^{-t/\epsilon} + (a + b)e^{-t} + \int_0^t e^{(s-t)} f(s)\, ds + O(\epsilon),$$

and we see once again that the solution $u(t)$ depends on two time scales.

12.2.1 Michaelis-Menten Enzyme Kinetics

Many of the equations governing chemical reactions exhibit the feature that different reactions take place at vastly different rates. For example, the storing of energy in a plant through photosynthesis and the releasing of that energy in burning occur at rates that are orders of magnitude different. It often occurs that if one process is significant, another process can be ignored because it is so slow as to be in nearly steady state, or so fast that it is already over.

As an example of this we consider the conversion of a chemical substrate S to a product P by enzyme catalysis. The reaction scheme

$$E + S \underset{k_{-1}}{\overset{k_1}{\rightleftharpoons}} ES \overset{k_2}{\rightarrow} E + P$$

was proposed by Michaelis and Menten in 1913. According to this reaction scheme, the enzyme E can combine at rate k_1 with S to form ES and ES can decompose into E and S at rate k_{-1} or into $E + P$ at rate k_2. The concentrations of E, S, ES and P are governed by the law of mass action which yields rate equations

$$\frac{d[E]}{d\tau} = -k_1[E][S] + k_{-1}[ES] + k_2[ES]$$

$$\frac{d[S]}{d\tau} = -k_1[E][S] + k_{-1}[ES]$$

$$\frac{d[ES]}{d\tau} = k_1[E][S] - k_{-1}[ES] - k_2[ES]$$

$$\frac{d[P]}{d\tau} = k_2[ES]$$

subject to initial conditions $[E] = E_0$, $[S] = S_0$, $[ES] = [P] = 0$ at time $\tau = 0$. Here square brackets denote concentrations. Notice that $[P]$ is found by direct integration once $[ES]$ is known, and

$$\frac{d[ES]}{d\tau} + \frac{d[E]}{d\tau} = 0,$$

so that

$$[ES] + [E] = E_0$$

for all time. We now introduce the nondimensional variables

$$u = [ES]/C_0, \qquad v = [S]/S_0, \qquad t = k_1 E_0 \tau, \qquad \kappa = (k_{-1} + k_2)/k_1 S_0,$$

$$\epsilon = C_0/S_0, \qquad \lambda = \frac{k_{-1}}{k_{-1} + k_2}, \qquad C_0 = E_0/(1 + \kappa)$$

to obtain

$$\epsilon \frac{du}{dt} = v - (v + \kappa)u/(1 + \kappa)$$

$$\frac{dv}{dt} = -v + (v + \kappa\lambda)u/(1 + \kappa)$$

subject to initial conditions $u(0) = 0$, $v(0) = 1$. It is typical that the initial concentration of substrate is large compared to the initial concentration of enzyme, so $\epsilon > 0$ is small. Notice also that λ and $\kappa/(1 + \kappa)$ are smaller than one, so that there are no big numbers to worry about other than $1/\epsilon$.

The naïve approximation is to set $\epsilon = 0$ so that

$$u = \frac{v}{v + \kappa}(1 + \kappa)$$

and

$$\frac{dv}{dt} = -\frac{\kappa v}{v + \kappa}(1 - \lambda).$$

This solution, called the QUASI-STEADY STATE APPROXIMATION (qss), is valid as long as $\epsilon\dot{u}$ remains small. However, since we cannot satisfy the two initial conditions with one first order equation, this cannot be uniformly valid for all time. The solution of this approximate equation is given implicitly by

$$v + \kappa\ln v = A - \kappa(1-\lambda)t$$

where A is not yet known.

To determine A and find how u and v behave initially, we introduce a fast time $\sigma = t/\epsilon$, and the original equations become

$$\frac{dU}{d\sigma} = V - (V+\kappa)U/(1+\kappa)$$

$$\frac{dV}{d\sigma} = \epsilon(-V + (V+\kappa\lambda)U/(1+\kappa))$$

$$U(0) = 0, \; V(0) = 1, \qquad U(\sigma) = u(\epsilon\sigma), \; V(\sigma) = v(\epsilon\sigma),$$

which are the boundary layer equations. Now, upon setting $\epsilon = 0$, we find the boundary layer equations reduce to

$$\frac{dU}{d\sigma} = V - (V+\kappa)U/(1+\kappa)$$

$$\frac{dV}{d\sigma} = 0, \qquad U(0) = 0, \; V(0) = 1,$$

These equations imply that during a short initial time interval V does not change, so $V(\sigma) = 1$, while

$$U(\sigma) = \left(1 - e^{-\sigma}\right).$$

In original variables

$$u(t) = 1 - e^{-t/\epsilon}.$$

The first order picture is now complete. In a short initial time segment u equilibrates rapidly to 1 while v remains fixed at $v = 1$. Thereafter v satisfies $v + \kappa\ln v = 1 - \kappa(1-\lambda)t$ and $u = \frac{v(1+\kappa)}{v+\kappa}$. A uniformly valid representation of this is given by

$$v + \kappa\ln v = 1 - \kappa(1-\lambda)t$$

$$u = \frac{v(1+\kappa)}{v+\kappa} - e^{-t/\epsilon},$$

to leading order in ϵ. Higher order corrections can be found using regular expansions in both time domains, but for most purposes do not give important information.

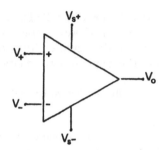

FIGURE 12.1 Schematic diagram of an operational amplifier.

12.2.2 Operational Amplifiers

In the previous example, the behavior of the solution could be described on two time scales, the initial layer during which there is very rapid equilibration followed by the outer region during which the motion is in a quasi-steady state.

One example in which only the quasi-steady state behavior is relevant to understanding the behavior of the system is an operational amplifier (op-amp). An op-amp is a complicated integrated circuit device with a very simple function. In circuit diagrams op-amps are depicted as in Figure 12.1, with five leads, v_+, v_-, v_0, which are two input voltages and the output voltage, respectively, and v_{s+}, v_{s-}, which supply power to the op-amp. (Most diagrams do not show the two supply voltages.) The job of an op-amp is to compare the voltages v_+, v_-, and if $v_+ > v_-$, it sets v_0 to v_{R+}, called the upper rail voltage, while if $v_+ < v_-$, it sets v_0 to v_{R-}, the lower rail voltage. Typically, the rail voltages are close to the supply voltages. It is a reasonable approximation to assume that the leads v_+ and v_- draw no current, and that the output v_0 can provide whatever current is necessary to maintain the appropriate voltage.

An op-amp cannot perform this function instantly, but should be described by the differential equation

$$\epsilon \frac{dv_0}{dt} = g(v_+ - v_-) - v_0.$$

The function $g(u)$ is continuous but very nearly represented by the piecewise constant function $g(u) = v_{R+}H(u) + v_{R-}H(-u)$, where $H(u)$ is the Heaviside step function. The number ϵ is a very small number, typically 10^{-6} sec., corresponding to the fact that op-amps have a "slew rate" of $10^6 - 10^7$ volts per second.

With no further interconnections between the leads of an op-amp, we have a comparator circuit. However, many other functions are possible. An example of how an op-amp is used in a circuit is shown in Figure 12.2. Here, the output lead v_0 is connected to the input lead v_- with the resistor R_2, and

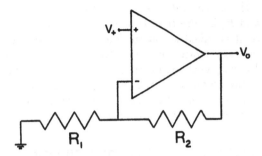

FIGURE 12.2   Amplifier circuit diagram.

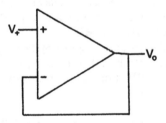

FIGURE 12.3   Voltage follower.

the lead v_- is connected to ground (voltage = 0) with the resistor R_1. From Kirkhoff's laws it follows that

$$v_- = \frac{R_1 v_0}{R_1 + R_2},$$

so that the equation describing the response of the op-amp becomes

$$\epsilon \frac{dv_0}{dt} = g\left(v_+ - \frac{R_1 v_0}{R_1 + R_2}\right) - v_0.$$

Although the input voltage v_+ is not constant, it is typical that it is changing on a time scale that is much slower than that of the op-amp, so we can treat v_+ as constant. Since $g(u)$ is v_{R+} for slightly positive arguments and is v_{R-} for slightly negative arguments, it follows that the op-amp equilibrates very rapidly to the voltage

$$v_0 = \left(1 + \frac{R_2}{R_1}\right) v_+,$$

as long as this lies between the rail voltages. In other words, since v_0 is a scalar multiple of v_+, this circuit is a linear amplifier within the range of the rail voltages.

Another easy circuit to understand is shown in Figure 12.3. Here R_2 is taken to be zero and R_1 is infinite so that $v_0 = v_+$ within the range of the rail voltages. This circuit is called a voltage follower because the output voltage follows the input voltage. Notice that in both of the circuits we have assumed that the input voltage v_+ is changing on a time scale that is much slower than the time scale for the response of the op-amp so that a quasi-steady state approximation is appropriate. Other examples of useful op-amp circuits are given in the exercises at the end of this chapter.

12.2.3   Slow Selection in Population Genetics

Suppose we have a population of organisms, one of whose characteristics is determined by some gene at a particular locus on the chromosome. We suppose that there are two alleles, denoted as type A or type B, and that the genotypes are the three diploids AA, AB, and BB.

We model the population of genotypes by letting N be the total number of individuals in the population and letting x_1, x_2, and x_3 denote the total number of AA, AB, and BB individuals respectively. The growth of each population is determined by

$$\frac{dx_i}{dt} = b_i p_i P - d_i x_i \qquad i = 1, 2, 3$$

where b_i and d_i are survival (at birth) and death rates, respectively, for the population x_i and p_i is the probability that random mating will produce genotype i. Simple probability arguments show that

$$p_1 = \left(\frac{2x_1 + x_2}{2P}\right)^2$$

$$p_2 = 2\left(\frac{2x_1 + x_2}{2P}\right)\left(\frac{2x_3 + x_2}{2P}\right)$$

$$p_3 = \left(\frac{2x_3 + x_2}{2P}\right)^2.$$

The total population is $P = x_1 + x_2 + x_3$ and we introduce the scaled variables $D = \frac{x_1}{P}$, $2H = \frac{x_2}{P}$, $R = \frac{x_3}{P}$, where D, H and R represent the dominant, heterozygote and recessive genotype proportions of the population. In terms of these variables, the total population satisfies

$$\frac{1}{P}\frac{dP}{dt} = \sum_{i=1}^{3} b_i p_i - (d_1 D + 2d_2 H + d_3 R) \equiv F(D, H, R)$$

where $p_1 = (D + H)^2$, $p_2 = 2(D + H)(R + H)$, $p_3 = (R + H)^2$. In addition,

$$\frac{dD}{dt} = b_1 p_1 - (d_1 + F(D, H, R)) D$$

$$\frac{dH}{dt} = b_1 p_2/2 - (d_2 + F(D, H, R)) H$$

$$\frac{dR}{dt} = b_3 p_3 - (d_3 + F(D, H, R)) R.$$

Of course, $D + 2H + R = 1$, so one of these four equations is superfluous.

We now suppose that the death rates for the three genotypes are only slightly different, and that the survival rates at birth are identical for the three genotypes. We incorporate this assumption into the model by taking $b_i = b$, $d_i = d + \epsilon \Delta_i$, $i = 1, 3$ and $\epsilon > 0$ small. With this specification the model equations become

$$\frac{1}{P} \frac{dP}{dt} = b - d - \epsilon [\Delta_1 D + \Delta_3 R]$$

$$\frac{dD}{dt} = b (p^2 - D) + \epsilon D [\Delta_1 D + \Delta_3 R - \Delta_1]$$

$$\frac{dp}{dt} = \epsilon [p (\Delta_1 D + \Delta_3 R) - \Delta_1 D]$$

where $p = D + H$, $R = 1 + D - 2p$.

In this system of equations, the total population P can be determined by quadrature if D and p are known. Thus, it is only necessary to solve the last two equations for D and p. This we can do with $\epsilon > 0$ very small using the two-scale method of this section.

First we make some important observations. Since ϵ is small, p is a slowly varying function. This suggests that D equilibrates rapidly to something close to p^2. The quasi-steady state approximation for this system is to take $D = p^2$ and substitute into the differential equation for p. But the qss approximation is valid only after some initial time interval of rapid transition. To find a solution representation that is valid for all time, we use two time scales. We take $s = t$ and $\tau = \epsilon t$ to be the fast and slow time scales, respectively, and take $r = D - p^2$. Using that $\frac{d}{dt} = \frac{\partial}{\partial s} + \epsilon \frac{\partial}{\partial \tau}$, we transform the equations for D and p into the equations

$$\frac{\partial p}{\partial s} = \epsilon p(p - 1) [\Delta_1 p + \Delta_3 (p - 1)] + \epsilon r [(\Delta_1 + \Delta_3) p - \Delta_1] - \epsilon \frac{\partial p}{\partial \tau}$$

$$\frac{\partial r}{\partial s} = -br + \epsilon \left[\Delta_1 \left(r + p^2 \right) \left(r - (p - 1)^2 \right) \right.$$

$$\left. + \Delta_3 \left(r - p^2 \right) \left(r + (p - 1)^2 \right) - \frac{\partial r}{\partial \tau} \right].$$

To lowest order in ϵ, the solutions are $p = p_0(\tau)$ and $r = r_0(\tau) e^{-bs}$, and the first order correction terms are governed by

$$\frac{\partial p_1}{\partial s} = p_0 (p_0 - 1) [\Delta_1 p_0 + \Delta_3 (p_0 - 1)]$$

$$+ r_0 e^{-bs} \left[(\Delta_1 + \Delta_3) p_0 - \Delta_1 \right] - \frac{\partial p_0}{\partial \tau}$$

$$\frac{\partial r_1}{\partial s} + b r_1 = \Delta_1 \left(r_0 e^{-bs} + p_0{}^2 \right) \left(r_0 e^{-bs} - (p_0 - 1)^2 \right)$$

$$+ \Delta_3 (r_0 e^{-bs} - p_0^2) \left(r_0 e^{-bs} + (p_0 - 1)^2 \right) - \frac{\partial r_0}{\partial \tau} e^{-bs}.$$

The function p_1 is bounded for all s if and only if

$$\frac{\partial p_0}{\partial \tau} = p_0 (p_0 - 1) (\Delta_1 p_0 + \Delta_3 (p_0 - 1))$$

which determines the slow variation of p_0. In addition, r_1 has no secular terms if and only if

$$\frac{\partial r_0}{\partial \tau} = r_0 (\Delta_1 - \Delta_3) \left(p_0{}^2 - (p_0 - 1)^2 \right).$$

For our purposes, the most important equation is that governing the slow variation of p. This equation is a nonlinear first order differential equation with cubic right hand side. It can be integrated exactly, but that is unnecessary to gain an understanding of its behavior. The cubic has three roots at $p_0 = 0, 1$ and $p_0 = \Delta_3/(\Delta_1 + \Delta_3) = \hat{p}$. Clearly, $\hat{p}$ is biologically relevant only if $0 < \hat{p} < 1$. The Δ_1, Δ_3 parameter plane can be divided into four regions in which p_0 has different behavior.

Region I: $\Delta_1 > 0$, $\Delta_3 > 0$. In this region, p_0 approaches $\hat{p}$ as $\tau \to \infty$ so that the heterozygote AB population is stable.

Region II: $\Delta_1 < 0$, $\Delta_3 > 0$. In this region, $\hat{p}$ is not biologically relevant and p_0 approaches 1, so that homozygote AA is the sole survivor since it has a selective advantage over the other populations.

Region III: $\Delta_3 < 0$, $\Delta_1 > 0$. In this region, $\hat{p}$ is also not biologically relevant and p_0 approaches 0, so that the homozygote BB is the sole survivor.

Region IV: $\Delta_3 < 0$, $\Delta_1 < 0$. In this region, $\hat{p}$ is unstable and the steady states $p = 0, 1$ are both stable. The outcome of the dynamics depends on initial data, but the surviving population will be homozygous, since they have a selective advantage over the heterozygote.

As a closing remark in this section, notice that these examples of nonlinear dynamics appear to have been done via differing techniques. In the enzyme kinetics model we found a qss equation and a boundary layer equation each valid in different temporal domains and then we matched the solutions. In the genetics example, we introduced the two times as independent time scales by setting $\frac{d}{dt} = \frac{\partial}{\partial \tau} + \epsilon \frac{\partial}{\partial \sigma}$. We found only one equation whose solution in the two time scales gave a uniformly valid approximation to the solution. No matching step was necessary. The point is that these are really the same methods, and both give the same result. It is simply a matter of personal preference how one actually calculates the approximate solutions.

12.3 Boundary Value Problems

12.3.1 Matched Asymptotic Expansions

Boundary value problems often show the same features as initial value problems with multiple scales in operation, and resulting boundary layers. The added difficulty is that we do not know a priori where the boundary layer is located.

As a beginning example, we consider a problem that we can solve exactly,

$$\epsilon y'' + y' + y = 0, \qquad y(0) = 0, \ y(1) = 1.$$

We find the naïve approximate solution by setting $\epsilon = 0$ to find $y' + y = 0$. It is immediately apparent that this reduced equation cannot be solved subject to the two boundary conditions. At most one of the boundary conditions can be satisfied by the solution $y_0(x) = Ae^{-x}$. It must be that $\epsilon y''$ is large somewhere and cannot be uniformly ignored. To find the appropriate boundary layer, we introduce a scaled space variable $\xi = (x - x_0)/\epsilon$ where x_0 is unknown but represents the "location" of the boundary layer. In terms of the "fast" variable ξ, the original equation becomes

$$Y'' + Y' + \epsilon Y = 0$$

where $Y(\xi) = y(\epsilon\xi)$, and, with $\epsilon = 0$, this equation reduces to $Y'' + Y' = 0$. The solution of the boundary layer equation is $Y_b = Be^{-\xi} + C$ where B and C are as yet unknown. The key observation is that Y is exponentially decaying for positive ξ, but exponentially growing for ξ negative. Since we hope to have a boundary layer that is influential only in a narrow region, we want it to be exponentially decaying, that is, we want $\xi > 0$. It follows that $x_0 = 0$, and the boundary layer should be located at $x = 0$. Thus, we take

$$Y_b = B\left(e^{-\xi} - 1\right)$$

as the lowest order representation of the solution near $x = 0$. The lowest order representation away from the boundary layer (called the "outer expansion") is

$$y_0 = e^{-(x-1)},$$

where the solution has been chosen to satisfy the boundary condition at $x = 1$. Now we need to choose the remaining unknown constant B.

The heuristic idea is to pick B so that y_0 and Y_b match, that is,

$$\lim_{x \to 0} y_0(x) = \lim_{\xi \to \infty} Y_b(\xi)$$

so that $B = -e$, and

$$y_0 = e^{-(x-1)}, \quad Y_b = e - e^{1-x/\epsilon}.$$

We form a composite approximation by summing the two approximations and subtracting the "common" part and find $y = e^{-(x-1)} - e^{1-x/\epsilon}$ (the common part is e since that is the value at which the two solutions agree).

This heuristic prescription to find B recognizes that in the limit as $\epsilon \to 0$, the boundary layer as $\xi \to \infty$ should agree with the outer solution as $x \to 0$. That is, there should be some common domain in which the boundary layer and the outer solutions have the same value.

This prescription to match the boundary layer with the outer solution works well for the leading order approximation, but needs to be made more precise to match higher order approximations. There are two ways to state the matching principle; they both give the same result.

Kaplan Matching Principle

Suppose $Y_b^n(\xi)$ is the nth order boundary layer expansion and $y_0^n(x)$ is the nth order outer expansion. Let $\eta(x, \epsilon)$ be an intermediate variable, that is $\lim_{\epsilon \to 0} \xi/\eta = \infty$ and $\lim_{\epsilon \to 0} x/\eta = 0$. Require that

$$\lim_{\epsilon \to 0} \frac{1}{\epsilon^n} \left(Y_b^n \left(\xi(\eta) \right) - y_0^n \left(x(\eta) \right) \right) = 0$$

independent of η.

Van Dyke Matching Principle

Suppose $Y_b^n(\xi)$ is the nth order boundary layer expansion and $y_0^n(x)$ is the nth order outer expansion. Require that the power series expansion of $Y_b^n(\xi(x))$ for fixed x be the same as the power series expansion of $y_0^n(x(\xi))$ for fixed ξ. (Expand in powers of ϵ the boundary layer solution in terms of the outer variable and the outer solution in terms of the boundary layer variable and require that the two expansions agree.)

The Kaplan matching principle works well on our simple problem if we take $\eta = x/\epsilon^{1/2}$. In terms of the intermediate variable η, $\xi = \epsilon^{-1/2}\eta$ and $x = \epsilon^{1/2}\eta$. We express the two approximate representations in terms of η as

$$Y_b = B \left(e^{-\eta/\sqrt{\epsilon}} - 1 \right), \qquad y_0 = e^{-(\sqrt{\epsilon}\eta - 1)}$$

and take limits as $\epsilon \to 0$ with η fixed. We learn that

$$\lim_{\epsilon \to 0} Y_b = -B = \lim_{\epsilon \to 0} y_0 = e,$$

which determines B.

To use the Van Dyke matching principle we express Y_b in terms of x and y_0 in terms of ξ and expand both as power series in ϵ. We obtain

$$Y_b = B\left(e^{-x/\epsilon} - 1\right) \sim -B + O(\epsilon)$$

and

$$y_0 = e^{(1-\epsilon\xi)} \sim e + O(\epsilon).$$

The two expansions agree if $B = -e$.

In this first example, the matching of the boundary layer with the outer solution is a relatively straightforward procedure. To understand the intricacies of the matching procedure, it is necessary to look at higher order terms. For this purpose we will discuss the solution of $\epsilon y'' + (1 + \alpha x)y' + \alpha y = 0$, $0 < x < 1$, $\alpha > -1$ with $y(0) = 0$, $y(1) = 1$. The usual starting point is the naïve approximation, that the solution should be a power series in ϵ, $y(x, \epsilon) = y_0(x) + \epsilon y_1(x) + \dots$ from which it follows that

$$(1 + \alpha x)y_0' + \alpha y_0 = [(1 + \alpha x)\, y_0]' = 0$$

$$[(1 + \alpha x)\, y_1]' = -y_0'',$$

and so on. The solutions are

$$y_0(x) \;=\; \frac{a}{1 + \alpha x},$$

$$y_1(x) \;=\; \frac{a\alpha}{(1 + \alpha x)^3} + \frac{b}{1 + \alpha x}$$

where a and b are as yet undetermined constants.

The solution as it stands can at best satisfy one of the two boundary conditions. We expect there to be a boundary layer in order to meet one of the conditions, but we are not yet sure where it should be located. We expect the independent variable in the boundary layer to be of the form $s = \frac{x - x_0}{\epsilon}$, and in terms of this variable, the differential equation becomes

$$Y'' + (1 + \alpha x_0 + \alpha \epsilon s)\, Y' + \epsilon \alpha Y = 0$$

and with $\epsilon = 0$ this reduces to

$$Y_0'' + (1 + \alpha x_0)\, Y_0' = 0.$$

Since we always want boundary layers to be decaying away from their end-points, we take $x_0 = 0$, and then for $s > 0$ there is an exponentially decaying boundary layer solution. Now, we seek a power series representation in ϵ for the boundary layer solution and find

$$Y_0'' + Y_0' = 0$$

$$Y_1'' + Y_1' + \alpha s Y_0' + \alpha Y_0 = 0.$$

Apparently we want the boundary layer solution to satisfy the condition on the left $y(0) = 0$ and the outer solution should satisfy the condition on the right $y(1) = 1$. Therefore,

$$Y_0(s) = A\left(1 - e^{-s}\right)$$

$$Y_1(s) = \alpha A s \left(\frac{1}{2}se^{-s} - 1\right) + B\left(1 - e^{-s}\right)$$

are the boundary layer solutions that satisfy $Y_0(0) = Y_1(0) = 0$, and

$$y_0 \;=\; \frac{1+\alpha}{1+\alpha x}$$

$$y_1 \;=\; \frac{-\alpha}{(1+\alpha)(1+\alpha x)} + \frac{(1+\alpha)\alpha}{(1+\alpha x)^3}$$

are the outer solutions that satisfy the conditions at the right $y_0(1) = 1$, $y_1(1) = 0$.

The last step is to determine the unknown constants A and B. To find A we can use a simple approach and require that $A = \lim_{s\to\infty} Y_0(s) = \lim_{x\to 0} y_0(x) = 1 + \alpha$. To this order, the uniformly valid expansion is

$$y(x) = (1+\alpha)\left(\frac{1}{1+\alpha x} - e^{-x/\epsilon}\right) + O(\epsilon).$$

This simple procedure for matching does not work to determine B, as the function $Y_1(s)$ does not have a bounded limit as $s \to \infty$. However, this should not be a cause for alarm since $Y_1(s)$ is a term of order ϵ, so the bad term only appears in the expansion as the product ϵs, a secular term. Secular terms usually come from expanding something like $e^{-\epsilon s}$ or $\frac{1}{1+\epsilon s}$ in powers of ϵs. There is probably some way to eliminate this secular term.

We invoke the Kaplan matching procedure. In the belief that the boundary layer expansion and the outer expansion are both valid in some common domain, we set $x = \eta t$, $s = \eta t/\epsilon$ where $\eta = \eta(\epsilon)$ and $\lim_{\epsilon\to 0} \eta(\epsilon) = 0$ and $\lim_{\epsilon\to 0} \frac{\epsilon}{\eta} = 0$. For example, $\eta = \epsilon^{1/2}$ works well. We now require that for all fixed t, the two expansions agree up to and including order ϵ, in the limit $\epsilon \to 0$.

In terms of η, the outer expansion is

$$y = \frac{1+\alpha}{1+\alpha\eta t} + \epsilon\alpha \left[\frac{1+\alpha}{(1+\alpha\eta t)^3} - \frac{1}{(1+\alpha)(1+\alpha\eta t)} \right] + \cdots$$

$$= (1+\alpha)(1-\alpha\eta t) + \epsilon\alpha \left(1+\alpha - \frac{1}{1+\alpha} \right) + \cdots.$$

Similarly, the boundary layer expansion is

$$y = (1+\alpha)\left(1 - e^{-\eta t/\epsilon}\right) + \alpha(1+\alpha)\eta t \left(\frac{t\eta e^{-\eta t/\epsilon}}{2\epsilon} - 1 \right)$$

$$+ \epsilon B \left(1 - e^{-\eta t/\epsilon}\right)$$

$$= (1+\alpha) - \alpha(1+\alpha)\eta t + \epsilon B + O(\epsilon^2).$$

To make the two expansions agree to order ϵ, we take

$$B = \alpha(1+\alpha) - \frac{\alpha}{1+\alpha},$$

and the terms common to both expansions are

$$(1+\alpha) - \alpha(1+\alpha)x + \epsilon\alpha(1+\alpha) - \frac{\epsilon\alpha}{1+\alpha}.$$

The uniformly valid composite expansion is

$$y(x) = (1+\alpha)\left(\frac{1}{1+\alpha x} - e^{-x/\epsilon} \right)$$

$$+ \epsilon\alpha(1+\alpha)\left(\frac{1}{2}\frac{x^2}{\epsilon^2} - 1 + \frac{1}{(1+\alpha)^2} \right) e^{-x/\epsilon} + O\left(\epsilon^2\right).$$

Notice that indeed the troublesome secular term is balanced exactly by the expansion of $\frac{1}{1+\alpha x}$.

The Van Dyke matching principle gives the same answer using a different calculation. We represent the boundary layer expansion in terms of the outer variable $x = \epsilon s$ as

$$Y_b = A\left(1 - e^{-x/\epsilon}\right) + \alpha A x \left(\frac{1}{2}\frac{x}{\epsilon}e^{-x/\epsilon} - 1 \right) + \epsilon B \left(1 - e^{-x/\epsilon}\right)$$

$$\sim A(1 - \alpha x) + \epsilon B + O\left(\epsilon^2\right)$$

and the outer expansion in terms of the boundary layer variable $s = x/\epsilon$ as

$$y = \frac{1+\alpha}{1+\alpha\epsilon s} + \frac{\epsilon(1+\alpha)\alpha}{(1+\alpha\epsilon s)^3} - \frac{\epsilon\alpha}{(1+\alpha)(1+\alpha\epsilon s)}$$

$$\sim (1+\alpha)(1-\alpha\epsilon s) + \epsilon\alpha\left(1+\alpha-\frac{1}{1+\alpha}\right) + O\left(\epsilon^2\right).$$

The identifications $A = (1+\alpha)$, $B = \alpha(1+\alpha-\frac{1}{1+\alpha})$ are apparent since $x = \epsilon s$.

Boundary layers can come in all kinds of shapes and locations. An example that shows some of the varieties that can occur is the nonlinear boundary value problem

$$\epsilon y'' + yy' - y = 0 \text{ on } 0 < x < 1$$

subject to the boundary conditions $y(0) = A$, $y(1) = B$. We begin by setting $\epsilon = 0$ and find that the outer solution satisfies $(y' - 1)y = 0$ so that one of

$$y = 0$$
$$y = x + A$$
$$y = x - 1 + B$$

holds, away from any boundary layers. For the boundary layer, we set $\sigma = \frac{x-x_0}{\epsilon}$ to obtain the boundary layer equation

$$\ddot{Y} + Y\dot{Y} - \epsilon Y = 0$$

where $Y(\sigma) = y(\epsilon\sigma + x_0)$. With $\epsilon = 0$ this equation reduces to

$$\ddot{Y} + Y\dot{Y} = 0$$

which can be integrated to

$$\dot{Y} + Y^2/2 = K^2/2.$$

There are two solutions of this equation, either

$$Y(\sigma) = K \tanh \frac{K}{2}(\sigma + \sigma_0)$$

or

$$Y(\sigma) = K \coth \frac{K}{2}(\sigma + \sigma_0).$$

Now we try to combine the three possible outer solutions with the two possible boundary layer solutions to find valid solutions of the boundary value problem. Our first attempts will be to use the hyperbolic cotangent boundary layer.

Region I: If $A > B - 1 > 0$ we take the outer solution

$$y(x) = x + B - 1$$

which satisfies the boundary condition at $x = 1$, and the boundary layer at $x = 0$ (ie. $x_0 = 0$) given by

$$Y(\sigma) = (B - 1) \coth\left(\left(\frac{B-1}{2}\right)(\sigma + \sigma_1)\right),$$

then $\lim_{\sigma \to \infty} Y(\sigma) = \lim_{x \to 0} y(x) = B - 1$ so that the two solutions match. To satisfy the boundary condition at $x = \sigma = 0$, we choose σ_1 to satisfy the transcendental equation

$$(B - 1) \coth\left(\frac{B-1}{2}\right) \sigma_1 = A$$

which has a unique solution provided $A > B - 1 > 0$. The uniformly valid solution is

$$y(x) = x + (B - 1) \coth\left(\frac{B-1}{2}\left(\frac{x}{\epsilon} + \sigma_1\right)\right) + O(\epsilon).$$

Region II: If $B < A + 1 < 0$, we take the outer solution

$$y(x) = x + A$$

which satisfies $y(0) = A$, and a boundary layer at $x = 1$ (ie. $x_0 = 1$) given by

$$Y(\sigma) = (A + 1) \coth\left(\left(\frac{A+1}{2}\right)(\sigma + \sigma_2)\right)$$

where σ_2 is chosen to satisfy the transcendental equation

$$B = (A + 1) \coth\left(\frac{A+1}{2}\right) \sigma_2$$

which is guaranteed to have a solution when $B < A + 1 < 0$. Here the matching condition $\lim_{x \to 1} y(x) = \lim_{\sigma \to \infty} Y(\sigma) = A + 1$ is satisfied. The uniformly valid representation of the solution is

$$y(x) = x - 1 + (A + 1) \coth\left(\frac{A+1}{2}\left(\frac{x-1}{\epsilon} + \sigma_2\right)\right) + O(\epsilon).$$

We have exhausted the possibilities for the hyperbolic cotangent boundary layer, and now use the hyperbolic tangent boundary layer. There are three ways to do this.

Region III: $0 < |A| < B - 1$. We take $y = x + B - 1$ to be the outer solution satisfying $y(1) = B$ and

$$Y(\sigma) = (B - 1) \tanh\left(\left(\frac{B-1}{2}\right)(\sigma + \sigma_3)\right)$$

to be the boundary layer at $x_0 = 0$. The number σ_3 is chosen to satisfy

$$(B - 1) \tanh\left(\frac{B-1}{2}\right)\sigma_3 = A,$$

so that $Y(0) = A$. This equation has a unique solution whenever $0 \le |A| < B-1$. The matching condition $\lim_{x\to 0} y(x) = \lim_{\sigma\to\infty} Y(\sigma) = B-1$ is satisfied and the uniformly valid representation of the solution is

$$y(x) = x + (B - 1) \tanh\left(\frac{B-1}{2}\left(\frac{x}{\epsilon} + \sigma_3\right)\right) + O(\epsilon).$$

Region IV: If $A+1 < 0$, $|B| < |A+1|$, we take the outer solution $y(x) = x+A$ and the boundary layer solution at $x = 1 (x_0 = 1)$

$$Y(\sigma) = (A + 1) \tanh\left(\left(\frac{A+1}{2}\right)(\sigma + \sigma_4)\right)$$

where to satisfy $y(1) = B$ we require $(A+1)\tanh(\frac{A+1}{2})\sigma_4 = B$. This transcendental equation has a unique solution whenever $|B| < |A + 1|$. The matching condition

$$\lim_{x\to 1} y(x) = \lim_{\sigma\to-\infty} Y(\sigma) = A + 1$$

is satisfied and the uniformly valid expansion is

$$y(x) = x - 1 + (A + 1) \tanh\left(\frac{A+1}{2}\left(\frac{x-1}{\epsilon} + \sigma_4\right)\right) + O(\epsilon)$$

Region V: If $-1 < A+B < 1$ and $B > A+1$ we use the hyperbolic tangent solution as an interior transition layer. On the left we take $y_-(x) = x + A$ and on the right we take $y_+(x) = x + B - 1$. At $x_0 = \frac{1-(A+B)}{2}$, $-y_-(x_0) = y_+(x_0) = \frac{B-A-1}{2} > 0$ so we take the transition layer

$$Y(\sigma) = \frac{B - A - 1}{2} \tanh\left(\frac{B-A-1}{2}\right)\sigma.$$

Notice that the matching conditions

$$\lim_{x\to x_0} y_-(x) = \lim_{\sigma\to-\infty} Y(\sigma) = \frac{A+1-B}{2}$$

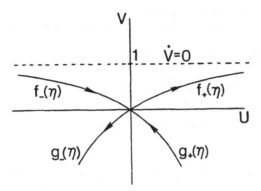

FIGURE 12.4 Phase portrait for $u' = v$, $v' = u(1 - v)$.

and

$$\lim_{x \to x_0} y_+(x) = \lim_{\sigma \to +\infty} Y(\sigma) = \frac{B - A - 1}{2}$$

are satisfied. The uniformly valid representation is

$$y(x) = x + \frac{B + A - 1}{2} + \frac{B - A - 1}{2} \tanh\left(\frac{B - A - 1}{4\epsilon}(2x - 1 + A + B)\right) + O(\epsilon).$$

This exhausts all the ways one can use these boundary layers, but there are values of A and B for which we do not yet have solutions. To get other solutions we must somehow incorporate the outer solution $y = 0$. To do so we must find a way to insert a "corner" layer that smoothly turns the corner between the solution $y = 0$ and a linear solution $y = x + \alpha$. For the corner layer we use the different scaling, $\eta = x/\epsilon^{1/2}$, $y = \epsilon^{1/2}u$ so that

$$\ddot{u} + u\dot{u} - u = 0.$$

We write this equation as a phase plane system

$$\dot{u} = v$$
$$\dot{v} = u(1 - v).$$

The phase portrait of this system (shown in Figure 12.4) has two potentially useful trajectories. The point $u = v = 0$ is the unique critical point for this system and is a saddle point. Emanating from $u = v = 0$ is a trajectory that goes to $v = 1$ as $u \to \infty$ and another trajectory from $v = 1$, $u = -\infty$ goes into the saddle point as $\eta \to \infty$. These two are the useful corner layer solutions. If we identify these trajectories by $u = f_+(\eta)$ and $u = f_-(\eta)$, respectively, then $\lim_{\eta \to -\infty} f_+(\eta) = 0$ and $\lim_{\eta \to +\infty} f_-(\eta) = 0$. Furthermore, one can always choose the origin of these functions so that $\lim_{\eta\to\infty}(f_+(\eta) - \eta) = 0$ and $\lim_{\eta\to-\infty}(f_-(\eta) - \eta) = 0$. Now we are able to find solutions in

Region VI: $0 < -A < 1 - B < 1$. We use the corner layers to write

$$y(x) = \sqrt{\epsilon}\, f_- \left(\frac{x + A}{\sqrt{\epsilon}}\right) + \sqrt{\epsilon}\, f_+ \left(\frac{x + B - 1}{\sqrt{\epsilon}}\right) + O(\epsilon)$$

which is the uniformly valid representation of the solution. Notice that for ϵ sufficiently small

$$y(0) = \sqrt{\epsilon}\, f_- \left(\frac{A}{\sqrt{\epsilon}}\right) + \sqrt{\epsilon}\, f_+ \left(\frac{B - 1}{\sqrt{\epsilon}}\right) \sim A$$

since $A < 0$ and $B - 1 < 0$, while

$$y(1) = \sqrt{\epsilon}\, f_- \left(\frac{A + 1}{\sqrt{\epsilon}}\right) + \sqrt{\epsilon}\, f_+ \left(\frac{B}{\sqrt{\epsilon}}\right) \sim B,$$

since $A + 1 > 0$ and $B > 0$.

This solution still does not fill out the A, B parameter space. We take another look at the phase plane Figure 12.4, and notice that there are two other trajectories into the critical point, one, say $u = g_+(\eta)$ with $\lim_{\eta\to\infty} g_+(\eta) = 0$ and $g_+ \geq 0$, and the other $u = g_-(\eta)$ with $\lim_{\eta\to-\infty} g_-(\eta) = 0$ and $g_-(\eta) \leq 0$. These trajectories are described by the implicit equation $u^2/2 + v + \ln(1-v) = 0$ where $v = \dot{u}$. For large v, this is approximately $u^2/2 \cong -v$ so that $u \sim \frac{2}{\eta + K}$, that is, u blows up at some finite value of η. This means that we can use $u = g_+(\eta)$ as a boundary layer near $x = 0$ and $u = g_-(\eta)$ as a boundary layer near $x = 1$. If $A > 0$ we can shift the independent variable so that $g_+(0) = A/\sqrt{\epsilon}$ and if $B < 0$ we can set $g_-(0) = B/\sqrt{\epsilon}$. This allows us to construct the following solutions.

Region VII: $A > 0$ and $0 < B < 1$. We shift the independent variable so that $\sqrt{\epsilon}\, g_+(0) = A$ and place a corner layer at $x_c = 1 - B$. The expansion takes the form

$$y(x) = \sqrt{\epsilon}\, g_+ \left(\frac{x}{\sqrt{\epsilon}}\right) + \sqrt{\epsilon}\, f_+ \left(\frac{x + B - 1}{\sqrt{\epsilon}}\right) + O(\epsilon).$$

Notice that the boundary conditions $y(0) = A$ and $y(1) = B$ are satisfied approximately to order ϵ for ϵ small.

Region VIII: $B < 0$ and $-1 < A < 0$. We shift the independent variable so that $\sqrt{\epsilon}\, g_-(0) = B$ and place a corner layer at $x_c = -A$. The expansion takes the form

$$y(x) = \sqrt{\epsilon}\, f_- \left(\frac{x + A}{\sqrt{\epsilon}}\right) + \sqrt{\epsilon}\, g_- \left(\frac{x - 1}{\sqrt{\epsilon}}\right) + O(\epsilon).$$

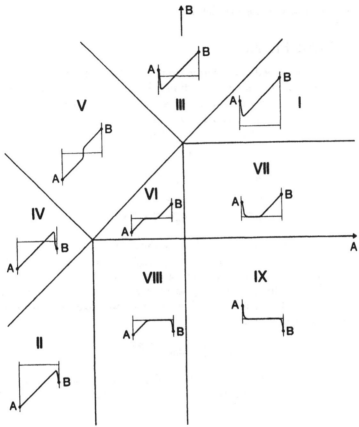

FIGURE 12.5   Qualitative behavior of boundary layer solutions in $A - B$ parameter space.

Region IX: $A > 0$, $B < 0$. We place boundary layers at both $x = 0$ and $x = 1$ and write

$$y(x) = \sqrt{\epsilon}\, g_+ \left(\frac{x}{\sqrt{\epsilon}} \right) + \sqrt{\epsilon}\, g_- \left(\frac{x - 1}{\sqrt{\epsilon}} \right) + O(\epsilon).$$

A summary of the nine regions and the shape of the solutions is shown in Figure 12.5.

It is not always possible to write down an explicit representation of the asymptotic expansion. In the previous example, we used functions $f_\pm(x)$ and $g_\pm(x)$ which we did not know exactly, but whose properties were understood. It is often the case that we can extract significant information about the solution without actually writing all of it down explicitly. In the dicussion that follows we show how important information can be obtained without being able to write down every detail.

12.3.2   Flame Fronts

We wish to find the speed of the flame front from the burning of a combustible gas. We suppose that a long (one dimensional) insulated tube contains a concentration Y_0 of a combustible gas at temperature (unburned) T_0. We must keep track of the concentration Y of the gas and temperature T. If the rate at which fuel is consumed is temperature dependent given by $K(T)$ then

$$\frac{\partial T}{\partial t} = D_T \frac{\partial^2 T}{\partial x^2} + BYK(T),$$

$$\frac{\partial Y}{\partial t} = D_Y \frac{\partial^2 Y}{\partial x^2} - YK(T).$$

These are "heat equations" modified to show the production of heat and the consumption of Y. The numbers D_T and D_Y are diffusion coefficients (assumed constant) of temperature and gas respectively. The number B is the heat of reaction for this gas and is positive when the reaction is exothermic, negative if it is endothermic. For this problem we take $B > 0$. The usual choice for the rate of reaction is

$$K(T) = k_0 e^{-E/RT}$$

where E is called the ACTIVATION ENERGY and R is the UNIVERSAL GAS CONSTANT. The function $K(T)$ is called an ARRHENIUS RATE FUNCTION.

The first step of the analysis is to rescale the equations in dimensionless variables. First, notice that in the absence of diffusion

$$\frac{dT}{dt} + B\frac{dY}{dt} = 0$$

so that $T + BY = T_0 + BY_0$. It follows that when the fuel is fully consumed, the temperature will be $T_1 = T_0 + BY_0$, where T_0 is the initial temperature, Y_0 the initial gas concentration. It is natural to introduce the rescaled variables

$$\theta = \frac{T - T_0}{T_1 - T_0}, \; y = Y/Y_0$$

so that $0 \leq \theta \leq 1, 0 \leq y \leq 1$. If we write the rate of reaction $K(T)$ in terms of θ, we find

$$K(T) = k_0 e^{-E/RT} = k_0 e^{-E/RT_1} \left(\exp\left[\frac{\theta - 1}{\epsilon} \frac{1}{1 + \gamma(\theta - 1)} \right] \right)$$

where $\gamma = 1 - T_0/T_1$ and $\epsilon^{-1} = E(T_1 - T_0)/RT_1^2$ is called the ZELDOVICH NUMBER. The Zeldovich number is typically on the order of ten so we assume that ϵ can be viewed as a small parameter. Finally we introduce dimensionless space and time scales

$$\tau = k_0 \epsilon^2 e^{-E/RT_1} t, \; \xi = \epsilon \left(\frac{k_0 e^{-E/RT_1}}{D_T} \right)^{1/2} x$$

in terms of which the model equations become

$$\frac{\partial \theta}{\partial \tau} = \frac{\partial^2 \theta}{\partial \xi^2} + \frac{1}{\epsilon^2} y f(\theta)$$

$$\frac{\partial y}{\partial \tau} = L^{-1} \frac{\partial^2 y}{\partial \xi^2} - \frac{1}{\epsilon^2} y f(\theta)$$

$$f(\theta) = \exp \left[\frac{\theta - 1}{\epsilon} \frac{1}{(1 + \gamma(\theta - 1))} \right]$$

where $L = D_T/D_Y$ is called the LEWIS NUMBER.

It is our goal to find an asymptotic representation of the solution of this rescaled equation, exploiting the smallness of ϵ. The solution we seek is a traveling wave moving from right to left, with the conditions $y(-\infty) = 1$, $\theta(-\infty) = 0$, $y(+\infty) = 0$, $\theta(+\infty) = 1$. To find a traveling wave solution, we introduce the traveling coordinate $s = v\tau + \xi$ and find

$$\theta'' - v\theta' + y f(\theta)/\epsilon^2 = 0$$

$$L^{-1} y'' - v y' - y f(\theta)/\epsilon^2 = 0,$$

whose derivatives are taken with respect to ξ.

It is important to understand the behavior of the function $f(\theta)/\epsilon^2$. Notice that $\lim_{\epsilon \to 0} f(\theta)/\epsilon^2 = 0$ for all $0 \le \theta < 1$, however $f(1)/\epsilon^2$ is unbounded as $\epsilon \to 0$. It appears that $f(\theta)/\epsilon^2$ is like a delta sequence when θ is restricted to lie in the interval $0 \le \theta \le 1$. Thus, we expect that the function $f(\theta)/\epsilon^2$ is important only when θ is nearly 1, where it is very important.

It is easy to construct outer solutions. There is the obvious constant solution $\theta = 1$, $y = 0$ which we take as the solution for $s > 0$. For $s < 0$ we suppose $\theta < 1$ so that $f(\theta)/\epsilon^2$ can be ignored (it is a transcendentally small term). Thus to leading order in ϵ we find

$$\theta_0'' - v_0 \theta_0' = 0$$

$$L^{-1} y_0'' - v_0 y_0' = 0$$

with solutions

$$\theta_0 = a e^{v_0 s}, \qquad y_0 = 1 - b e^{v_0 L s}$$

for $s < 0$. Notice that with $v_0 > 0$, $\theta(-\infty) = 0$, $y(-\infty) = 1$. We can match this solution to the solution for $s > 0$ by taking $a = 1$, $b = 1$. Thus, we find

$$\theta_0 = \begin{cases} e^{v_0 s} & s < 0 \\ 1 & s > 0 \end{cases}$$

$$y_0 = \begin{cases} 1 - e^{v_0 s} & s < 0 \\ 0 & s > 0 \end{cases},$$

where the velocity v_0 is not yet known.

To determine v_0 we must examine more closely what happens in a neighborhood of $s = 0$, where the function $f(\theta)/\epsilon^2$ has its most profound effect. To do so we introduce the rescaled variables $\eta = s/\epsilon$, $y = \epsilon u$, $\theta = 1 + \epsilon\phi$, and the traveling wave equations become

$$\phi'' - v\epsilon \phi' + uf(1 + \epsilon\phi) = 0,$$

$$L^{-1}u'' - v\epsilon u' - uf(1 + \epsilon\phi) = 0.$$

To leading order in ϵ these equations are

$$\phi_0'' + u_0 e^{\phi_0} = 0$$

$$L^{-1}u_0'' - u_0 e^{\phi_0} = 0.$$

It is clear that $\phi_0'' + L^{-1}u_0'' = 0$ so that by applying boundary conditions at $+\infty$, $\phi_0 + L^{-1}u_0 = 0$. We substitute $u_0 = -L\phi_0$ to obtain the equation for ϕ_0,

$$\phi_0'' - L\phi_0 e^{\phi_0} = 0.$$

The solution of this equation that we want will be negative and will approach zero as $\eta \to +\infty$. If we multiply this equation by ϕ_0' and integrate once we find

$$\frac{1}{2}\phi_0'^2 - L(\phi_0 - 1)e^{\phi_0} = L.$$

As a result, as $\eta \to -\infty$, it must be that $\phi_0' \to \sqrt{2L}$. We do not actually need an exact formula for $\phi_0(\eta)$.

This solution forms a corner layer to be inserted between the two outer solutions. To make everything match we must apply a matching condition. In particular, we require $\lim_{s \to 0^-} \theta_0'(s) = \lim_{\eta \to -\infty} \phi_0'(\eta)$, or $v_0 = \sqrt{2L}$. To see this more precisely, we use the Kaplan Matching principle. We represent $\phi_0(\eta)$ for large negative values of η as $\phi_0(\eta) \sim -K + \phi_0'(-\infty)\eta + \dots$. We introduce the intermediate variable $y = s/\epsilon^{1/2}$, and expand $e^{v_0 s}$ and $1 + \epsilon\phi_0(\eta)$ in powers of ϵ for y fixed. We find

$$e^{v_0 s} = 1 + \epsilon^{1/2}v_0 y + O(\epsilon) = 1 + \epsilon\phi_0(\eta) = 1 + \epsilon^{1/2}\phi_0'(-\infty)y + O(\epsilon).$$

To match these we must take $v_0 = \phi_0'(-\infty) = \sqrt{2L}$, as proposed.

To summarize, we have patched together the two pieces of the outer solution representation using a corner layer, and by requiring that the corner layer match the outer solution correctly, we are able to determine $v_0 = \sqrt{2L}$.

This solution representation shows two things. First, the actual flame reaction takes place in a very thin region near the origin. Ahead of the flame, temperature is rising and fuel concentration is decreasing as the gas and heat diffuse into and out of, respectively, the flame region. The region ahead of the flame is called the preheat zone.

The velocity of the flame was found to be $v_0 = \sqrt{2L}$ in the dimensionless variables. In terms of original unscaled variables, the traveling wave coordinate s is given by

$$s = \epsilon \left(\frac{k_0}{D_T} \right)^{1/2} e^{-E/2RT_1} (x + V_0 t)$$

where

$$V_0 = D_T \left(\frac{2k_0}{D_Y} \right)^{1/2} \epsilon e^{-E/2RT_1}$$

is the velocity in original space time coordinates. This formula for V_0 contains useful information on how the velocity of the flame front depends on the parameters of the problem, such as initial concentration of the gas, the activation energy of the gas, and so on.

12.3.3 Relaxation Dynamics

We end this chapter with two examples of interest in neurophysiology. These examples show that shock layers do not necessarily occur at boundaries, and that one need not have an explicit representation of the solution to understand its behavior.

We study the FitzHugh-Nagumo system

$$\epsilon \frac{du}{dt} = f(u) - v$$

$$\frac{dv}{dt} = u - \gamma v$$

where $f(u) = u(u-1)(\alpha - u)$, and $\epsilon \ll 1$. This system of equations describes the evolution of voltage and current in the circuit of Figure 11.5, with the nondimensional variable u related to the voltage and v related to the current i_2. The parameter α is related to the input current I_0 and voltage source v_0. Notice that if $\gamma = 0$ (which corresponds to $R = 0$ in Figure 11.5) and $\alpha = -1$, this system of equations is equivalent to the van der Pol equation.

Like the van der Pol equation, the FitzHugh-Nagumo system was originally proposed as a model describing the behavior of excitable neural (and

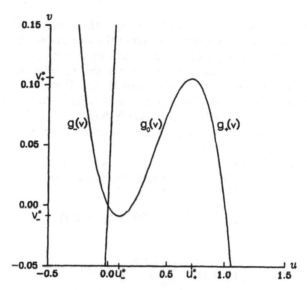

FIGURE 12.6 Nullclines $v = f(u)$ and $u = \gamma v$ for FitzHugh-Nagumo dynamics.

cardiac) cells. We will examine the equation in two cases, the excitable case with $0 < \alpha < 1/2$, and the oscillatory case with $-1 < \alpha < 0$. We will always assume that $\frac{4}{\gamma} > (\alpha - 1)^2$. This curious restriction guarantees that there is one and only one critical point for the differential equation system.

Case 1

Excitable dynamics $0 < \alpha < 1/2$. We wish to solve (approximately) the FitzHugh-Nagumo system subject to some fixed initial conditions $u(0) = u_0$, $v(0) = v_0$ and see what happens as a function of u_0, v_0. We begin by examining the reduced equations. If we set $\epsilon = 0$, we obtain the reduced "outer" equation

$$\frac{dv}{dt} = u - \gamma v, \; v = f(u).$$

This is an implicit differential equation for v. The algebraic equation $v = f(u)$ has three solution branches which we denote by

$$u = g_-(v), \quad u = g_0(v) \quad \text{and} \quad u = g_+(v)$$

(see Figure 12.6). The lower branch $g_-(v)$ exists only for $u < u_-^*$ and $g_+(v)$ exists only for $u > u_+^*$ where $32u_{\pm}^* = \alpha + 1 \pm \sqrt{\alpha^2 - \alpha + 1}$. The middle branch $g_0(v)$ exists in the range $u_-^* \le u \le u_+^*$. Thus, $g_-(v)$ exists only for $v > v_-^* = f(u_-^*)$ and $g_+(v)$ exists only for $v < v_+^* = f(u_+^*)$.

Depending on which branch we choose, the outer equation can have three representations, namely

$$\frac{dv}{dt} = g_\pm(v) - \gamma v$$

on the branches $u = g_\pm(v)$ or

$$\frac{dv}{dt} = g_0(v) - \gamma v$$

on the middle branch. Notice that on the upper branch and on the middle branch, v increases monotonely since

$$g_+(v) - \gamma v > 0, \ g_0(v) - \gamma v > 0,$$

whereas, on the lower branch, $g_-(0) = 0$ so that $g_-(v) - \gamma v$ changes sign at $v = 0$.

The boundary layer equation is found by making the change of variables $\sigma = t/\epsilon$ and then setting ϵ to zero, to obtain

$$\frac{dU}{d\sigma} = f(U) - V, \ \frac{dV}{d\sigma} = 0$$

where $U(\sigma) = u(\epsilon\sigma)$, $V(\sigma) = v(\epsilon\sigma)$. These equations are also relatively easy to decipher. We must have $V(\sigma) = V_0$, a constant, and with the constant V_0 chosen, U equilibrates to a stable root of $f(U) = V_0$ according to

$$\frac{dU}{d\sigma} = f(U) - V_0.$$

The middle branch $g_0(v)$ is an unstable equilibrium of this equation and the upper and lower branches $g_\pm(v)$ are stable. Thus, if $U(0) > g_+(V_0)$, $U(\sigma)$ equilibrates to $g_+(V_0)$ as $\sigma \to \infty$, while if $U(0) < g_-(V_0)$, $U(\sigma)$ equilibrates to $g_-(V)$ as $\sigma \to \infty$.

We now have enough information to piece together the solution of the initial value problem. Starting from initial data u_0, v_0, we use the boundary layer solution

$$\frac{dU}{d\sigma} = f(U) - v_0, \ V = v_0$$

which equilibrates (quickly on the $t = \epsilon\sigma$ time scale) to a root of $f(u) = v_0$. If $f(u_0) > v_0$, u increases while if $f(u_0) < v_0$, u decreases. Thus, if $v_0 > v_+^*$ or if $v_-^* < v_0 < v_+^*$ and $u_0 < g_0(v_0)$, the solution equilibrates rapidly to $u = g_-(v_0)$. After the initial layer evolution, u and v evolve according to the outer dynamics $\frac{dv}{dt} = g_-(v) - \gamma v$, $f(u) = v$ (starting from initial data $u = g_-(v_0)$, $v = v_0$) and eventually they return to

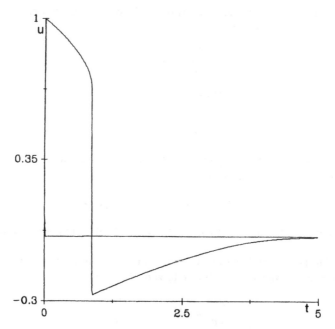

FIGURE 12.7 Response of FitzHugh-Nagumo equations to superthreshold stimulus, calculated for $\alpha = .1$, $\gamma = 2$, $\epsilon = .02$, and with initial data $u(0) = 0.11$, $v(0) = 0$.

rest at the unique equilibrium $u = v = 0$. This trajectory is called a SUBTHRESHOLD TRAJECTORY.

On the other hand, if $v_0 < v_-^*$ or if $v_-^* < v_0 < v_+^*$ and $u_0 > g_0(v_0)$, then the initial layer solution equilibrates rapidly to $u = g_+(v_0)$. Subsequently, u and v evolve according to the outer dynamics

$$\frac{dv}{dt} = g_+(v) - \gamma v, \ f(u) = v.$$

As long as possible, v and u move along the branch $u = g_+(v)$ with v increasing. However, at $u = u_+^*$, $v = v_+^*$ this outer branch ceases to exist although v is supposedly still increasing. We must insert a new transition layer starting with v slightly larger than v_+^* and u nearly u_+^*. This transition layer will satisfy the layer equation

$$\frac{dU}{d\sigma} = f(U) - \left(v_+^* + \delta\right), \ 0 < \delta \ll 1,$$

and U will equilibrate rapidly to $g_-(v)$, after which the dynamics on the lower branch

$$\frac{dv}{dt} = g_-(v) - \gamma v, \ f(u) = v$$

take over. On the lower branch, u and v eventually return to rest.

This last trajectory is different from the first in that it spent an amount of time

$$T = \int_{v_0}^{v_+^*} \frac{dv}{g_+(v) - \gamma v},$$

on the upper, or excited, branch. This trajectory is called a SUPER-THRESHOLD TRAJECTORY. Plots of $u(t)$ for the superthreshold response are shown in Figure 12.7, for $\alpha = .1$, $\gamma = 2$, $\epsilon = .02$, and initial data $u(0) = .11$ and $v(0) = 0$.

The critical curve separating the two types of responses is called the threshold. These two possibilities of response, namely rapid return to rest after subthreshold stimuli or an excursion on the excited branch following superthreshold stimuli, are typical of excitable media.

Case 2

Oscillatory dynamics, $\alpha < 0$. With $\alpha < 0$ the previous analysis works with slightly different details. The principal change is that the unique steady state lies on the middle branch $u = g_0(v)$. Thus, when we start from some arbitrary initial point (not on the middle branch $g_0(v)$) we rapidly equilibrate, keeping v fixed, to one of the stable outer branches $g_\pm(v)$. Now, there are no equilibria on these branches, so on $g_+(v)$, v is continually increasing and on $g_-(v)$, v is continually decreasing. However, these branches have only finite extent and therefore the outer dynamics fail to be valid after a finite amount of time, and a new transition must be inserted. The transition layer allows u to rapidly switch from near $u = u_+^*$, $v = v_+^*$ to $u = g_-(v_+^*)$, $v = v_+^*$ after which dynamics on $u = g_-(v)$ are followed, or from $u = u_-^*$, $v = v_-^*$ to $u = g_+(v_-^*)$, $v = v_-^*$ after which dynamics on $u = g_+(v)$ are followed. The resulting solution continues to cycle forever in a periodic fashion. The period of oscillation can be approximated by the amount of time that is spent on the outer branches,

$$T = \int_{v_-^*}^{v_+^*} \frac{dv}{g_+(v) - \gamma v} + \int_{v_+^*}^{v_-^*} \frac{dv}{g_-(v) - \gamma v} + O(\epsilon).$$

These oscillations are called RELAXATION OSCILLATIONS. A plot of the periodic oscillation of $u(t)$ is shown in Figure 12.8, with $\alpha = -.1$, $\epsilon = .02$, $\gamma = 2$.

Now that we understand the FitzHugh-Nagumo dynamics we wish to understand the behavior of periodic traveling wave solutions along an infinitely

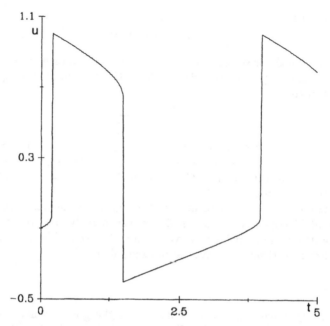

FIGURE 12.8 Relaxation oscillations of the FitzHugh-Nagumo equations, calculated for $\alpha = -0.1$, $\gamma = 2$, $\epsilon = .02$.

long excitable cable modeled by the FitzHugh-Nagumo dynamics

$$\epsilon\frac{\partial u}{\partial t} = \epsilon^2\frac{\partial^2 u}{\partial x^2} + f(u) - v$$

$$\frac{\partial v}{\partial t} = u - \gamma v.$$

Here we have added a diffusive term to the voltage equation in the FitzHugh-Nagumo model. One way to motivate the addition of this term is to consider an infinitely long array of discrete circuit elements coupled resistively (see for example the derivation of the telegrapher's equation in Chapter 8 and Figure 8.2). In the limit of a continuous medium we find diffusive coupling.

To find traveling solutions we put the equation into a traveling coordinate system by setting $s = x - ct$, where c represents the speed of propagation. In this coordinate system, the equations become

$$\epsilon^2 u'' + \epsilon c u' + f(u) - v = 0$$

$$cv' + u - \gamma v = 0.$$

We find the outer and boundary layer equations in the usual way. We set $\epsilon = 0$ to find the outer solution

$$cv' + u - \gamma v = 0, \qquad v = f(u),$$

which, except for the scalar c, is the same as the outer equation found in the last example. The boundary layer equation is found by introducing a fast variable $\sigma = s/\epsilon$ and in terms of this variable the equations are

$$\ddot{U} + c\dot{U} + f(U) - V = 0$$
$$c\dot{V} + \epsilon(U - \gamma V) = 0,$$

where $U(\sigma) = u(\epsilon \sigma)$, $V(\sigma) = v(\epsilon \sigma)$. Setting $\epsilon = 0$, we find the boundary layer equation

$$\ddot{U} + c\dot{U} + f(U) - V = 0, \quad \dot{V} = 0.$$

First we solve the boundary layer equation to see what kinds of boundary layers are possible. We look for solutions of this equation with the property that they provide a transition between the two stable branches of the outer equation (this is the matching condition). Thus, we seek a solution $U(\sigma)$ such that $\lim_{\sigma \to \pm \infty} U(\sigma) = g_{\pm}(V_0)$. We can calculate this trajectory explicitly when $f(u)$ is a cubic. In the more general case when $f(u)$ is shaped like a cubic, with three zeros, one can prove that such a trajectory always exists with a unique value of $c = c(V_0)$.

For V_0 fixed between v_-^* and v_+^*, the function $F(u) = f(u) - V_0$ can be rewritten as

$$F(u) = (u - g_- (V_0)) (u - g_+ (V_0)) (g_0 (V_0) - u)$$

or more succinctly

$$F(u) = (u - z_1) (u - z_3) (z_2 - u)$$

where $z_1 < z_2 < z_3$. To solve the boundary layer equation, we try a solution of the form

$$U' = A (U - z_1) (U - z_3)$$

and substitute into the governing boundary layer equation. We learn that we must have

$$A^2 (2U - z_1 - z_3) + cA - U + z_2 = 0$$

for all U, or that

$$A^2 = 1/2$$

and

$$2cA = z_1 - 2z_2 + z_3.$$

To satisfy the conditions as $\sigma \to \pm\infty$ we must have $A < 0$ so that

$$c = \frac{-1}{\sqrt{2}}(z_1 - 2z_2 + z_3) = \frac{-1}{\sqrt{2}}(g_-(V_0) - 2g_0(V_0) + g_+(V_0))$$

and

$$U(\sigma) = \frac{z_1 + z_3}{2} + \frac{z_3 - z_1}{2} \tanh\left((z_3 - z_1)\sigma/2\sqrt{2}\right)$$

is the boundary layer solution. The important thing to notice is that the speed c of the boundary layer is uniquely determined by the level V_0 at which the transition is to be positioned. It is not difficult to see that $c = c(V_0)$ is an odd function about $V_0 = \frac{1}{2}(v_-^* + v_+^*) \equiv \hat{V}$, that $c(\hat{V}) = 0$, and that $c(V) > 0$ for $V < \hat{V}$. Of course, $c(V)$ is only defined for $v_-^* \le V \le v_+^*$.

Now we can construct a periodic solution. We fix $V_0 < \hat{V}$ and use a transition layer between $g_-(V_0)$ and $g_+(V_0)$ with parameter $c = c(V_0)$. We set $V_1 = 2\hat{V} - V_0$, and note that $c(V_1) = -c(V_0)$. We use the outer solution on $u = g_+(v)$ from V_0 to V_1 and at V_1 insert another transition layer, this one going from $g_+(V_1)$ to $g_-(V_1)$. Finally we use the outer solution on the lower branch $u = g_-(v)$ from V_1 to V_0 to complete the cycle. Notice that the second transition layer has the opposite orientation from the first transition layer, that is, the second transition jumps down from the branch $g_+(V)$ to $g_-(V)$ while the first transition jumps up. Since the speed of a down jump is the negative of the speed of an up jump at a fixed level V, the speed c for both transitions is exactly the same.

For each fixed V_0 there is a different speed $c = c(V_0)$ and wavelength (spatial period of the solution) given by

$$\Lambda = \int_{V_0}^{V_1} \frac{dv}{g_+(v) - \gamma v} + \int_{V_1}^{V_0} \frac{dv}{g_-(v) - \gamma v}.$$

The relationship between Λ and c, in this case parameterized by V_0, is the dispersion relation for these periodic waves.

Notice that if $-1 < \alpha < 0$, these solutions make sense for all V_0, with $v_-^* < V_0 < \hat{V}$, but for $0 < \alpha < 1$ there is a critical point lying on the branch $u = u_-(v)$ and we must restrict V_0 to be non-negative. If we take $V_0 = 0$, the wavelength of the solution is unbounded and there is only one pulse. This solution is called a SOLITARY PULSE, and the corresponding speed is called the solitary speed.

This construction produces solutions for all speeds less than the solitary speed for ϵ sufficiently small. However, for fixed ϵ but small speed, the solution construction is invalid because the width of the boundary layer ϵ becomes comparable to the width of the domain of the outer solution, which is not permitted. To find periodic solutions with small speed, a different expansion procedure is necessary (see Problem 12.3.13).

Further Reading

General introductions to singular perturbation theory are

1. J. D. Cole, *Perturbation Methods in Applied Mathematics*, Blaisdell, Waltham, Ma, 1968.

and its revision

2. J. Kevorkian and J. D. Cole, *Perturbation Methods in Applied Mathematics*, *Applied Mathematical Sciences Series* 34, Springer-Verlag, Berlin, 1981.

As its title implies

3. M. Van Dyke, *Perturbation Methods in Fluid Dynamics*, Academic Press, New York, 1964,

is devoted to problems related to fluid dynamics, but it also contains a good explanation of matching principles. The book

4. R. E. O'Malley, Jr., *Introduction to Singular Perturbations*, Academic Press, New York, 1974

is more technical in nature, as is

5. D. R. Smith, *Singular Perturbation Theory*, Cambridge University Press, Cambridge, 1985

which proves the validity of these methods. This last reference also has an extensive bibliography on singular perturbation methods.

The proof of the averaging theorem in a general setting uses only the implicit function theorem and can be found in

6. J. K. Hale, *Ordinary Differential Equations*, Wiley-Interscience, New York, 1969.

Books discussing the modeling of flame fronts are

7. F. A. Williams, *Combustion Theory*, Addison-Wesley, Reading, Mass., 1965.

8. J. Buckmaster and G. S. S. Ludford, *Theory of Laminar Flames*, Cambridge University Press, New York, 1982.

Problems for Chapter 12

Section 12.1

1. Find two terms of the multiscale expansion of

$$u'' + 2\epsilon u' + u = 0 \qquad u(0) = a, \ u'(0) = 0.$$

2. Find two terms of the multiscale expansion of the general solution of van der Pol's equation

$$u'' + \epsilon u' \left(u^2 - 1\right) + u = 0$$

In particular, determine w_2 where $w = 1 + \epsilon^2 w_2 + O(\epsilon^3)$ is the frequency of oscillation.

Remark: This is a good problem on which to practice your REDUCE skills.

3. Use the two variable method to find the transient behavior of solutions of

$$u'' + u' \left(u^2 - \mu\right) + u = 0$$

in a neighborhood of $\mu = 0$. Discuss the behavior of solutions when $\mu = \mu(\epsilon t)$ is a slowly varying function of time as well.

4. Find the long time behavior of the following nonlinearly damped mass-spring systems

a. $u'' + \epsilon(u')^3 + u = 0$

Generalize this calculation to $u'' + \epsilon(u')^{2\nu+1} + u = 0$.

b. $u'' + \epsilon|u'|u' + u = 0$.

5. Use the multiscale technique to calculate solutions of Mathieu's equation

$$u'' + (a + b\cos t)u = 0$$

for b small and a near n^2, $n = 1, 2, \ldots$. In which regions of parameter space is the solution decaying to zero or unbounded? Where is the solution exactly periodic?

6. Use the multiscale technique to find solutions of the initial value problem

$$u'' + \epsilon(\cos t) \left(u'\right)^2 + u = 0 \qquad t > 0$$

$$u(0) = 0, \ u'(0) = -1$$

7. a. Find approximate solutions to the equation

$$u'' + \epsilon \cos(\epsilon t)(u')^3 + u = 0$$

   b. Generalize this technique to solve $u'' + 2\epsilon q(\epsilon t)(u')^3 + u = 0$ for $t \geq 0$.

8. Find the leading order term of the two variable expansion for the solution of

$$u'' + \epsilon \left[2\epsilon + (u')^2\right] u' + u = 0, \quad t > 0,$$

$$u(0) = a, \quad u'(0) = b.$$

9. Find approximate solutions for the weakly coupled oscillators

$$u'' + k^2 u = \epsilon k^2 v, \quad v'' + k^2 v = \epsilon k^2 u, \quad t > 0$$

   and

$$u(0) = u_0, \quad v(0) = v'(0) = u'(0) = 0, \quad 0 < \epsilon \ll 1$$

   Interpret your solution. In particular how does the amplitude of oscillation vary between u and v?

Section 12.2

1. Find a uniformly valid solution of the initial value problem

$$\epsilon u'' + u' + u = f(t)$$

   by introducing the two independent time scales $\tau = t$, $\sigma = t/\epsilon$ and setting $\frac{d}{dt} = \frac{d}{d\tau} + \frac{1}{\epsilon}\frac{\partial}{\partial \sigma}$. Compare your solution with that given in the text.

2. Solve the enzyme kinetics problem by introducing two independent time scales $\tau = t$, $\sigma = t/\epsilon$ and setting $\frac{d}{dt} = \frac{\partial}{\partial \tau} + \frac{1}{\epsilon}\frac{\partial}{\partial \sigma}$. Compare your solution with that given in the text.

3. Find the leading terms of the solution of

$$\epsilon u'' + (1+t)u' + u = 1 \quad \text{for} \quad t > 0$$

   and

$$u(0) = 0, \quad u'(0) = \frac{1}{\epsilon}.$$

4. Find the solution of

$$u' = v$$

$$\epsilon v' = -u^2 - v$$

   for $t > 0$ with $u(0) = 1$, $v(0) = 0$, to leading order in ϵ.

5. Find the relationship between v_- and v_0 for an op-amp with a resistor connecting v_+ with v_0.

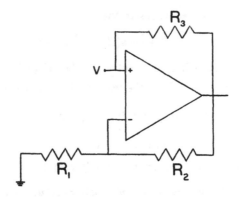

FIGURE 12.9   Nonlinear current device.

6. Find the current-voltage relationship between the voltage v_+ and the current through the resistor R_3 for the op-amp circuit shown in Figure 12.9. What happens if the input voltages are allowed to exceed the rail voltages?

7. Show that the op-amp circuit shown in Figure 12.10 is equivalent to a capacitor and resistor in series, parallel to an inductor and resistor in series. Find the equivalent inductance of the circuit.

8. Find the equations describing the circuit shown in Figure 12.11. Set the rail voltages in the right most op-amp to be smaller than the rail voltages in the other two op-amps and assume that the operating range of the circuit is below the range of the higher rail voltages. Show that this circuit

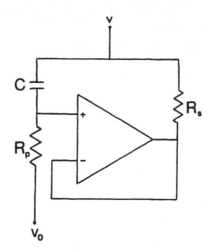

FIGURE 12.10   Simulated inductor circuit.

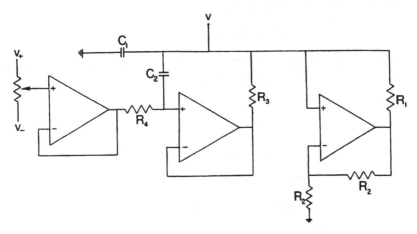

FIGURE 12.11   Piecewise linear FitzHugh-Nagumo circuit.

is an analog for the FitzHugh-Nagumo system with piecewise linear, rather than cubic, function $f(u)$.

Section 12.3

1. Compare the exact solution of $\epsilon y'' + y' + y = 0$ with $y(0) = 0$, $y(1) = 1$ with the zeroth order asymptotic solution given in the text.

2. Find the approximate solution of

$$\epsilon \frac{d^4 u}{dt^4} - (1+t)\frac{du}{dt} = 1 \qquad 0 < t < 1$$

   subject to $u' = u = 0$ at $t = 0$, 1.
   Hint: There are boundary layers at $t = 0$ and $t = 1$. What scaling should be used in the boundary layers?

3. Find a uniformly valid approximation to the solution of

$$\epsilon \frac{d^4 u}{dt^4} - (1+t^2)\frac{du}{dt} = 1 \qquad 0 < t < 1$$

   subject to $u'' = u' = u = 0$ at $t = 0$ and $u'(1) = 0$.

4. Classify the boundary layer solutions of

$$\epsilon u'' = (u^2 - 1)(u^2 - 4) \qquad 0 < x < 1$$

   subject to $u'(0) = 0$, $u(1) = \beta$ as a function of the parameter β.

5. Find two boundary layer solutions of

$$\epsilon u'' - uu' + u^3 = 0, \qquad 0 < x < 1$$

with $u(0) = 1/2$, $u(1) = -1/2$, one with a boundary layer at $x = 0$ and one with a boundary layer at $x = 1$.

6. Find two distinct solutions of

$$\epsilon u'' + u' - (u - 2)^2 = 0, \qquad 0 \le x \le 1$$

subject to $u' - u = 1$ at $x = 0$ and $u' + \frac{1}{3}u = 1$ at $x = 1$.

7. Find three solutions of

$$\epsilon u'' - uu' + u^3 = 0, \qquad 0 < x < 1$$

subject to $u(0) = 2/3$, $u(1) = -1/2$.
Hint: Two of the solutions have boundary layers at the ends of the interval and one has an interior layer centered at $x = 1/4$.

8. Find a solution of $\epsilon u'' = 1 - (u')^2, 0 < x < 1$ with $u(0) = \alpha$, $u(1) = \beta$, $|\alpha - \beta| < 1$.
Hint: The solution has a corner layer.

9. Draw a sketch of the velocity of a flame front in a combustible gas as a function of the initial gas concentration.

10. What is the effect of heat loss on the velocity of a flame front in a combustible gas?

11.   **a.** Approximate the period of oscillation of the equation

$$\epsilon \frac{du}{dt} = g(u) - v$$

$$\frac{dv}{dt} = u$$

where $g(u) = -u + H(u - \alpha)$, H is the Heaviside function, and $\alpha < 0$, in the limit $\epsilon \ll 1$.

   b. Approximate the dispersion curve for the system of equations

$$\epsilon \frac{du}{dt} = \epsilon^2 \frac{d^2 u}{dx^2} + g(u) - v$$

$$\frac{dv}{dt} = u$$

with $g(u)$ as in **a.**

12. Estimate the period of oscillation of the van der Pol oscillator

$$\epsilon\frac{du}{dt} = u\left(1 - u^2\right) - v$$

$$\frac{dv}{dt} = u.$$

13. Find slow speed periodic traveling wave solutions of the equations

$$\epsilon u_t = \epsilon^2 u_{xx} + f(u) - v$$

$$v_t = u$$

using regular perturbation arguments.

Hint: Set $s = x - ct$ with $c = \epsilon^{1/2}\mu$. Find a regular expansion in powers of $\epsilon^{1/2}$.

14. The Belousov-Zhabotinsky reaction is an oxidation-reduction reaction involving bromomalonic acid and cerium ions which has the potential to be excitable or self oscillatory. A satisfactory model of this chemical reation is given by

$$\epsilon\frac{du}{dt} = u - u^2 - \hat{f}v\frac{u - q}{u + q}$$

$$\frac{dv}{dt} = u - v.$$

Discuss the behavior of this system with $0 < f < 3$, q a very small fixed number and ϵ very small. (Typical values are $q = 2 \times 10^{-4}$, $\epsilon = 10^{-2}$.) When is the system excitable and when is it oscillatory? Estimate the period of oscillation.

REFERENCES

1. M. J. Ablowitz and H. Segur, **Solitons and the Inverse Scattering Transform**, SIAM, Philadelphia, 1981.

2. M. Abramowitz and I. A. Stegun, **Handbook of Mathematical Functions**, Dover, New York, 1965.

3. K. Aki and P. G. Richards, **Quantitative Seismology, Theory and Methods: Vol. 2**, Freeman, 1930.

4. R. Askey, **Orthogonal Polynomials and Special Functions**, SIAM, Philadelphia, 1975.

5. K. E. Atkinson, **A Survey of Numerical Methods for the Solution of Fredholm Integral Equations of the Second Kind**, SIAM, Philadelphia, 1976.

6. G. K. Batchelor, **An Introduction to Fluid Dynamics**, Cambridge University Press, Cambridge, 1967.

7. M. S. Bauer and D. S. Chapman, "Thermal Regime at the Upper Stillwater Damsite," Uinta Mountains Utah, *Tectonophysics* **128** 1–20 (1986).

8. K. G. Beauchamp, **Applications of Walsh and Related Functions**, Academic Press, London, 1984.

9. P. G. Bergmann, **Introduction to the Theory of Relativity**, Prentice-Hall, New York, 1942.

10. N. Bleistein and R. A. Handelman, **Asymptotic Expansions of Integrals**, Holt, Rinehart and Winston, New York, 1975.

11. J. Buckmaster and G. S. S. Ludford, **Theory of Laminar Flames**, Cambridge University Press, New York, 1982.

12. G. F. Carrier, M. Krook, and C. E. Pearson, **Functions of a Complex Variable**, McGraw-Hill, New York, 1966.

13. Y. Choquet-Bruhat and C. DeWitt-Moretti, **Analysis, Manifolds and Physics**, North-Holland, Amsterdam, 1982.

14. S. N. Chow and J. K. Hale, **Methods of Bifurcation Theory**, Springer-Verlag, New York, 1982.

15. R. V. Churchill, **Fourier Series and Boundary Value Problems**, McGraw-Hill, New York, 1963.

16. J. D. Cole, **Perturbation Methods in Applied Mathematics**, Blaisdell, Waltham, MA, 1968.

17. E. T. Copson, **Theory of Functions of a Complex Variable**, Oxford at the Clarendon Press, 1935.

18. E. T. Copson, **Asymptotic Expansions**, Cambridge University Press, Cambridge, 1967.

19. A. M. Cormack, "Representation of a Function by its Line Integrals, with Some Radiological Applications," *Journal of Applied Physics* **34**, 2722–2727 (1963), and *Journal of Applied Physics* **35**, 2908–2912 (1964).

20. R. Courant and D. Hilbert, **Methods of Mathematical Physics**, Volume I, Wiley-Interscience, New York, 1953.

21. C. deBoor, **A Practical Guide to Splines**, Applied Mathematical Sciences, Volume 27, Springer-Verlag, New York, 1978.

22. J. J. Dongarra, J. R. Bunch, C. B. Moler, and G. W. Stewart, *LINPACK User's Guide*, SIAM, Philadelphia, 1979.

23. W. Eckhaus and A. Van Harten, **The Inverse Scattering Transformation and the Theory of Solitons**, North Holland, Amsterdam, 1981.

24. A. Erdelyi, **Asymptotic Expansion**, Dover, New York, 1956.

25. H. Flaschka, "On the Toda Lattice, II—Inverse Scattering Solution," *Prog. Theor. Phys.* **51**, 703–716 (1974).

26. J. N. Franklin, **Matrix Theory**, Prentice-Hall, Englewood Cliffs, New Jersey, 1968.

27. G. E. Forsythe, M. A. Malcolm, and C. B. Moler, **Computer Methods for Mathematical Computations**, Prentice-Hall, 1977.

28. B. Friedman, **Principles and Techniques of Applied Mathematics**, Wiley and Sons, New York, 1956.

29. P. R. Garabedian, **Partial Differential Equations**, Wiley and Sons, New York, 1964.

30. K. O. Geddes, G. H. Gonnet, and B. W. Char, *MAPLE User's Manual*, Department of Computer Science, University of Waterloo, Waterloo, Ontario, Canada N2L 3G1.

31. I. M. Gelfand and S. V. Fomin, **Calculus of Variations**, Prentice-Hall, Englewood Cliffs, 1963.

32. I. M. Gelfand and B. M. Levitan, "On the Determination of a Differential Equation from its Spectral Function," *American Mathematical Society Transl.*, Ser. 2, pp. 259–309, (1955).

33. I. M. Gelfand and G. E. Shilov, **Generalized Functions**, Volume I, Academic Press, New York, 1972.

34. H. Goldstein, **Classical Mechanics**, Addison-Wesley, Reading, 1950.

35. D. Gottlieb and S. A. Orszag, **Numerical Analysis of Spectral Methods; Theory and Applications**, SIAM, Philadelphia, 1977.

36. J. K. Hale, **Ordinary Differential Equations**, Wiley-Interscience, New York, 1969.

37. P. R. Halmos, **Finite Dimensional Vector Spaces**, Van Nostrand Reinhold, New York, 1958.

38. H. F. Harmuth, **Transmission of Information by Orthogonal Functions**, Springer-Verlag, New York, 1972.

39. The Hatfield Polytechnic, *Symposium on Theory and Applications of Walsh Functions*, 1971.

40. A. C. Hearn, *REDUCE User's Manual, Version 3.1*, Rand Publication CP78, The Rand Corporation, Santa Monica, CA 90406.

41. P. Henrici, **Applied and Computational Complex Analysis**, Wiley and Sons, 1986.

42. H. Hochstadt, **The Functions of Mathematical Physics**, Wiley-Interscience, New York, 1971.

43. H. Hochstadt, **Integral Equations**, Wiley-Interscience, New York, 1973.

44. F. Hoppensteadt, **Mathematical Theories of Populations: Demographics, Genetics and Epidemics**, SIAM, Philadelphia, 1975.

45. P. Horowitz and W. Hill, **The Art of Electronics**, Cambridge Press, Cambridge, MA, 1980.

46. G. Iooss and D. Joseph, **Elementary Stability and Bifurcation Theory**, Springer-Verlag, New York, 1981.

47. E. Jahnke and F. Emde, **Tables of Functions**, Dover, New York, 1945.

48. O. Kappelmeyer and R. Häenel, **Geothermics with Special Reference to Applications**, Gebruder Borntraeger, Berlin, 1974.

49. J. P. Keener, "Analog Circuitry for the van der Pol and Fitzhugh-Nagumo Equations," *IEEE Transactions on Sys., Man., Cyber*, SMG-13, 1010–1014, 1983.

50. J. B. Keller, "Inverse Problems," *American Mathematical Monthly* **83**, 107–118 (1976).

51. J. B. Keller and S. Antman, **Bifurcation Theory and Nonlinear Eigenvalue Problems**, Benjamin, New York, 1969.

52. J. Kevorkian and J. D. Cole, **Perturbation Methods in Applied Mathematics**, Applied Mathematical Sciences Series 34, Springer-Verlag, Berlin, 1981.

53. S. Kida, "A vortex filament moving without change of form," *J. Fluid Mech.* **112**, 397–409 (1981).

54. D. J. Kortweg and G. deVries, "On the Change of Form of Long Waves Advancing in a Rectangular Canal, and on a New Type of Long Stationary Waves," *Phil. Mag.* **39**, 422–443, 1895.

55. G. L. Lamb, Jr., **Elements of Soliton Theory**, Wiley-Interscience, New York, 1980.

56. L. D. Landau and E. M. Lifshitz, **Mechanics**, Pergamon Press, Oxford, 1969.

57. C. L. Lawson and R. J. Hanson, **Solving Least Squares Problems**, Prentice-Hall, Englewood Cliffs, New Jersey, 1974.

58. P. Lax, "Integrals of Nonlinear Equations of Evolution and Solitary Waves," *Comm. Pure Appl. Math.* **21**, 467–490 (1968).

59. G. A. Lion, "A Simple Proof of the Dirichlet-Jordan Convergence Theorem," *American Mathematical Monthly* **93**, 281–282 (1986).

60. W. V. Lovitt, **Linear Integral Equations**, Dover, New York, 1950.

61. W. S. Massey, **Algebraic Topology: an Introduction**, Harcourt, Brace and World, New York, 1967.

62. L. M. Milne-Thomson, **Theoretical Aerodynamics**, MacMillan, London, 1966.

63. C. Moler, *MATLAB User's Guide*, May 1981, Department of Computer Science, The University of New Mexico, Albuquerque, N. M.

64. *muMATH-83 Reference Manual*, The Soft Warehouse, P. O. Box 11174, Honolulu, Hawaii 96828.

65. J. D. Murray, **Asymptotic Analysis**, Springer-Verlag, New York, 1984.

66. F. Natterer, **The Mathematics of Computerized Tomography**, Wiley and Sons, New York, 1986.

67. A. W. Naylor and G. R. Sell, **Linear Operator Theory in Engineering and Science**, Holt, Rinehart, and Winston, New York, 1971.

68. F. W. J. Olver, **Asymptotics and Special Functions**, Academic Press, New York, 1974.

69. R. E. O'Malley, Jr., **Introduction to Singular Perturbations**, Academic Press, New York, 1974.

70. F. M. Phelps III, F. M. Phelps IV, B. Zorn, and J. Gormley, "An experimental study of the brachistochrome," *Eur. J. Phys.* **3**, 1–4 (1984).

71. M. Pickering, **An Introduction to Fast Fourier Transform Methods for Partial Differential Equations, with Applications**, Wiley and Sons, New York, 1986.

72. W. H. Press, B. P. Flannery, S. A. Teukolsky, and W. T. Vetterling, **Numerical Recipes: The Art of Computing**, Cambridge University Press, Cambridge, 1986.

73. R. H. Rand, **Computer Algebra in Applied Mathematics: an Introduction to MACSYMA**, Pitman, Boston, 1984.

74. R. H. Rand and D. Armbruster, **Perturbation Methods, Bifurcation Theory and Computer Algebra**, *Applied Mathematical Sciences* 65, Springer-Verlag, New York, 1987.

75. G. F. Roach, **Green's Functions: Introductory Theory with Applications**, Van Nostrand Reinhold, London, 1970.

76. T. D. Rossing, "The Physics of Kettledrums," *Scientific American*, November 1982.

77. W. Rudin, **Principles of Mathematical Analysis**, McGraw-Hill, New York, 1964.

78. W. Rudin, **Real and Complex Analysis**, McGraw-Hill, 1974.

79. B. T. Smith, J. M. Boyle, J. J. Dongarra, B. S. Garbow, Y. Ikebe, V. C. Klema, C. B. Moler, **Matrix Eigensystem Routines—EISPACK Guide**, Lecture Notes in Computer Science, Volume 6, Springer-Verlag, 1976.

80. D. R. Smith, **Singular Perturbation Theory**, Cambridge University Press, Cambridge, 1985.

81. I. Stakgold, **Boundary Value Problems of Mathematical Physics**, Macmillan, New York, 1968.

82. I. Stakgold, "Branching of Solutions of Nonlinear Equations," *SIAM Rev.* **13**, 289, 1971.

83. I. Stakgold, **Green's Functions and Boundary Value Problems**, John Wiley and Sons, New York, 1979.

84. F. Stenger, "Numerical Methods Based on Whittaker Cardinal, or Sinc Functions," *SIAM Rev.* **23**, 165–224, 1981.

85. G. W. Stewart, **Introduction to Matrix Computations**, Academic Press, New York, 1980.

86. J. J. Stoker, **Nonlinear Elasticity**, Gordon and Breach, New York, 1968.

87. J. J. Stoker, **Differential Geometry**, Wiley-Interscience, New York, 1969.

88. G. Strang, **Linear Algebra and its Applications**, Academic Press, New York, 1980.

89. G. Strang and G. Fix, **Finite Element Analysis and Applications**, Wiley and Sons, New York, 1968.

90. A. F. Timin, **Theory of Approximation of Functions of a Real Variable**, Macmillan, New York, 1963.

91. E. C. Titchmarsh, **Eigenfunction Expansions Associated with Second-Order Differential Equations**, Oxford University Press, Oxford, 1946.

92. F. G. Tricomi, **Integral Equations**, Wiley and Sons, New York, 1957.

93. M. Van Dyke, **Perturbation Methods in Fluid Dynamics**, Academic Press, New York, 1964.

94. M. Van Dyke, **An Album of Fluid Motion**, The Parabolic Press, Stanford, 1982.

95. R. Wait and A. R. Mitchell, **Finite Element Analysis and Applications**, Wiley and Sons, New York, 1986.

96. J. L. Walsh, "A Closed Set of Normal Orthogonal Functions," *Am. J. Math* **45**, 5–24 (1923).

97. H. F. Weinberger, **A First Course in Partial Differential Equations**, Blaisdell, New York, 1965.

98. H. F. Weinberger, **Variational Methods for Eigenvalue Approximation**, SIAM, Philadelphia, 1974.

99. G. B. Whitham, **Linear and Nonlinear Waves**, Wiley and Sons, New York, 1974.

100. J. H. Wilkinson, **The Algebraic Eigenvalue Problem**, Oxford University Press, London, 1965.

101. F. A. Williams, **Combustion Theory**, Addison-Wesley, Reading, MA, 1965.

102. E. Zauderer, **Partial Differential Equations of Applied Mathematics**, Wiley and Sons, New York, 1983.

103. A. H. Zemanian, **Distribution Theory and Transform Analysis**, McGraw-Hill, New York, 1965.

INDEX